MONOGRAPHS ON THE PHYSICS AND CHEMISTRY OF MATERIALS

MONOGRAPHS ON THE PHYSICS AND CHEMISTRY OF MATERIALS

Strong solids (*Third edition*) A. Kelly and N. H. Macmillan

Optical spectroscopy of inorganic solids B. Henderson and G. F. Imbusch

Quantum theory of collective phenomena G. L. Sewell

Experimental high-resolution electron microscopy (*Second edition*) J. C. H. Spence

Experimental techniques in low-temperature physics Guy K. White

Principles of dielectrics B. K. P. Scaife

Surface analytical techniques J. C. Rivière

Basic theory of surface states Sydney G. Davison and Maria Stęślicka

Acoustic microscopy Andrew Briggs

Dynamic light scattering W. Brown

Quasicrystals: a primer (Second edition) C. Janot

Interfaces in crystalline methods A.P. Sutton and R.W. Balluffi

Atom probe field ion microscopy M.K. Miller, A. Cerezo, M.G. Hetherington, and G.D.W. Smith

Rare earth iron permanent magnets J.M.D. Coey

Statistical physics of fracture and breakdown in disordered solids B.K. Chakrabarti and L.G. Benguigui

Quasicrystals

A Primer

SECOND EDITION

C. JANOT
*Institut Laue–Langevin, Grenoble,
France*

CLARENDON PRESS · OXFORD

Oxford University Press, Great Clarendon Street, Oxford OX2 6DP
Oxford New York
Athens Auckland Bangkok Bogota Bombay
Buenos Aires Calcutta Cape Town Dar es Salaam
Delhi Florence Hong Kong Istanbul Karachi
Kuala Lumpur Madras Madrid Melbourne
Mexico City Nairobi Paris Singapore
Taipei Tokyo Toronto Warsaw
and associated companies in
Berlin Ibadan

Oxford is a trade mark of Oxford University Press

Published in the United States by
Oxford University Press Inc., New York

First published 1992
Second edition 1994
First published in paperback 1997

A catalogue record for this book is available from the British Library

Library of Congress Cataloging in Publication Data
Janot, C. (Christian), 1936–
Quasicrystals: a primer / C. Janot.—2nd ed.
(Monographs on the physics and chemistry of materials; 50)
Includes bibliographical references and index.
I. Quasicrystals. I. Title. II. Series.
QC173.4.Q36J36 1994 548'.8—dc20 94–28803
ISBN 0 19 851778 5 (hbk)
ISBN 0 19 856551 8 (pbk)

Printed in Great Britain by
Bookcraft (Bath) Ltd
Midsomer Norton, Avon

Preface to the second edition

This revised edition of the book is intended to be an improvement on the original. Basically, the book remains the 'primer' it was intended to be, and has not been transformed into 'the' up-to-date relation of 'everything about quasicrystals'. Many other books, cited in the text, have achieved that purpose.

However, since quasicrystal science is a subject of great interest, progress in experimental achievements and fundamental understanding has been rather explosive during the past two years. New quasicrystals have been discovered and prepared to such a level of perfection that they can successfully compete with the mosaic spread of the best periodic crystals. The somewhat 'amateur' method of preparing quasicrystals which prevailed for some time has now been improved. Detailed and carefully explored phase diagrams are available. They clearly establish the composition and temperature domains for the existence of several systems such that their quality can be optimized. Convenient and efficient characterization methods have also been identified and systematically applied. As an obvious consequence, structural concepts have been developed or even revised, characteristic properties have been identified in minute detail, and quasicrystals have been definitively established as a new and very interesting state of matter. I have attempted to integrate all these novel developments at the introductory level of the book and accordingly, all chapters except the first have undergone some modification. In particular, Chapters 5 and 6 have been substantially amended.

It is my great pleasure to thank again the many people who have helped me with this work, particularly those who provided me with new figures, sometimes even prior to publication of their data.

Grenoble
February 1994

C. J.

Preface to the first edition

Books were just a commodity that had to be produced like jam or bootlaces

George Orwell
(*Nineteen eighty four*)

In his pessimistic picture of the year 1984, George Orwell did not anticipate the discovery by Dany Shechtman *et al.*[1] of another sort of Big Brother monster in the world of crystallography! Rapidly solidified aluminium–manganese alloys produced diffraction patterns consisting of sharp peaks revealing a three-dimensional icosahedral symmetry. The pattern was considered highly surprising, because an icosahedron has six five-fold symmetry axes and it is well known that five-fold symmetry axes cannot exist in periodic structures; sharp peaks were 'normally' associated with periodic crystal atomic structures.

In fact, what is really surprising is that Shechtman's discovery so startled the world of science. Early this century, mathematicians[2] had introduced quasiperiodic and almost periodic functions with Fourier transforms into dense sets of sharp Dirac-like peaks. Moreover, modulated structures had been suggested as early as 1927 by Dehlinger[3] and incommensurate modulated phases had been discovered in 1936.[4] Their description in terms of high-dimensional space groups were even formalized before 1980 by Janner and Janseen.[5–8]

Appropriately, Levine and Steinhardt[9] also published in 1984 the study they had completed on a new class of ideal atomic structures. The new structures were highly ordered, but quasiperiodic rather than periodic. This was really the beginning for quasiperiodic crystals, or quasicrystals. These ideal quasicrystals had diffraction patterns with sharp Bragg peaks arranged in disallowed crystallographic patterns. In particular, quasicrystals with icosahedral symmetry were shown to be possible. It was thus tempting to think that the AlMn alloys as observed by Shechtman *et al.* were merely icosahedral quasicrystals.

However, although they obviously had the merit of initiating the subject, AlMn 'quasicrystals' suffer from some major drawbacks: (i) they are not equilibrium compounds and are obtained via rapid quenching procedures; (ii) the diffraction peaks are rather sharp, but not really Bragg-like, and in fact exhibit finite widths; and (iii) they cannot be grown as large single (quasi)crystals for the purpose of proper X-ray or neutron diffraction studies. This is probably why the most serious competition for the ideal quasicrystal model was originally the icosahedral glass model, in which icosahedral clusters are randomly aggregated within certain rules.

The next prominent milestone in the still rather short history of quasicrystals was the year 1986, with the discovery of the Al_6Li_3Cu icosahedral phase[10] which formed (i) as an equilibrium compound and (ii) as single grains approaching millimetre sizes and with perfectly faceted triacontahedral shape. Moreover, the diffraction peaks still had finite widths, but did not scale with the reciprocal space vector in the way suggested by the icosahedral glass model. The experimental observations were consistent with the quasicrystal model in which strains (phason strain as we explain in the book) are quenched during solidification. The strained quasicrystal model then generated so-called 'random tiling' structures which also stimulated the interesting suggestion that the icosahedral phase is favoured for entropic reasons.

But this was not the end of it. Again two years later, in 1988, quasicrystals of the AlFeCu system were obtained[11], followed in 1990 by AlPdMn quasicrystals.[12] They form by regular slow casting procedures, they behave as stable phases, single grains with dodecahedral shapes are easily grown, and the widths of the diffraction peaks are only limited by the instrument resolution. In short, they are apparently perfect (phason-less) quasicrystals whose structure can actually be described within the quasicrystal model and the high-dimensional periodic representation. Of course, other models, including glass-like aggregates, random tilings, multiple twinnings, or large unit cell crystals, can be continually altered so as to be reasonably consistent with the data. But more and more complex modifications are required when samples improve. The 'glass structure' must be artificially modified so that its intrinsic disorder does not dominate the peak widths and in fact finally vanishes; multiple twinnings must be taken at such a scale that quasiperiodicity is actually restored; random tilings have to be designed with complicated ordering algorithms; large unit cells in crystal models have to be so large that they would imply a physically implausible range of interaction. By contrast, data and theory continue to converge on the quasicrystal model with possible long wavelength phason strain and/or defects.

Meanwhile, many other quasicrystals with various 'disallowed' crystallographic diffraction patterns have emerged and have initiated a tremendous burst of theoretical and experimental research activity to determine whether the new alloys are indeed quasicrystals, and if so, what new properties they possess.

To summarize the 'state of the art'[13] it can be stated that structures are reasonably understood, though remaining a controversial issue on some aspects which sometimes do not appear necessarily relevant. As far as properties are concerned, they have just emerged from their infancy. For a long time (on the quasicrystal time-scale!) properties such as electrical resistivity, magnetism, vibrational density of state, mechanical parameters, etc. have been measured on poor quality quasicrystals. Unfortunately, the corrupting effects of quasiperiodicity departure are more dramatic than periodicity breakings in crystals. Thus, the resulting property data have been frustratingly untypical, ranging from bad crystal to amorphous-like behaviour. Presently, the experimental situation is being tremendously improved, though still lacking the proper use of the special selection rules that should result from long-range quasiperiodic order. This is certainly a necessary path toward further understanding. Indeed, these established selection rules for regular periodic crystals have until now commanded the lion's share of attention from physicists studying the crystalline state, although the non-crystalline state is widely prevalent in condensed matter physics and in the universe. Physicists do prefer simple things!

This book is intended to be a 'commodity' (in Orwell's words!) and the approach is substantially descriptive, so as to make it accessible to a wide circle of readers. Further documentation of the subject is available in a number of published volumes.[14–40]

The book is divided into six chapters. In Chapter 1, the definition and geometrical properties of quasiperiodic structures are introduced. Their typical features are described and compared to the other states of solid matter, namely periodic crystals, glassy materials, and incommensurate structures. Methods of generating quasiperiodic structures are also exemplified.

Elementary solid state physics, including the basis for crystallography, has been described in many texts such as those of Kittel[41] and Omar[42]. The treatment presented here follows closely that given by H.P. Myers.[43]

Chapter 2 is of two-fold interest, with a 'catalogue' of the quasicrystals produced to date and an analysis of the necessary characterization steps to ascertain a good true quasiperiodic order.

Chapter 3 details the cut procedure that allows a three-dimensional quasiperiodic structure to be deduced from higher dimensional periodic arrangements. The basis of quasicrystallography is given in this chapter, including technical sections on indexing in high-dimensional space, the cut algorithm, etc.

Chapter 4 reports on experimental studies of real quasicrystal structures with a critical analysis of typical results. Atomic models are also described.

Chapter 5 departs from ideal quasicrystals and makes a presentation of the 'defects' in a general sense. This includes phonons, or atomic vibrations, for which experimental data and theoretical approaches are still very scarce. I have tried to schematize the physics and the qualitative aspects of propagating localized and/or critical modes, without the burden of a heavy mathematical formalism, which can be found in original specialized papers. It is also shown that distortions of the quasilattice may arise not only from phonon-like strain but also from some sorts of atomic rearrangements coming from the complementary ingredient of the high-dimensional structure, and which are called phason strain. Dislocations in quasicrystals are introduced as combined phonon-like plus phason-like strain fields.

The last chapter addresses the questions of growability, stability, and electronic properties of quasicrystals. There are still pending problems and the reader may remain unsatisfied by the few qualitative developments presented in the chapter. Obviously, there is still room for work to come.

Each chapter ends with problems which the reader is encouraged to go through. These generally include complementary developments to questions that may have been too briefly addressed in the course of the offered sections. They also propose some particular illustrations or numerical applications. A computer program that generates a three-dimensional Penrose lattice by the cut procedure is included.

The book has been signed with my name only. Marc de Boissieu would have certainly deserved to be a co-author to account for his large contribution to the work. It has been his own decision that he had not done enough. I personally know that, without him, the project would have never been finalized.

Many other colleagues and friends have, willingly or not, contributed to the book by discussions and collaborations. A non-exhaustive list would certainly include Jean-Marie Dubois, Roland Currat, Claude Benoit, Marc Audier, Marko Jaric, Claire Berger, Jens-Boie Suck, Alan Goldman, Leonid Levitov, Chris Henley, Philip Gaskell, and An Pang Tsai.

I would like to thank the Directors of the Institut Laue–Langevin (Grenoble) for encouragement and support.

I would express my gratitude to Mrs Françoise Giraud, Marie-Rose Guillermet, and Marie-Eve Meyer for their unstinting cooperation in typing the manuscript, and Mrs Irmgard Stadler for preparing the drawings. I am especially grateful to F. Giraud who has perfectly 'orchestrated' the whole process and managed to deal with my occasional gloomy feelings.

I acknowledge the kind permission given by various publishers for the reproduction of material published earlier by them.

Finally, I thank my wife for her loving and patient acceptance of boring weekends and busy holidays!

Writing this book has, overall, been a very pleasant exercise. It makes me realize further what a fortunate scientist I was to be living at the same time as the quasicrystal discovery and how grateful we must be to Dan Shechtman and his co-workers who gave us this wonderful toy!

Grenoble C. J.
August 1991

References

1. Shechtman, D., Blech, I., Gratias, D., and Cahn, J.W. (1984). Metallic phase with long-range orientational order and no translational symmetry. *Phys. Rev. Lett.* **53**, 1951–3.
2. Esclandon, C. (1902). *C.R. Acad. Sci. Paris* **135**, 891–3.
3. Dehlinger, U. (1927). Über die Verbreiterung der Debyelinien bei kaltbearbeiteten Metallen. *Z. Krist.* **65**, 615–31.
4. Johansson, C.H. and Linde, J.O. (1936). Röntgenographische und elektrische Untersuchungen des CuAu-systems. *Ann. Phys. Lpz.* **25**, 1–48.
5. Janner, A. and Janssen, T. (1977). Symmetry of periodically distorted crystals. *Phys. Rev.* **B15**, 643–58.
6. Janner, A. and Janssen, T. (1979). Superspace groups. *Physica* **A99**, 47–76.
7. Janner, A. and Janssen, T. (1980). Symmetry of incommensurate crystal phases. I: Commensurate basic structures. *Acta Cryst.* **A36**, 399–408.
8. Janner, A. and Janssen, T. (1980). Symmetry of incommensurate crystal phases. II: Incommensurate basic structures. *Acta Cryst.* **A36**, 408–15.
9. Levine, D. and Steinhardt, P.J. (1984). Quasicrystals: a new class of ordered structures. *Phys. Rev. Lett.* **53**, 2477–80.
10. Dubost, B., Lang, J.M., Tanaka, M., Sainfort, P., and Audier, M. (1986). Large AlCuLi single quasicrystals with triacontahedral solidification morphology. *Nature* **324**, 48–50.
11. Tsai, A.P., Inoue, A., and Masumoto, T. (1988). New stable quasicrystals in Al–Cu–M (M = V, Cr or Fe) systems. *Trans. JIM* **29**, 521–4.
12. Tsai, A.P., Inoue, A., Yokoyama, Y., and Masumoto, T. (1990). New icosahedral alloys with superlattice order in the AlPdMn system prepared by rapid solidification. *Phil. Mag. Lett.* **61(1)**, 9–14.
13. Steinhardt, P.J., and DiVincenzo, D. (eds) (1991). *Quasicrystals: the state of the art*, Directions in Condensed Matter Physics Vol. 11. World Scientific, Singapore.
14. Venkataraman, G., Sahoo, D., and Balakrishnan, V. (1989) *Beyond the crystalline state* Springer, Berlin.
15. Steinhardt, P.J. and Ostlund, S. (1987). *The physics of quasicrystals.* World Scientific, Singapore.
16. Jaric M.V. (ed.) (1988–90). *Aperiodicity and order.* Vol. 1 (1988), Vol. 2 (1989), and Vol. 3 (1990). Academic, New York.
17. Jaric, M.V. and Lundqvist, S. (eds) (1990). *Quasicrystals.* World Scientific, Singapore.
18. Fujiwara, T. and Ogawa, T. (eds) (1990). *Quasicrystals.* Springer, Berlin.
19. Gratias, D. and Michel, L. (eds) (1986). Proceedings of the International Workshop on 'Aperiodic crystals'. *J. Phys. (France)* **43**, Colloque C3, supplément No. 7.
20. Janot, C. and Dubois, J. M. (eds) (1988). *Quasicrystalline materials.* World Scientific, Singapore.
21. Toledano, J.C. (ed.) (1990). Geometry and thermodynamics: common problems of quasicrystals, liquid crystals and incommensurate systems. *Nato Asi Series B: Physics* **229**. Plenum, New York.
22. Yacaman, M.J., Romeu, D., Castaño, V., and Gómez, A. (eds) (1990).

Quasicrystals and incommensurate structures in condensed matter. World Scientific, Singapore.
23. Baake, M. (1993). *Selected topics in the theory of quasicrystals*. World Scientific, Singapore.
24. Delacour, J. and Levy, J.C.S. (1989). *Systèmes à mémoire, une approche multidisciplinaire*. Masson, Paris.
25. Godrèche, C. (1988). *Du cristal à l'amorphe*. Editions de Physique, Paris.
26. Grünbaum, B. and Sheppard, G.C. (1987). *Tilings and patterns*. W.H. Freeman, New York.
27. Hargittai, I. (1992). *Fivefold symmetry*. World Scientific, Singapore.
28. Hargittai, I. (1990). *Quasicrystals, networks and molecules of fivefold symmetry*. VCH, New York.
29. Jena, P., Khanna, S.N., and Rao, B.K. (1992). *Physics and chemistry of finite systems: from clusters to crystals I and II*. NATO ASI Series C, Vol. 374. Plenum, New York.
30. Kuo, K.H. (1987). *Quasicrystals*. Materials Science Forum, Vols. 22–24. Trans. Tech. Publications, Switzerland, Aedermannsdorf.
31. Kuo, K.H. and Ninomiya, T. (1990). *Quasicrystals*. World Scientific, Singapore.
32. Levy, J.C.S. (1993). *Nouvelles structures de matériaux*. Masson, Paris.
33. Luck, J.M., Moussa, P., and Waldschmidt, M. (1990). *Number theory of Physics*. Springer Proceedings in Physics, Vol. 47. Springer Verlag, Berlin.
34. Perez-Mato, J.M., Zuniga, F.J., and Madariaga, G. (1991). *Methods of structural analysis of modulated structures and quasicrystals*. World Scientific, Singapore.
35. Sadoc, J.F. (1990). *Geometry in condensed matter physics*. World Scientific, Singapore.
36. Scott, J.F. and Clark, N.A. (1987). *Incommensurate crystals, liquid crystals and quasicrystals*. NATO ASI Series B, Physics, Vol. 166. Plenum, New York.
37. Senechal, M. (1990). *Crystalline symmetries: an informal mathematical introduction*. Adam Hilger, Bristol.
38. Strandburg, K.L. (1992). *Bond-orientational order in condensed matter systems*. Springer Verlag, Berlin.
39. Waldschmidt, M. and Luck, J.M. (1992). *From number theory to physics*. Springer Verlag, Berlin.
40. Yacaman, M.J. and Torres, M. (1993). *Crystal–quasicrystal transitions*. Elsevier, Amsterdam.
41. Kittel, C. (1976). *Introduction to solid state physics* (*Fifth edition*). Wiley, New York.
42. Omar, M.A. (1975). *Elementary solid state physics*. Addison-Wesley, Reading, MA.
43. Myers, H.P. (1990). *Introductory solid state physics*. Taylor and Francis, London.

Contents

1

How to fill space with atoms in condensed matter states

Good order is the foundation of all good things
Edmund Burke

1.1 Introduction

For many years condensed matter has been identified with solid crystalline substances, and the physics of condensed matter is to a very great extent the physics of crystals. Crystals exhibit perfect translational symmetry which introduces interesting selection rules which can be applied usefully in the interpretation of experiments as well as in theoretical modelling.

However, interest in 'non-crystalline' materials has grown considerably in the past few years. 'Non-crystalline' must be understood in the broadest sense of lacking strict translational symmetry. This may range from liquids or amorphous solids to incommensurate structures. The former may exhibit little order, sometimes restricted to a short-range arrangement of atoms. The latter can be described as underlying periodic lattices with actual atom positions displaced with respect to these lattice sites; the displacement is also periodic in space, but the lattice and displacement periods are not in a rational ratio.

Quasiperiodic structures, or quasicrystals for short, are non-crystalline materials with perfect long-range order, but with no three-dimensional periodicity ingredient whatsoever, not even the underlying lattice of the incommensurate structures.

The structure of condensed matter is mostly investigated via diffraction experiments (electrons, X-rays, or neutrons) which produce intensity distributions related to the Fourier transform of the structure. Long-range order means a finite number (possibly very large) of Fourier components, with sharp diffraction peaks.

In this chapter, the basic aspects of order in condensed matter are briefly reviewed and quasiperiodicity is pictorially introduced.

1.2 Periodic structures

1.2.1 Lattices, cells, bases, and space groups

In practice, the atomic arrangement in condensed matter is never perfect. But this aspect is neglected in crystallography and crystals are described by reference to perfect infinite arrays of geometrical points called **lattices**. In a lattice, every point has identical surroundings; all lattice points are equivalent and the crystal lattice therefore exhibits perfect translational symmetry. Relative to any arbitrarily chosen origin, any other lattice point has the position vector:

$$\boldsymbol{r} = n_1\boldsymbol{a} + n_2\boldsymbol{b} + n_3\boldsymbol{c}. \tag{1.1}$$

The number n_i are integers and the vectors $\boldsymbol{a}$, $\boldsymbol{b}$, $\boldsymbol{c}$ are fundamental units of the translational symmetry. The volume associated with a single lattice point is unique, but there is a choice regarding the vectors $\boldsymbol{a}$, $\boldsymbol{b}$, $\boldsymbol{c}$ (Fig. 1.1). This volume is called the **primitive cell.**

It can be demonstrated that there are only 14 different periodic ways of arranging identical points in a 3-dim space. These 14 arrays are called **Bravais lattices** (Fig. 1.2 and Table 1.1). The volume unit depicting the lattice is apparently not always a primitive cell. This is because it is often convenient to use a larger volume called the crystallographic unit cell, or **unit cell** for short. In general the unit cell illustrates the symmetry in a more obvious manner and/or favours the use of orthogonal axes.

The 14 Bravais lattices are determined by the allowable spatial symmetries of periodic arrays of points. However, a group of atoms can be associated

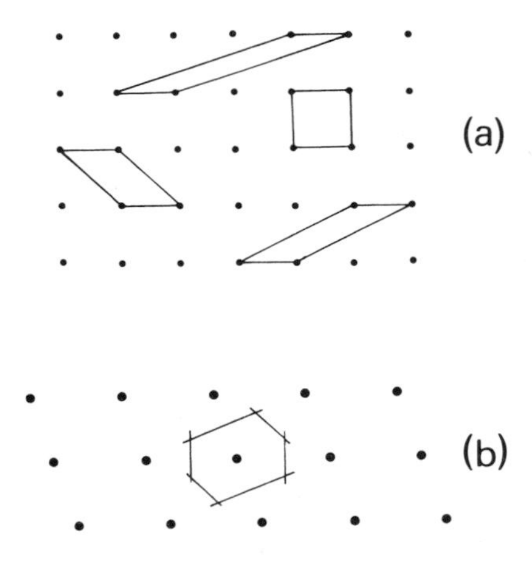

FIG. 1.1 (a) The primitive cell has arbitrary shape but a well determined volume. (b) Voronoi cells or Wigner–Seitz cells are the open volume in between planes bisecting atomic bonds around one atom.

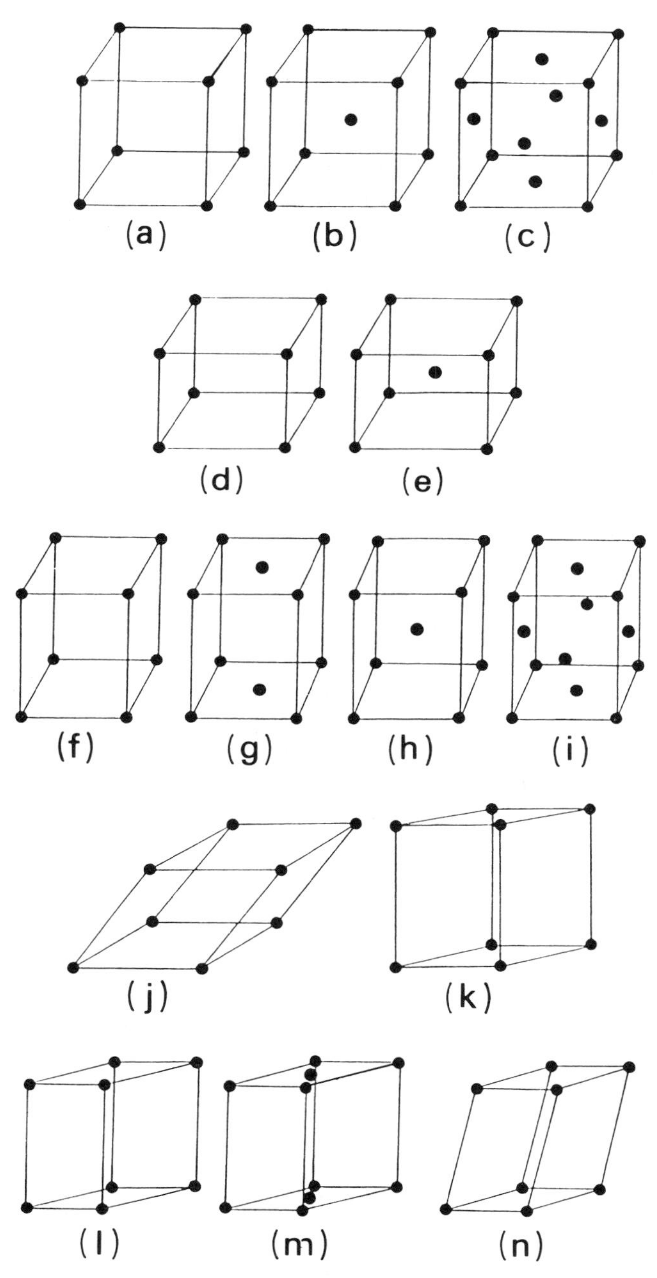

FIG. 1.2 The 14 Bravais lattices in 3-dim spaces, described in terms of unit cells showing clearly the symmetries: cubic systems (a, b, c); tetragonal systems (d, e); orthorhombic systems (f, g, h, i); rhombohedral systems (j, k); monoclinic and triclinic systems (l, m, n).

Table 1.1 The seven crystal systems and fourteen Bravais lattices

System	Conventional unit cell[†]	Bravais lattice
Triclinic	$a \neq b \neq c$ $\alpha \neq \beta \neq \gamma$	P (primitive)
Monoclinic	$a \neq b \neq c$ $\alpha = \beta = 90° \neq \gamma$	P C (base-centred)
Orthorhombic	$a \neq b \neq c$ $\alpha = \beta = \gamma = 90°$	P C I (body-centred) F (face-centred)
Tetragonal	$a = b \neq c$ $\alpha = \beta = \gamma = 90°$	*P* I
Cubic	$a = b = c$ $\alpha = \beta = \gamma = 90°$	P I F
Trigonal	$a = b = c$ $120° > \alpha = \beta = \gamma \neq 90°$	R (rhombohedral primitive)
Hexagonal	$a = b \neq c$ $\alpha = \beta = 90°, \gamma = 120°$	P

[†] The angles α, β, and γ are those between the base vectors, and are defined according to the usual geometrical convention so that α is the angle between $\boldsymbol{b}$ and $\boldsymbol{c}$ (cyclic).

with each lattice point. This immediately opens new possibilities for structural arrangements in which the position vector is now written:

$$\boldsymbol{r}_j = n_1\boldsymbol{a} + n_2\boldsymbol{b} + n_3\boldsymbol{c} + \boldsymbol{R}_j, \tag{1.1a}$$

$\boldsymbol{R}_j$ being a position vector, within the unit cell, relative to the lattice point (n_1, n_2, n_3). The group of atoms which is associated with every lattice point is called a **basis**. A crystal structure is therefore specified by its Bravais lattice *and* a basis. Geometrically, the introduction of the basis induces new **symmetry elements** such as rotations or reflections. These symmetry elements belong to the so-called **point group** symmetry of the basis. Each operation or combination of operations of the point group must turn the atomic arrangement of the crystal into itself. In all, there are 230 different symmetry patterns available for three-dimensional periodic structures by the combination of translational and point symmetry operations. These combinations are called **space groups**. Each crystal structure, however complicated with regard to symmetry, is limited to one of the 230 space groups.

1.2.2 Atomic planes, rows, and indices

A Bravais lattice may also be considered as a stacking of two-dimensional arrays, i.e. planes. There is an infinite number of different families of parallel planes associated with any Bravais lattice (Fig. 1.3). Each family is characterized by a point arrangement and density, and also by a specific interplanar spacing (the larger the interplanar spacing, the greater the density, owing to uniqueness of the specific volume per lattice point). **Miller indices** have been introduced to describe lattices planes and direction in an exact manner.

The Miller indices allow all crystallographic planes to be described within the unit cell; they are derived by the following prescription:

(1) for the plane of interest, determine the intercepts x, y, z on the coordinate axes;

(2) express the intercepts in terms of the basic vectors of the unit cell:

$$x/a,\ y/b,\ z/c.$$

(these are necessarily rational numbers);

(3) form the reciprocals a/x, b/y, c/z;

(4) express as the triplet of the smallest integers hkl, written (hkl).

The distance from the origin to the plane (hkl) lying within the unit cell is the interplanar spacing d_{hkl}. Negative Miller indices are written $\bar{h}$ (pronounced 'bar-h'). A set of symmetrically equivalent planes is written as $\{hkl\}$, including negative indices.

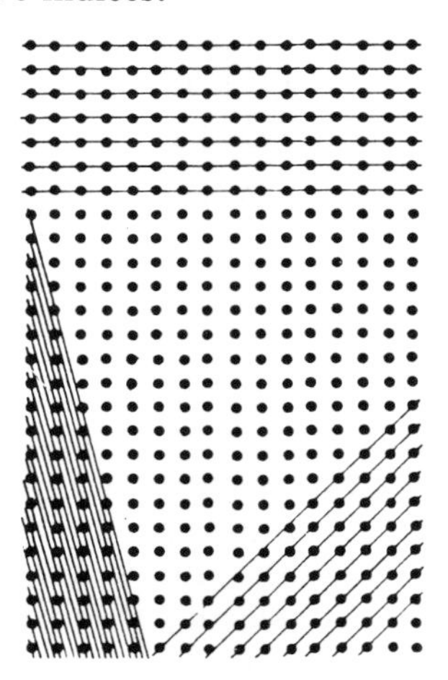

FIG. 1.3 A lattice is also an assembly of identical planes (in 3-dim). There is an infinite number of different families of parallel planes in a given Bravais lattice.

The positions of lattice or basis points are denoted by their coordinates within the unit cell expressed in terms of base vectors (0,0,0 for the origin site, $\frac{1}{2}$, $\frac{1}{2}$, $\frac{1}{2}$ for the body-centred position in a cubic lattice, etc.). Directions, or rows of atoms, in the lattice are specified by the coordinates of the lattice point that is the nearest to the origin in the chosen direction. A direction is written $[uvz]$ and a set of symmetrically equivalent directions $\langle uvz \rangle$. In cubic lattices (and only in these) directions can be defined in terms of the normals to the lattice planes: $[hkl]$ indicates the normal direction to the plane (hkl).

1.2.3 *The reciprocal lattice*

A family of crystal planes (hkl) is completely characterized by the normal to the planes, which can be denoted by a unit vector $\boldsymbol{n}_{hkl}$ and the interplanar spacing d_{hkl}. Thus the structure of a crystal might be described by suitable tabulations of $\boldsymbol{n}_{hkl}$ and d_{hkl}.

Experience has shown that the best way to do this is to define the so-called **reciprocal space** formed by the vectors

$$\boldsymbol{G}_{hkl} = 2\pi \boldsymbol{n}_{hkl}/d_{hkl}. \tag{1.2}$$

There is obviously one point in the reciprocal space for each family of planes in the **direct** lattice. Base vectors $\boldsymbol{A}$, $\boldsymbol{B}$, $\boldsymbol{C}$ can be defined for the reciprocal lattice so that:

$$\boldsymbol{G}_{hkl} = h\boldsymbol{A} + k\boldsymbol{B} + l\boldsymbol{C}, \tag{1.3}$$

in which

$$\boldsymbol{A} = 2\pi \boldsymbol{b} \times \boldsymbol{c}/\boldsymbol{a}\cdot(\boldsymbol{b} \times \boldsymbol{c}) \tag{1.4}$$

and $\boldsymbol{B}$ and $\boldsymbol{C}$ are given by cyclic permutations.

The following relationships are straightforward:

$$\begin{aligned} \boldsymbol{A}\cdot\boldsymbol{a} &= 2\pi \text{ (and cyclic permutations)} \\ \boldsymbol{A}\cdot\boldsymbol{B} &= \boldsymbol{A}\cdot\boldsymbol{C} = 0. \end{aligned} \tag{1.5}$$

It is easy to demonstrate that eqns (1.3), (1.4), and (1.5) are fully compatible with the definition (1.2) of the reciprocal lattice. However, when looked at carefully, eqn (1.3) seems to overdetermine the reciprocal lattice with respect to the number of physically significant planes in the direct space. For the sake of illustration of this statement, let us consider a simple cubic direct lattice. In the prescription for establishing Miller indices it has been said that the triplet of the smallest indices must be formed. Thus, should all the planes (200), (300), (400), . . ., (n00) reduce to (100)? The (100) planes of simple cubic lattice are those limiting the faces of the unit cell (Fig. 1.4); then the

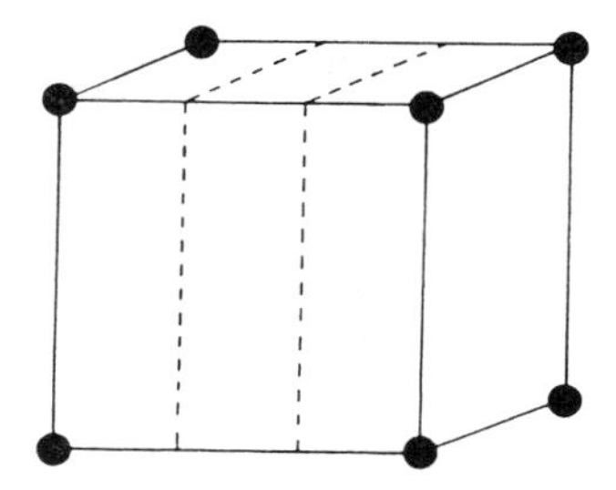

FIG. 1.4 In a primitive (cubic) lattice only one third of the planes (300) pass through lattice points.

(300) planes would be planes parallel to (100) but with an interplanar spacing one third that of the (100) spacing. Most of the planes (300) or (n00) do not pass through lattice points and can hardly have physical significance. We conclude that the reciprocal lattice may appear as a rather artificial concept, with most of its points associated with unphysical planes of the direct lattice. Actually, the construction is justified by its usefulness, in particular when diffraction by a crystal is concerned.

1.2.4 Experimental determination of crystal structures

Crystal structures are determined by diffraction of electrons, X-rays, or neutrons. The physical basis for diffraction experiments lies in the interference effects produced by phase differences between rays *elastically* scattered from different atoms in the crystal. Each atom in the sample may be considered to be a source of secondary spherical waves whose strength is controlled by the scattering power of the atom f_j (atomic form factor for electrons and X-rays, scattering length for neutrons). The sample is irradiated by a collimated beam of monochromatic (or monochromatized) rays with a wavevector $\boldsymbol{k}$; the scattered rays may be considered as families of parallel beams with wavevector $\boldsymbol{k}'$, with $|\boldsymbol{k}'| = |\boldsymbol{k}|$. Consider first the simple case illustrated in Fig. 1.5. Two atoms labelled A and B scatter the incoming beam; A is chosen as the zero of coordinates and B has position vector $\boldsymbol{\rho}$ relative to A. The phase difference caused by the displaced position of atom B relative to atom A is

$$\Delta = \boldsymbol{\rho} \cdot (\boldsymbol{k}' - \boldsymbol{k}) = \boldsymbol{\rho} \cdot \Delta \boldsymbol{k}. \tag{1.6}$$

If any atom j in the sample at vector ρ_j relative to atom A is now considered, the signal amplitude in direction $\boldsymbol{k}'$ arising from the whole sample is

$$\sum_j \frac{\Phi_0 f_j}{r_j} \exp\{\mathrm{i}(\boldsymbol{k}' \cdot \boldsymbol{r}_{\mathrm{A}} - \omega t + \boldsymbol{\rho}_j \cdot \Delta \boldsymbol{k})\}. \tag{1.7}$$

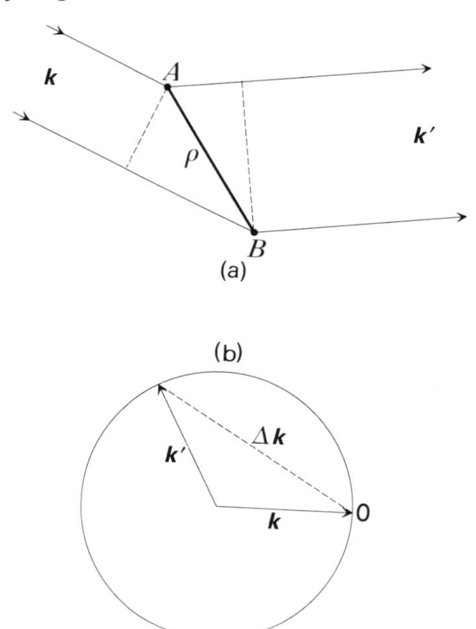

FIG. 1.5 (a) An incoming plane wave $\boldsymbol{k}$ is scattered by two 'atoms' A and B. One of the scattered waves has wavevector $\boldsymbol{k}'$. The scattering vector $\Delta\boldsymbol{k} = \boldsymbol{k}' - \boldsymbol{k}$ can be described by the Ewald construction (b): a sphere of radius $2\pi/\lambda$ is drawn in reciprocal space with the vector $\boldsymbol{k}$ appropriately oriented; the permitted diffracted waves are then determined by the reciprocal lattice points lying on this sphere.

Φ_0 is the amplitude of the incident beam. Actually, the only term in eqn (1.7) which changes significantly from atom to atom and thus can produce diffraction effects is

$$\sum_j f_j \exp\{\mathrm{i}(\boldsymbol{\rho}_j \cdot \Delta\boldsymbol{k})\}. \tag{1.8}$$

which, for a rectangular parallelepiped sample with size $\boldsymbol{Ma}$, $\boldsymbol{Nb}$, $\boldsymbol{Pc}$ containing MNP atoms has the magnitude:

$$\frac{\sin\frac{1}{2}\boldsymbol{Ma}\cdot\Delta\boldsymbol{k}}{\frac{1}{2}\boldsymbol{Ma}\cdot\Delta\boldsymbol{k}}\cdot\frac{\sin\frac{1}{2}\boldsymbol{Nb}\cdot\Delta\boldsymbol{k}}{\frac{1}{2}\boldsymbol{Nb}\cdot\Delta\boldsymbol{k}}\cdot\frac{\sin\frac{1}{2}\boldsymbol{Pc}\cdot\Delta\boldsymbol{k}}{\frac{1}{2}\boldsymbol{Pc}\cdot\Delta\boldsymbol{k}}. \tag{1.9}$$

Each factor in eqn (1.9) is just the expression governing the amplitude of light reflected from a linear grating. Absolute maxima are determined by

$$\sin\tfrac{1}{2}\boldsymbol{a}\cdot\Delta\boldsymbol{k} = 0$$

and cyclic permutation, i.e.

$$\Delta \boldsymbol{k} = n_1 \boldsymbol{A} + n_2 \boldsymbol{B} + n_3 \boldsymbol{C}, \tag{1.10}$$

n_i being integers.

This is nothing else than a vector of the reciprocal lattice and represents a family of plane $(n_1 n_2 n_3)$ in the direct lattice. Thus, each diffracted beam is associated with a particular family of crystal planes (see Fig. 1.5 for the corresponding Ewald construction) and gives a very sharp peak. If 2θ is the angle of the diffracted beam relative to the incident one, then

$$|\Delta \boldsymbol{k}| = 2|\boldsymbol{k}|\sin\theta = |\boldsymbol{G}_{n_1 n_2 n_3}| = \frac{2\pi}{d_{n_1 n_2 n_3}}$$

which leads to Bragg's law

$$2d\sin\theta = \lambda. \tag{1.11}$$

An equivalent expression for Bragg's law is obtained if we note that the projections of $\boldsymbol{k}$ and $\boldsymbol{k}'$ on $\Delta \boldsymbol{k} = \boldsymbol{G}_{hkl}$ (see Fig. 1.5) add in absolute values to $|\boldsymbol{G}_{hkl}|$, i.e.

$$\boldsymbol{k}' - \boldsymbol{k} = \boldsymbol{G}_{hkl}$$

which also gives

$$2\boldsymbol{G}_{hkl} \cdot \boldsymbol{k} \pm |\boldsymbol{G}_{hkl}|^2 = 0 \tag{1.12}$$

This means that waves with vector $\boldsymbol{k}$ originating at a reciprocal space site and ending in a plane perpendicular to the mid-point of $\boldsymbol{G}_{hkl}$ (or vice versa) are totally diffracted. These special $\boldsymbol{k}$ vectors define the so-called Brillouin zone which will be discussed and used at length later in the book.

Bragg's law is usually discussed in terms of reflections as a result of constructive interferences of rays diffracted from successive planes of a given family (kkl). From Fig. 1.6 it is easy to demonstrate that the general condition for Bragg reflections is:

$$2d_{hkl}\sin\theta = n\lambda,$$

where n is the order of diffraction. Equation (1.12) can be rewritten as

$$2(d_{hkl}/n)\sin\theta = \lambda,$$

which shows that the nth-order diffraction in planes (hkl) is equivalent to first-order diffraction in planes $(nh\,nk\,nl)$. Thus, if we choose to describe diffraction by crystals as always arising in first order, then we must introduce planes of the form $(nh\,nk\,nl)$ and the redundancy problems arising from the definition (1.3) of the reciprocal lattice is now resolved. Using the result that

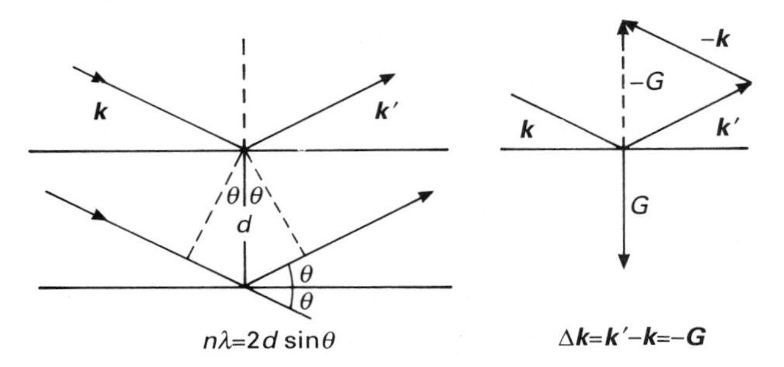

FIG. 1.6 Diffraction by lattice points in terms of Bragg reflection from parallel planes.

the wavevector difference $\Delta\boldsymbol{k}$ must be a vector of the reciprocal lattice for a net diffraction signal, eqn (1.8) becomes

$$\sum_j f_j \exp\{\mathrm{i}(\boldsymbol{\rho}_j \cdot \boldsymbol{G}_{hkl})\}. \tag{1.13}$$

This is called the structure factor, F_{hkl}, associated with the planes (hkl). Substituting eqn (1.3) for $\boldsymbol{G}_{hkl}$ and using the coordinates $u_i v_j w_j$ of the atoms at vectors ρ_j in units of the cell vectors, we find

$$F_{hkl} = \sum_j f_j \exp\{\mathrm{i}2\pi(hu_j + kv_j + lw_j)\}. \tag{1.14}$$

To illustrate this, let us consider the simple cubic lattice with only one atom at 000. Equation (1.14) gives

$$F_{hkl} = f\exp(\mathrm{i}\,2\pi\,0) = f,$$

which means that all values of h, k, l are permitted. The possible values of the Bragg angle θ are deduced from

$$\sin^2\theta = \frac{\lambda^2}{4d_{hkl}^2},$$

with

$$d_{hkl}^2 = a^2/(h^2 + k^2 + l^2).$$

The possible values of $(h^2 + k^2 + l^2)$ are 0, 1, 2, 3, 4, 5, 6,, 8, 9, 10, 11, 12, 13, 14,, 16, etc. A similar analysis can be carried out for a body-centred cubic lattice (atoms at 000 and $\frac{1}{2}\,\frac{1}{2}\,\frac{1}{2}$). If the atoms are identical:

$$F_{hkl} = \begin{cases} 2f \text{ when } h + k + l \text{ is even} \\ 0 \text{ when } h + k + l \text{ is odd} \end{cases}$$

and diffracted beams exist only for Bragg angles corresponding to $h^2 + k^2 + l^2 = 0, 2, 4, 6, 8, 10, 12, 14, 16, 18, 20, \ldots$. If the two atoms in the cell are different (CsCl structure) all the simple cubic diffracted beams arise, but with amplitude

$$F_{hkl} = \begin{cases} f_1 + f_2 \text{ when } h + k + l \text{ is even} \\ f_1 - f_2 \text{ when } h + k + l \text{ is odd} \end{cases}$$

Conditions on *hkl* and/or combinations of them are known as **extinction rules**. As a further example, for a face-centred cubic lattice h, k, l must be all odd or all even.

Extinction rules of the cubic structures are summarized in Table 1.2 and illustrated in Fig. 1.7.

To summarize the main important points of the above sections, it can be said that:

(1) diffraction by a periodic crystal gives sharp Bragg peaks;

(2) each Bragg peak is associated with a plane family, and then with (hkl) Miller indices and a point $\boldsymbol{G}_{hkl}$ in the reciprocal lattice. The scattering vector $\Delta \boldsymbol{k} = \boldsymbol{k} - \boldsymbol{k}'$ is equal to $\boldsymbol{G}_{hkl}$ (written also $\boldsymbol{Q}$ for short hereafter);

Table 1.2

Bravais	Miller indices allowed	First reflection
Simple cubic	all	(100)
Body-centred cubic	$h + k + l$ even	(110)
Face-centred cubic	h,k,l all odd or all even	(111)

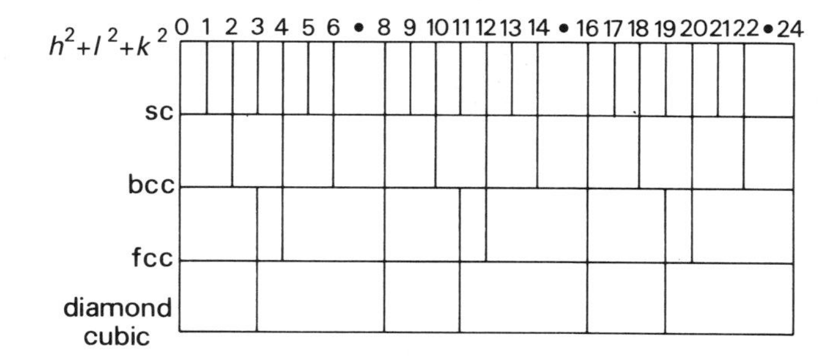

FIG. 1.7 Qualitative relative positions and extinction rules for the diffracted beams in cubic systems. The more atoms in the basis, the fewer Bragg peaks are produced.

(3) intensities of the Bragg peak are given by the second power of the amplitude of the structure factors F_{hkl}, which according to eqn (1.13) are the Fourier components of the atomic distribution in reciprocal space.

The latter statement suggests a way that may lead from diffraction data to structure. The drawbacks to be overcome are truncation effects and phase reconstruction problems, which, most of the time, preclude a direct calculation of the atomic density via the inverse Fourier transform of the experimental structure factor. Indeed, diffraction experiments cannot scan the entire reciprocal space and only a few of the complete set of Bragg peaks are actually available; moreover, only intensity can be measured, and then the phases of the structure factors cannot be extracted from data.

In practice, the space group of a structure under investigation is first determined by indexing the experimental positions of the Bragg peaks. Integrated intensities of these Bragg peaks are then compared with the equivalent Fourier components of a structure model compatible with the determined space group. Various iterative algorithms are currently used to fit the model to the data conveniently.

1.2.5 The notion of forbidden symmetries

As stated previously, there are only 230 space groups permitted. This means that any arbitrarily chosen point group symmetry cannot readily be combined with translational operations. Without demonstrating this general statement at length, an illustration taken from 2-dim space tilings may be useful for the understanding of what startled the world of crystallography so much with the advent of materials showing five-fold symmetry. As illustrated in Fig. 1.8, a 2-dim space can easily be tiled by rectangular, square, triangular, or hexagonal tiles. This is obviously due to these regular polygons having vertex angles equal to integral fractions of 2π. It is then possible to arrange lattice points in such a way that each of them is regularly surrounded by six triangular tiles (vertex angle equal to $2\pi/6$), four square tiles ($2\pi/4$), or three hexagonal tiles ($2\pi/3$). This is the illustration of the fact that two-, three-, four- and six-fold rotations are permitted symmetry operations for building the 230 3-dim space groups by combination with the 14 translational Bravais lattices.

Conversely, five-fold rotations and rotations of order over six-fold are forbidden by periodic space tiling. For a pentagonal tile, for instance, the vertex angles are equal to 108° which is contained 3.333 . . . times in 2π; it is then possible to distribute only three pentagons, plus an angular gap of 36°, around a given 'lattice point' (Fig. 1.8). Suppose also, as in Fig. 1.9, that we have anyhow a 2-dim Bravais lattice based on a pentagonal cell.

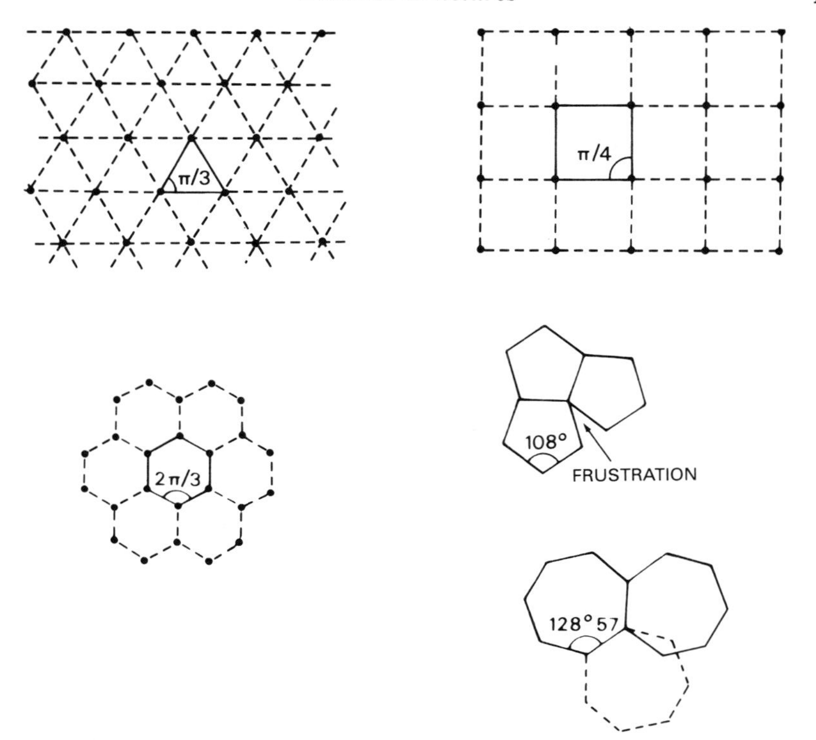

FIG. 1.8 Permitted tilings (triangle, square, hexagon) and impossible tilings (pentagon, heptagon, . . .) of the two-dimensional space.

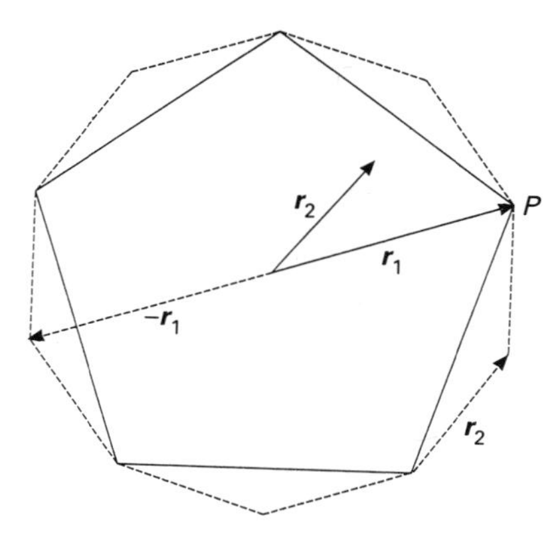

FIG. 1.9 Pentagonal and decagonal cells are not compatible with periodicity.

Having a lattice point at $\boldsymbol{r}_1$ and at the body centre imposes the need to have another point at $-\boldsymbol{r}_1$, which is not the case. This may be satisfied by choosing a decagonal cell (dotted line in Fig. 1.9), but now points translated from the central point by the edge vector $\boldsymbol{r}_2$ are not lattice points as they should be. Five-fold symmetry is definitely not compatible with translational operations.

Heptagons are not acceptable tiles either. Their vertex angles are equal to $128°57'$, which is contained 2.8 times in 2π; attempted tiling of the 2-dim space leaves angular gaps as large as 102°86 (Fig. 1.8).

More generally, vertex angles of a regular polygon with n edges are equal to $\pi(n-2)/n$. Periodic tiling is possible if these angles are integral fraction of 2π, i.e. $2n/(n-2)$ is an integer. Beyond pentagons and heptagons already mentioned, this number is 2.666 . . ., 2.571 . . ., 2.5, 2.444 . . . for successive polygons of higher orders, and it remains always greater than 2 (e.g. 2.083 33 . . . for tiles with 50 edges!).

1.3 Liquids, glasses, and amorphous alloys

1.3.1 Description of 'disordered' systems

As explained in the previous section, disorder in condensed matter is usually not considered in regular crystallography. However, real systems may depart from perfect translational order in many respects, as schematized in Fig. 1.10. First of all, a geometrically perfect crystal (Fig. 1.10(*a*)) may be chemically ordered (Fig. 1.10(*d*)) or chemically disordered (Fig. 1.10(*g*)). But the geometry itself may be disordered due to fluctuations in the length of the chemical bonds (bond disorder: Fig. 1.10(*b*) and/or in the number of bonds around 'lattice' sites (topological disorder: Fig. 1.10(*c*). All sorts of disorder may coexist, resulting in deeply chaotic materials, such as certain liquids or amorphous solids (Fig. 1.10(*e*), (*f*), (*h*), and (*i*)).

The leading question is now to describe the atomic structure of systems in which translational operation are absent. This may be done by introducing a set of vectors $\{\boldsymbol{r}_i\}$ where $\boldsymbol{r}_i$ refers to the position of the nucleus of the ith atom in real space. There are only a few systems where it is legitimate, as in an ideal gas, to treat $\boldsymbol{r}_i$ as an independent random variable within the total volume, any pair distance being equally probable. Thus it may be stated that disorder in (condensed) matter has nothing to do with complete randomness and the effects of physical/chemical packing constraints on the probability distribution of the set $\{\boldsymbol{r}_i\}$ must be considered as the main problem to deal with.

Another concern is the question of how physical properties 'propagate' through the system. In perfectly ordered crystals all physical attributes are strictly periodic. Any observable quantity (electron density, potential, etc.) has the mathematical property

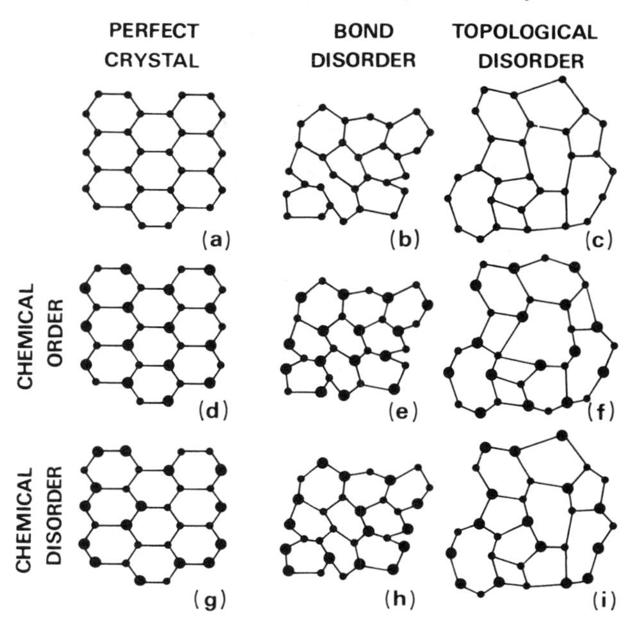

FIG. 1.10 A perfect periodic crystal (a) may be chemically ordered (d) or disordered (g); it may also contain bond disorder (b) and more complex topological disorder (c); both chemical and geometrical disorders may coexist (e, f, h, i).

$$F(\boldsymbol{r}) = F(\boldsymbol{r} + \boldsymbol{R}) \tag{1.15}$$

for all lattice vectors $\boldsymbol{R}$ and for any position vector $\boldsymbol{r}$. In disordered condensed systems, atomic potential (density) and chemical constraints certainly prevent 'short-range' properties from drastically changing while 'interstitial regions' can vary considerably. Thus, eqn (1.15) may be turned into a weakened form:

$$F(\boldsymbol{r} + \boldsymbol{r}_i) \simeq F(\boldsymbol{r}) \tag{1.16}$$

for each 'permitted' value of $\boldsymbol{r}_i$ and for $|\boldsymbol{r}|$ smaller than, say, something of the order of an 'atomic radius'. Thus the system contains many similar regions, which might be reminiscent of related crystals, but which are packed in a somewhat 'random' manner. This is due to the relative magnitudes of the energy ingredients entering in condensed matter. To turn a collection of isolated atoms into a solid or a liquid, interatomic distances have to be decreased from infinity of a couple of (Å), with energy changes of the order of several eV. An additional energy of tenths of 1 eV may correspond to the generation of a favourable chemical and/or short range order. Forcing the system to periodicity, or long-range order, requires only an additional few hundredths of 1 eV, typically.

In the absence of periodic order, the general formula (1.1) giving lattice sites in crystals must be forgotten. A catalogue of the set of vectors $\{\boldsymbol{r}_i\}$ giving atomic centres is not the best description because of the indefinitely large number of atoms in the system. Thus statistical parameters have to be introduced. Formally, these parameters are the s-body densities $n(\boldsymbol{r}_1, \ldots, \boldsymbol{r}_s)$ defined by the probability $\mathrm{d}P$ of finding atoms centred at $\boldsymbol{r}_1, \boldsymbol{r}_2, \ldots, \boldsymbol{r}_s$, such as

$$\mathrm{d}P = n(\boldsymbol{r}_1, \boldsymbol{r}_2, \ldots, \boldsymbol{r}_s)\, \mathrm{d}\boldsymbol{r}_1 \mathrm{d}\boldsymbol{r}_2, \ldots, \mathrm{d}\boldsymbol{r}_s.$$

Assuming that the average density of atoms per unit volume is spatially constant, i.e. $n(\boldsymbol{r}_1) \equiv n \equiv N/V$, allows to introduce normalized atomic distribution functions:

$$G(\boldsymbol{r}_1, \boldsymbol{r}_2, \ldots, \boldsymbol{r}_s) = \frac{1}{n^s} n(\boldsymbol{r}_1, \boldsymbol{r}_2, \ldots, \boldsymbol{r}_s), \tag{1.17}$$

which measure the probability of finding s atoms at given positions $\boldsymbol{r}_1, \boldsymbol{r}_2, \ldots, \boldsymbol{r}_s$ relative to all hypothetical possible arrangements of these s atoms in all other possible assemblies of the same average density.

Almost all the directly observable evidence about atomic arrangements in condensed matter is contained in the pair distribution function $G(\boldsymbol{r}_1, \boldsymbol{r}_2)$, except for possible measurements using X-ray near edge structure (XANES) which have access to higher-body distribution functions. Because of the assumed spatial homogeneity $G(\boldsymbol{r}_1, \boldsymbol{r}_2)$ depends only on the separation vector $\boldsymbol{r}_{12} = \boldsymbol{r}_2 - \boldsymbol{r}_1$, i.e.

$$G(\boldsymbol{r}_1, \boldsymbol{r}_2) \equiv G(\boldsymbol{r}_{12}) \equiv G(\boldsymbol{r}) = \frac{1}{n^2} n(\boldsymbol{r}).$$

Thus $G(\boldsymbol{r})$ measures the probability density of finding two atoms separated by a vector $\boldsymbol{r}$, relative to all possible assemblies having the same average density n. To illustrate this, $G(\boldsymbol{r})$ can be calculated for typical systems (Fig. 1.11):

1. In an ideal gas, any pair distance is equally probable, beyond some value related to atomic radius

$$G(\boldsymbol{r}) = 1 \qquad \text{for } |\boldsymbol{r}| \geqslant d.$$

2. In a perfect single crystal, pair distances are restricted to vectors of the lattice. Thus $G(\boldsymbol{r})$ is a set of delta functions at the lattice sites, i.e.

$$G(\boldsymbol{r}) = \frac{1}{n} \sum_{\boldsymbol{R}} \delta(\boldsymbol{r} - \boldsymbol{R}).$$

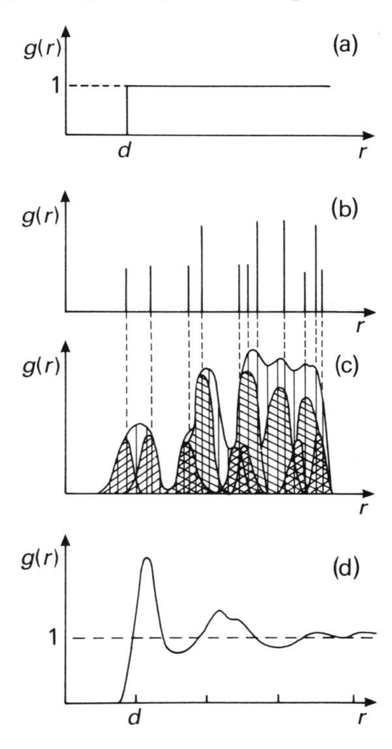

FIG. 1.11 Schematics for the radial distribution function of (a) an ideal gas (b) a perfect ideal crystal, (c) a 'real' crystalline system powdered, and (d) an amorphous alloy.

In a powder specimen, there are similar grains in all possible orientations. The pair distribution function may be averaged over all directions of the lattice vector $\boldsymbol{R}$, which gives the radial distribution function $g(r)$:

$$g(r) = \sum_{R} \frac{n(R)}{n} \delta(r - R),$$

where $n(R)$ is the number of differently oriented lattice vectors (multiplicity) that have the same length. The peaks in $g(r)$ can be significantly broadened by local disorder, thermal vibration, and/or instrument resolution effects (Fig. 1.11(*c*)).

3. Absence of long-range order with only relatively weak fluctuations in the nearest-neighbour distances will have the effects of broadening in the short-distance peaks and of smearing out the distant peaks of $g(r)$

into a uniform continuum averaging to 1 at some large pair distance (Fig. 1.11(*d*)). Obviously, maps of the atomic arrangements cannot be constructed by formal analytical operations on $g(r)$ but can only be guessed at and shown to be not inconsistent with it. A sample of $g(r)$ for different atomic arrangements as shown in Fig. 1.12 suggests however, that valuable insight into the structure of condensed matter can be gained by measuring the radial distribution function.

For complex systems with different atomic species, $g(r)$ is a weighted sum of partial radial distribution functions. For instance, in a binary alloy $g(r)$ is written

$$g(r) = c_1^2 g_{11}(r) + 2c_1c_2g_{12}(r) + c_2^2 g_{22}(r) \tag{1.18}$$

in which c_i is the concentration of atoms i and g_{ij} the distribution function of pairs ij. It is easy to see the usefulness of measuring *partial* radial distribution functions if the structure of real systems is to be deciphered.

Generally speaking, what we can learn from $g(r)$ is shown schematically in Fig. 1.13. The positions of the first maxima correspond to the average pair distance of first, second, . . . neighbours. The widths of the peaks measure fluctuations in these distances. The area under each peak gives the atom

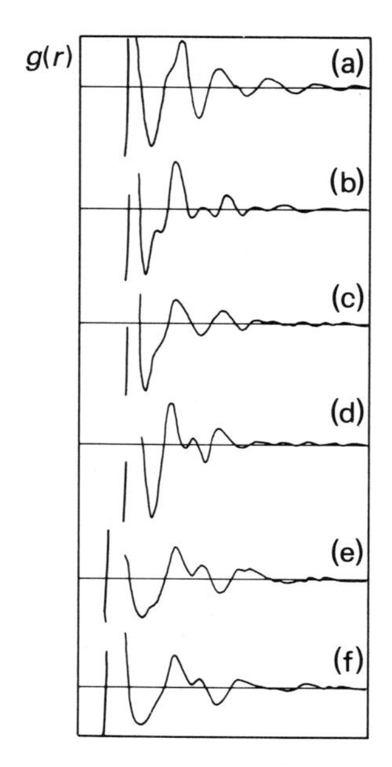

FIG. 1.12 Samples of radial distribution functions for assemblies of microcrystalline systems with (a) bcc, (b) fcc, (c) hcp, (d) A15 symmetries, or (e) a dense random packing of hard spheres, compared with (*f*) a typical experimental pattern for amorphous alloys.

numbers in the successive coordination shells. Finally, the coherence length, or range, of order L can be defined as the distance at which $g(r)$ oscillations around 1 are dying out.

The question is how to measure $g(r)$ in diffraction experiments.

1.3.2 Diffraction with disordered systems

The structure factor of an assembly of atoms can be expressed by eqn (1.8), which may be rewritten as

$$F(\boldsymbol{Q}) = \sum_j f_j \exp\{\mathrm{i}(\boldsymbol{r}_j \cdot \boldsymbol{Q})\}.$$

$\boldsymbol{Q} = \boldsymbol{k} - \boldsymbol{k}'$ has been substituted for $\Delta \boldsymbol{k}$ and $\boldsymbol{r}_j$ for $\boldsymbol{\rho}_j$ for convenience. As intensities only are obtained from diffraction data, the experimentally meaningful parameter is the interference function

$$S(\boldsymbol{Q}) = \frac{1}{N} \sum_{i,j} \exp\{\mathrm{i}\boldsymbol{Q}(\boldsymbol{r}_i - \boldsymbol{r}_j)\}. \tag{1.19}$$

(written here for an assembly of N identical atoms).

$S(\boldsymbol{Q}) = 1$ for all $\boldsymbol{Q}$ value in the special case of a gas-like disorder and is different from zero for $\boldsymbol{Q} = \boldsymbol{G}_{hkl}$ in perfect periodic crystals. Thus, in a partially ordered structure, the extent to which $S(\boldsymbol{Q})$ deviates from unity

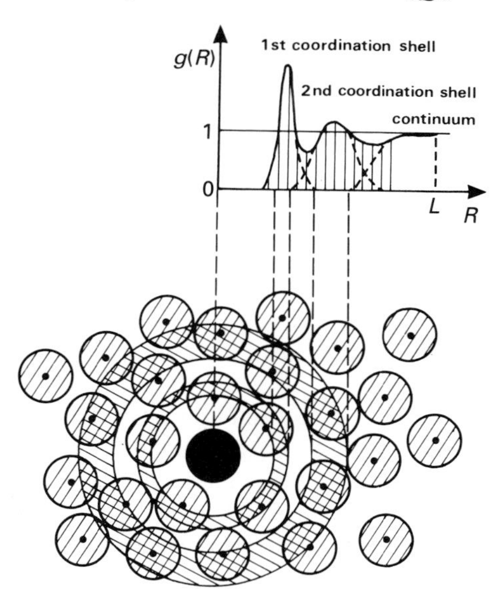

FIG. 1.13 Radial distribution functions give average pair distances (peak positions), average fluctuations in these distances (peak width), average coordination numbers (peak area), and coherence length (limit of density oscillations).

is a measure of the residual order. Calculation of $S(\mathbf{Q})$, or $S(Q)$, over all sites of a macroscopic specimen of atomic density $n = N/V$, in the basic assumptions of homogeneity and macroscopic isotropy, is simply an average calculation which ends with

$$S(Q) = 1 + n\int_0^\infty \{g(r) - 1\} \frac{\sin Qr}{Qr} 4\pi r^2 \mathrm{d}r. \qquad (1.20)$$

By Fourier inversion of eqn (1.20), the radial distribution function $g(r)$ is obtained as deduced from the experimental data $S(Q)$, i.e.

$$g(r) = 1 + \frac{1}{(2\pi)^3 n}\int_0^\infty \{S(Q) - 1\} \frac{\sin Qr}{Qr} 4\pi Q^2 \mathrm{d}Q. \qquad (1.21)$$

Despite its analytical simplicity the Fourier inversion of an observed interference function is not that easy from a practical point of view. The atomic density n has to be known beforehand and errors arise from truncation of the integral at large and small values of Q. As in crystallography, it is often more efficient to make a direct comparison with a model in calculating the interference functions by an easy Fourier transform of the model into 'reciprocal space'. Equations (1.20) and (1.21) also apply to partial interference functions in the case of many-fold atomic systems. In the total $S(Q)$, the $S_{ij}(Q)$ partials are weighted by a composition term $c_i c_j$ and a scattering term $f_i f_j$. This allows us to determine experimentally the partials S_{ij} when contrast variation measurements are possible (changes of $f_i f_j$ by isotope or isomorphous substitution and anomalous scattering effects near an absorption edge).

Typical structure factors, as measured in different systems, are pictured in Fig. 1.14. In disordered systems, one fairly well-defined ring is surrounded with fainter rings outside it. Roughly speaking, it can be said that radiation samples density fluctuation over pieces of material whose size scales with Q^{-1}. Thus at very small Q, substantial volumes of matter are looked at and $S(Q)$ is small due to average homogeneity. As Q increases, density fluctuations are detected and $S(Q)$ reaches a maximum for Q^{-1} comparable to the interatomic distances. Beyond this point, further oscillations about unity are mainly due to the sharp cut-off in $g(r)$ at distances smaller than the close interatomic spacing.

Between perfect periodic crystals and liquids there is a place for any level of disorder. Disorder in a periodic crystal may be viewed as atoms displaced out of the regular atomic planes (hkl). These displaced atoms no longer contribute to the structure factors F_{hkl} and the intensities of the Bragg peaks at $\mathbf{G}_{hkl}$ are reduced in proportion. The less dense the plane (hkl), the more affected are the corresponding Bragg intensities. This means that Bragg peaks with large Q values die out more quickly when disorder increases. Now, the displaced atoms scatter the incident beam out of the reciprocal

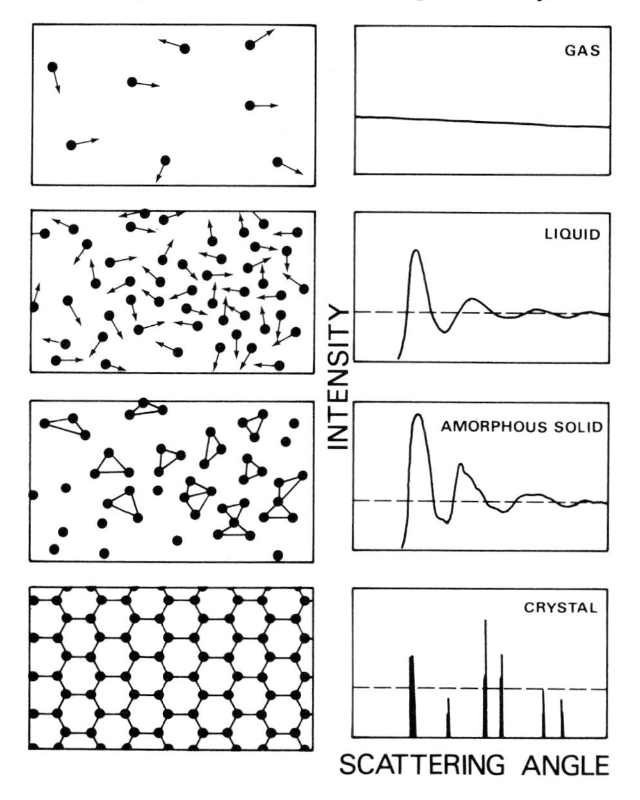

FIG. 1.14 Pictorial representation of atom distribution in matter with their expected typical diffraction patterns.

lattice points, yielding the so-called diffuse scattering in between the Bragg peaks. For liquids and amorphous solids the diffraction pattern may be considered as modulated diffuse scattering, the modulations arising from effects of residual order. For a gas-like system, the diffracted signal is totally flat diffuse scattering. It can then be inferred that, for any kind of structure, fluctuations in short-range order relative to an average long-range structure will mainly show up in diffuse scattering, which as such is a very valuable experimental tool.

1.4 Quasiperiodicity: another type of long-range order

1.4.1 A one-dimensional example of non-periodic long-range order

As introduced in Section 1.2, long-range order in an ideal infinite periodic crystal is invariant under a lattice of translation vectors defined in eqn (1.1). Because of the periodicity, the density may be expanded in a Fourier series, with non-zero values of the Fourier components only for vectors of the reciprocal lattice as defined by eqn (1.3).

However, periodicity is not a requisite for long-range order. For the sake of illustration of this statement, let us consider the very simple one-dimensional periodic structure made of lattice points equally spaced on a straight line. The density of these 1-dim lattice points is given by

$$\rho(x) = \sum_n \delta(x - na), \tag{1.22}$$

with a the lattice parameter. The Fourier components of such a structure are a series of sharp peaks defined by a single integer (Miller index) h according to

$$F_h = \sum_h \delta(Q_h - 2\pi h/a). \tag{1.23}$$

If a one-dimensional structure is built by superimposing two series of the sort expressed by eqn 1.22), but with two different repetition lengths, the density now becomes

$$\rho(x) = \sum_{n,m} \delta(x - na) + \delta(x - \alpha ma). \tag{1.24}$$

This structure is still long-range ordered, in any case, but is not periodic if α, the ratio of the two periods, is not a rational number, i.e. there is no possible coincidence in space between the two periodic components of the whole structure. Each of these periodic components can be expanded in a Fourier series similar to eqn (1.23), which add to give the set of Fourier components of the non-periodic structure (1.24) as follows:

$$F_{h,h'} = \sum_{h,h'} \delta\{Q_{hh'} - 2\pi/a\,(h + h'/\alpha)\}. \tag{1.25}$$

$F_{hh'}$ is now defined by two independent integers h, h', but is still a sharp function, as for a periodic structure. This means that a reciprocal space for $\rho(x)$ (eqn 1.24) may still be an interesting notion, but the density in this reciprocal space is higher than for the regular periodic structure, and may even be very large because the multiples of the irrational number α modulo one cover the unit interval in a dense manner.

Non-periodic long-range ordered structures can be built using any non-random procedure to generate atomic arrangements. A very useful one-dimensional example is the so-called **Fibonacci chain**. There are several ways to obtain this sequence of intervals. One is to start with a finite sequence of two segments, one short, S, and one large, L, and to operate with the iterative rule $S \to L$ and $L \to LS$ to build successive strings with increasing length. Not all starting sequences are allowed. For an appropriate choice, for example just L, infinite repetition of the operation gives an infinite sequence of L, S which is obviously perfectly ordered at long distance. At any point in the chain, the type of segment (L or S) that must be found is uniquely determined by the chosen starting sequence. According to the construction rule, the growing strings in the successive iterations are

$$
\begin{array}{l}
L \\
LS \\
LSL \\
LSLLS \\
LSLLSLSL \\
LSLLSLSLLSLLS \\
\vdots \\
\text{etc.}
\end{array}
$$

If the ratio $L/S = \tau$ is an irrational number, the sequence has no repetition distance. The canonical Fibonacci sequence corresponds to $\tau = 2\cos 36° = (1 + \sqrt{5})/2 = 1.618\,034\ldots$, the so-called golden mean. This golden mean is an interesting irrational number in the context of icosahedral symmetries (e.g. the distance ratio (centre of vertex)/(centre to mid-edge) in a regular pentagon is equal to $\tau/2$). Interesting properties, sometimes very strange, of τ are mentioned in the problem section of this chapter.

It can be demonstrated that the frequency ratio of occurrence of L and S is also the golden mean τ, or at least that this ratio has a limit equal to τ for the chain grown *ad infinitum*. The distance D from the origin of the first L segment to the end of the nth segment, through the (L, S) sequence, can be calculated and is equal to:

$$D = S\{n + (1/\tau)E\lfloor (n + 1)/\tau \rfloor \}, \tag{1.26}$$

in which $E\lfloor (n + 1)/\tau \rfloor$ is the integer part in $(n + 1)/\tau$ (e.g. $E\lfloor (N + 1/\tau \rfloor = 0, 1, 1, 2, 3, 3, 4, 4, 5, 6, 6$ *for* $n = 0 \ldots 10$, respectively). The existence of eqn (1.26) is evidence of the perfect long-range order in the sequence. It is also easy to verify that each successive string as obtained in a given iteration is simply the concatenation of the strings obtained in the two preceding iterations. Thus, if f_n denotes a given string, $f_{n+1} = f_n f_{n-1}$, with $f_1 = S$, $f_2 = L$ and $n = 3, 4, \ldots$. The Fibonacci chain will be used in the next section to illustrate the point that rather complicated non-periodic structures may have a well defined set of Fourier vectors, as long as they have a proper long-range order.

1.4.2 The sharp diffraction peaks of a Fibonacci chain

The complete approach to the Fourier transform analysis of a Fibonacci chain has elegantly been presented by Levine and Steinhardt.[1,2] Actually, the diffraction pattern of a Fibonacci structure is one of its most distinctive features. The pattern consists of a set of Bragg peaks that densely fill the reciprocal space. This result can intuitively be understood by considering the atomic positions given by eqn (1.26), which can be renormalized to

$$x_n = n + 1/\tau E \lfloor (n + 1)/\tau \rfloor . \qquad (1.27)$$

(This corresponds to a Fibonacci chain with segment length equal to 1 and τ). The first term alone corresponds to a periodic spacing equal to one. The second term increases by τ^{-1} each time n is increased by τ. Thus, x_n can be divided into a sum of two functions that describe periodic spacing but with incommensurate periods, in a way somewhat similar to the example of eqn (1.24). If the first term only were kept, the diffraction pattern would consist of Bragg peaks spaced periodically in reciprocal space with some fundamental period Q_1. Because the second term is incommensurate, it leads to Bragg peaks at some incommensurate reciprocal space period Q_2. The full pattern then consists of the union of the two sets of peaks, plus peaks at linear combinations of Q_1 and Q_2. As the two are incommensurate, the peaks fill the reciprocal space densely, though in a countable non-fractal way.

The above intuitive statements can be made more quantitative by computing the Fourier components of the density distribution described by eqn (1.27). These Fourier components are written classically according to eqn (1.8):

$$F(Q) = \lim_{N \to \infty} \frac{1}{N} \sum_n \exp(iQx_n), \qquad (1.28)$$

where we are summing over the N atomic positions in the Fibonacci chain. It can be demonstrated that with x_n positions given by eqn (1.27), non-zero Fourier components are obtained at vectors $Q_{hh'}$ such that

$$Q_{hh'} = \frac{2\pi\tau^2}{\tau^2 + 1}(h + h'/\tau), \qquad (1.29)$$

where h and h' are integers.

Before substituting x_n and $Q_{hh'}$ into the expression for $F_{hh'}$ eqn (1.28), eqn (1.27) may be re-expressed as

$$x_n = n(1 + 1/\tau^2) + 1/\tau^2 - 1/\tau F\{ (n + 1)/\tau\},$$

where the curly brackets indicate the fractional part and we have used the identity $y = E\lfloor y \rfloor + F\{y\}$; terms with factor n have also been regrouped. The exponent in eqn (1.28) is now given by

$$\mathrm{i}Q_{hh'}x_n = \mathrm{i}2\pi n(h + h'/\tau) + \frac{\mathrm{i}2\pi}{\tau^2 + 1}(h + h'/\tau) - \frac{\mathrm{i}2\pi\tau}{\tau^2 + 1}(h + h'/\tau)\,F\{(n+1)/\tau\}.$$

Just adding and substracting h'/τ gives

$$\mathrm{i}Q_{hh'}x_n = \mathrm{i}2\pi[nh + h'((\mathrm{n}+1)/\tau)] - \mathrm{i}2\pi h'/\tau + \text{2nd term} + \text{3rd term (unchanged)}.$$

Now $-\mathrm{i}\,2\pi h'/\tau$ can combine with the former second term and $h'((n+1)/\tau)$ in the first term can be split as

$$h'((n+1)/\tau) = h'E\lfloor(n+1)/\tau\rfloor + h'F\{(n+1)/\tau\}$$

Regrouping the terms properly finally gives

$$\mathrm{i}Q_{hh'}x_n = \mathrm{i}2\pi(nh + h'E\lfloor(n+1)/\tau\rfloor) - \frac{\mathrm{i}2\pi}{1+\tau^2}(\tau h' - h) + \frac{\mathrm{i}2\pi\tau}{\tau^2+1}(\tau h' - h)F\{(n+1)/\tau\} \tag{1.30}$$

The first term in the final expression is an integer times $\mathrm{i}2\pi$, and therefore only contributes a factor of unity upon exponentiation in $F_{hh'}$ eqn (1.28). The second term is independent of n and so only contributes an overall phase factor to $F_{hh'}$. The last term is n-dependent and contributes to the sum in $F_{hh'}$ in an important way. Since $F\{(n+1)/\tau\}$ is a fractional part, it varies over the [0, 1] interval and the last term in eqn (1.30) lies between zero and $\mathrm{i}X \equiv (\mathrm{i}2\pi\tau/(\tau^2+1))(\tau h' - h)$ which, in turn, is *uniformly and densely* distributed over the interval $[0, \mathrm{i}X]$ owing to τ being an irrational number. This enables us to approximate the sum over n in eqn (1.28) by an integral in order to calculate the contribution of the last term in eqn (1.30) to $F_{hh'}$. Thus, eqn (1.28) now becomes

$$F_{hh'} = \exp\left(-\frac{\mathrm{i}2\pi}{1+\tau^2}(\tau h' - h)\right)\frac{1}{X}\int_0^X \exp(\mathrm{i}y)\,\mathrm{d}y$$

or

$$F_{h,h'} = \frac{\sin\left(\dfrac{\pi\tau}{\tau^2+1}(\tau h' - h)\right)}{\left(\dfrac{\pi\tau}{\tau^2+1}(\tau h' - h)\right)}\exp\left(\mathrm{i}\pi\frac{\tau-2}{\tau+2}(\tau h' - h)\right). \tag{1.31}$$

Finally, the Fourier transform of the Finonacci chain can be written:

$$F(Q) = \sum_{h,h'} F_{hh'} \delta(Q - Q_{hh'}). \tag{1.32}$$

The diffraction peaks are indexed by two integers, even though the structure is one-dimensional. The peaks are real sharp peaks, as for a regular periodic structure, but they form a very dense pattern since any h' value can be associated with a given h. Long-range ordered structures without periodicity, presenting such a singularly continuous Fourier spectrum, are called quasiperiodic. The corresponding materials are quasiperiodic crystals or quasicrystals for short. The appearance of more Miller indices than the physical dimensionality is typical of incommensurate crystals and quasicrystals.

According to eqn (1.31), the brightest spots occur for $Q_{hh'}$ with small $\tau h' - h$ or h/h' close to τ; that is when (h, h') are successive Fibonacci integers (1, 1), (2, 1), (3, 2), (5, 3) .... Outside this sequence, the peak intensities decrease strongly; for instance when the difference $h' - h$ is 1, 2, 3 ... times as large as that of two successive Fibonacci integers, the intensity is about 10 per cent, 5 per cent, 0.5 per cent ... only of that the closest brightest spot (Fig. 1.15). This result also applies to real 3-dim quasicrystals and explains why diffraction patterns exhibit isolated bright spots in an otherwise dark field (see Chapter 2).

Another interesting property of the Fibonacci Fourier transform is its self-similarity, i.e. any small part of it, however small, will always show a unique sequence of peaks after proper rescaling. It is easy to show that the sequence of Fourier vectors $Q_{hh'}$ as expressed in eqn. (1.29) is invariant when multiplied by any power of τ. This can also be seen by magnifying any part of Fig. 1.15.

In the following sections various procedures for building quasiperiodic structures will be briefly presented.

1.4.3 Orientational order in quasicrystals

One dimensional samples of quasi-periodic structures, as described in Section 1.4.2, cannot suggest aspects of a quasicrystal that are related to their 'point group' symmetry. The basic ingredients that must be injected into the generation of a quasicrystal can be summarized as follows:

1. *Quasiperiodic translational order*. The density function is quasiperiodic, and thus can be expressed as a finite sum of periodic functions with some incommensurate periods.
2. *Minimal separation between atoms*. This distinguishes the quasicrystal from a single set of sites obtained by superimposing two periodic lattices with periods whose ratio is irrational.

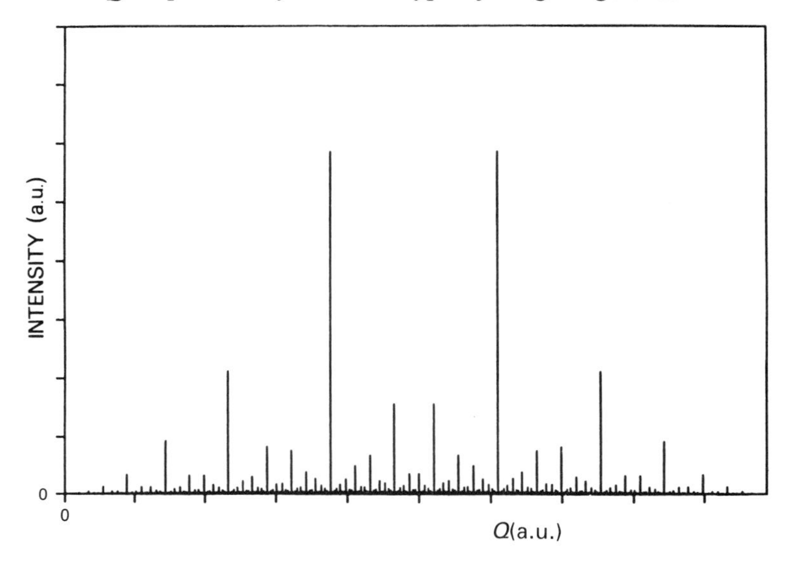

FIG. 1.15 Fourier spectrum of a Fibonacci chain calculated with 987 inflation steps.

3. *Orientational order*. The bond angles between neighbouring atoms, or clusters, have long-range correlations and are oriented, on the average, along a set of 'star' axes that define the orientational order.

A technique called the generalized dual or multigrid method[3,4,5] is the straightforward formal approach for obtaining quasicrystal packings with arbitrary orientational symmetry. A **grid** is a countable infinite set of infinite non-intersecting curves (in 2-dim) or surfaces (in 3-dim). An example of a very simple grid is shown in Fig. 1.16(*a*) in the form of a set of parallel equidistant straight lines. In 2-dim, an ***N*-grid** is a set of N grids such that each curve in the ith grid intersects each curve in the jth grid at exactly one point for each $i \neq j$. A two-dimensional 2-grid is shown in Fig. 1.16(*b*), with two sets of parallel equidistant straight lines at an angle with respect to each other. In 3-dim, an N-grid is a set of N grids such that any triplet of surfaces in the ith, jth and kth grids, respectively (for $i \neq j \neq k$), intersect at exactly one point. Associated with an N-grid is a star of vectors $\boldsymbol{e}_i$; $\boldsymbol{e}_1$ and $\boldsymbol{e}_2$ are indicated for the 2-grid of Fig. 1.16(*b*). This star of vectors defines the rotational symmetries of the lattice, or of the space tiling, resulting from the N-grid. For instance, the 3-grid shown in Fig. 1.17(*a*) with three star vectors at $2\pi/3$ from each other generates a tiling with a three-fold axis. This 3-grid is called **singular** because lines belonging to three different grids pass through

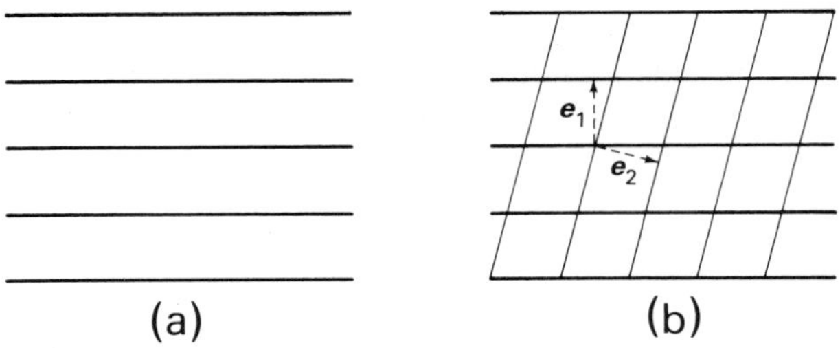

FIG. 1.16 (a) An example of a grid as a set of parallel equidistant straight lines and (b) a 2-grid with star axes ($\boldsymbol{e}_1$, $\boldsymbol{e}_2$).

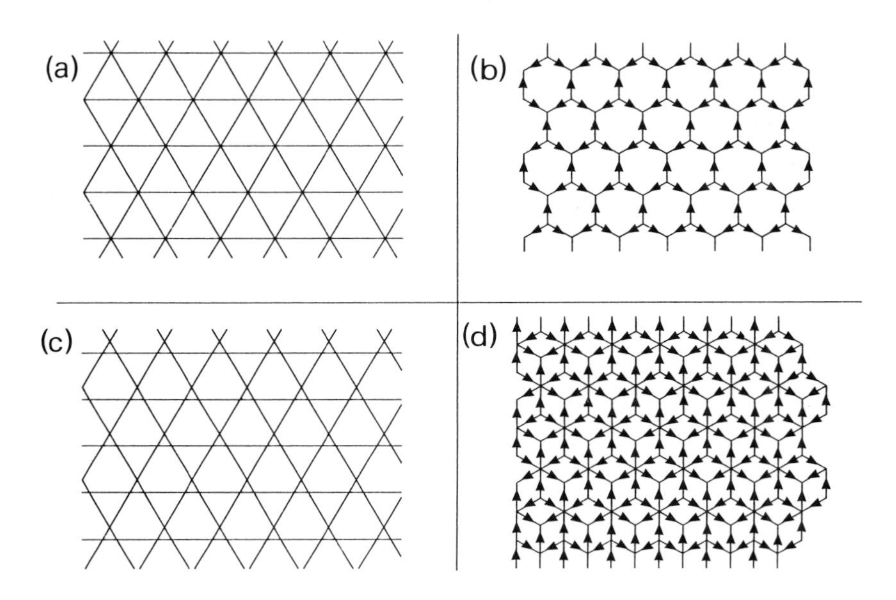

FIG. 1.17 Examples of 3-grids: singular (*a*) or regular (*c*), with their associated dual lattice points (*b* and *d*, respectively).

one point. If three lines of an *N*-grid never pass through one point, the *N*-grid is called regular. An example of a regular 3-grid with parallel grids and three-fold orientational symmetry is shown in Fig. 1.17(*c*). A regular *N*-grid (see the figure) may contain very small cells, too small for the vertices to satisfy the condition of minimal separation between atoms. One way to escape this difficulty is to replace the *N*-grid by its dual lattice. The dual transformation associates with each open region or cell of the *N*-grid a point in the dual space. In a 2-dim *N*-grid, the cells are polygons whose dual is a point placed at their mass centre. Duals of 3-grids are shown in Fig. 1.17(*b*) and (*d*). In a 3-dim *N*-grid, the cells are polyhedra whose duals are obtained by joining the mass centres of the faces; as an example, Fig. 1.18 presents regular polyhedra which are dual to each other.

Even the dual lattice of a *N*-grid may contain atom separations that are too short. This is illustrated in Fig. 1.19, which shows a **regular** 5-grid (or pentagrid) formed from five *periodically* spaced grids oriented normal to the five symmetry axes of a regular pentagons. Because five-fold rotations are not compatible with periodicity, this 5-grid does not generate a finite number of Voronoi cell shapes and the dual lattice cannot fulfil the condition of minimal separation between atoms (see Fig. 1.1(*b*) for the definition of a Voronoi cell). Conversely, a similar but singular 5-grid would fulfil this condition.

Now, the generalized dual multigrid method can be used to generate quasicrystal lattice. The grids must be quasiperiodic. For instance, parallel straight lines in 2-dim or parallel planes in 3-dim can be arranged in such a way that the spacings between consecutive lines or planes form a Fibonacci sequence. Then a 2-dim pentagrid or a 3-dim icogrid are formed by such

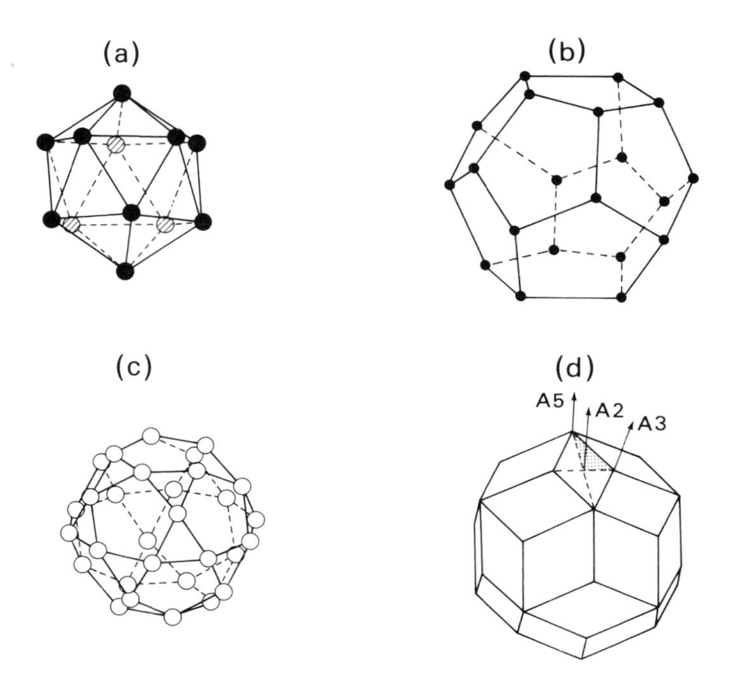

FIG. 1.18 Regular polyhedra with icosahedral symmetry: (a) icosahedron, (b) dodecahedron, (c) icosidodecahedron, and (d) triacontahedron; (a) and (b) on the one hand and (c) and (d) on the other are dual to each other.

FIG. 1.19 Regular periodic pentagrid with infinite number of Voronoi cell shapes.

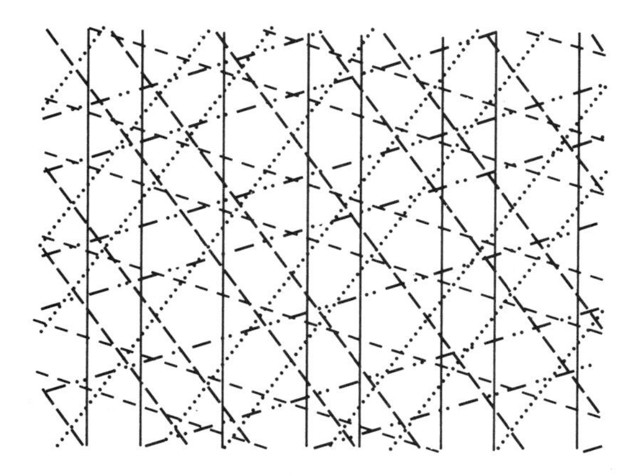

FIG. 1.20 Example of a 2-dim quasilattice obtained by superimposing five sets of Ammann lines (in Fibonacci sequences) oriented as the edges of a regular pentagon.

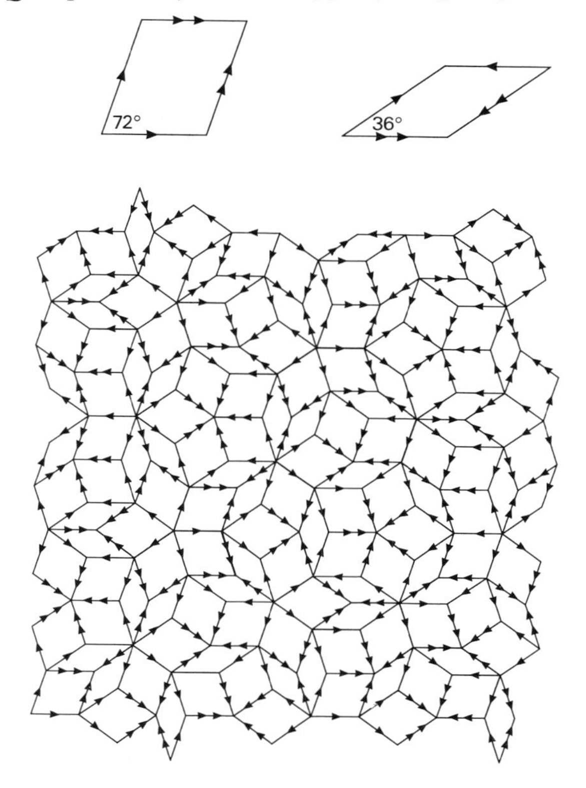

FIG. 1.21 Two-dimensional Penrose tiling with two lozenges as unit cells and edge matching rules (see text).

Fibonacci grids oriented normal to the five symmetry axes of a regular pentagon, or to the six five-fold axes of a regular icosahedron. The 2-dim Fibonacci pentagrid, or Ammann grid, is shown in Fig. 1.20. There are only a finite number (eight actually) of cell shapes, which ensures good characteristics in the dual lattice. The resulting connected space-filling packing of unit cells is a so-called 2-dim Penrose lattice (Fig. 1.21).[6] The forms of the tiles for a Penrose lattice are here two lozenges with equal edges, one with angles of 36° and 144° (skinny lozenge) and the other one with angles of 72° and 108° (fat lozenge). The dual operation that relates a 2-dim Penrose lattice to an Ammann grid is illustrated in Fig. 1.22. This example demonstrates the usefulness of the dual multigrid method in generating a quasicrystal lattice: the desired long-range quasiperiodic order is injected via the grid sequence; the chosen orientational order results from the selected star

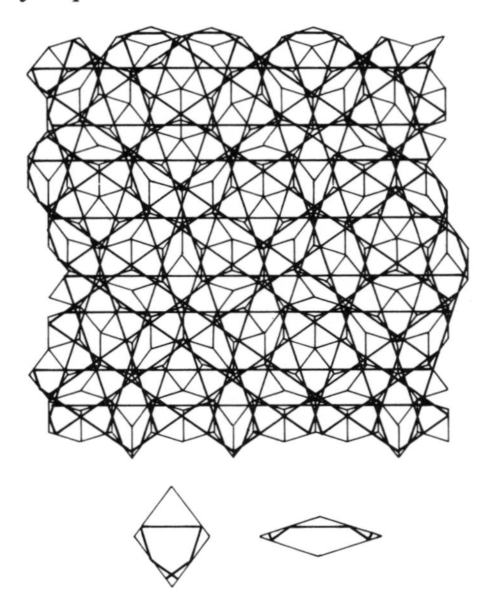

FIG. 1.22 Illustration of the dual operation that relates a 2-dim Penrose lattice to an Ammann grid.

of vectors for the N-grid; and the dual transformation maps the set of intersection of the N-grid into a connected space-filling packing of unit cells.

The 3-dim quasilattice generated from the 3-dim icogrid is also a (3-dim) Penrose lattice.[7] The forms of the tiles are now two rhombohedra, one oblate and the other one prolate, as shown in Fig. 1.23. An interesting result is that the computed Fourier transform of a generated 3-dim Penrose tiling compares qualitatively well with experimental electron diffraction pattern obtained from real quasicrystalline materials (Fig. 1.24). In particular, the five-fold orientational symmetries show up very clearly and bright spots are very sharp.

1.4.4 Direct quasiperiodic space tiling procedures

The quasiperiodic structures described so far have several properties which can usefully be analysed with the aim of building these structures directly. Self-similarity is one of these properties of interest, as illustrated in Fig. 1.25 for the case of the Fibonacci chain. Self-similarity implies that deflation/inflation operations transform the structure into itself within rescaling effects. The Fibonacci chain transforms into another Fibonacci chain with different tile sizes via the 'substitution' rule $L \to LS$ and $S \to L$ (Fig. 1.25).

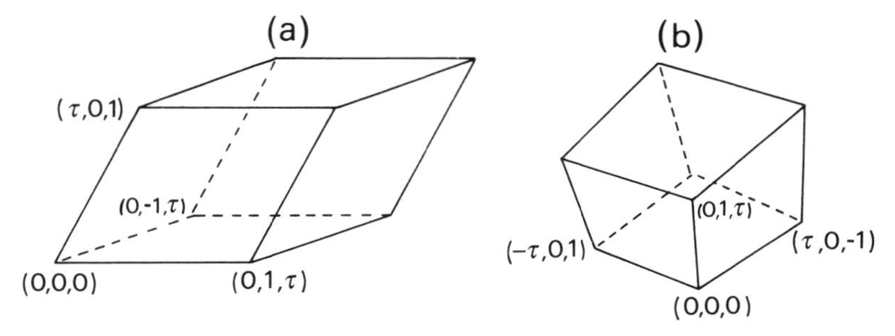

FIG. 1.23 Schematic of (a) prolate and (b) oblate rhombohedra that can be used for 3-dim Penrose tiling. Vertex coordinates are given in a frame of three orthogonal 2-fold axes of a regular icosahedron.

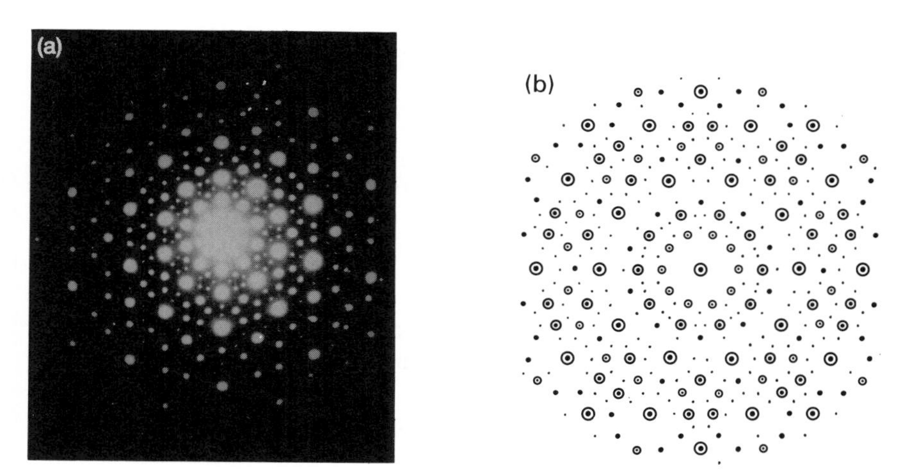

FIG. 1.24 Electron diffraction pattern of an AlMn quasicrystal (5-fold zone axis) (a) compared with the computed Fourier pattern of a 3-dim Penrose tiling (b) and as obtained perpendicular to a 5-fold axis.

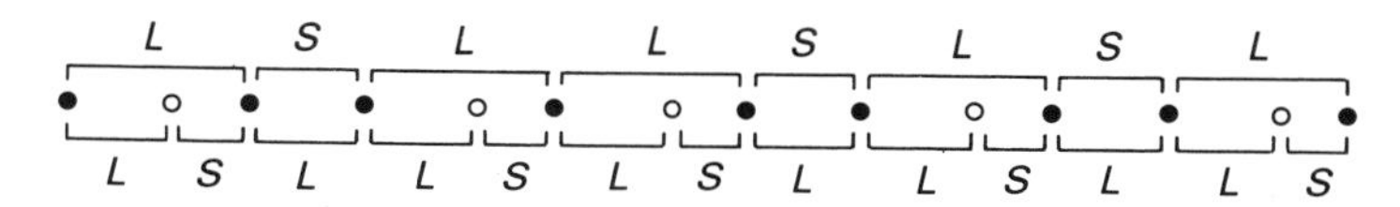

FIG. 1.25 The Fibonacci chain is a self-similar structure. The original sequence of L and S segments is indicated by solid disks. The deflated chain consists of all of the sites, solid and open disks. Both sequences are Fibonacci chains, identical within rescaling.

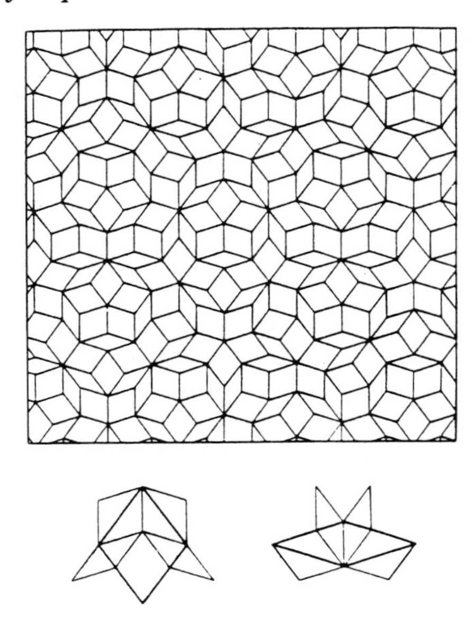

FIG. 1.26 Deflation rules for the two rhombic unit cells of a 2-dim Penrose tiling. Iteratively deflating an initial fat lozenge generates a growing portion of a Penrose tiling.

Thus, starting with a single line segment L and repeating the deflation interatively, we can generate a Fibonacci chain containing an arbitrarily large number of points.

Penrose tiling can also be generated by an inflation-deflation procedure. This can be deduced formally from the fact that Penrose lattice (2-dim and 3-dim) are related to Fibonacci sequences via the dual multigrid method. The self-similar properties of a 2-dim Penrose lattice are depicted in Fig. 1.26. Again here, deflating properly an initial 'fat' lozenge iteratively generates a growing piece of the quasiperiodic structure. It is worth pointing out that quasiperiodic structures are self-similar, but self-similarity generally does not ensure quasiperiodicity, even though it imposes some sort of long-range order.

One can now wonder whether it is possible to build directly a quasiperiodic lattice, given a set of tiles. The answer is no, and the number of tilings, ranging from periodic structures to various random tilings, for the two Penrose tiles (or any other set of tiles) is non-denumerable. In order to force quasiperiodicity one needs **matching rules**. For instance, different arrows are put on the edges of the tiles and it is required that tiles only touched by edges arrowed in the same way[6] (Fig. 1.21). Matching rules also exist for a 3-dim Penrose tiling[7, 8, 9]. These matching rules, although being a very strict prescription for space tiling, act in a somewhat subtle way. For a given number of unit cells, there are many different clusters that can be constructed without violation of the matching rules. But the number of these possible clusters, in strict accordance with the matching rules and without any holes and/or tears, decreases as the clusters are grown. Thus one finds it very difficult to build a perfect Penrose tiling adding tiles one by one, since many blind alleys are available at each step even if the matching rules are obeyed. In other words, a perfect Penrose tiling obeys the edge-matching rules but these edge-matching rules do not allow the tiling to grow without defects. The point will be analysed further in Chapter 6.

A comprehensive account of the properties of two-dimensional tilings may be found in the book by Grünbaum and Shepard.[10]

1.4.5 Quasiperiodicity as generated by projection or cut from higher dimensional space

As already mentioned, the typical features of quasiperiodic structures arise from the fact that their orientational symmetries are not compatible with space groups which are accepted in the relevant physical space. But these 'forbidden' point group symmetries may be accepted by periodic tiling at the cost of increasing dimensionality.

As explained in Section 1.2.5, the only rotational operations compatible with translations in a 2-dim or a 3-dim space are 2-fold, 3-fold, 4-fold, and 6-fold rotations. This means that the usable elementary tiles, or unit cells, are restricted in 3-dim to those corresponding to the 14 Bravais lattices (Fig. 1.2). Clearly, the polyhedra of interest are fully defined by *three* basis vectors that can be accommodated by the three dimensions of the space.

If now, say, 5-fold rotations are considered relevant 'unit cell' may be a regular pentagon in 2-dim or a regular icosahedron in 3-dim. A regular pentagon has five vertices, which are completely defined if and only if *four* of the vectors from centre to vertex are specified. Thus, one may anticipate that at least four space dimensions are required to accommodate periodic tilings made of tiles having the rotational symmetries of a pentagon. Similarly, a regular icosahedron is defined by *six* of the vectors from centre to vertex and, consequently, a 6-dim space (at least) is requested if one is to accommodate a periodic lattice with icosahedral symmetries.

The rotational symmetries of an icosahedron are the six five-fold axes already mentioned, plus 10 three-fold axes (from centre to triangular faces), 15 two-fold axes (from centre to middle edges) and the inversion operation. These operations constitute the icosahedral point group symmetries $m\bar{3}\bar{5}$. If the inversion is excluded, we are left with the point group 532. It can be demonstrated that there are 16 possible space groups in 6-dim formed by combining the two point groups with the three *6-dim cubic* Bravais lattices (primitive or simple (P), face-centred (FC), and body centred (BC). Now we are going to introduce the so-called **projection** and/or **cut procedures.**[11–14] These methods allow us to generate 3-dim quasiperiodic structures by merging our 3-dim physical space $R3_{\text{par}}$ into an n-dim hyperspace (R_n). For instance R_n is 6-dim (at least) for icosahedral symmetries; it is tiled by a 6-dim cubic Bravais lattice whose six edge directions project in $R3_{\text{par}}$ on to the six-fold axes of a regular icosahedron; the complementary space of $R3_{\text{par}}$ into $R6$ is called $R3_{\text{perp}}$; the 6-dim point group is isomorphic with the 3-dim one; and the point group rotations map the 6-dim Bravais lattice on itself, but also map the 3-dim subspaces $R3_{\text{par}}$, $\mathbf{R3}_{\text{perp}}$ into themselves. This constraint induces severe selections for the relevant 6-dim space group and, as we shall see later, allows proper indexing of experimental diffraction patterns.

To justify the relevance of high-dimensional periodic structures for quasicrystals, let us consider the expression of local mass density in a solid. In an ordinary periodic crystal, the mass density follows the translational invariances of the crystal and can be written as a Fourier series or sum of density waves:

$$\rho(\boldsymbol{r}) = \frac{1}{V}\sum_{\boldsymbol{G}} \rho(\boldsymbol{G})\exp(\mathrm{i}\boldsymbol{G}\cdot\boldsymbol{r})$$

where V is the unit cell volume and $\boldsymbol{G}$ are vectors of the reciprocal lattice, depending on only *three* integer linearly independent basis vectors $\boldsymbol{a}_i^*$ such that

$$\boldsymbol{G} = h\boldsymbol{a}_1^* + k\boldsymbol{a}_2^* + l\boldsymbol{a}_3^*$$

($\boldsymbol{G} = 0$ if and only if all h, k, $l = 0$). The $\boldsymbol{a}_i^*$ are said to span the reciprocal lattice. They are related to the 'basis' vector $\boldsymbol{a}_i$, which defines the unit cell of the crystal in physical space as previously explained.

In a quasicrystal, the Fourier transform of the mass density is again a Fourier series and the wavevectors in the Fourier sum also form a discrete reciprocal lattice. However, the number of integer linearly independent basis vectors required to 'span' the reciprocal lattice exceeds the spatial dimension, and the point symmetry of the reciprocal lattice is incompatible with periodic translational order. For example, six basis vectors are required to span the reciprocal lattice for three-dimensional quasicrystals with icosahedral symmetry:

$$\boldsymbol{G} = n_1\boldsymbol{a}_1^* + n_2\boldsymbol{a}_2^* + n_3\boldsymbol{a}_3^* + n_4\boldsymbol{a}_4^* + n_5\boldsymbol{a}_5^* + n_6\boldsymbol{a}_6^*$$

where the $\boldsymbol{a}_i^*$ can be selected to point along the five-fold axes of an icosahedron. Of course, the vectors $\boldsymbol{G}$ must be expressed in a three-dimensional space. In a well-chosen cubic coordinate system, the $\boldsymbol{a}_i^*$vectors are of the form $(\pm 1, \pm \tau, 0)$ (and permutations). Thus all the vectors $\boldsymbol{G}$ have cubic coordinates of the form $(h + \tau h', k + \tau k', l + \tau l')$ where h, h', k, k', l, l' are integral numbers (selected with extinction rules). This manifold may be the simplest possible definition of a quasicrystal. The expression of the reciprocal vectors $\boldsymbol{G}$ in terms of their cubic coordinates shows that, owing to the irrationality of τ, the quasiperiodic structures may exhibit some unusual properties when compared with regular crystals.

1. The $\boldsymbol{G}$ vectors form a dense pseudocontinuous set, and Bragg peaks are expected to show up 'everywhere' in the reciprocal space ($h + \tau h'$ are not integral numbers and their fractional parts densely fill the interval [0,1]).
2. There is not a 'smallest' basic $\boldsymbol{G}$ value as a consequence of property 1; this lack will be a difficulty when indexing experimental diffraction patters.
3. Quasiperiodic structures will obey some inflation rules, again due to the peculiar properties of the golden mean τ. For instance, if all h, h', k, k', l, l' integers are allowed by extinction rules, τ inflated structures are identical within rescaling (note that $\tau^{n+1} = \tau^n + \tau^{n-1}$).
4. The dense reciprocal space of a quasicrystal may be given a **periodic image** in a high-dimensional space. For instance, the equation giving $\boldsymbol{G}$ can be considered as the description of a six-dimensional periodic reciprocal lattice whose Fourier transform would generate a six-dimensional periodic mass density distribution. Projection and cut operation will then relate the physical three-dimensional description to its six-dimensional image.

To illuminate the high dimensional representation of quasiperiodic structures, perhaps less formally, let us take the simple example of a function which is periodic in 2-dim:

$$\rho(xy) = \sum_{n,m} \delta(x - na)\delta(y - ma).$$

This function is pictured in Fig. 1.27 as a distribution of density points at the vertices of a square lattice. Its Fourier components are distributed at the vectors of a square lattice with $2\pi/a$ spacing in reciprocal space and we may write

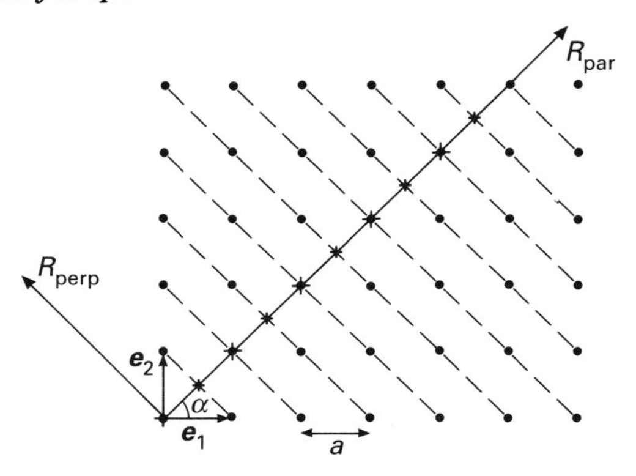

FIG. 1.27 A 2-dim square lattice (•) projects on a 1-dim space as a 1-dim periodic structure (×) as long as the 'slope' of the 1-dim space into the 2-dim one is rational with respect of lattice rows.

$$F_{hh'} = \sum_{h,h'} \delta(Q - Q_{hh'}),$$

with hh' the trivial integer Miller indices.

The lattice sites of the 2-dim square structure (Fig. 1.27) can be projected on to any 1-dim subspace as a straight line R_{par} (in Fig. 1.27) at an angle α with the horizontal rows of the square lattice. If the slope of R_{par} is *rational* with respect to rows of the square lattice, the projected 1-dim structure is a **discrete periodic** set of sites, since R_{par} passes repeatedly through lattice sites of the square lattice, and full rows of sites in 2-dim project at one single point in 1-dim. Conversely, if the slope of R_{par} is *irrational*, the projected 1-dim structure is no longer periodic and exhibits unphysically short atomic distances; it can even be demonstrated that the projected set of points is as dense as a singularly continuous distribution. One way to relax the density problem is to restrict projections on R_{par} to those points of the square lattice that are confined within a strip parallel to R_{par} and having a cross-section in the complementary space R_{perp} equal to that of the square unit cell. This is shown in Fig. 1.28. The projected structure in now made of two 'tiles' of length a cos$\alpha = L$ and a sin$\alpha = S$, respectively. Note that the strip width is equal to $\Delta = a(\cos\alpha + \sin\alpha)$. With the additional constraint that $\cos\alpha/\sin\alpha = \tau$, the golden mean, it is easy to check that the distribution of (L, S) segments over R_{par} now obeys a Fibonacci sequence.

Thus the strip/projection scheme provides an easy way to generate any non-periodic long-range ordered structure, when the high dimensional

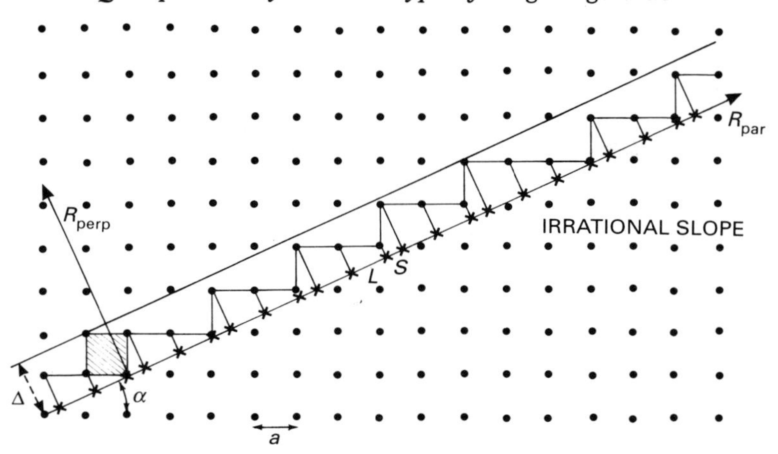

FIG. 1.28 A 1-dim illustration of the strip/projection method. The slope of R_{par} is irrational. Sites of the portion of the square lattice inside the strip project on R_{par} in a perfectly ordered non-periodic structure made of two tiles of incommensurate sizes. With a slope of R_{par} equal to τ, the Fibonacci chain is obtained.

periodic structure compatible with the required symmetry operations is known.

This scheme is also adequate when Fourier expansion of the structure is concerned. As already mentioned, the Fourier components of a square lattice are equally intense peaks distributed at the vectors $Q_{hh'}$ of the reciprocal square lattice. Now only a strip Δ of the 2-dim structure must be considered, which means that density has to be multiplied by 1 within the strip and 0 outside, say by a 'window' function $W(x_{perp})$ acting only in the complementary space. The Fourier pattern of the strip is then the convolution product of the Fourier pattern of the infinite square lattice by the Fourier transform $G(Q_{perp})$ of $W(x_{perp})$. The result is shown in Fig. 1.29. The Fourier components are still features centred at the reciprocal lattice points of the square structure; these features still have no extension in the reciprocal space R^*_{par} associated with R_{par} but oscillate in the reciprocal space R^*_{perp} associated with R_{perp}, according to

$$G(Q_{perp}) \cong \Delta \left(\frac{\sin Q_{perp}\Delta}{2}\right) \left(\frac{Q_{perp}\Delta}{2}\right)^{-1}, \tag{1.33}$$

with

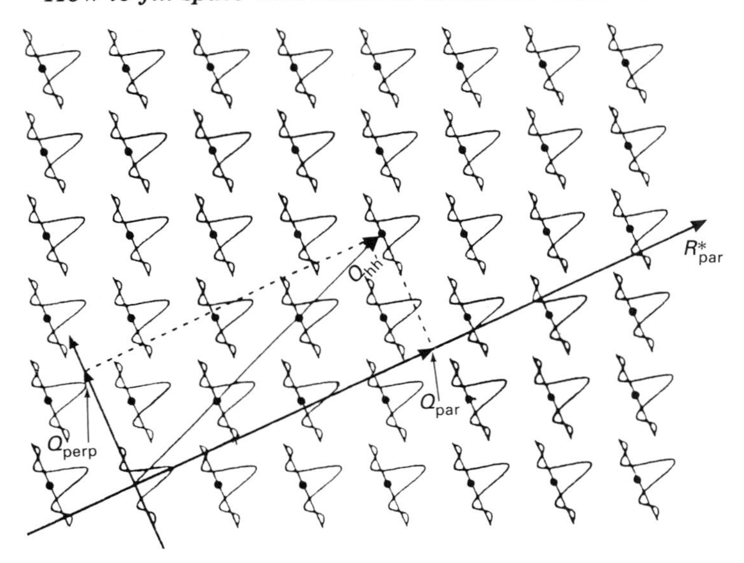

FIG. 1.29 Fourier pattern of the 2-dim strip shown in Fig. 1.28. Positions and intensities of the diffraction peaks of the Fibonacci chain correspond to the intersections by R^*_{par} of the full set of 2-dim diffraction features.

$$Q_{perp} = \frac{2\pi}{a} \frac{1}{(2+\tau)^{\frac{1}{2}}} (h - h'\tau).$$

Because of the non-rational slope of R^*_{par} into R^* (2-dim), all these Fourier components, though being centred at a $Q_{hh'}$ in 2-dim, have non-zero values on R^*_{par}, at positions given by

$$Q_{par} = \frac{2\pi}{a} \frac{1}{(2+\tau)^{\frac{1}{2}}} (h + h'\tau). \tag{1.34}$$

As the Fourier transform of a projection is a cut (and vice versa), the Fourier pattern (positions and magnitudes) of the 1-dim Fibonacci chain is then given by these features distributed at Q_{par} positions along R_{par}. The interested reader will readily verify that eqns (1.33) and (1.34) are identical to those directly obtained in Section 1.4.2. It will surely be appreciated how simple the derivation using the strip/projection method is when compared with the intricate calculations carried out directly in the physical space. Indexing by two integers is now straightforward and perfectly understandable. The dense singularity continuous aspect of the Fourier patterns is also

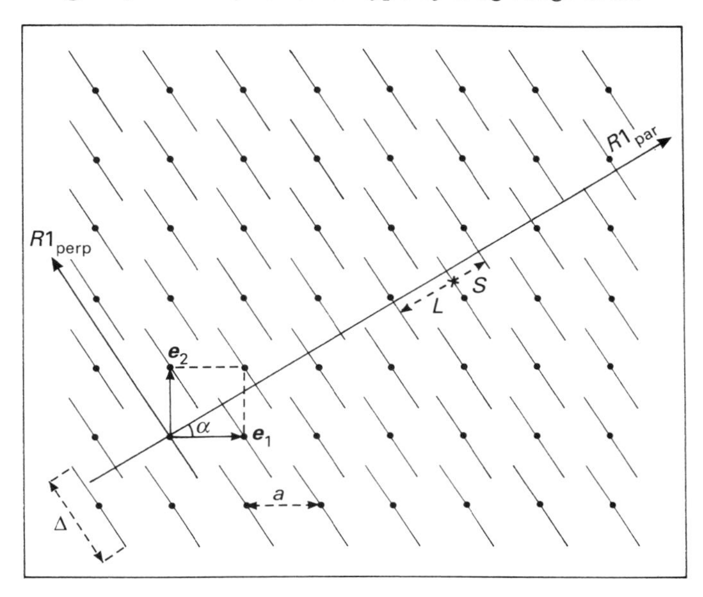

FIG. 1.30 Illustration of the cut-projection method. The basis associated with the 2-dim square lattice is a single line segment perpendicular to the physical 1-dim space. The cut of this 2-dim structure by R_{par} is a Fibonacci chain.

fully demonstrated, components with large values of Q_{perp} giving no significant contributions, as already pointed out.

A closely related method for generating quasilattices is the so-called cut procedure or embedding method.[15, 16] This approach again uses the fact that the Fourier transform of a projection is a cut and vice versa. But now the cut operation is in direct space and the projection in reciprocal space. The quasiperiodic structure in the physical space is obtained as 3-dim irrational cuts of extended objects which are periodically distributed in a n-dim space ($n > 3$). The main advantage of this method is that quasicrystals can be described by symmetry (space group), metric (lattice parameters), and distribution of density in an n-dim unit cell, as for regular structures. Again, a low dimensional example is used in Fig. 1.30 for an easy illustration of the method. In this example, a 1-dim quasiperiodic chain corresponds to the cut of a periodic square lattice arrangement of segments, called A_{perp} hereafter; these segments have been chosen flat in the complementary space. The density distribution in the 2-dim $\boldsymbol{R}$ space can be written:

$$\rho(\boldsymbol{r}) = \delta(\boldsymbol{r} - \boldsymbol{r}^{\text{lat}}) * A_{\text{perp}}, \tag{1.35}$$

where r^{lat} denotes the vertex positions in the 2-dim square lattice and * is a convolution product; the $\rho(r_{par})$ density distribution of the 1-dim quasicrystal is the cut of $\rho(r)$ by R_{par}. The length Δ of the A_{perp} segment is equal to $n_1 a^2$, in which n_1 is the average number of points per unit length of the quasicrystal; this results from density conservation via the cut procedure. In Fig. 1.30, the slope of R_{par} in R has been chosen to generate the Fibonacci chain ($\cos\alpha/\sin\alpha = \tau$); Δ is also equal to $a(\cos\alpha + \sin\alpha)$. The Fourier pattern of the 2-dim periodic density eqn (1.35) is now given by the expression

$$F(Q) = \sum_{hh'} \delta(Q - Q_{hh'})G(Q_{perp})/a^2, \tag{1.36}$$

in which $Q_{hh'}$ are the vectors of the reciprocal lattice associated with the 2-dim square lattice; they have two components $(2\pi/a)h$, $(2\pi/a)h'$). As previously introduced for the strip projection methods, Q scans the reciprocal space R^*, which is the sum of the reciprocal spaces R^*_{par} and R^*_{perp}; Q has physical and complementary components Q_{par} and Q_{perp} respectively; and $G(Q_{perp})$ is the Fourier transform of the segment A_{perp} and is expressed by a formula identical to eqn (1.33). $F(Q)$ is obviously indexable by the two integers (h, h') (Fig. 1.31) and so are the Fourier components $F(Q_{par})$ of the 1-dim quasicrystal density, as obtained via projection of $F(\mathbf{Q})$ on to R_{par} (Fig. 1.32), that is:

$$F(Q_{par}) = \sum_{hh'} \delta(Q_{par} - Q^{hh'}_{par})G(Q^{hh'}_{perp})/a^2, \tag{1.37}$$

with $Q^{hh'}_{par}$ and $Q^{hh'}_{perp}$ given also by eqns (1.33) and (1.34).

The unit cell of the n-dim periodic structure contains all the information needed to build the related quasicrystal, as illustrated in Fig. 1.32. This information is fully given by the symmetry features of the n-dim unit cell and the size and shape of the A_{perp} hypersurface. These A_{perp} hypersurfaces are piecewise connectable flat volumes in the complementary space as long as simple quasiperiodic tilings are concerned: straight line segments for a 1-dim Fibonacci sequence, regular pentagons for a 2-dim Penrose lattice, and triacontahedra for a 3-dim Penrose lattice.

The cut procedure is certainly the more generic and simpler way to generate quasiperiodic structures. So far, we have restricted the derivation to specific cases with $A3_{perp}$ hypersurfaces which are flat in the complementary space (no physical or parallel components). This is by no means a requisite and Fig. 1.33 presents a situation in which a small part of A_{perp} has been shifted parallel to the physical space. The result is changes in the length of some of the tiles, in an ordered fashion, but the same long-range order is

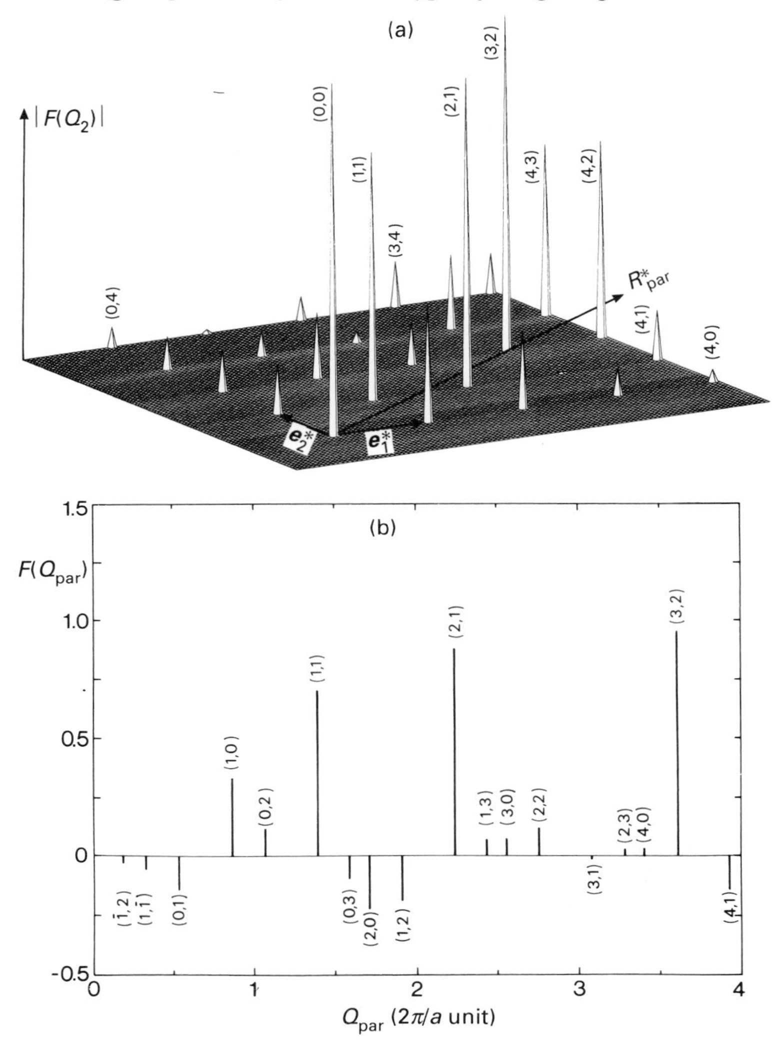

FIG. 1.31 Schematic drawing of the diffraction pattern corresponding to the 2-dim structure of Fig. 1.30. (a) Bragg peaks are positioned at the vector of a square reciprocal lattice, with intensity modulation according to $G(Q_{\text{perp}})$. (b) When projected on R^*_{par} (physical reciprocal space), these Bragg peaks give the 1-dim Fourier pattern of the Fibonacci chain.

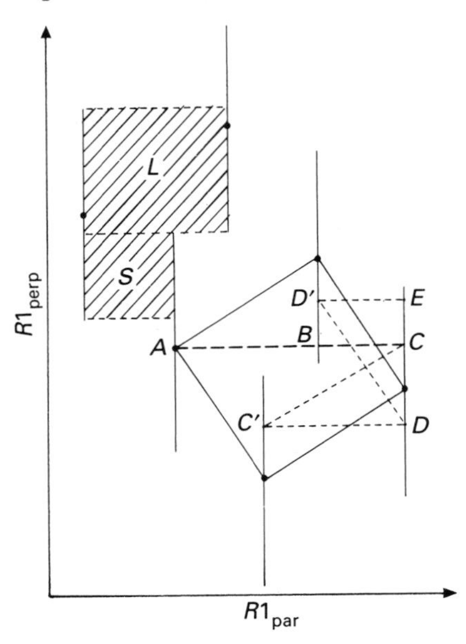

FIG. 1.32 The unit cell of the n-dim structure, with its decoration, contains all the information necessary to build the related quasicrystal (successive 1-dim tiles are AB, BC, $C'D$, with C' equivalent to C, $D'E$, with D' equivalent to D, etc . . .). Long and short segments appear in the infinite 1-dim structure in proportion to the hatched areas (acceptance volume).

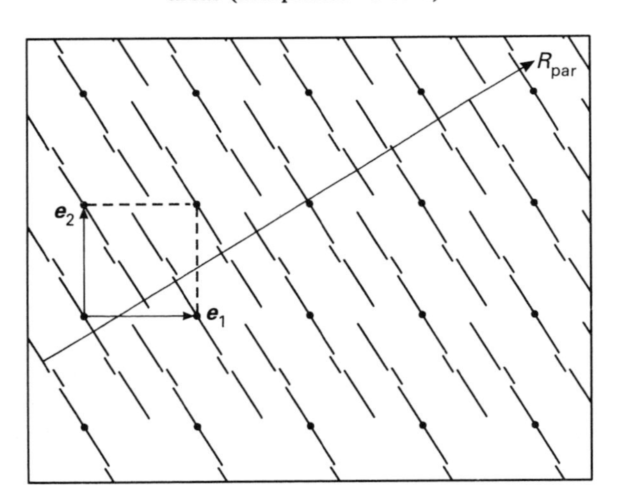

FIG. 1.33 Examples of a simple parallel component in the A_{perp} atomic surfaces. The same long-range order is maintained, but tile sizes are modified as compared to the situation of Fig. 1.30.

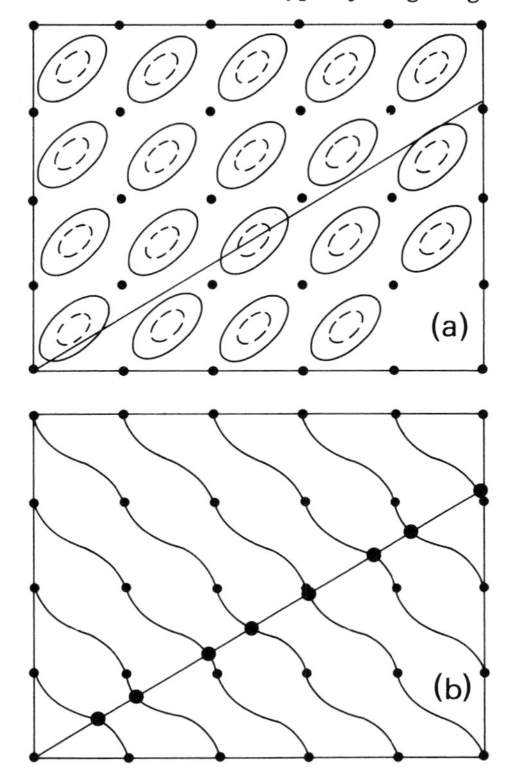

FIG. 1.34 One-dimensional quasiperiodic structure as a cut of 2-dim 'crystals'. The basis associated with the 2-dim Bravais lattice points determines the QC structure: (a) general case; (b) modulated line of δ-functions.

strictly maintained. Even more general situations are presented in Fig. 1.34. Actually, the $A3_{\text{perp}}$ hypersurfaces have only to be invariant under the operations of the point group symmetry of interest. For instance, if one decides that a small spherical hole must be dug out for some reason, with its centre on one five-fold axis of a triacontahedral $A3_{\text{perp}}$ (3-dim Penrose tiling), this implies that 12 similar such holes are actually dug out (six five-fold axes plus the inversion centre).

Of course, the 1-dim ↔ 2-dim relations are the easy ones. The reader may have some worries about 6-dim structures. In particular, we have not been accustomed by everyday life to 'see' in more than three dimensions! A convenient way to visualize a high-dimensional structure is to draw 2-dim slices of

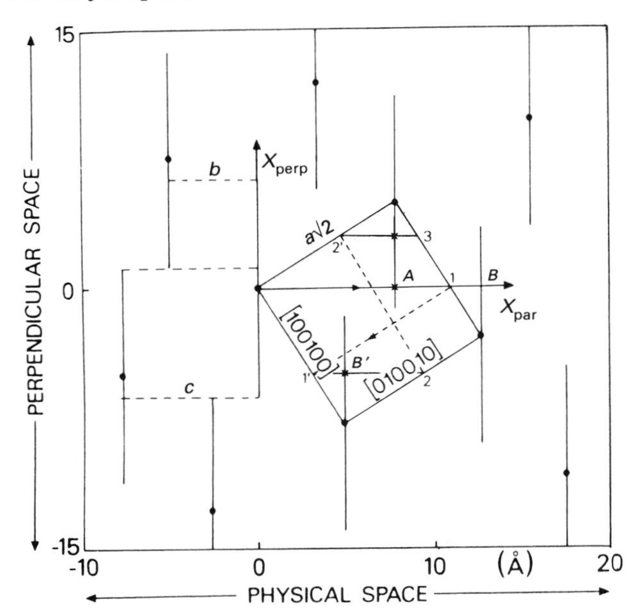

FIG. 1.35 Visualization of a 6-dim cubic image and its cut by the physical space. The figure represents a 2-dim slice of the 6-dim space containing two-fold axes, X_{par} and X_{perp}, in the physical and complementary spaces, respectively. The 6-dim cubic unit cell is decorated with triacontahedral hypersurfaces which show up in the figures as line segments along X_{perp} around sites (solid discs). Atom positions along X_{par} and occurrences of atomic distances are determined as in Fig. 1.32.

the whole, (say 6-dim) space, parallel to a plane containing one axis in $R3_{par}$ and one axis in $R3_{perp}$. An example is presented in Fig. 1.35. The 6-dim image is a primitive cubic lattice decorated with a regular triacontahedron having the same cross-section as the cubic unit cell when projected on to the $R3_{perp}$ subspace. The lattice parameter of the 6-dim cubic lattice has been chosen equal to 6.5 Å, corresponding to a cut 3-dim Penrose tiling with edges of 4.6 Å (see problems). The 6-dim space slice shown in Fig. 1.35 has coordinates axis x_{par} and x_{perp} along two-fold axis in $R3_{par}$ and $R3_{perp}$, respectively. The triacontahedra at the lattice points show up as straight line segments parallel to x_{perp}. These segments are intersected by x_{par} at positions such as A, B, etc. . ., which are the physical positions for atoms in $R3_{par}$ on this two-fold axis. In this direction, two nearest-neighbours distances show up: $b = 4.83$ Å and $c = 7.82$ Å, corresponding to the short and long diagonals of the rhombohedron faces of the cut 3-dim Penrose tiling. The relative frequency of occurrence for a distance is given by the acceptance

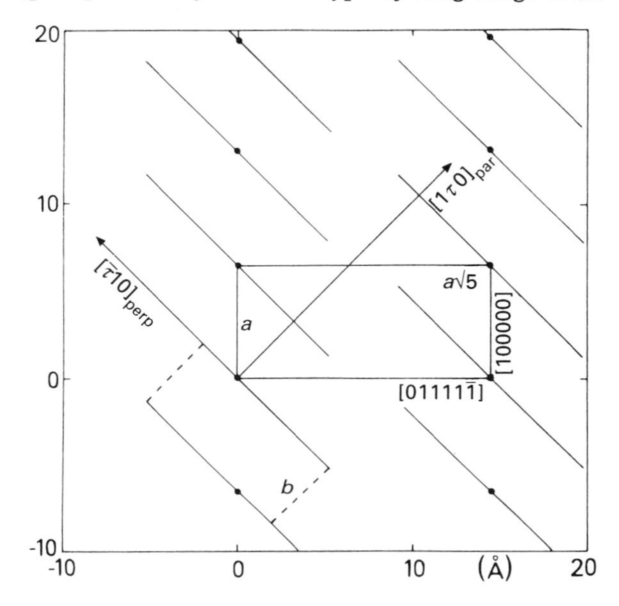

FIG. 1.36 The same representation as in Fig. 1.35, but for a slice of the 6-dim space containing 5-fold axes.

volume, which is the common parts of the projections on to $R3_{perp}$ of the two corresponding $A3_{perp}$ hypersurfaces (see Fig. 1.35). Any direction can be treated the same way, as exemplified in Figs. 1.36, and 1.37 for three-fold and five-fold axis slices.

The embedding method, as applied to generalized quasiperiodic structures is a trivial extension of the high-dimensional representation of incommensurate structures.[15, 16] This remark may help in somewhat 'de-dramatizing' the need to 'think' in more than three dimensions when quasiperiodic structures are concerned. The purpose of the next section is to advocate this point and also to underline similarities and differences between quasicrystals and more usual incommensurate structures.

1.4.6 Modulated crystals and quasicrystals

In Sections 1.2 and 1.3, periodic structures and 'disordered' amorphous systems have briefly been described as the two extreme cases of atomic arrangements in condensed matter. The former is probably the simplest kind of perfect long-range order that can be imagined, while the latter tries to propose models for the most chaotic solids.

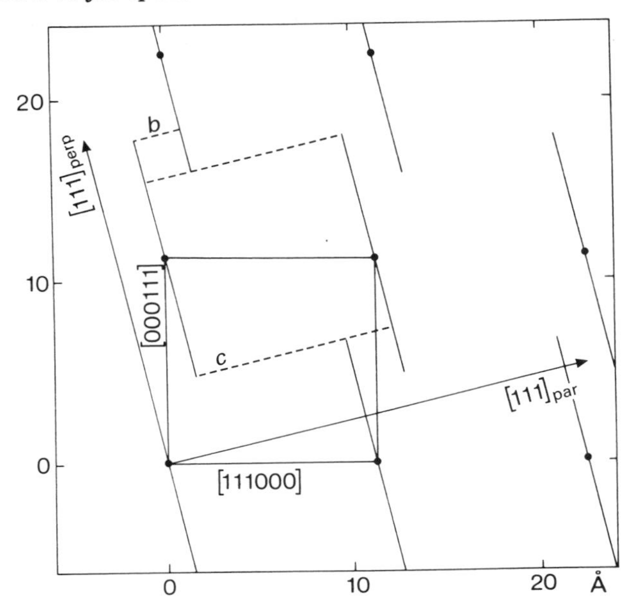

FIG. 1.37 The same as in Figs. 1.35 and 1.36, but with 3-fold axes.

Incommensurate structures have been mentioned only briefly in the introduction of this chapter. Actually, many incommensurate systems may be viewed as *modulated* crystals. This implies the existence of a basis 3-dim periodic structure to which the modulation is applied. The word modulation must be understood here in a generic sense, and stands for any distortion or static perturbation characterized by its own independent periodicity. An example of a simple modulated structure is shown in Fig. 1.38: the basis periodic structure is a 2-dim square lattice and the modulation is produced by displacements of the atoms from their normal site, parallel to the horizontal rows of the square lattice and equal to a sinusoidal function of the site positions. Thus, in the unmodulated square lattice atom positions are at vectors $\boldsymbol{r}$ such that

$$\boldsymbol{r}_n = a(n_1\boldsymbol{e}_1 + n_2\boldsymbol{e}_2),$$

($\boldsymbol{e}_1, \boldsymbol{e}_2$ being convenient orthogonal unit vectors) and these positions transform into

$$\boldsymbol{r}'_n = a\{\, (n_1 + \varepsilon \sin q\, n_1 a)\boldsymbol{e}_1 + n_2\boldsymbol{e}_2 \,\}$$

for the modulated structure; q appears as the distortion wavevector.

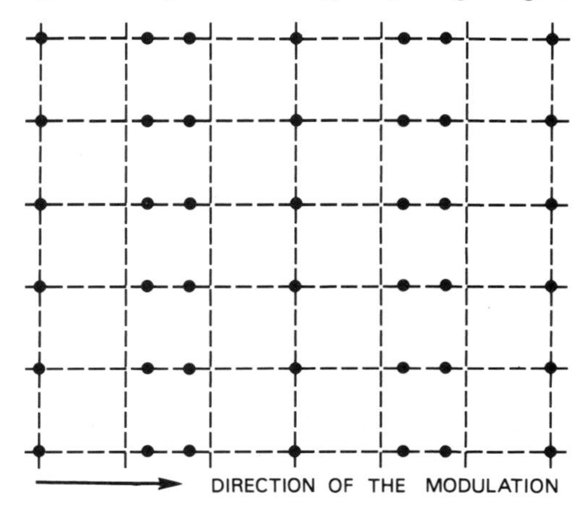

FIG. 1.38 Example of a 2-dim square lattice with a 1-dim modulation. For the sake of simplicity, the modulation period has been taken commensurate with the lattice.

This distortion wavevector q can be given the form $2\pi/\alpha a$. If α is a rational number the structure remains periodic, but with a larger unit cell (Fig. 1.39(b)). The corresponding structure is then referred to as a **commensurate modulated structure** or superstructure. Conversely, if α is an irrational number, the periodicity is lost and the structure is known as an **incommensurate modulated structure** (Fig. 1.39(c)). The Fourier transform of the modulated structure can be derived from trivial arithmetics:

$$F(\boldsymbol{Q}) = \sum_{r_n} \int e^{i\boldsymbol{Q}\cdot\boldsymbol{r}}\, \delta[\boldsymbol{r} - \boldsymbol{r}_n - \tilde{\boldsymbol{e}}_1\, \epsilon\, a \sin(qn_1 a)]\, d\boldsymbol{r}$$

$$= \sum_{m=-\infty}^{+\infty} J_m(Q\epsilon) \sum_{\boldsymbol{G}} \delta(\boldsymbol{Q} - \boldsymbol{G} - m\boldsymbol{q})$$

in which $\boldsymbol{G}$ are vectors of the reciprocal lattice for the unmodulated crystal and $J_m(Q\epsilon)$ are Bessel functions of integer order m. Thus Bragg peaks still appear at positions typical of the undistorted structure, with structure factors weighted by $J_0(Q\epsilon)$ (Fig. 1.40). Obviously, if the modulation is commensurate, q produces an exact partition of the intervals between fundamental reflections and the satellites belong to a finite set. Conversely, the satellites form an infinite set if the modulation is incommensurate. Owing to the decaying value of $J_m(Q\epsilon)$ when m increases, only a small number of satellites are observed in practical cases. The analysis also applies to a 3-dim modulated structure.[17–19]

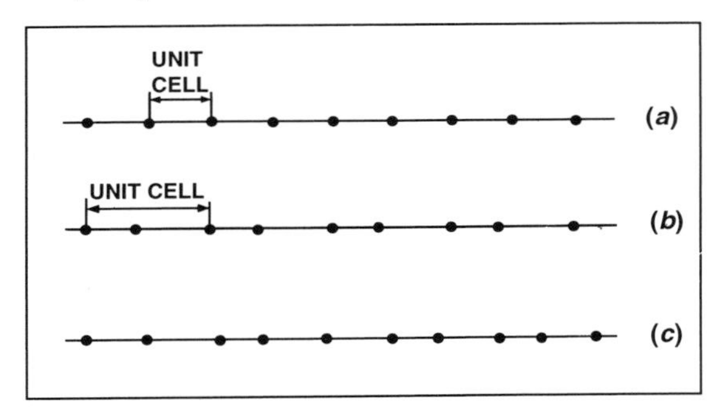

FIG. 1.39 A periodic chain of atoms (a) transforms into a longer-period structure (b) upon commensurate modulation but into a non-periodic arrangement (c) upon incommensurate modulation.

An n-dim modulated structure can be viewed as the intersection of an $(n + 1)$-dim periodic structure with the n-dim physical space. This property is generic and independent of the physical nature of the system. To illustrate this, let us consider for instance the simple case of an incommensurate structure resulting from the sinusoidal modulation of a periodic chain, periods of the chain and of the sine function being irrationally related. The position (or origin or phase) of the modulation function with respect to the chain may be chosen arbitrarily. When this phase is changed, atom positions in the chain are 'rearranged' into a new configuration having the same total equilibrium free energy as before. A continuous shift of the modulation function with respect to the chain generates an infinite number of 'indistinguishable' incommensurate structures which can be vizualized simultaneously by piling them up along an axis perpendicular to the chain (Fig. 1.41).[20] In the then defined perpendicular space, a given point of the chain has 'equivalent' positions distributed on a periodic profile which is nothing other than the modulation function. The incommensurate structure, in all its 'indistinguishable varieties', then appears as a cut of a periodic high-dim (2-dim here) structure by the physical space. The perpendicular space, which describes phases of the modulation function, is called the phases space, or phason space. Any translation in the high-dim space with a non-zero component only in the perpendicular space is called a 'phason mode', or phason for short, the term phonon being restricted to translation parallel to the physical space. Though all these concepts can easily be extended to quasicrystals, there are some important qualitative differences between incommensurate phases and quasicrystals.[21] Most of the incommensurate structures corres-

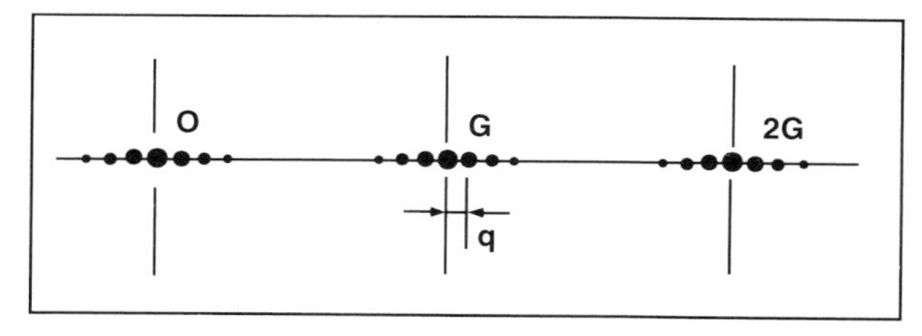

FIG. 1.40 Schematic representation of the diffraction pattern for a modulated chain of atoms as shown in Figs. 1.39(b) or 1.39(c).

pond to 1-dim or 2-dim modulations and the overall symmetry is rather low. The associated high-dim space has then dimension 4 or 5 and the cut that generates the physical space has low symmetry. The relative locations of the atoms are distributed in space according to a modulation law that is a continuous function of the perpendicular space coordinates; the atomic surfaces in the high-dim space are then continuous surfaces. The existence of an average underlying periodic lattice also gives meaning to the notions of average unit cell, average atomic distances, fundamental on satellite reflections, etc.

Conversely, for quasicrystals the high-dim space may have very large dimension. For instance, the associated high-dim space for icosahedral quasicrystals has a minimal dimension of six. Moreover, the overall symmetry is the highest possible in 3-dim. There is no underlying periodic lattice and, consequently, no 'average' parameters can be sensibly defined. The most dramatic difference comes from the atomic surfaces in the high-dim space, which are not continuous anymore, but appear as piecewise discontinuous objects, mostly parallel to the complementary space (see Fig. 1.30). For the conservation of atoms with respect to the R_{perp} variables, these atomic surfaces must connect in such a way that there is no 'annihilation creation' of atoms under any R_{perp} translation (this is called the **closeness** condition). The requisite is fulfilled with the example of Fig. 1.30, as can easily be checked by moving the R_{par} line, keeping its direction constant. Conversely, if the A_{perp} objects had been designed shorter (larger), some perpendicular translations of R_{par} would have resulted in atoms being temporarily annihilated (created). Needless to say, the atomic objects in the high-dim space cannot intersect, since no two atoms should come infinitely close to each other (**hard core** conditions).

The reciprocal space, or equivalently the diffraction pattern, of a quasicrystal differs from that of a modulated crystal by the absence of 'funda-

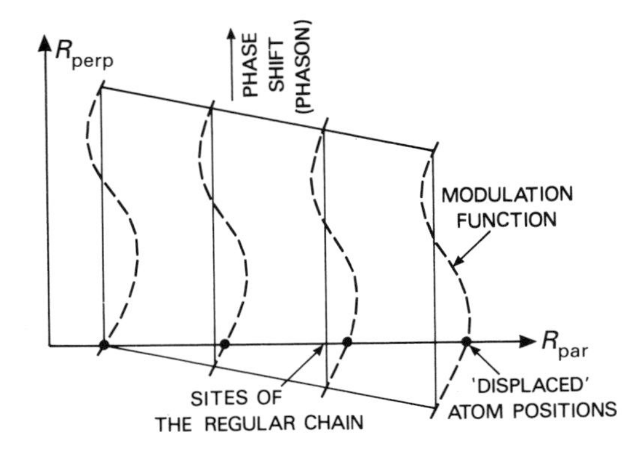

FIG. 1.41 Continuous phase shifts of the modulation function generate 'indistinguishable' structures. All these structures can be visualized when piled up together in a 'perpendicular' space. This is the foundation of the cut, or embedding, methods.[20]

mental reflections' as opposed to 'satellite reflections': there is no subset of indices corresponding to a would-be average periodic lattice.

Finally, it must be noted that all these differences between modulated crystals and quasicrystals are essentially valid for actual *3-dim* physical systems. In 1-dim models, distinctions are somewhat smeared out; a Fibonacci chain, for instance, is totally equivalent to a periodic chain modulated by a function with period τ times as large as that of the basis chain.

In conclusion, there are certainly many procedures that can be used to generate ordered geometrical arrangements of points in space, without the perequisite of periodicity. Quasiperiodicity is a special class of such a non-periodic long-range order. Quasiperiodic structures can be defined, and easily constructed, as irrational cuts of object which are periodically distributed in an n-dim space ($n > 3$). There is no regular reciprocal lattice attached to these quasiperiodic structures, but the Fourier pattern remains meaningful. It contains a dense singularly continuous set of sharp features indexable with as many integers as the dimensions of the n-dim periodic space. A relatively small fraction of these Fourier components have significant intensities.

Experimental 'quasicrystallography', as developed later on in this book, is based on the concepts presented in this first chapter.

1.5 Problems

1.1 Determine the five two-dimensional Bravais lattices.

1.2 Show that the volume of the primitive cell is unique for a given lattice, even if its shape can be changed.

1.3 Show that $d_{hkl} = a/(h^2 + k^2 + l^2)^{\frac{1}{2}}$ for 3-dim cubic systems. How is the interplanar spacing for a 6-dim cubic system written?

1.4 Determine the permitted Bragg reflections for bcc and fcc Bravais lattices (in three dimensions).

1.5 Consider two square lattices decorated with line segments equal to $a(\cos\alpha + \sin\alpha)$. One lattice has a single decoration at the origin site, while the other has the origin and square centre decorated. A cut of these structures by a 1-dim space at angle α with the horizontal rows of the square lattices generate two 1-dim quasiperiodic structures ($\cos\alpha/\sin\alpha = \tau$). Compare the permitted Bragg reflections for the two cases.

1.6 Properties of the golden mean τ:
(a) show that $\tau^2 - \tau - 1 = 0$;
(b) given the Fibonacci series $u_0 = 0$, $u_1 = 1$, $u_2 = 1$, $u_3 = 2$, $u_4 = 3, \ldots, u_n = u_{n-1} + u_{n-2}, \ldots$, show that u_{n+1}/u_n converges to τ as n increases (approximants of τ).
(c) show that τ^n also form a Fibonacci series, i.e.: $\tau^n = \tau^{n-1} + \tau^{n-2}$ and calculate $\tau^2, \tau^3, \tau^4, \ldots$ as linear combination $m\tau + l$.

1.7 Show that the frequencies of S and L segments in a Finonacci string f_n are $v(S, n) = u_{n-2}$ and $v(L, n) = u_{n-1}$, respectively. Demonstrate that $v(L, n)/v(S, n) = \tau$ at the limit of large n values. Also use the cut method (Fig. 1.32) to obtain this result. Show that periodic repetitions of substrings are excluded. Justify eqn (1.26).

1.8 Compare the Fourier pattern of a Fibonacci chain with that of its periodic approximant 5/3, both being generated by the cut method.

1.9 Generate a 2-dim N-grid with octagonal symmetry and its dual connected space filling packing of unit cells.

1.10 With the example of a regular triangular unit cell, show that a periodic lattice can be generated by a deflation procedure.

1.11 Show that, in the cut method, the sizes of atomic volumes in the n-dim periodic structure are determined by the density and composition of the 3-dim quasicrystal. Take examples of monatomic and binary systems.

1.12 Show that the Fourier transform of a projection is a cut. Use the example of 2-dim ↔ 1-dim relation.

References

1. Levine, D. and Steinhardt, P. J. (1984). Quasicrystals: a new class of ordered structures. *Phys. Rev. Lett.* **53**, 2477–80.
2. Levine, D. and Steinhardt, P. J. (1986). Quasicrystals. I. Definition and structure. *Phys. Rev.* **B34**, 596–616.
3. Kramer, P. and Neri, R. (1984). On periodic and non-periodic space fillings of E^m obtained by projections. *Acta Cryst.* **A40**, 580–7.
4. De Bruijn, N. G. (1981). Sequences of zeros and ones generated by special production rules. *Math. Proc.* **A84**, 27–37.
5. De Bruijn, N. G. (1981). Algebraic theory of Penrose's non-periodic tilings of the plane. *Math. Proc.* **A84**, 39–66.
6. Penrose, R. (1974). The role of aesthetics in pure and applied mathematical research. *Bull. Inst. Math. Appl.* **10**, 266–71.
7. Mackay, A. L. (1982). Crystallography and the Penrose pattern. *Physica* **A114**, 609–13.
8. Socolar, J. E. S. and Steinhardt, P. J. (1986). Quasicrystals. II. Unit-cell configurations. *Phys. Rev.* **B34**, 617–47.
9. Katz, A. (1988). Matching rules for the 3-dim Penrose tilings. In *Quasicrystalline materials*. (ed. C. Janot and J.M. Dubois). pp. 195–204. World Scientific, Singapore.
10. Gruenbaum, B. and Sheppard, C. G. (1986). *Tilings and patterns*. Freeman, New York.
11. Elser, V. (1986). The diffraction pattern of projected structures. *Acta Cryst.* **A42**, 36–43.
12. Katz, A. and Duneau, M. (1986). Quasiperiodic patterns and icosahedral symmetry. J. Physique **47**, 181–96.
13. Katz, A. and Duneau, M. (1985). Quasiperiodic patterns. *Phys. Rev. Lett.* **54**, 2688–91.
14. Kalugin, P. A., Kitaev, A. Y., and Levitov, L. S. (1985). $Al_{0.86}Mn_{0.14}$: a six-dimensional crystal. *JETP Lett.* **41**, 145–9.
15. Bak, P. (1986). Icosahedral crystals from cuts in six-dimensional space. *Scripta Met.* **20**, 1199–1204.
16. Janssen, T. (1986). Crystallography of quasi-crystals. *Acta Cryst.* **A42**, 261–71.
17. de Wolf, P. M. (1974). The pseudo-symmetry of modulated crystal structures. *Acta Cryst.* **A30**, 777–85.
18. Janner, A. and Janssen, T. (1977). Symmetry of periodically distorted crystals. *Phys. Rev.* **B15**, 643–58.
19. Janner, A. and Janssen, T. (1979). Superspace groups. *Physica* **A99**, 47–76.

20. Currat, R. and Janssen, T. (1988). Excitations in incommensurate crystal phases. *Solid State Physics* **41**, 201–302.
21. Katz, A. (1990). On the distinction between quasicrystals and modulated crystals. In *Quasicrystals* (ed. M. V. Jaric and S. Lundqvist). pp. 200–17. World Scientific, Singapore.

2

Real quasicrystals: preparation and characterization

And now for something completely different
Monty Python's Flying Circus

2.1 Introduction

Even if Chapter 1 has been convincing enough for the reader to be persuaded that quasiperiodic arrangements of atoms are a possible state of matter, this by no means implies that real materials do fulfil such a type of order. This is the purpose of Chapter 2: to prove this point by describing how quasicrystals can be prepared.

The message of the chapter is also that quasicrystals may be easy to prepare but that preparation of *good* quasicrystals requires great care. This poses the basis problem of the necessary characterization of the quasicrystalline state as a fundamental prerequisite for any measurement of properties.

The quasicrystals of the 'first generation', namely those of the AlMn-like systems, demonstrated that quasiperiodic order can form 'in nature'. But they were of rather poor quality, probably because of being prepared via a rapid quenching procedure. Investigations of structure and properties of quasicrystals were dogged for a while by the lack of stable systems and/or of sufficiently large single grains. But these investigations jumped ahead when perfect stable icosahedral phases were grown in alloys such as AlCuFe and AlPdMn. Existing reversible phase transitions from quasicrystalline states to periodic crystals have also provided good experimental evidences to differentiate a true icosahedral symmetry from the pseudo-icosahedral symmetry that may be mimicked by some special arrangements of microcrystals. At present, quasicrystalline phases may be considered just as particular intermetallic compounds among others.

2.2 Preparation methods

2.2.1 The melt spinning technique

Many quasicrystals are obtained by rapid solidification of a liquid. The procedure is similar to the production of metallic glasses, where cooling rates

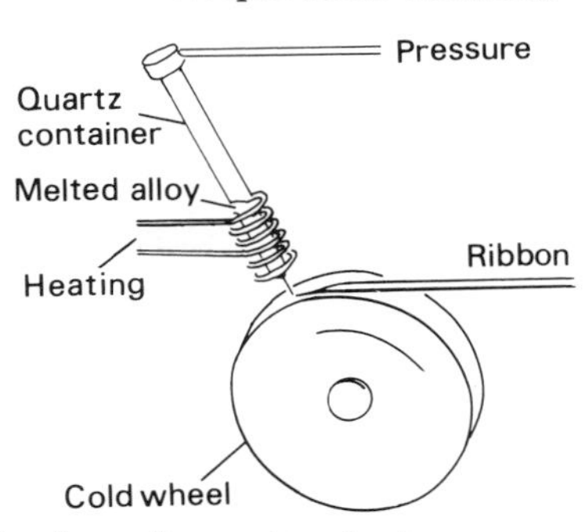

FIG. 2.1 Schematic view of a melt spinning apparatus. The alloy is melted in the quartz container and then squirted on the rotating wheel through a small hole. The quenching rate depends on several parameters: ejection speed of the metal, angle between the wheel and the container, speed rotation, and dexterity of the experimentalist (!).

of 10^5 to $10^9\,\mathrm{K\,s^{-1}}$ are necessary to avoid the nucleation of high-temperature equilibrium phases. There are several techniques which permit varying the rate of supercooling at the nucleation state, such as melt spinning or planar flow casting. In these processes, molten alloys are squirted on to a rotating wheel (Fig. 2.1). The liquid, when dropped on the wheel, is quenched at a rate of the order of $10^6\,\mathrm{K\,s^{-1}}$ and the sample is obtained as ribbons a few μm thick and a few mm wide. The quenching rate can be changed to some extent by varying the rotation speed of the wheel. The ribbons generally contain single grains of quasicrystal with sizes of about 1 μm across (Figs. 2.2 and 2.3).[1] Such a size is suitable for electron diffraction (see Section 2.3) but single crystal X-ray and neutron studies are not possible. Single phase quasicrystals may be produced, but with difficulty and with poor reproducibility, because the parameters involved in the procedure are somewhat uncontrolled. The $Al_{80}Mn_{20}$ initial quasicrystal, for instance, is obtained in coexistence with other crystalline phase. But the addition of a small quantity of silicon (about 5 per cent) favours quasicrystal formation and leads to an almost single phase.

Faster quenching rates can be obtained by using the splat cooling technique in which a droplet of liquid alloy is squeezed between an anvil and a piston. However, the quantity of sample obtained is obviously much smaller than with melt spinning, and the formation parameters are even more difficult to control.

2.2.2 Other production techniques for metastable quasicrystals

All the production methods relevant to metastable alloys and glasses have been also applied to quasicrystals. They are all based on disorder generation

FIG. 2.2 Typical microstructure of AlMnSi metallic ribbons obtained by melt spinning (courtesy of M. Audier).

FIG. 2.3 Single grain of quasicrystal as observed by electron microscopy (courtesy of Professor Nissen).[1,9]

at the atomic level. This is generally done by a solid state reaction. A typical method is the multilayer deposition technique in which alternating layers of Al and Mn are deposited on a substrate, the thickness being of the order of 1000 Å. Once the multilayer with the right average composition is obtained, the sample is bombarded by high-energy ions of inert gases (e.g. Xe^{2+}). An amorphous quasicrystalline or crystalline state is obtained, depending on the energy of the ions and on the sample temperature. Here disorder is introduced by the kinetic energy of the ions (like a ball in a bowling game) and is also driven by the temperature of the sample (the atoms become more mobile as the temperature increases). A schematic view of this process is shown in Fig. 2.4.[3] The samples obtained are quite small ($2 \times 2\,\mathrm{mm}^2$ and 1000 Å thick) and are ideal for electron microscopy studies. Using this technique, single quasicrystalline phases can be obtained in the AlMn system, allowing the study of the various phase transformations between quasicrystalline, amorphous, and crystalline states.[2-5] The AlMn quasicrystalline state can also form by direct ion implanation of Mn in an oriented Al matrix.[6]

Mechanical alloying (also called ball milling) can also produce amorphous or crystalline states. Powders of the different elements are alloyed by the

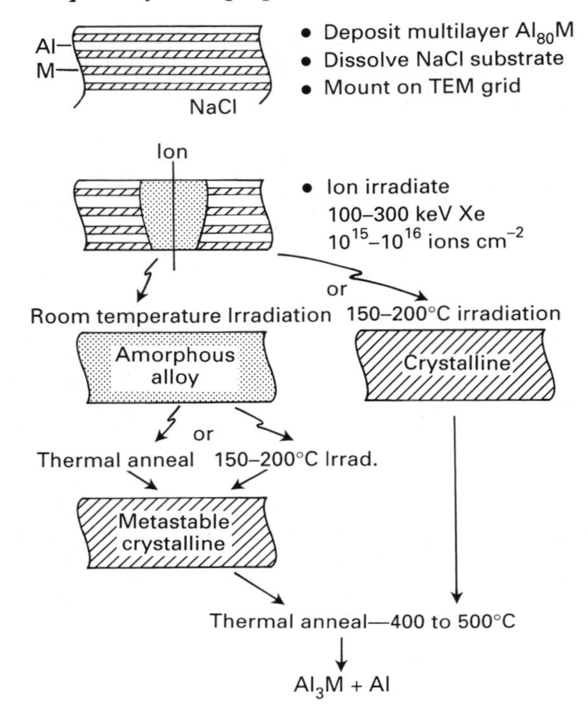

FIG. 2.4 Schematic view of the ion mixing preparation of the AlMn quasicrystal (from Lilienfeld *et al.*).[3]

kinetic energy of balls vibrating in a steel container. Eckert *et al.* obtained amorphous or quasicrystalline samples in the AlCuMn system.[7]

Apart from solid state reaction preparation, two other techniques should be considered: the evaporation technique and the laser or electron melting of thin layers. In the former method, a 'fog' of small droplets of liquid alloy is produced, and quenched. The 'smoke particles' present interesting external shapes and structures, with typical sizes in the range 500 to 3000 Å. Ishimasa *et al.*[8,9] have prepared dodecagonal quasicrystalline phase in the NiCr system (Fig. 2.5) using laser or electron melting of a small sample of thin layers; quenching results from radiative effects and the thermal capacity of the rest of the sample.[10]

2.2.3 Conventional casting

Conventional casting (i.e. slow cooling from the melt) supposes the quasicrystalline state to be stable, at least in some composition and temperature

FIG. 2.5 Electron diffraction of a dodecagonal NiCr quasicrystal (courtesy of Professor Nissen).[8, 9]

range. This technique has the advantage of enabling the growth of large single grains whose structures can be determined by quantitative X-ray or neutron single crystal diffraction. The Al_6Li_3Cu alloy was the first icosahedral phase found to be stable. The phase diagram of the system is not simple and the non-congruent formation of the quasicrystalline phase precludes the use of traditional crystal growth techniques. However, using a geode-like preparation, Dubost *et al.*[11] were able to prepare beautiful single quasicrystals with triacontahedral shape (Fig. 2.6). This opened the route for precise structure determination. Since then, other quasicrystals have been obtained by conventional slow casting in the GaMgZn,[12] AlPdMn, and AlCu (Fe, Os, Ru) systems,[13–15] (all icosahedral phases) and the decagonal AlCuCo[16] and AlCoNi systems.[17] However, one can see that stable quasicrystals to date are still very scarce.

The task of understanding the quasicrystal growth process is far from having been achieved, and requires a knowledge of the atomic structure. General aspects of quasicrystal growth will be presented in Chapter 6.

The first point to be understood in crystal growth of equilibrium phases (crystal or quasicrystal) is what is going on when a liquid of a given composi-

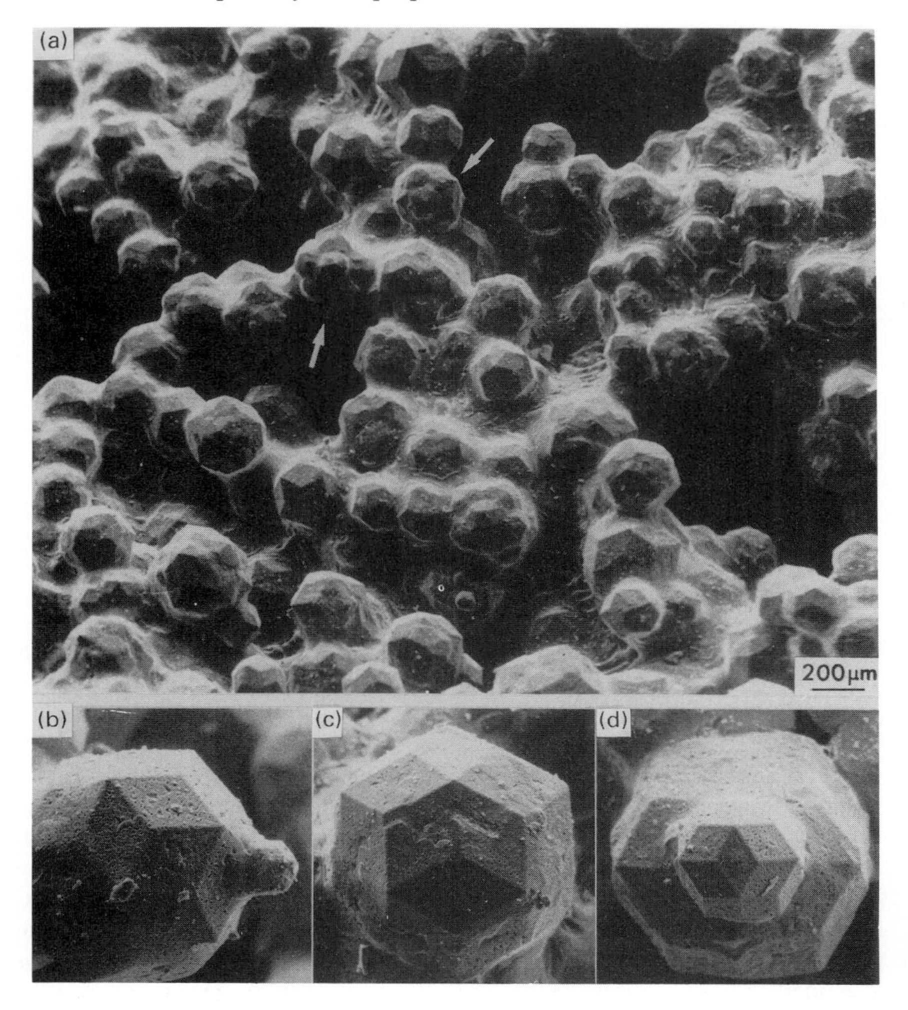

FIG. 2.6 (a) Inside of a geode: the wall of the shrinkage cavity is covered with small triacontahedral single grains. (b) to (d) Enlarged view of a single grain of AlLiCu quasicrystal (courtesy of M. Audier).[11]

tion is cooled down: this is called the solidification route and must be sought in the equilibrium phase diagram. The example of a simple binary system will be used for the sake of illustration (Fig. 2.7). If the system is a definite compound, say A_3B, two different cases may occur. The solid may be in equilibrium with a liquid of the same composition (such as a pure element) and

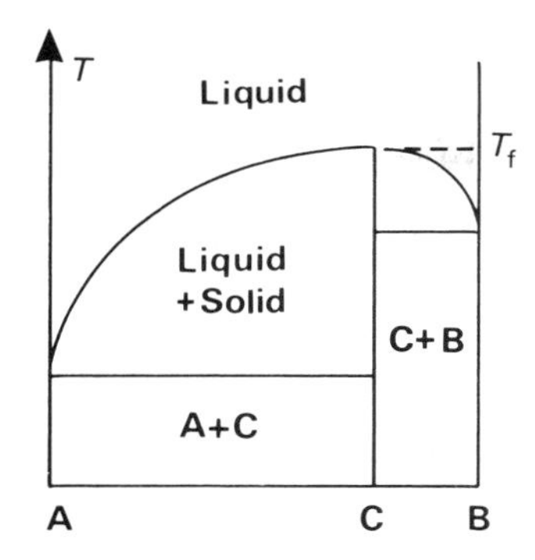

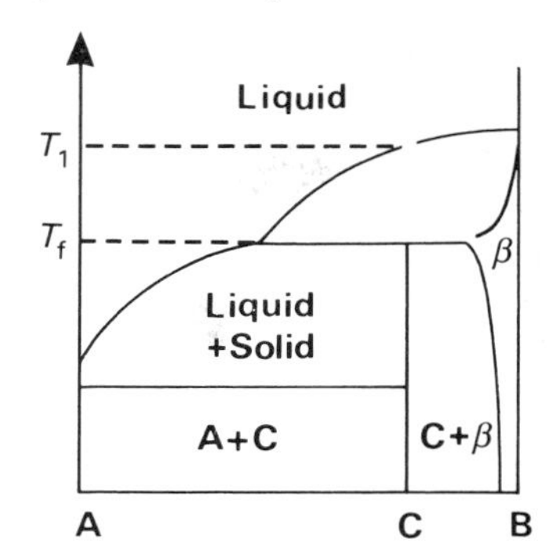

FIG. 2.7 Examples of equilibrium phase diagrams for binary systems showing congruent (left) and non-congruent (right) melting.

the fusion is called congruent. But the liquid may have a different composition and the compound decomposes during cooling: this is non-congruent melting. Non-congruent melting will always result in an ingot with a mixture of phases and makes the growth of large single quasicrystals very difficult.

The ternary AlLiCu phase diagram was carefully revisited by Dubost *et al.*[18] Several crystalline phases exist in the region of the icosahedral phase: the bcc R-phase, whose chemical composition is almost the same as the icosahedral one, the hexagonal T_1-phase, and the Al_2Cu phase. Unfortunately, the solidification of the T_2-icosahedral phase is not congruent. Starting from an aluminium enriched alloy it is possible to avoid the formation of the R-phase and the icosahedral phase growth first, but then two eutectics, T_2 + Al and T_1 + Al, will form. To obtain large single crystals one must produce a geode-like growth which avoids the formation of the two eutectics with a low melting point. With the geode-like technique, single grains result from free solidification with separation of the solid dendrites from the residual liquid in internal shrinkage cavities of a solidifying ingot. Cooling then causes the inside of the ingot to be coated with single grains (triacontahedra) in a way similar to that found in natural geode.[19]

2.3 Characterization of quasicrystalline samples

2.3.1 Electron, X-ray, and neutron interactions with matter

Single crystal diffraction is the only way to characterize a quasicrystal sample properly. X-rays, electrons, and neutrons have a wave-like behaviour, giving rise to interference when diffracted by materials. However, the interaction mechanisms are not similar.

The scattering of X-rays by atoms is actually due to interaction with atomic electrons. The electrons are accelerated by the electric field of the X-ray photon, and secondary X-rays, with the same wavelength and phase as the incident radiation, are emitted. The amplitude of the diffracted X-rays is proportional to the number of electrons in the atom, and is highly dependent on the diffraction angle (this is usually called the atomic form factor). The typical range of the X-ray wavelength is 0.7–2 Å. Interaction with material is quite strong and sizes required for samples are of the order of 0.1 mm, depending on absorption.

Neutrons interact with the nuclei of the atoms. The neutron is momentarily captured, forming a compound nucleus, and is then remitted. The phase difference between the incident neutron wave and the diffracted one can be 0 or π, which means that the coherent scattering length (the equivalent of the atomic scattering factor) is either positive or negative. Moreover, the coherent scattering length does not have any systematic dependence on the number of protons and neutrons present in the nucleus. In particular, atoms which have almost the same number of electrons can have very different coherent scattering lengths, whereas for X-rays the factors are almost the same. This is also true for isotopes, which makes neutron diffraction particularly powerful when isotopes exist. (For instance Lithium has two stable isotopes: ^{6}Li, with a coherent scattering length $b = +0.2$ and ^{7}Li with $b = -0.22$ (in 10^{-12} cm). Depending on the isotope that is present, the 'lighting' of the atomic structure will be different. This has been widely used in atomic structure determination. The interaction of matter with neutrons is quite weak compared with the interaction with X-rays, and sample sizes have to be much larger if any signal is to be detected. A typical crystal size for neutron diffraction purposes would be of the order of 1 cm. Neutron wavelengths range from 0.7 Å to 3.5 Å typically, but can extend to much larger values when 'cold neutrons' are used (several tens of Å).

Electrons are scattered by the atomic potential. They are strongly absorbed by matter and specimens have to be very thin for transmission experiments. The wavelength associated with the electron in a standard electron microscope is much shorter than the wavelengths of X-rays and neutrons (for a 200 kV electron microscope the wavelength is 2.54 10^{-2} Å). The first and most useful characterization is done via so-called selected area electron diffraction (SAED). In this mode the diffraction pattern is obtained by transmission geometry and the thickness of the sample has to be of the order of a few hundreds of Å. The area involved in the diffraction process has a typical size of $10^3 \times 10^3$ Å^2 (i.e. $0.1 \times 0.1\ \mu m^2$). This is what makes this technique so powerful: only very small single grains are needed for single crystal diffraction purposes. There is also the well-known possibility of high-resolution electron microscopy images (HREM), which allow us to see, in a first crude approximation, the projection of the atoms on to a flat image.

2.3.2 Electron diffraction

As stated above, this is the first and most important way of characterizing a quasicrystalline sample because of the small size of the single grain needed. If an electron beam irradiates a sample, the diffraction pattern observed in the detector (fluorescent screen or photographic plates) is found by the Ewald construction (see Chapter 1, Fig. 1.5). The wavelength is very short and the radius $1/\lambda$ of the Ewald sphere is very large when compared to typical distances in reciprocal space. For instance, a cubic crystal with a 3 Å lattice parameter will have points regularly spaced in reciprocal space and a reciprocal lattice constant $a^* = 1/a = 0.33$ Å^{-1}. This is to be compared with $1/\lambda \sim 40$ Å^{-1}. Thus where the diffraction pattern is formed, the Ewald sphere is almost a plane. Moreover, due to the thinness of the sample, points in reciprocal space are broadened and look rather like small line segments (Fig. 2.8). It is thus easy to understand that the observed diffraction pattern (i.e. the intersection of the Ewald sphere with the reciprocal lattice) will correspond directly to a reciprocal plane passing through the origin. Moving the sample around a symmetry axis will correspond to a rotation of the reciprocal lattice around the point I (Fig. 2.8). Thus an electron diffraction pattern corresponds to a *section* of the reciprocal space.

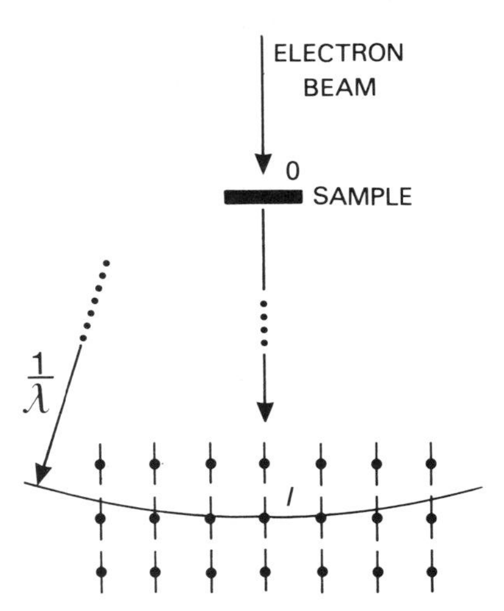

FIG. 2.8 The Ewald construction. The sample is in 0 and the origin of the reciprocal lattice in I. The sample is very thin which gives rise to broadening of the peaks (small line segments).

2.3.2.1 Crystalline samples As explained in Chapter 1, the intensity distribution in the diffraction pattern is driven by two ingredients: (i) the lattice and space group, which tell us where Bragg reflections are in the reciprocal space, and (ii) the atomic decoration of the unit cell which influences the intensity of each Bragg spot.

To illustrate the point, let us take a simple face-centred cubic lattice, such as aluminium. There is a single atom located at the origin in the unit cell. All other atomic positions are generated by applying the fcc lattice symmetries, i.e. translations ($\frac{1}{2}$ 0 0), (0 $\frac{1}{2}$ 0) and (0 0 $\frac{1}{2}$). The decorated unit cell is represented in Fig. 2.9. It can easily be shown that the reciprocal lattice is a centred cubic lattice (I). This means that not all reflections of the primitive cubic lattice are allowed, but some extinctions exist. More precisely, only the reflections for which (h, k, l) are all odd or all even are allowed (see Chapter 1). In the kinematical approximation, computation of the intensity of a given Bragg spot is trivial: all the spots have the same intensity. However, this is only a poor approximation of the real electron diffraction pattern, because dynamical effects, such as multiple scattering, have to be taken into account. Such calculations are highly dependent on apparatus and specimen parameters and are generally difficult to handle. This is why electron diffraction is mainly used for symmetry and lattice parameter determination. X-ray and neutron diffraction are preferred for quantitative intensity measurements and structure specifications.

Let us now return to the electron diffraction patterns of the fcc lattice. The corresponding sections of the reciprocal space are of special interest when they show symmetry elements. These sections are called zone axes, and are planes orthogonal to a given direction in the reciprocal space. Starting from

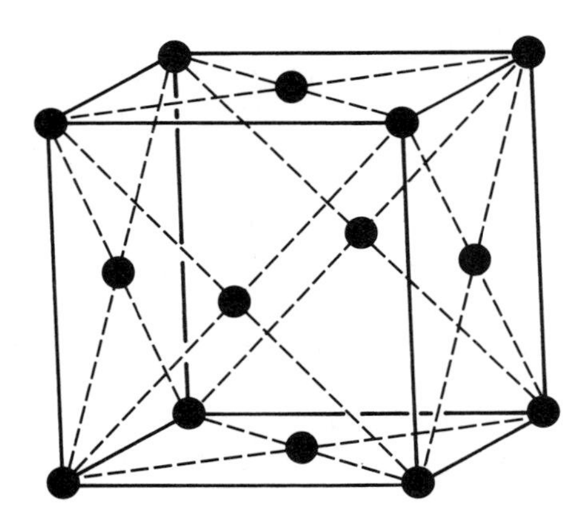

FIG. 2.9 The fcc lattice of aluminium. The unit cell is decorated by only one atom, located at (0, 0, 0). Other positions are obtained via the symmetry operations and translations of the fcc lattice.

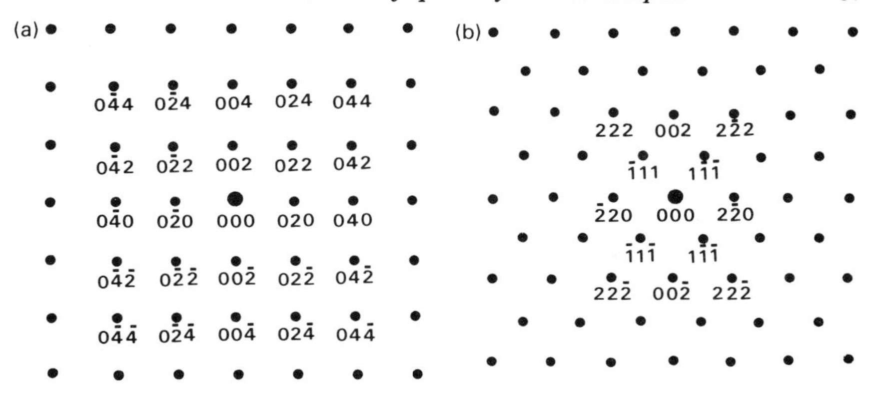

FIG. 2.10 (a) [100] zone axis of fcc Al. Bragg spots are regularly sited on a regular square lattice. (b) [101] zone axis obtained after a 45° rotation around [010].

the [100] zone axis (Fig. 2.10(a)), one can then tilt the sample. Obviously, there are two four-fold axes, [010] and [001], in the diffraction pattern. Moving the sample around the [001] axis is equivalent to a rotation of the reciprocal space (see the Ewald construction). After a 45° rotation the vector [110] is orthogonal to the diffraction plane and the diffraction pattern shows only a two-fold rotational symmetry. The two vectors which define the plane are (002) and ($2\bar{2}0$) (due to extinction effects, (001) and ($1\bar{1}0$) do not exist). By geometrical construction the point ($2\bar{2}2$) is also in the plane, and thus also the ($1\bar{1}1$) point, which is the centre of the cube. The [$\bar{1}10$] zone axis is thus typical of the I lattice (Fig. 2.10(b)). A simple way to visualize the different axes and their respective orientation is to work out the so-called stereographic projection. For a given orientation of the lattice, an imaginary sphere is drawn with its centre at the origin of the lattice. The different symmetry axes (or symmetry planes) intersect this sphere at points (lines) which are then projected on to the horizontal plane. This is shown in Fig. 2.11 for a cubic lattice. The different axes, which will be perpendicular to the sample when moving it around the [010] direction, are then easily found: they all lie on the horizontal line [010]–[$0\bar{1}0$] (Fig. 2.11). Their relative angles are also easily determined on the stereographic projection.

For a given symmetry point group, each zone axis has a precise number of symmetry elements. Thus, to specify a given point group completely, one must check their respective orientations. With the example of cubic lattice, turning by 45° around [010] axis (four-fold symmetry), leads to the [$\bar{1}01$] zone, which is a two-fold axis. Then, when turning around the two-fold axis lying in this plane, the [$\bar{1}11$] direction is aligned with the beam after a 45° rotation and the diffraction pattern shows a three-fold zone axis (see also the stereographic projection).

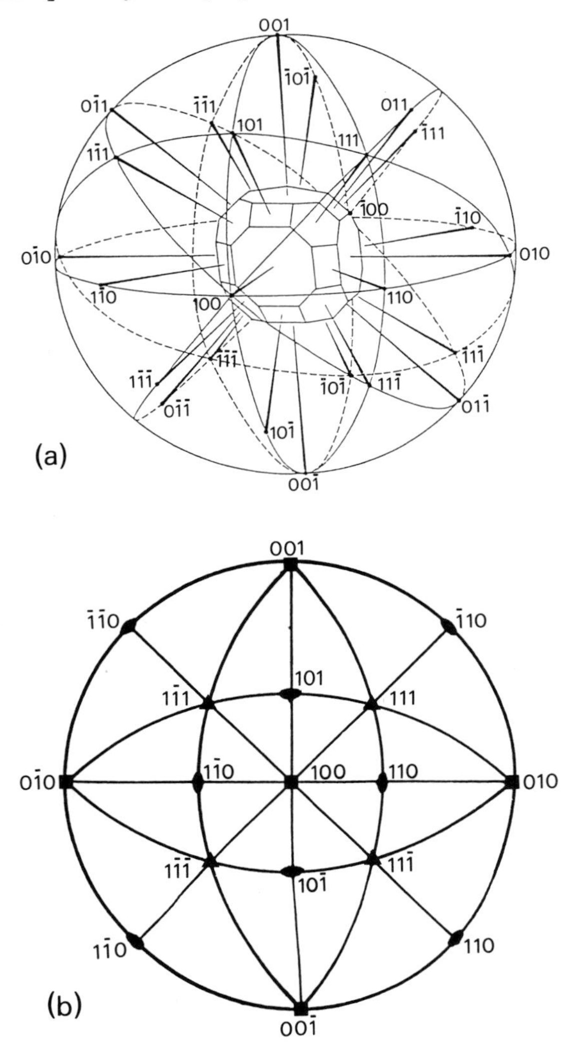

FIG. 2.11 (a) Schematic view of the reciprocal space for the cubic I lattice. A sphere is drawn. (b) Stereographic projection of the [100] zone axis of the cubic lattice. The intersections of the different axes with the sphere are projected along the plane.

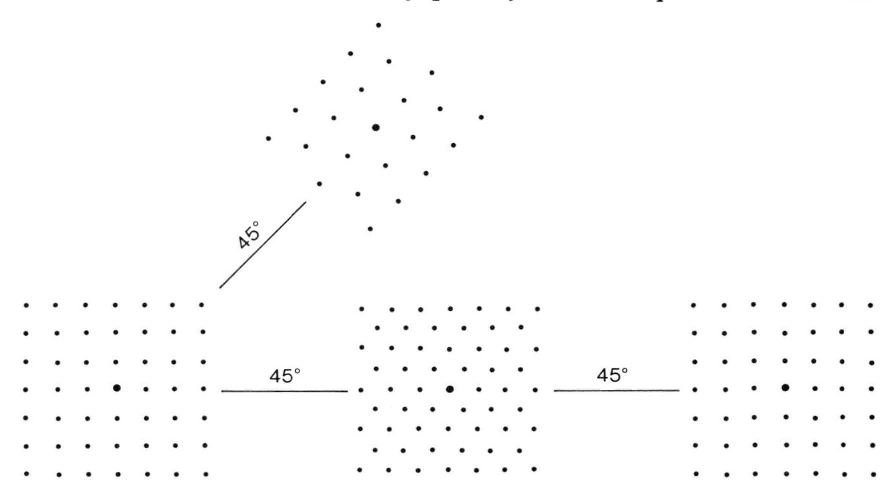

FIG. 2.12 The different zone axes of fcc aluminium.

Extinctions of the I lattice are only visible once the indexation is done. This, for instance, is the case in the [101] zone axis.

These different steps are generally presented in one figure, where the rotation angles are redrawn. All the symmetry elements are globally shown and demonstrate the existence of a given point group or space group when extinctions show up (Fig. 2.12).

2.3.2.2 Icosahedral quasicrystals Let us consider now the icosahedral quasicrystal whose diffraction pattern startled the scientific community so

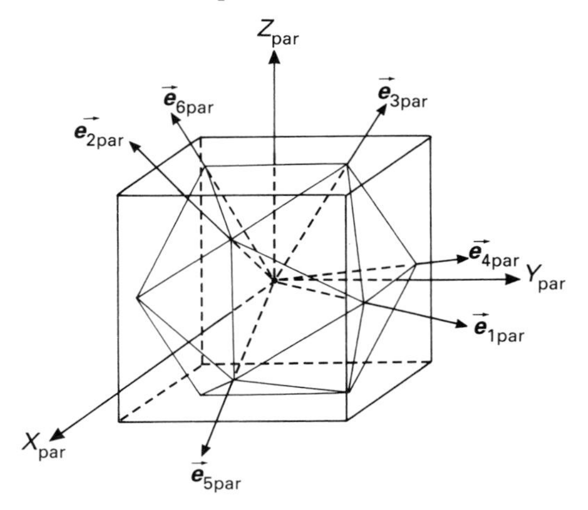

FIG. 2.13 An icosahedron. Three 2-fold axes form an orthogonal basis.

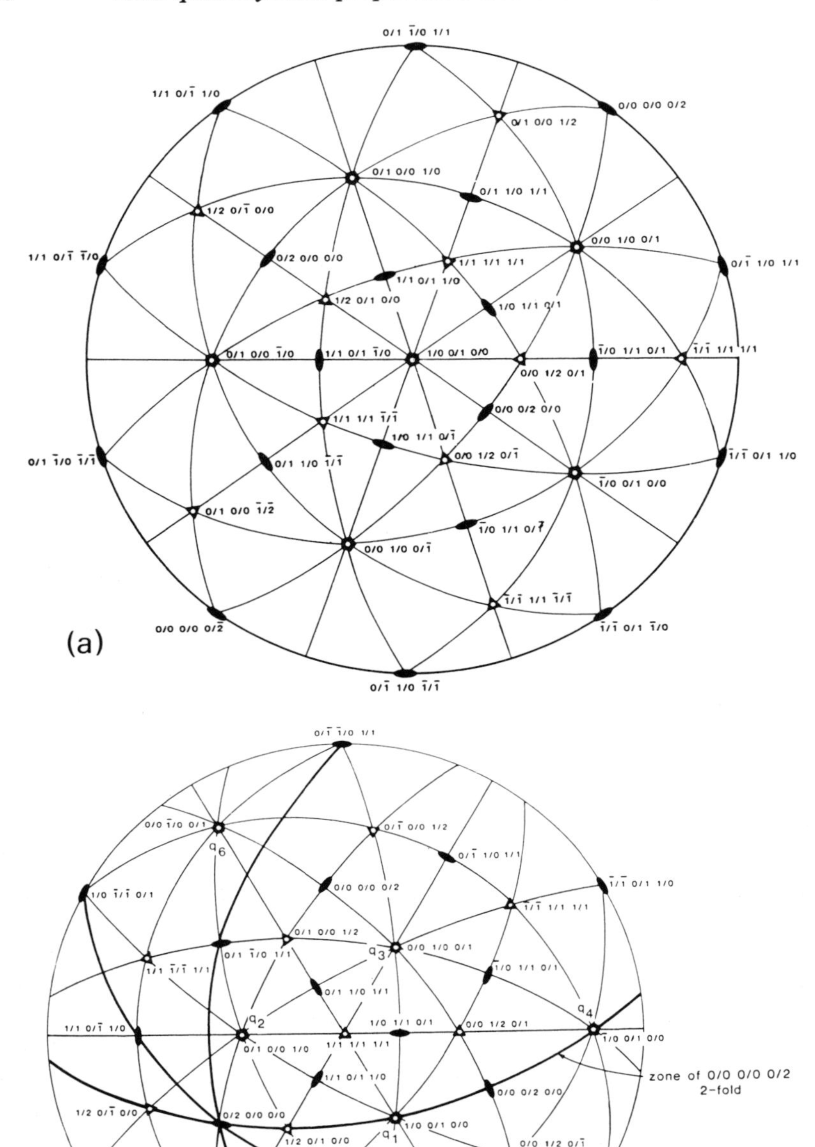
(a)
(b)
zone of 0/0 0/0 0/2
2-fold
zone of 0/0 1/0 0/1
5-fold
zone of 0/0 1/2 0/1
3-fold

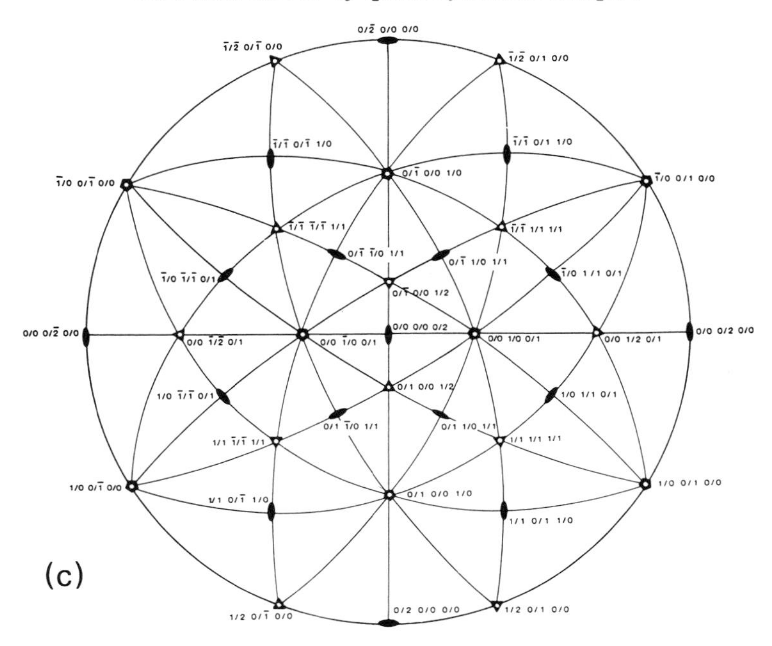

FIG. 2.14 Stereographic projection of the icosahedron. Indexing is done according the scheme proposed by Cahn *et al.*[20] (a) 5-fold zone axis; (b) 3-fold zone axis; (c) 2-fold zone axis.

much. At this point we strongly suggest that readers hand-build their own icosahedra and dodecahedra (models are proposed in the problem section of the chapter). This is very helpful to visualize the different symmetry elements and their respective orientations.

An icosahedron has six 5-fold axes, ten 3-fold axes and fifteen 2-fold axes. Three of the 2-fold axes are orthogonal and can be chosen as a basis. An icosahedron inside a cube is shown in Fig. 2.13. The stereographic projections of the icosahedron along 2-fold, 3-fold, or 5-fold zone axes are shown in Fig. 2.14. Indexing of the different directions is done using the Cahn and Gratias scheme[20] (see Chapter 3 for details). For now, we are mainly interested in the respective positions of the symmetry axes in order to characterize the icosahedral point group completely. There are two icosahedral point groups Y or (235) and Y_h or $m\,\overline{3}\overline{5}$ which possess the same symmetry elements as Y plus an inversion centre. Electron diffraction patterns do not allow us to distinguish these two point groups from each other. Indeed, the cut of the reciprocal space corresponds to what is called the zero-order Laue zone, i.e. a slice of the reciprocal space going through the origin. Because of the Friedel law ($F(hkl) = F^*(-h - k - l)$) the zone is necessarily centro-symmetric: each Bragg spot has a corresponding symmetrical spot with the same intensity. Convergent beam electron diffraction imaging is a technique allowing the distinction of the two point groups, but is beyond the scope of this book.

From Fig. 2.14 the orientations of the various symmetry axes are easily found. For instance, the 5-fold zone axis contains ten 2-fold axes at 36° angles from each other. (The axes perpendicular to the 5-fold axis lie on a great circle.) Starting from this zone axis one can turn clockwise around the 2-fold axes, e.g. the axis labelled ($0/1, \bar{1}/0, 1, 1$) (at the top in Fig. 2.14(*a*)). The successive axes that will show up are 2-fold and then 5-fold, after 31.72° and 63.45° rotations respectively. Doing the same operation, but around a 2-fold axis located at 36° from the preceding one will bring successively 3-fold, 2-fold, and 3-fold axes at angles 37.38°, 58.29°, and 79.2°. Thus the complete characterization of the icosahedral sample must fulfil the orientation relationship of the zone axis. This is exemplified in Figs. 2.15 and 2.16, where different electron diffraction patterns of the AlCuFe icosahedral phase are shown. Let us consider the 5-fold zone axis. There is no periodic distribution of spots, which is a direct consequence of the icosahedral symmetry. Instead, sequences of spots whose distances from the centre are in the τ ratio are observed (see Fig. 2.15). To ensure perfect icosahedral symmetry, the symmetry elements visible on the electron diffraction pattern have to be checked carefully. The Bragg spots must be sharp, i.e. show no broadened or shaped peaks. Rows of spots must be well aligned when looking at a symmetry axis at grazing angle. Before the discovery of the perfect AlCuFe quasicrystal, all the published SAED showed some departure from the ideal pattern. The most common one is a shift of some spots, turning a row from a straight line into a zigzag feature. This is typical of a special kind of defect in quasicrystals called phasons (see Chapters 4 and 5). These shifts can also arise from the presence in the sample of a crystalline phase with large unit cells and pseudo-icosahedral symmetry. Careful examination of the SAED is thus crucial to ensure the quasicrystallinity of the sample.

Finally, one should note that the 2-fold zone axis is the most interesting one, since it contains 2-fold, 3-fold, and 5-fold axes plus mirror planes, though not exhibiting the very nice and once puzzling 5-fold symmetry pattern.

2.3.3 High-resolution electron microscopy

High-resolution microscopes have been improved spectacularly in the past decade and this technique is nowadays more common. Resolution can reach the atomic level (i.e. 2–3 Å), and it is very tempting to try to 'see' the atoms of a quasicrystal. In order to understand better what is seen on a high-

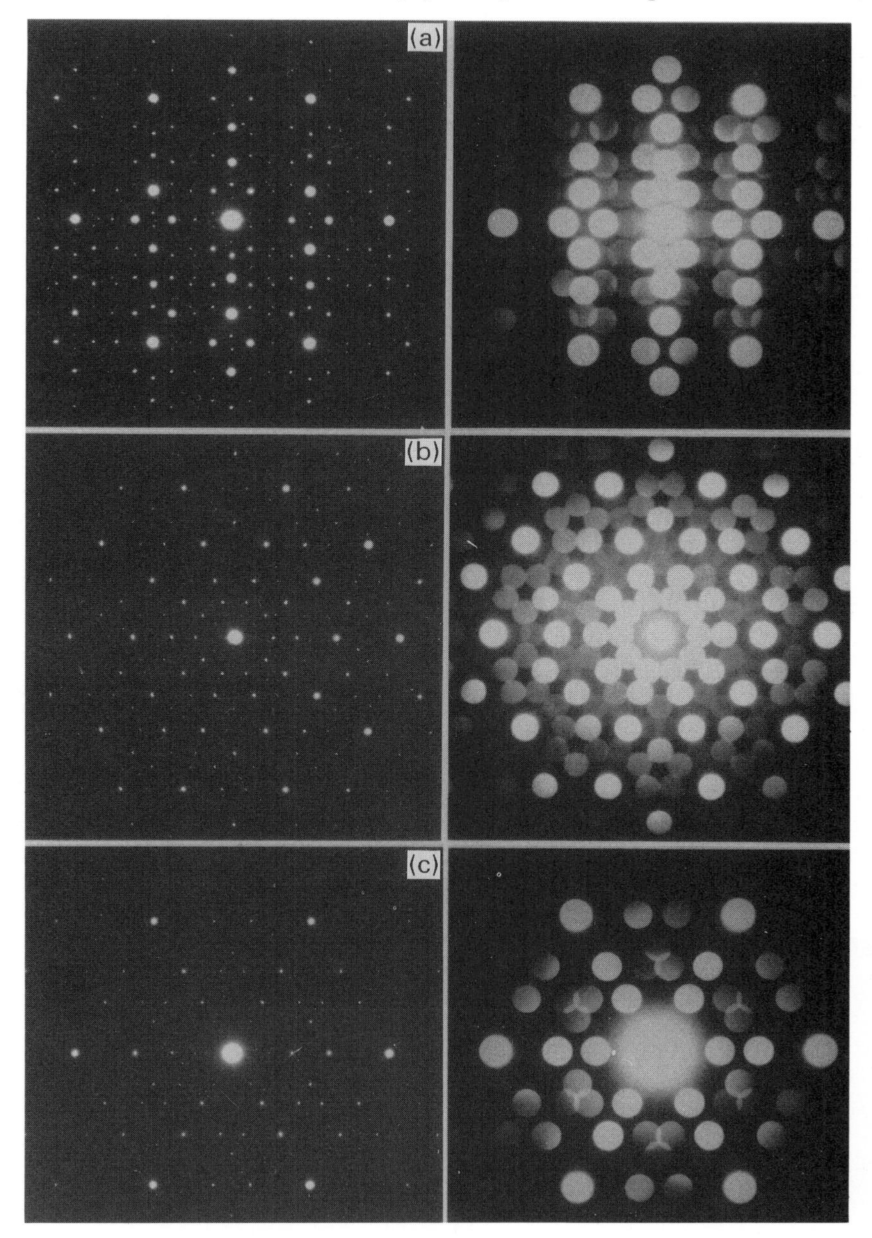

FIG. 2.15 SAED patterns of the AlMnSi icosahedral phase: (a) 2-fold; (b) 5-fold; (c) 3-fold zone axis. (Courtesy of M. Audier).

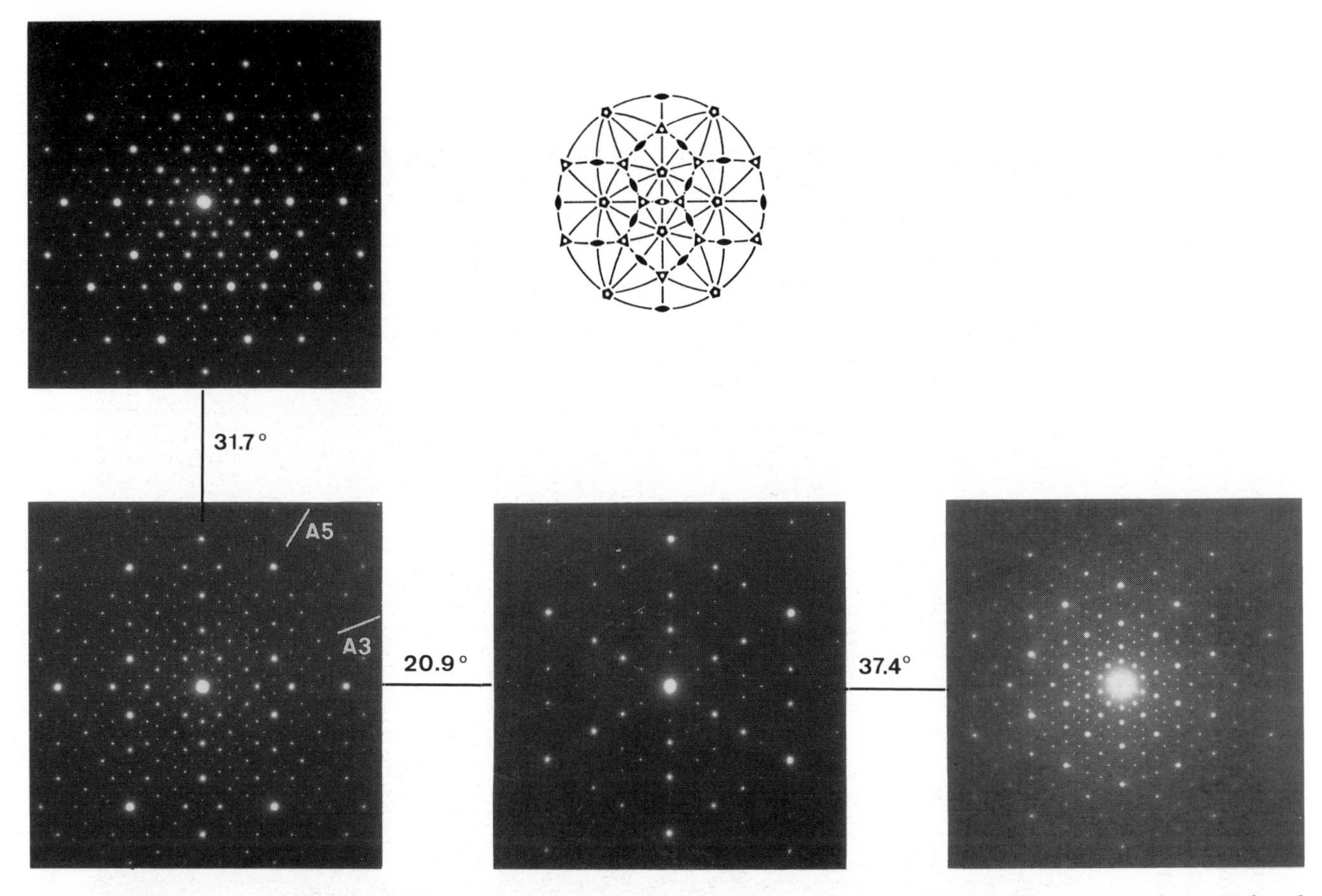

FIG. 2.16 SAED patterns of the AlCuFe icosahedral phase showing the relative orientation of the different axes when rotating the sample around a 2-fold axis. (Courtesy of M. Audier).

resolution image let us recall the basic principle of the image formation. The electron microscope provides an image which is the Fourier transform of the amplitudes of the diffraction pattern. The light and dark dots are related to the sample in direct space. In a first and very crude approximation the image formation can be summarized by the following sequence:

$$\text{Sample} \xrightarrow{\text{FT}} \text{Fourier amplitude} \xrightarrow{\text{FT}} \text{Image},$$

FIG. 2.17 High resolution micrograph of a crystalline $Al_{59}Cu_5Li_{26}Mg_{10}$ Z-phase. The unit cell is hexagonal with $a = c/2 = 14.3$ Å. The HREM is observed perpendicular to the c axis. (Courtesy of M. Audier).

where FT means Fourier transform, and the Fourier amplitude corresponds to the structure factor, i.e. phase *and modulus* of the Bragg spot. In doing this, we have neglected the dynamical effects and the transfer function of the microscope which convolutes the results.

Only one layer of reciprocal space is Fourier transformed for the image formation. The mathematical operation for selecting a layer is called a cut. If the whole reciprocal space could be transformed, a complete three-dimensional image of the sample would be obtained. It can be shown that Fourier transforming a layer (cut) is equivalent to *projecting* the three-dimensional image on to a plane. Thus a Fourier transform of a cut may be considered as a projection. This property is very useful when computing diffraction patterns of quasicrystals. An example of a high-resolution electron micrograph (HREM) from a crystalline sample is shown in Fig. 2.17. Periodicity and lattice planes are clearly visible. However, the white dots correspond to columns of either atoms or holes depending on several para-

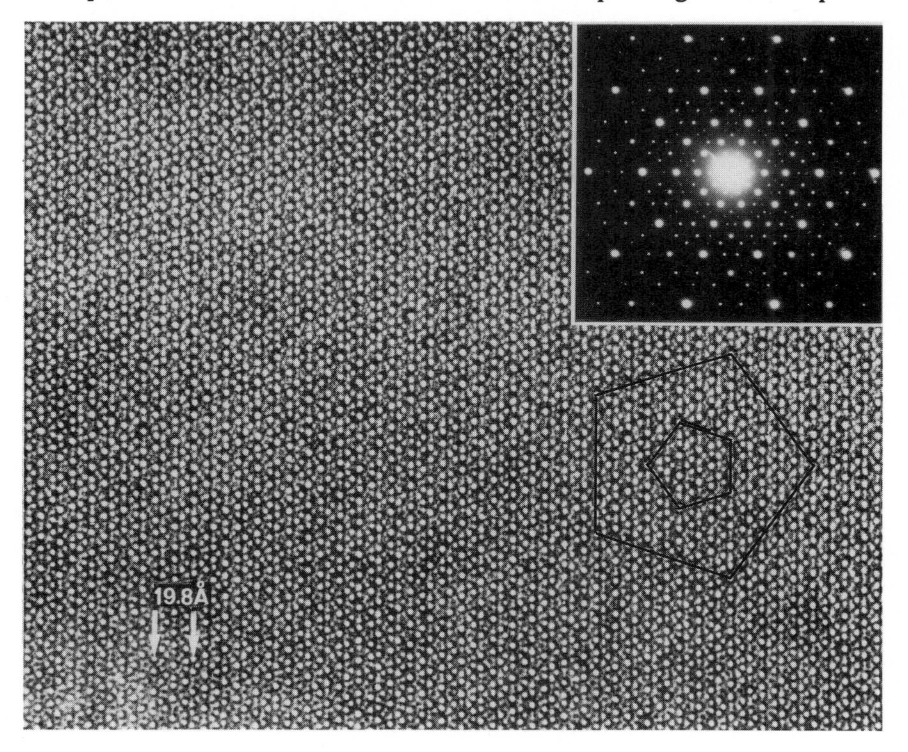

FIG. 2.18 HREM of the icosahedral AlCuFe sample corresponding to a 5-fold zone axis. Pentagonal arrangements are outlined. The succession of lattice planes, when looked at from a grazing angle, is not periodic: the mean distances between planes are related by τ, the golden mean. (Courtesy of M. Audier).

meters such as the thickness of the sample and the focusing of the electron microscope. Except for simple cases, the link between the image and the structure (even if the structure is known) is not straightforward, and dynamical effects and apparatus parameters have to be taken into account in the calculations.

What does the HREM of a quasicrystal look like? A 5-fold zone axis of an AlCuFe icosahedral phase is shown in Fig. 2.18. What are the characteristics of this HREM? First of all, it is obvious that no periodicity can be defined. There are white dots which are very well aligned and which can be interpreted as rows of atoms (or holes) and lattice planes. The succession of the lattice planes is not periodic and their mean distances are related by the golden mean τ. It is difficult to find evidence of quasiperiodic stacking. One way of showing this long-range order is to Fourier Transform a series of dark lines simulating the lattice plane stacking. This is shown in Fig. 2.19: the diffractogram consists of Bragg spots, which are the signature of an ordered structure (i.e. the planes are not stacked randomly). Indexing of the diffractogram is only possible if two integers are used (their position is in a τ ratio), which shows that the stacking is quasiperiodic. Finally, several pentagonal motifs are visible (outlined in Fig. 2.18). In the figure the two main aspect of a quasiperiodic structure are observed, namely long-range orientational order and long-range quasiperiodic translation order.

At such a small scale, periodicity, if it exists, should be visible. This point, which has been the subject of a famous controversy, will be discussed in more detail in Section 2.5.

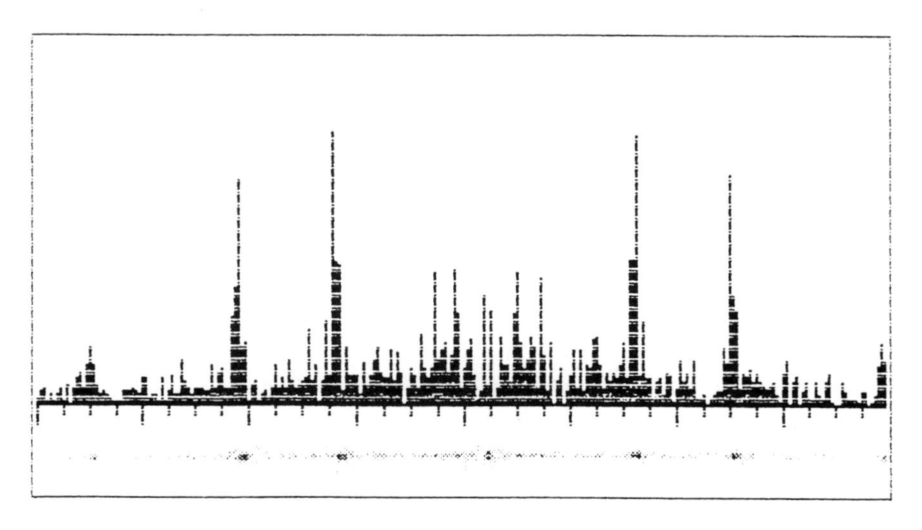

FIG. 2.19 Fourier transform of the intense planes of an HREM (Fig. 2.18). The observed order is not periodic.

2.3.4 Neutron and X-ray diffraction

As stated previously, a careful characterization needs a single crystal experiment to be done. Quite large single grains have to be used for X-ray and neutron experiments. This is why relatively few results have been published to date.

AlLiCu samples were the first quasicrystals grown as single grains. They were large enough to be suitable for X-ray and even neutron diffraction experiments. Figure 2.20 shows X-ray precession photographs taken along different zone axes.[21] This technique allows us to scan slices of reciprocal space and gives a similar picture (for the zero-order Laue zone) to that obtained with electron diffraction. Higher order zones can also be scanned in such an experiment. Again, symmetry axes and cell parameters have to be checked. In Fig. 2.20 the symmetry requirements are shown to be perfectly fulfilled: all the spot positions are consistent with respect to the icosahedral symmetry.

This leads to the possibility of quantitative four-circle measurement. In these experiments the diffracted beam is collected by a detector. Quantitative information is easily obtained on peak shape, integrated intensity, and symmetry-related reflections. To be fully consistent with a quasicrystalline state, once the cell parameters have been determined, the positions and also the intensities of equivalent reflections have to be checked.

When single crystals are not available, powder diffraction is the only way to get quantitative information on the diffracted intensity. In terms of characterization of a quasicrystalline sample, this is the least constraining technique. The only condition to be fulfilled is to match the line position with the indexing scheme; as orientation information is lost, only the length of the $\boldsymbol{Q}$ vector of a given reflection is measured and compared to the calculated one. Parasitic crystalline phases are generally well evidenced by this technique. Bancel *et al.*[22] were the first to provide a completely indexed diffraction pattern. This problem will be addressed more completely in Chapter 3.

2.4 The various families of quasicrystals and their order perfection

The icosahedral point group is not the only one which is incompatible with periodicity. In fact, a large variety of quasicrystals have already been obtained. Some of them are quasiperiodic in a plane, and present periodicity in one direction. Symmetries found so far correspond to 10-fold (decagonal), 8-fold (octogonal), and 12-fold (dodecagonal) rotation. Even a one-dimensional quasicrystal was discovered recently: the diffraction pattern shows a periodic plane, and a perpendicular quasiperiodic row of spots. Figure 2.21 shows an example of a decagonal electron diffraction pattern obtained in the AlCuCoSi alloy.

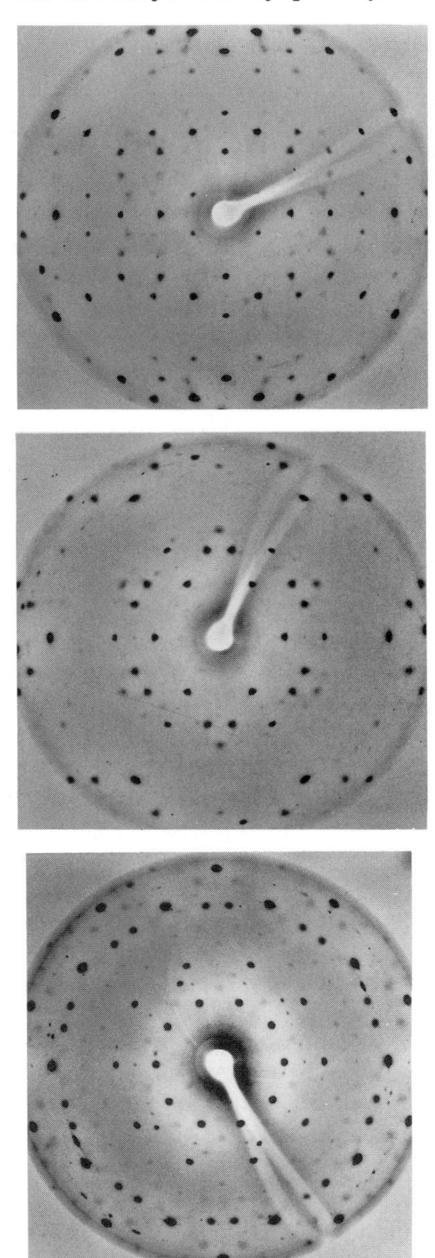

FIG. 2.20 X-ray precession photograph taken with an AlLiCu icosahedral single grain. From top to bottom, 2-fold, 3-fold, and 5-fold zone axes (from Denoyer *et al.*[21].

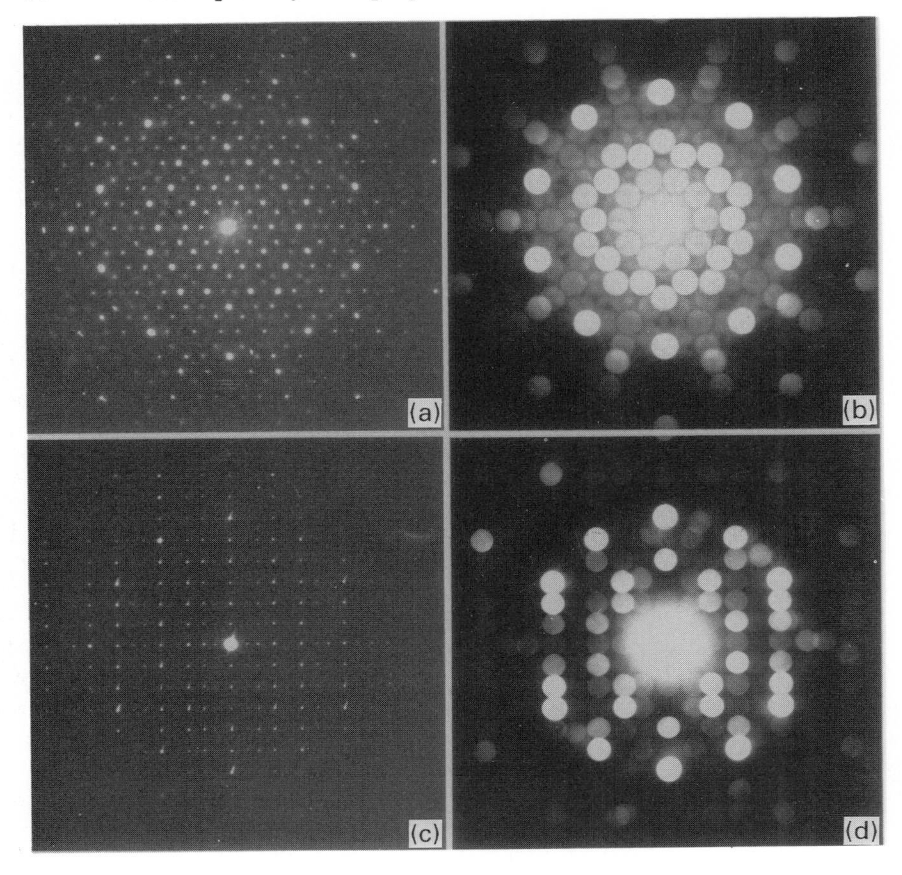

FIG. 2.21 Examples of a decagonal diffraction pattern: (a), (b) 10-fold zone axis; (c), (d) after a $\pi/2$ rotation, a plane containing a periodic direction is visible. (Courtesy of M. Audier).

To date, there is no general rule which can be used to predict which alloys and which compositions are susceptible to producing a quasiperiodic sample. However, it has been observed that the existence of large unit cell crystalline Frank–Kasper phases is a favourable circumstance for quasicrystal formation.

A list of some of the quasiperiodic samples obtained so far is given in Table 2.1 (From Steurer[23].)

Table 2.1 (a) Systems with one-dimensional quasicrystalline (Fibonacci) phases

GaAs–AlAs	artificial	[24]
Mo–V	artificial	[25]
Al–Pd		[26]
$Al_{80}Ni_{14}Si_6$		[27]
$Al_{65}Cu_{20}Mn_{15}$		[27]
$Al_{65}Cu_{20}Co_{15}$		[27]

(b) Systems with octogonal phases. The second column gives the translation period along the eight-fold axis

$V_{15}Ni_{10}Si$	6.3 Å	[28]
$Cr_5Ni_3Si_2$	6.3 Å	[28]
Mn_4Si	6.2 Å	[29]
$Mn_{82}Si_{15}Al_3$	6.2 Å	[30]
Mn–Fe–Si	—	[31]

(c) Systems with decagonal phases. The second column gives the translation period along the 10-fold axis and the third column lists the closely related crystalline phases

Al_5Ir	16 Å	Al_3Ir	[32]
Al_5Pd	16 Å	Al_3Pd	[32]
Al_5Pt	16 Å	Al_3Pt	[32]
Al_5Os	16 Å	$Al_{13}Os_4$	[33]
Al_5Ru	16 Å	$Al_{13}Ru_4$	[22]
Al_5Rh	16 Å	Al_5Rh_2	[31]
Al_4Mn	12 Å	$Al_{11}Mn_4$,	[34]
		μ-Al_4Mn	[35]
Al_4Fe	16 Å	$Al_{13}Fe_4$	[36]
$Al_{77.5}Co_{22.5}$	16 Å	$Al_{13}Co_4$	[37]
Al_4Ni	4 Å	—	[38]
$Al_6Ni(Si)$	16 Å	$Al_9(Ni,Si)_2$	[38]
Al–Cr(Si)	12 Å	$Al_{45}Cr_7$	[33]
$Al_{79}Mn_{19.4}Fe_{2.6}$	—	—	[39]
$Al_{65}Cu_{20}Mn_{15}$	12 Å	$Al_{11}Mn_4$	[40]
$Al_{65}Cu_{20}Fe_{15}$	12 Å	$Al_{13}Fe_4$	[40]
$Al_{65}Cu_{20}Co_{15}$	4,8,12 and 16 Å	$Al_{13}Co_4$	[40]
$Al_{75}Cu_{10}Ni_{15}$	4 Å	—	[40,41]
V–Ni–Si	—		[36]
$Al_{70}Co_{20}Ni_{10}$	4 Å	—	[17]

Table 2.1 Continued

(d) Systems with dodecagonal phases. The second column gives the translation period along the twelve-fold axis

$Cr_{70.6}Ni_{29.4}$	—	[9,10]
V_3Ni_2	4.5 Å }	[42]
$V_{15}Ni_{10}Si$	4.5 Å }	

(e) Systems with icosahedral phases. The quasilattice constants or the reciprocal ones are listed in the second column. Quasicrystal structures of the Al–Mn–Si or (Al,Zn)–Mg types are denoted by A or B, respectively.

$Al_{86}Mn_{14}$	4.60 Å	A }	[43]
$Al_{86}Fe_{14}$	—	— }	[22]
$Al_{85}Cr_{15}$	4.65 Å	A	[44,45]
Al_4Ru	—	—	[46]
$Al_{78}Re_{22}$	—	—	[47]
Al_4–V	—	— }	
Al–Mo	4.75 Å	A }	[48]
Al–W	—	— }	
Al–$(Cr_{1-x}Fe_x)$	—	— }	
Al–$(Mn_{1-x}Fe_x)$	—	— }	[49]
$Al_{62}Cr_{19}Si_{19}$	4.60 Å	A	[50]
$Al_{60}Cr_{20}Ge_{20}$			[51]
Al–Cr–Ru	—	—	[47]
Al–Mn–(Cr,Fe)	—	—	[52]
$Al_{73}Mn_{21}Si_6$	4.60 Å	A	[53]
$Al_{55}Mn_{20}Si_{25}$	—	A	[54]
$Al_{75.5}Mn_{17.5}Ru_4Si_3$	—	A	[55]
$Al_{74}Mn_{17.6}Fe_{2.4}Si_6$	4.59 Å	A	[56]
$Al_{75}Mn_{15}Cr_5Si_5$	—	—	[57]
$Al_{60}Ge_{20}Mn_{20}$	—	A	[58]
$Al_{70}Fe_{20}Ta_{10}$	—	—	[59]
$Al_{65}Cu_{20}Mn_{15}$	—	—	[58]
$Al_{65}Cu_{20}Fe_{15}$	4.45 Å	stable	[13]
$Al_{65}Cu_{20}Cr_{15}$	stable	—	[60]
$Al_{65}Cu_{20}V_{15}$	4.59 Å	—	[60]
$Al_{65}Cu_{20}Ru_{15}$	4.53 Å	stable }	[14]
$Al_{65}Cu_{20}Os_{15}$	4.51 Å	stable }	
$Al_{70}Pd_{20}Mn_{10}$	—	stable	[61]
Al_6CuLi_3	5.04 Å	B	[62,63]
Al_6CuMg_4	5.21 Å	B	[64]
$Al_{51}Cu_{12.5}(Li_xMg_{36.5-x})$	5.02–5.07 Å	B	[65]
Al_6AuLi_3	5.11 Å	B }	[48]
$Al_{51}Zn_{17}Li_{32}$	5.11 Å	— }	

Table 2.1 Continued

$Al_{50}Mg_{35}Ag_{15}$	5.23 Å	B	[66]
Al–Ni–Nb	–	–	
$(Al,Zn)_{49}Mg_{32}$	5.15 Å	B	[67]
$(Al,Zn,Cu)_{49}Mg_{32}$	5.15 Å	B	[68]
$Ga_{16}Mg_{32}Zn_{52}$	5.09 Å stable	B	[12]
Ti_2Fe	$q = 2.82(2)$ Å^{-1}	–	[69]
Ti_2Mn	$q = 2.78(2)$ Å^{-1}	–	
Ti_2Co	$q = 2.76(2)$ Å^{-1}	–	
Ti–Ni	–	–	[70]
$Ti_2(Ni,V)$	–	A	[70,71]
Nb–Fe	–	–	[33]
Mn–Ni–Si	–	–	[72]
$V_{41}Ni_{36}Si_{23}$	–	–	[73]
$Pd_{58.8}U_{20.6}Si_{20.6}$	5.14 Å	–	[74]

Systems with icosahedral symmetry seem to be the most common family of quasicrystals. This family keeps expanding by the discovery of new systems. All these icosahedral quasicrystals can be roughly assigned to three different groups mainly based upon structural details and the perfection of their long-range order.

The first group includes the AlMn-like[43] icosahedral phases, which obviously had the merit of initiating the subject. As described in the previous sections of this chapter, they are metastable compounds obtained via rapid quenching procedures and they cannot be grown as sizeable single grains. Their diffraction patterns exhibit peaks with finite widths which do not simply scale with Q_{par} and/or Q_{perp}. Large amounts of diffuse scattering are often observed. Their average structure is related to a simple (primitive) cubic lattice in the 6-dim image. Obviously they are strongly disordered (see Chapter 4) and they may be composed of a more or less dense packing of icosahedral atomic clusters (icosahedral glass). In support of this idea, most of their properties (electrical resistivity, magnetic spin-glass-like behaviour, etc.)[75] are reminiscent of those observed in glassy or amorphous alloys. This first group is now of historical interest only, and should be discarded for quantitative investigations of the icosahedral state.

The second group corresponds to materials whose local atomic order is based on Bergman-like clusters similar to those observed in Frank–Kasper crystalline phases (see Chapter 4). The archetype of this group belongs to the AlLiCu systems[11, 62, 63] and has the composition $Al_{5.70}Cu_{1.08}Li_{3.22}$. As previously explained, they form at equilibrium via slow cooling from the melt. By using the geode-like preparation, it is possible to grow single grains approaching millimetre size with faceted triacontahedral shapes (see Fig. 2.6). Their diffraction pattern exhibits shaped peaks whose widths scale with Q_{perp}, revealing extended phason disorder. Their average structure is also related to a simple (primitive) cubic lattice in the 6-dim image, with a well-defined chemical order. They are at the origin of the so-called random-tiling models and have stimulated the suggestion that the icosahedral order may be favoured for entropic reasons. This point will be discussed later in the book. Quasicrystals belonging to this group are denoted by B in Table 2.1. It has become increasingly clear that the large phason disorder generally observed in these quasicrystals is related to their very low rate of formation, involving numerous intermediate states. The fraction of the final product which remains as trapped statistical domains of these intermediate states generates a phason disorder which cannot be totally eliminated but is subjected to preparation procedures[76]. An exception within this second group is the newly discovered AlMgLi quasicrystal (approximately 20–40 per cent Mg and 20–30 per cent Li)[77]; its 6-dim periodic image is not primitive but a face-centred cubic lattice. Provided that a higher degree of structural perfection can be achieved, the members of this second group are of great interest with respect to properties specification because they are made of 'good' normal metals. Thus they do not complicate the intrinsic quasi-periodicity by interfering with the particular intricacies of transition metals.

The third group is of greatest interest and contains icosahedral quasicrystals which can reach a very high level of perfection. Their local atomic order is based on a Mackay-like atomic cluster (see Chapter 4). Their 6-dim image is face-centred cubic, resulting from a modified primitive lattice via chemical order superstructure. Their structural perfection is revealed by the absence of any diffuse scattering and Bragg diffraction peaks whose width is due to instrument resolution only. Their properties (see Chapters 5 and 6) cannot be explained by either the usual schemes of periodic crystals or those of amorphous matter. Obviously, they form a new solid state in their own right. However, their perfection depends critically on composition and preparation procedures. This point will be addressed in some detail later in this chapter via examples of unusual phase transitions and the presentation of phase diagrams. They are also stable phases which can be grown as large single grains with dodecahedral faceting (Fig. 2.22). Notable members of this 'perfect quasicrystal' group, in their best compositions, are $Al_{0.62}Cu_{0.255}Fe_{0.125}$[78], $Al_{0.695}Pd_{0.225}Mn_{0.08}$[79, 80], $Al_{0.64}Cu_{0.22}Ru_{0.14}$[81], $Al_{0.71}Pd_{0.20}Re_{0.09}$[82]. The first two of this selection have been investigated quite extensively, particularly with respect to their structures, the thermodynamics of their stability and formation, and the characterization of their quality. The AlPdMn system in particular has become a sort of quality reference, fixing some limit to the properties that must be accepted in establishing the typicality of structure and properties. Fortunately, the icosahedral phase of the AlPdMn system forms peritectically just below the melting points (see phase diagram in Fig. 2.34 below), which means that the reaction growth is almost congruent. This also reduces the probability of formation via several very different intermediate states and the subsequent production of phason-like defects.

The ultimate characterization of the range of positional order in an atomic assembly is usually made via measurements of the **mosaic spread** and the **anomalous transmission** of X-rays due to dynamic diffraction (the Bormann effect). The mosaic spread is a parameter which measures the average misorientation within a structure with respect to perfect order. The misorientations are revealed experimentally by broadening of the Bragg peaks which can be measured using a high-resolution X-ray diffraction microprobe[83] or by γ diffraction.[84] The results for the best AlPdMn quasicrystals are very impressive: the mosaic spread can be less than 0.001° full width at half maximum which is about one-third of the intrinsic Darwin width of Si(111)!! The anomalous transmission is observed through *thick samples* whose structural perfection allows multiple (dynamic) scattering effects. The usual kinematic diffraction theory, which is applicable in most cases, treats diffraction as a single-scattering process: photons experience a single scattering interaction before they exit from the sample. Positive interferences of these single-scattering events produce the Bragg peaks. Some photons are not scattered at all and may emerge from the sample as the transmitted beam, absorption permitting; there is still a transmitted beam even when the sample is out of the Bragg position, except, of course, if the sample is too thick. Actually, the single-scattering process is an approximation of the true diffraction phenomena, which cannot apply to very-long-range spatial perfection. In this case, multiple-scattering events are generally permitted. Then, as illustrated in Fig. 2.23, a Bragg situation will produce *two diffracted beams*, one deflected at an angle of 2Θ with the incident beam (as for kinematic diffraction), and the other making a coherent contribution to the transmitted beam. This is the so-called anomalous transmission which is the only transmitted beam through a thick sample and which vanishes when the sample is turned out of the Bragg position. Indeed, this has been observed with AlPdMn quasicrystals[83] demonstrating the high perfection of the sample over length scales of the order of several millimetres. As we shall explain later, it is hard to imagine that a random tiling structure could reveal such a dynamic diffraction effect.

Now, these perfect quasicrystals are not easily obtained. Frequently, parasitic crystalline phases form instead of or mixed with the quasicrystal. We must be able to identify these different possible states.

2.5 Quasicrystals versus twinned crystals

One of the first objections to the quasicrystalline model has been the so-called twinned model, proposed by Pauling, which led to a famous controversy.[85, 86] The basic idea is that a set of small crystals, with relative

FIG. 2.22 Single grains of the AlCuFe alloy. The shape is a dodecahedron (courtesy of M. Audier).

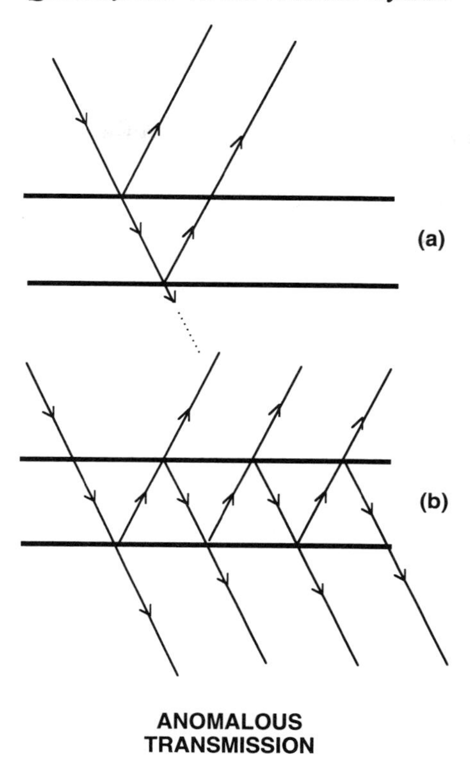

FIG. 2.23 Schematic illustration, restricted to interference between two atomic planes, of the differences between (a) kinematic and (b) dynamic diffraction. This is repeated at all atomic planes below those shown.

orientations following icosahedral symmetry, will reproduce the observed diffraction pattern. In the following, rather than studying in detail the original proposition by Pauling, we will look at the AlCuFe icosahedral phase, which can exist in two forms: either perfectly icosahedral or in a microcrystalline state. We will see that the latter state presents a structure similar to the one proposed by Pauling, but can be distinguished without ambiguity from the perfect icosahedral state.

Beforehand, it may be worth remembering what the first diffraction patterns looked like. They all exhibited some departure from perfect icosahedral symmetry. The main feature consisted of a misalignment in a spot row: instead of being on a perfect straight line, they zig-zagged. In the quasicrystalline scheme, this was interpreted as a kind of typical disorder called phason. But for others, it was the signature of an underlying periodicity. In fact, large periodic unit cells can simulate pseudo-icosahedral symmetry. Is it possible to distinguish the two schemes experimentally?

2.5.1 The AlCuFe microcrystalline state

The $Al_{62}Cu_{25.5}Fe_{0.125}$ alloy gives rise to very nice single grains in the form of dodecahedrons (Fig. 2.22) when slowly cooled from the melt. However, for the slightly different composition $Al_{63.5}Cu_{24}Fe_{12.5}$ it has been shown that at room temperature the phase is not in the icosahedral state, but instead is in a microcrystalline state.[87–90] This is exemplified in Fig. 2.24, which shows a 5-fold high resolution micrograph of the sample: when viewed at glancing angles, periodic fringes are clearly visible within small domains whose typical sizes are about 150 Å. These domains are not oriented randomly, but follow an icosahedral symmetry, as shown in Fig. 2.24(*b*). Looking at various zone axes, M. Audier and P. Guyot[87] showed that the unit cell to be considered is rhombohedral with parameters $a = 18.86$ Å and $\alpha = 63.43°$. (In fact the unit cell is larger, because of a superstructure order.) They are oriented according to a stellate polyhedron (Fig. 2.24(*b*)). Simulating the corresponding diffraction pattern is not an easy task. We will restrict ourselves to the 5-fold zone axis. First we have to compute the diffraction pattern of one grain, the final diffraction pattern being the superposition of five such grains, rotated by successive angles of 72°. The pseudo 5-fold axis can be obtained either by setting a [001] axis vertical, or by looking at the irrational zone $[\bar{1}\bar{1}\tau]$, or rather at one of its rational approximants (for instance $[\bar{8}\bar{8}13]$). As the lattice constant is fairly large, there is a high density of reflections in the reciprocal space. Obviously, only a few of them will be strong enough to show up, because of the atomic decoration. In order to get a realistic structure factor the unit cell has been decorated by an icosahedral atomic cluster, and the kinematical diffraction pattern has been computed.[91] The resulting picture is shown in Fig. 2.25. A pseudo 5-fold symmetry is clearly visible. However, spots are not very well aligned. For instance, there is a misalignment of about 1.4° between the spots 530 and 850 (Fig. 2.25(*a*)) and 3.8° between 233 and 355 (Fig. 2.25(*b*)). In general, strong spots are relatively well aligned, misalignment being more pronounced for low-intensity spots. Alternatively, starting from the icosahedral state, it is possible to generate a crystalline state by deformation of the periodic image; the deformation is along the phason space, and will lead to spot displacement in reciprocal space which scales with Q_{perp}. Generally, weak reflections are associated with a large Q_{perp} component, so that misalignment of weak reflections is enhanced.

The final diffraction pattern is the superposition of five of these diffraction patterns properly rotated. The result is shown in Fig. 2.26: the observed typical triangular shape of the spots is well reproduced by the simulation.

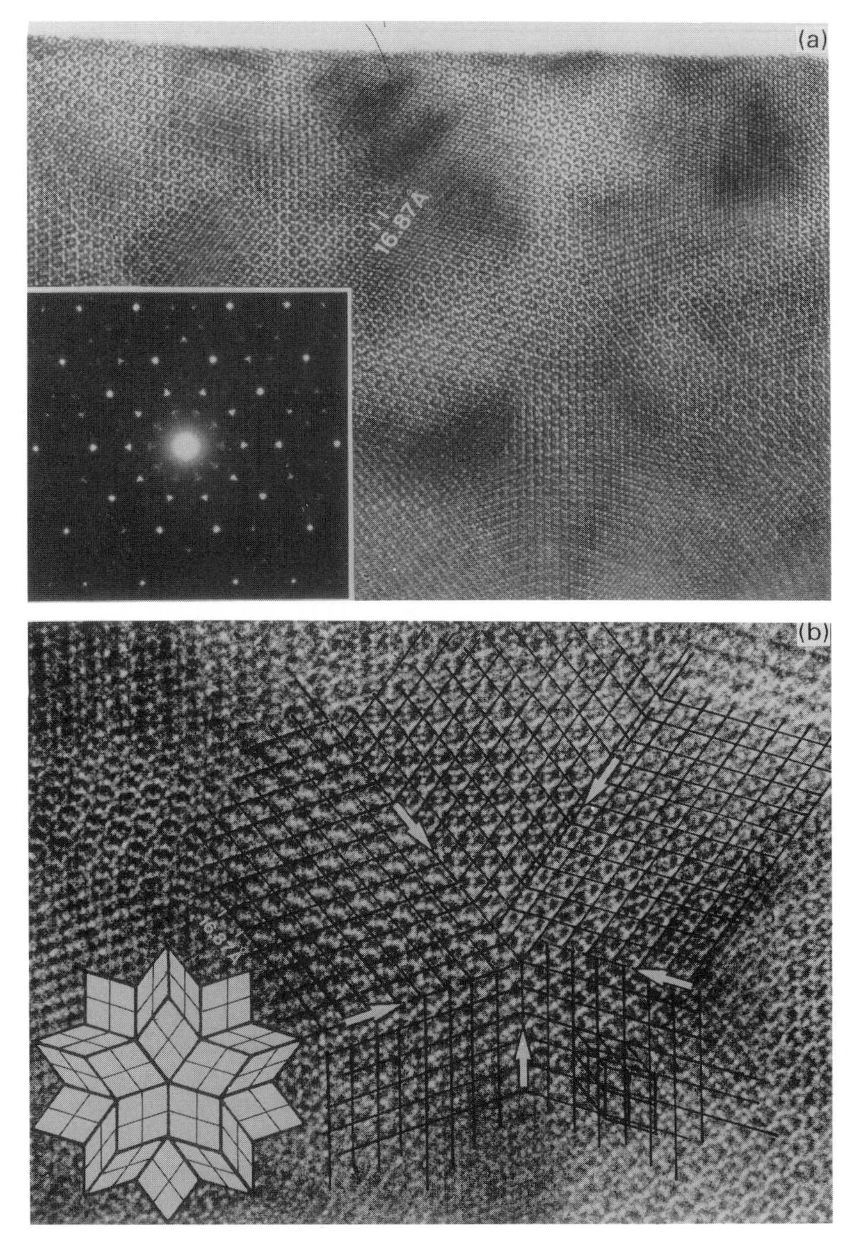

FIG. 2.24 (a) HREM of the microcrystalline state in the AlCuFe system. Looking at grazing angles periodic fringes are clearly visible. (b) Magnification of (a). Five different domains are visible. The lattice periodicity is outlined. The rhombohedral unit cell is projected on to the 5-fold axis, which leads to a periodicity $a \sin(63.43) = 16.9$ Å. (The 5-fold axis is parallel to one of the edges of the rhombohedral unit cell.) (Courtesy of M. Audier).

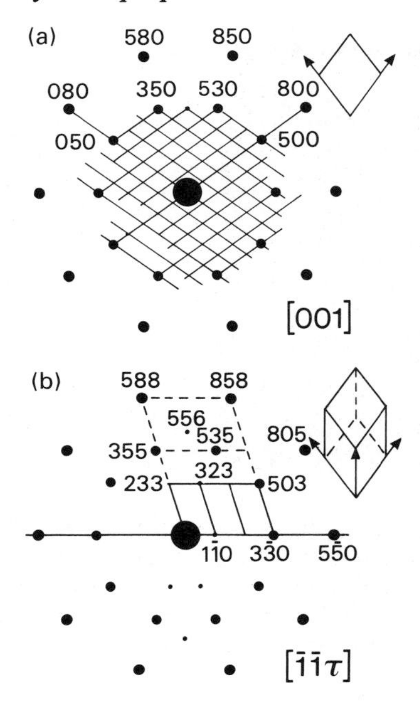

FIG. 2.25 Simulation of a diffraction pattern of a rhombohedral unit cell decorated by an icosahedral atomic cluster, taken along a 5-fold axis. (Courtesy of M. Audier).

One should note that if only the centroid of the reflections is considered, a perfect icosahedral symmetry is restored. Thus only a careful study allows us to say that this is not an icosahedral state but a microcrystalline state.

2.5.2 The AlCuFe perfect icosahedral state

When heating up the $Al_{63.5}Cu_{24}Fe_{12.5}$ dodecahedral single grain a phase transformation occurs. The periodic domains disappear, leading to a perfect icosahedral state.[87, 90] This is clearly identified either in a diffraction pattern or in HREM. For instance, the diffraction pattern shown in Fig. 2.15 exhibits very well-defined Bragg spots, which no longer have the triangular shape typical of the microcrystalline state. Moreover, a lot of very low-intensity spots are now visible, which are the signature of a very low phason disorder. This has also been demonstrated in powder diffraction patterns: in the microcrystalline state, lines are broadened because of the distribution of the reciprocal lattice vectors around one mean value. In the perfect icosahedral state, lines have a width which is only limited by the instrument resolution.[89, 90, 92, 93] A careful study using a synchrotron high-resolution

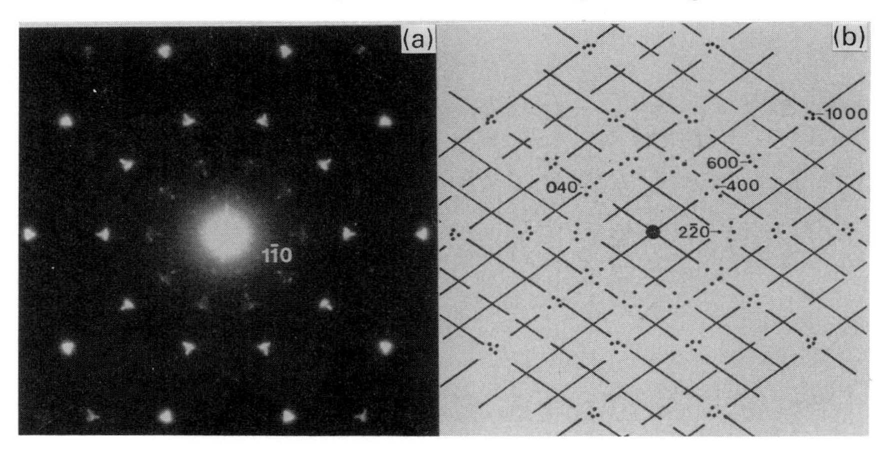

FIG. 2.26 (a) Experimental diffraction pattern of the AlCuFe microcrystalline state along the 5-fold axis. (b) Simulation of the spot positions with the superposition of five periodic reciprocal lattice rotated by 72°. For the sake of simplicity only one periodic lattice is shown. The observed triangular shape of the spots (*a*) is well reproduced in the simulation pattern (*b*). (Courtesy of M. Audier).

powder diffractometer does not show any significant broadening of the lines. The perfect high-resolution image has already been shown (Fig. 2.18) and is obviously different from that of Fig. 2.24: the image shows a uniform contrast, and the white spots form perfect straight lines which are in a quasiperiodic sequence. If phason disorder had existed, it would have shown up as jags in the rows (see Chapter 5), which is obviously not the case. Thus the icosahedral state can be distinguished without ambiguity from any microcrystalline state. These microcrystalline states are reminiscent of the icosatwin model proposed by Pauling,[85, 86] i.e. a large periodic unit cell with domains that are oriented according to an overall icosahedral symmetry. Of course, it is always possible to find a crystalline unit cell that would match Bragg peak positions of the perfect icosahedral state, but the unit cell has to be at least of the order of 500 Å and the concept of periodicity does not have any real sense at this scale. This point will be emphasized further later on.

2.6 Phason-induced phase transition and phase diagram in the AlFeCu system

As stated in the previous section, the composition $Al_{63.5}Fe_{12.5}Cu_{24}$ does not produce a stable quasicrystal since a microcrystalline rhombohedral phase is

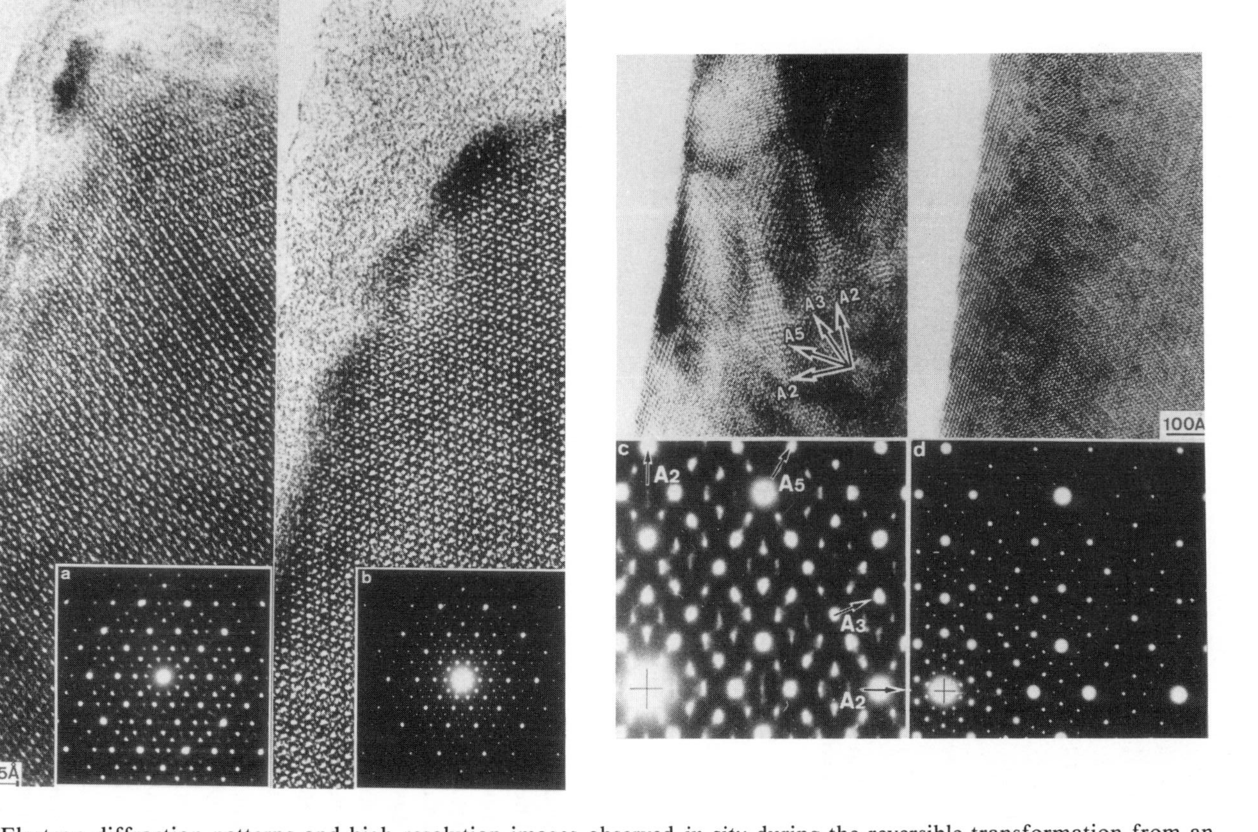

FIG. 2.27 Electron diffraction patterns and high resolution images observed *in situ* during the reversible transformation from an AlFeCu microcrystalline rhombohedral state (a), (c) to an icosahedral phase (b), (d) in 5-fold (a), (b) and 2-fold (c), (d) zone axis[87] (courtesy of M. Audier).

observed at room temperature. This means that the quasiperiodic structure is not the ground state of the system for that particular composition. However, both phases (icosahedral and rhombohedral) are equilibrium compounds in their respective temperature ranges of stability and can be transformed reversibly into each other. This is illustrated in Fig. 2.27 which shows electron diffraction patterns and high resolution images corresponding to the same sample area but at different temperatures.

The phase transition has been observed *in situ*.[87] Below 500°C the sample is totally in the microcrystalline state and this rhombohedral order disappears at about 750°C. Above 865°C the system progressively turns into a mixture of a cubic plus monoclinic structures before total melting. However, this is not the end of the story. A detailed study of the rhombohedral microcrystal ↔ quasicrystal transition has demonstrated that transient states form during this transformation following a partially reversible mechanism. Starting from the high-temperature range in which the icosahedral phase is stable, the transformation occurs roughly between 700°C and 500°C via two successive unstable states, first a modulated icosahedral structure and then a mixture of two pentagonal phases (planar quasiperiodic tilings stacked periodically along the third direction). The reverse transition, from rhombohedral microcrystals into the icosahedral quasicrystal occurs more simply via the pentagonal transient states without involving the previous modulated quasicrystal. Both modulated icosahedral and pentagonal orders are unstable, and the transition proceeds at any given temperature (within the proper range) as time goes on. For instance, the modulation period increases.[94, 95] Electron diffraction patterns and high-resolution images of the modulated and pentagonal phases are shown in Fig. 2.28 and 2.29 respectively.

The modulation (Fig. 2.28) preserves the overall icosahedral symmetry and shows up via satellite reflections in the diffraction pattern. Each main icosahedral Bragg peak is actually flanked by 12 satellites aligned on 5-fold directions (Fig. 2.28) and some periodic franges appear in the high resolution image. Menguy *et al.*[96] have interpreted the formation of this modulated quasicrystal as resulting from phason shear 'waves' (see Chapter 3 for details) deforming the high-dimensional image in E_{perp} periodically along the 5-fold axis of this image.[97] As the transformation proceeds, the modulation induces periodicities along the 5-fold directions which generates a mixture of two pentagonal phases (Fig. 2.29) and finally the microcrystalline rhombohedral stable phase, still mimicking an overall 5-fold symmetry (Fig. 2.24–2.26). The transitions appear to be driven by atomic diffusion involving both phason jumps and local chemical composition changes. The microcrystalline domains of the rhombohedral phase obviously result from the shear strain 'waves' mentioned above whose combination preserves the overall pseudo-icosahedral symmetry through all the transition stages.

The phase diagram of the ternary system AlFeCu has systematically been

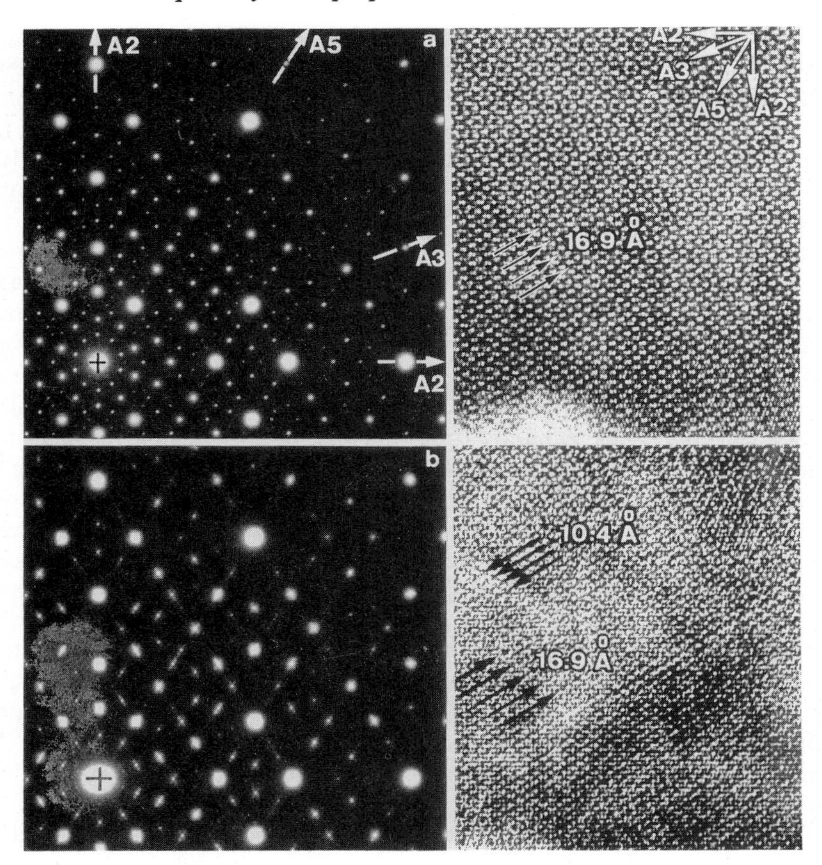

FIG. 2.28 Electron diffraction patterns and high resolution images of the perfect icosahedral AlFeCu phase (a) compared with the modulated icosahedral compound and (b) in a 2-fold zone axis (from Menguy,[94] courtesy of M. Audier). In (b) satellites are visible along 5-fold directions.

revisited, in particular in the vicinity of the icosahedral phase composition.[98] An icosahedral section of this diagram, measured at 700°C, is shown in Fig. 2.30. The extent of the icosahedral phase occurs within the boxed area labelled i, that is over a triangle with vertices roughly corresponding to compositions $Al_{63.6}Cu_{23.7}Fe_{12.7}$, $Al_{62.3}Cu_{24.5}Fe_{13.2}$, and $Al_{59.7}Cu_{29.8}Fe_{10.5}$. When the annealing temperature is reduced to the lowest possible value for atomic diffusion still to be active, this triangle reduces to a tiny area of compositions close to $Al_{62.3}Cu_{24.9}Fe_{12.8}$ which corresponds to the perfect icosahedral quasicrystal stable down to low temperature. Above 700°C, region i also progressively shrinks to be replaced by the monoclinic λ and

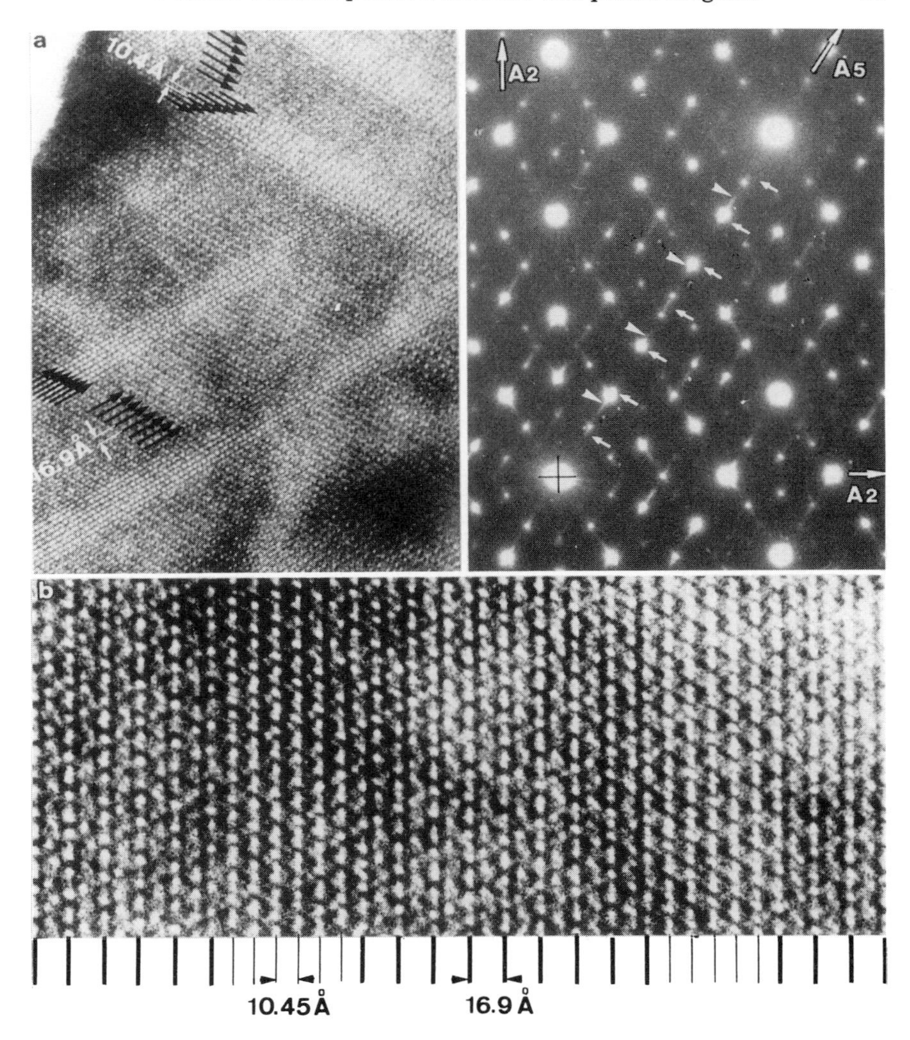

FIG. 2.29 Mixed state of two pentagonal structures: (a) high-resolution image and electron diffraction patterns and (b) enlarged part of the high resolution image showing the two periods. Arrows outline reflection spots which are typical of these two periods (from Menguy[94], courtesy of M. Audier).

cubic β phases which melt above 1100°C. In Fig. 2.30, the composition areas for the periodic approximants P (pentagonal), R (rhombohedral), and O (orthorhombic) are also boxed.

As already mentioned, the AlCuRu system is rather similar to the AlFeCu system. Its phase diagram has been the subject of only preliminary

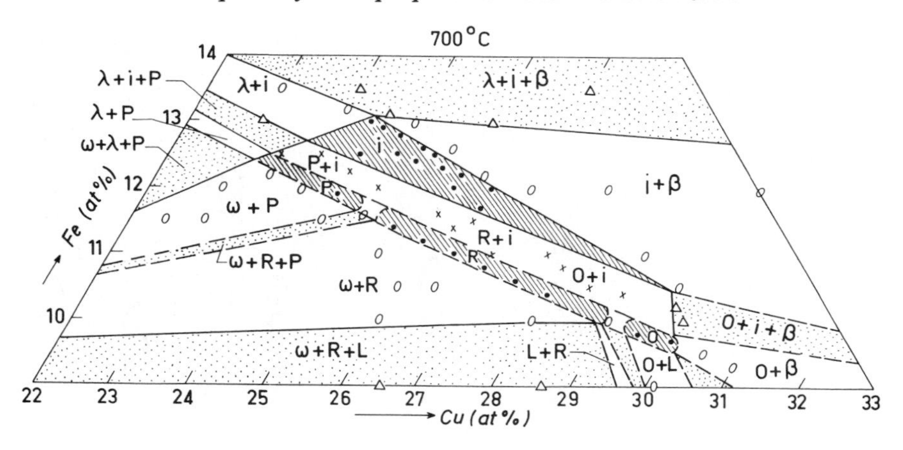

FIG. 2.30 Isothermal section at 700°C of the AlFeCu phase diagram: i, icosahedral phase; P, pentagonal phase; R, rhombohedral phase; O, orthorhombic phase. Mixed symbols indicate multiphase areas: Δ ternary; O binary; • single phase; x at least binary. Broken lines indicate that area borders are not accurately determined because of a similar composition on each side (courtesy of F. Faudot and Y. Calvayrac who kindly provided this diagram prior to publication).

investigations so far.[82] However, the composition at which the quasicrystal is a structural ground state also appears to be critically imposed.

2.7 A phase diagram for the AlPdMn system

Alloys of the AlPdMn system can be obtained either at equilibrium via regular casting methods or in quenched states via melt spinning techniques.[15, 16] For a very specific composition, namely $Al_{70.5}Pd_{21}Mn_{8.65}$, a perfect icosahedral phase forms in equilibrium conditions and remains stable down to room temperature.[60, 99, 100] Large single grains can also be obtained for single crystal diffraction measurements with both neutrons and X-rays.[101] Again, a slight departure from the optimized stoichiometry results in multiphase materials and/or non-perfect icosahedral order.

Using melt spinning techniques to produce (metastable?) materials[102] results in a family of AlPdMn quasicrystals or crystals with various structures.

A first interesting scan of the phase diagram at constant Mn content (about 10 at per cent), by changing the Al/Pd ratio in the formula $Al_{90-x}Pd_xMn_{10}$, shows clearly the crucial role of Pd. At low Pd content, below 15 at per cent, typically, the icosahedral phase is polluted by the presence of some residual fcc aluminium. Moreover, the structure of the

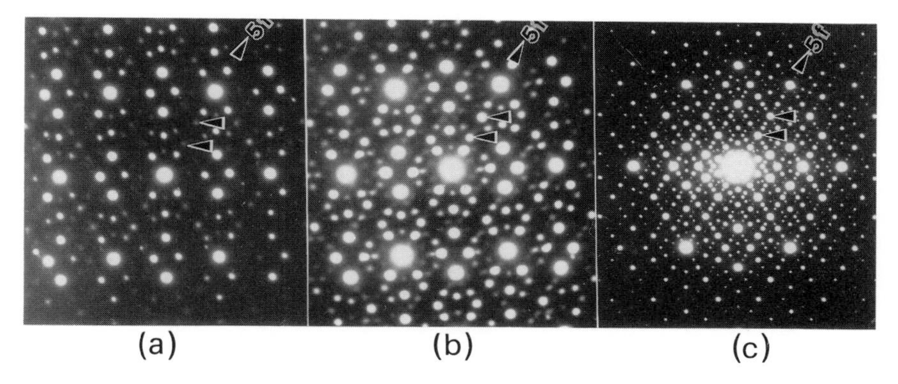

FIG. 2.31 Selected area electron diffraction patterns in two-fold zone axis for alloys with compositions: (a) $Al_{80}Pd_{10}Mn_{10}$ (unperfect SI structure); (b) $Al_{77}Pd_{13}Mn_{10}$ (unperfect FCI structure); (c) $Al_{70}Pd_{20}Mn_{10}$ (perfect FCI structure) (courtesy of An Pang Tsai).

icosahedral phase itself is also dependent on the Pd content. The point is illustrated in Fig. 2.31 showing electron diffraction patterns of two-fold zone axes for samples with compositions $Al_{80}Pd_{10}Mn_{10}$, $Al_{77}Pd_{13}Mn_{10}$, and $Al_{70}Pd_{20}Mn_{10}$. Clearly, additional spots show up (some of them are indicated by arrows in the figures) for samples with the higher Pd content. We will see in subsequent chapters that these additional spots, when indexed into the 6-dimensional scheme of the icosahedral structure, are evidences for a phase transition from a 6-dim simple primitive cubic (SI) to a 6-dim face centred cubic (FCI) Bravais lattice. Simultaneously with the appearance of the extra spots, the width of the diffraction peaks decreases drastically, as expected from better long-range order. We conclude that increasing the Pd concentration with respect to Al at the ideal fixed Mn content (i) stabilizes the FCI structures, which can actually be viewed as a superstructure ordering of a SI lattice, and (ii) improves the perfection of the icosahedral structure. This suggests strong chemical order effects on the structure,[102] which is actually confirmed by diffraction data.[101] The FCI structure is preserved in a small domain around the ideal composition, but again with a cost to be paid in structure perfection. The lattice parameter of the 6-dim cubic structure also has its smallest value when the alloy has the ideal composition. Now, if the phase diagram is scanned along Pd/Mn relative contents, other structural changes are observed. For instance, in the series of alloys with compositions $Al_{70}Pd_{30-x}Mn_x$, decagonal phases form at both ends of the concentration range. For a Pd content larger than 25 at per cent the periodic repetition distance of the quasiperiodic atomic planes is of the order of 16 Å, as for the previously mentioned $Al_{75}Pd_{25}$ decagonal quasicrystal. For less

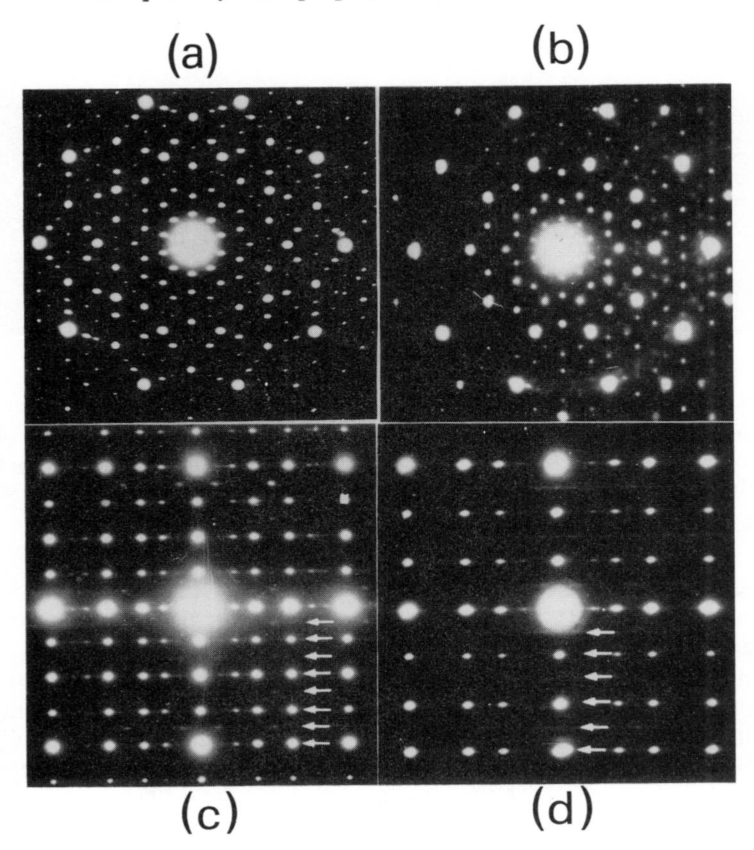

FIG. 2.32 Selected area electron diffraction patterns in the 10-fold (a), (b) and 2-fold (c), (d) zone axes for $Al_{70}PD_8Mn_{22}$ (b), (d) and $Al_{70}Pd_{25}Mn_5$ (a), (c) decagonal phases (courtesy of An Pang Tsai).

than about 10 at per cent Pd, the decagonal phase is AlMn-like with a repetition distance of 12 Å. Electron diffraction patterns of the 10-fold and the 2-fold zone axes are shown in Fig. 2.32 for alloys with compositions $Al_{70}Pd_8Mn_{22}$ and $Al_{70}Pd_{25}Mn_5$. Six- and eight-layer structures are revealed for the former and the latter respectively.

The global concentration Pd + Mn = 30 at per cent seems to be very critical for the formation of quasicrystalline phases. On either side of the corresponding line in the phase diagram, crystalline phases have a tendency to form. Finally, it is interesting to note that the small region around the ideal composition, in which the FCI quasicrystal phase can be observed, elongates on a line corresponding to alloy composition with 1.75 electrons per atom

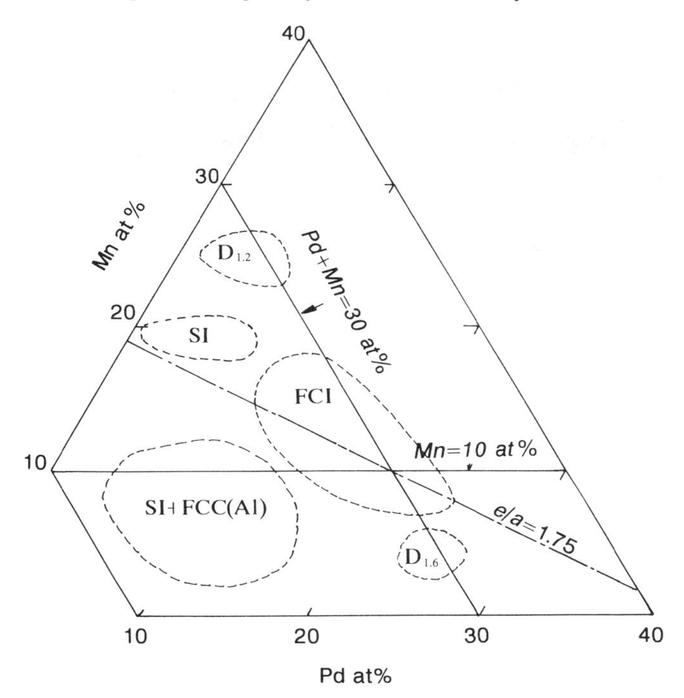

FIG. 2.33 A phase diagram of the rapidly solidified AlPdMn system: SI, FCI, $D_{1.2}$, and $D_{1.6}$ are for simple (primitive) icosahedral, face centred icosahedral, and decagonal (with 1.2 and 1.6 nm periodicity) structures, respectively (courtesy of An Pang Tsai).

(average) (see Fig. 2.33). Thus, electronic structure ingredients are expected to play a rather strong role in the stabilization mechanism of the FCI quasicrystalline phase.

The above conclusions drawn from samples produced by the melt spinning technique have been somewhat modified and made more accurate by metallurgical investigations of alloys produced by slow solidification.[103] The liquidus projection of the ternary AlPdMn phase diagram has been obtained in the vicinity of the icosahedral phase. This is shown in Fig. 2.34.

As expected from the almost congruent melting process, the icosahedral phase area is preserved up to liquidus temperature and is only slightly reduced when lower isothermal sections are considered. Many crystalline compounds, as well as other quasicrystals (modulated icosahedral or decagonal), can form beyond the icosahedral phase. The ideal composition appears to correspond to manganese compositions significantly smaller than initially suggested, i.e. in the vicinity of $Al_{69.5}Pd_{22.5}Mn_8$. Large single-grain

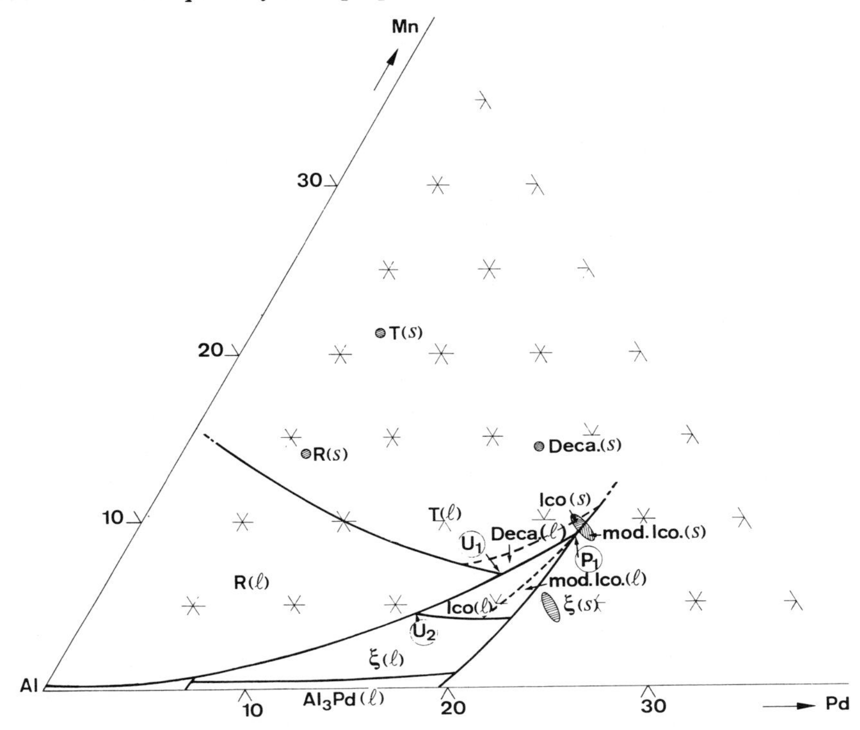

FIG. 2.34 Liquidus projection of the AlPdMn system: l, liquid state; s, solid state; R, T, ξ, rhombohedral phases; Deca, Ico, mod.Ico, quasicrystals with decagonal, icosahedral, and modulated icosahedral structure respectively.[103]

samples of good quasicrystals have been grown with this system using Bridgman and Czochralski procedures.[79, 103]

Figure 2.35 shows an optical micrograph of a single grain of icosahedral phase obtained by Czochralski growth.[103] The composition of the master alloy must be taken as far as possible from the ternary quasiperitectic points U_1, U_2 (see Fig. 2.34) if formation of the parasitic phase, namely $Al_{72.1}Pd_{20.7}Mn_{7.2}$, is to be avoided. The seed quasicrystal is extracted from an oriented ingot resulting from a Bridgman growth. The Czochralski growth temperature is in the range 830–888°C, using an induction-melted alloy in a cold copper crucible.

According to Fig. 2.34 the following phases can form in competition with the icosahedral quasicrystal:[103]

- the orthorhombic ξ phases $Al_{73}Pd_{23}Mn_4$ or $Al_{74}Pd_{21.6}Mn_{4.4}$;
- an orthorhombic T phase $Al_{71}Pd_6Mn_{23}$;

- an orthorhombic R phase $Al_{78.6}Pd_{5.7}Mn_{15.6}$;
- a decagonal quasicrystal (Deca) $Al_{67.5}Pd_{14}Mn_{18.5}$;
- an Al_3Pd_2 compound;
- a modulated icosahedral quasicrystal (on the Pd-rich side).

All these phases can nucleate either simultaneously or one after the other with well-defined orientation relationships.[103]

2.8 Conclusion

Six years after their discovery by Shechtman *et al.*, it is striking to see how large the number of quasicrystalline systems is. In the light of the AlCuFe example, their characterization has to be handled very carefully: electron diffraction and HREM are the first and unavoidable tests. This is clearly the only way to distinguish perfect quasicrystalline states from microcrystalline states without ambiguity. Then other techniques, such as single crystal X-ray or neutron diffraction (when the sample is large enough), or powder diffraction can be employed. Composition and/or temperature domains of existence are very critical for some systems. This is of course of importance when measuring properties which are supposed to be deeply influenced by quasi-periodicity versus periodicity. At present, the most promising system seems to be the AlPdMn icosahedral quasicrystal whose perfection reaches that of the best available periodic crystals.

FIG. 2.35 Optical micrograph of a single grain of the icosahedral AlPdMn quasicrystal obtained by Czochralski growth (scale in centimetres) (courtesy of M. Boudard).

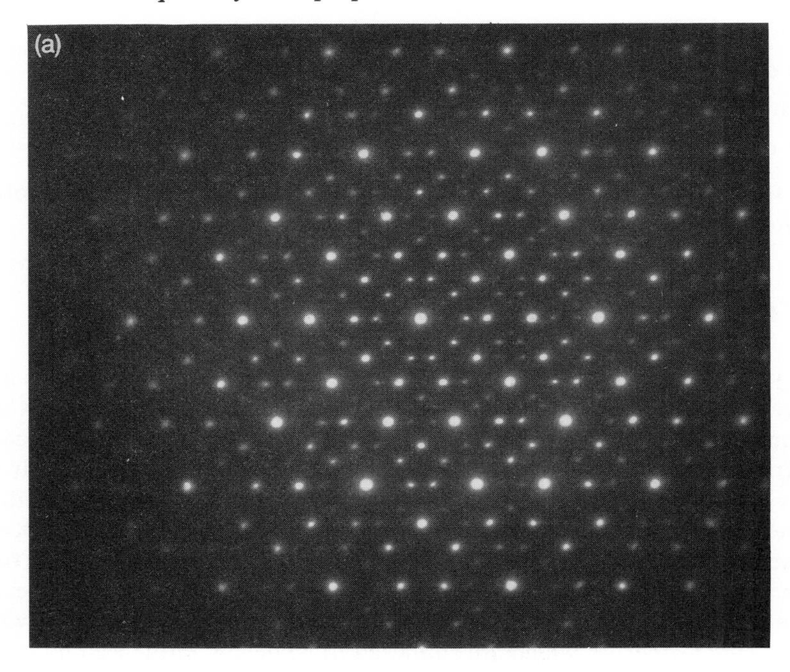
(a)

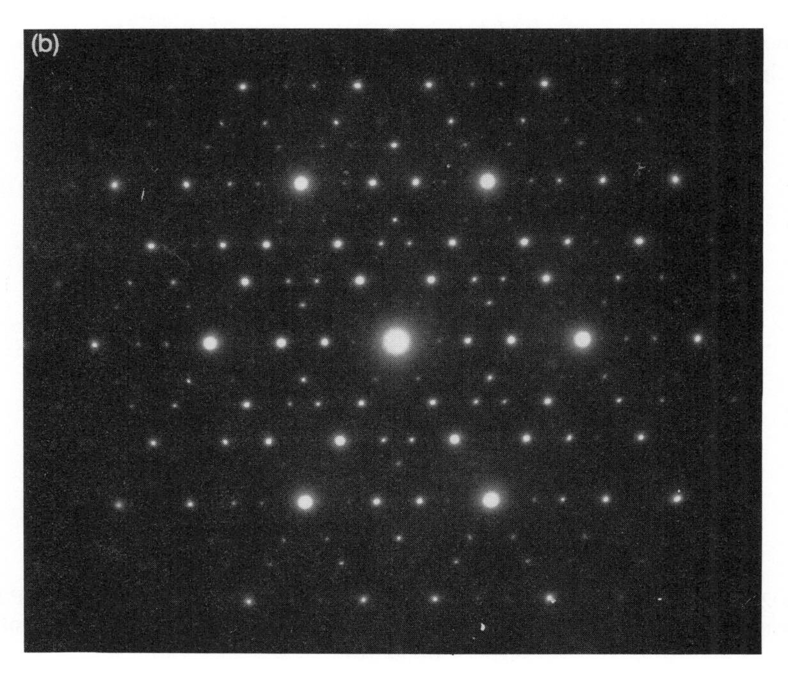
(b)

FIG. 2.36 See Problem 2.1.

2.9 Problems

2.1 The pictures in Fig. 2.36 show several electron diffraction patterns observed with quasicrystal samples.
   (a) Determine the zone axis for these patterns.
   (b) Which patterns correspond to good quasicrystals? Justify your conclusions.
   (c) Propose an interpretation of the structure corresponding to the 'bad quasicrystal' patterns.

2.2 Use the cut-and-fold models to make your icosahedron and your dodecahedron (Figs. 2.37 and 2.38 respectively).

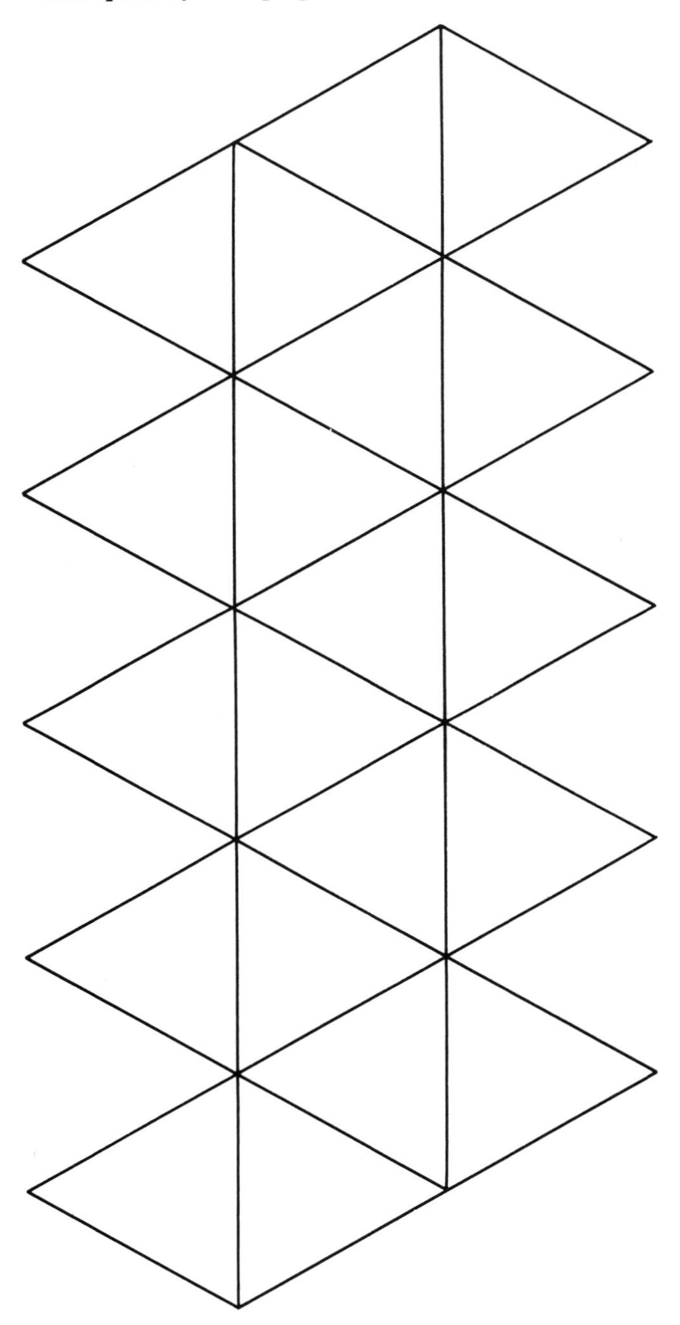

FIG. 2.37 Cut-and-fold model for an icosahedron.

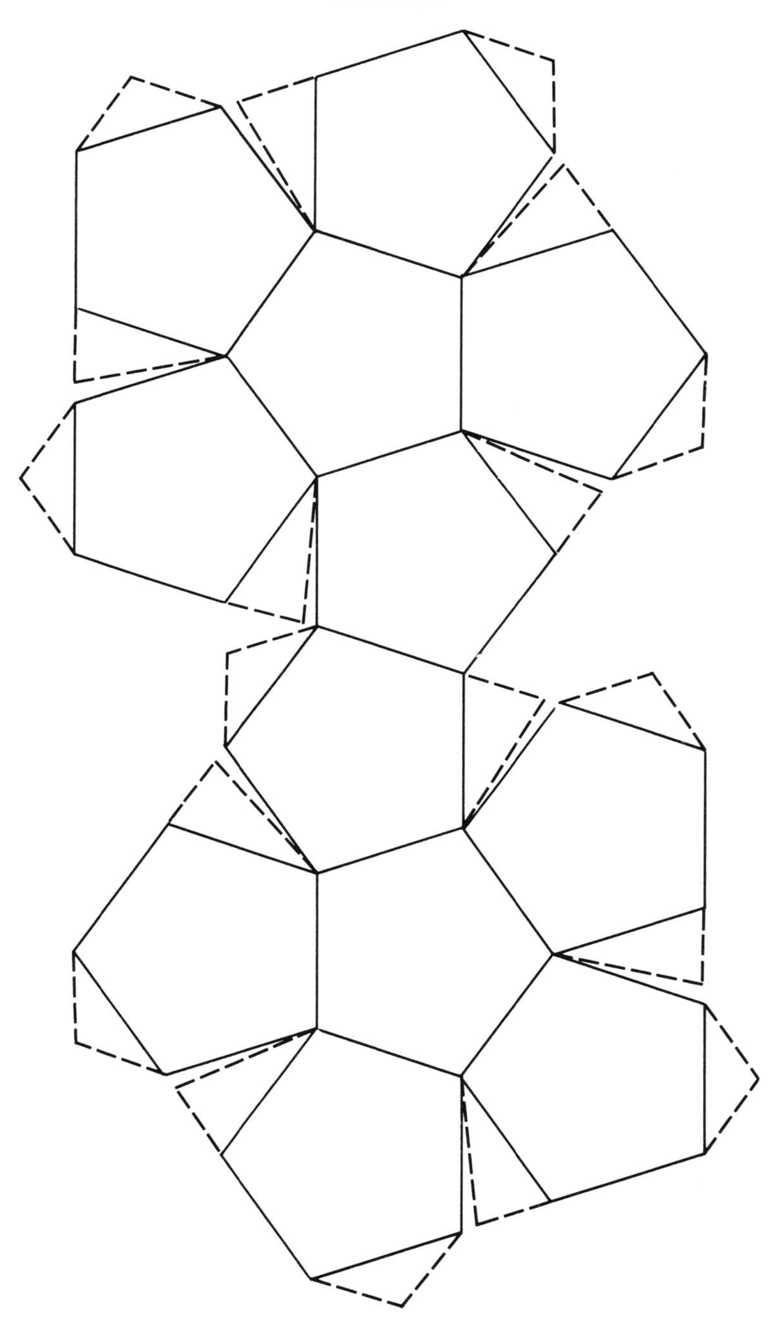

FIG. 2.38 Cut-and-fold model for a dodecahedron.

References

1. Nissen, H.U., Wessicken, R., Beeli, C., and Csanady, A. (1988). AlMn quasicrystal aggregates with icosahedral morphology symmetry. *Phil. Mag.* **B57**, 587–97.
2. Lilienfeld, D.A., Nastasi, M., Johnson, H.H., and Ast, D.G. (1985). Amorphous-to-quasicrystalline transformation in the solid state. *Phys. Rev. Lett.* **55**, 1587–90.
3. Follstaedt, D.M. and Knapp, J.A. (1987). Icosahedral phase formation by ion beam alloying methods. *Nucl. Inst. Meth.* **B24/25**, 542–7.
4. Lilienfeld, D.A., Hung, L.S., and Mayer, J.W. (1987). Ion induced metastable phases. *Nucl. Inst. Meth.* **B19/20**, 1–7.
5. Mayer, J., Urban, K., and Fidler, J. (1987). Phase transitions between the quasicrystalline, crystalline and amorphous phases in Al-16 at% V. *Phys. Stat. Sol.(a)* **99**, 467–73.
6. Budai, J.D. and Aziz, M.J. (1986). Formation of icosahedral AlMn by ion implantation into oriented crystalline films. *Phys. Rev.* **B33**, 2876–8.
7. Eckert, J., Schultz, L., and Urban, K. (1990). Phase transitions and quasicrystal formation in AlCuMn induced by ball milling. *Europhys. Lett.* **13**, 349–54.
8. Ishimasa, T., Nissen, H.V., and Fukano, Y. (1988). Electron microscopy of crystalloid structure in NiCr small particles. *Phil. Mag.* **A58**, 835–63.
9. Ishimasa, T., Fukano, Y., and Nissen, H.-U. (1988). The crystalloid structure in Ni–Cr small particles. In *Quasicrystalline materials* (ed. C. Janot and J.M. Dubois) pp. 168–77. World Scientific, Singapore.
10. Aziz, M.J. and Budai, J.D. (1986). Precipitation of icosahedral Al–Mn during pulsed laser melting. *J. Mater. Res.* **1**, 401–4.
11. Dubost, B., Lang, J.M., Tanaka, M., Sainfort, P. and Audier, M. (1986). Large AlCuLi single quasicrystals with triacontahedral solidification morphology. *Nature* **324**, 48–50.
12. Ohashi, W. and Spaepen, F. (1987). Stable Ga–Mg–Zn quasi-periodic crystals with pentagonal dodecahedral solidification morphology. *Nature* **330**, 555–6.
13. Tsai, A.P., Inoue, A., and Masumoto, T. (1987). A stable quasicrystal in Al–Cu–Fe system. *Japan J. Appl. Phys.* **26**, L1505–7.
14. Tsai, A.P., Inoue, A., and Masumoto, T. (1988). New stable icosahedral Al–Cu–Ru and Al–Cu–Os alloys. *Japan. J. Appl. Phys.* **27**, L1587–90.
15. Tsai, A.P., Inoue, A., Yokoyama, Y., and Masumoto, T. (1990). New icosahedral alloys with superlattice order in the AlPdMn system prepared by rapid solidification. *Phil. Mag. Lett.* **61(1)**, 9–14.
16. He, L.X., Wu, Y.K., Meng, X.M., and Kuo, K.H. (1990). Stable AlCuCo decagonal quasicrystals with deca-prism solidification morphology. *Phil. Mag. Lett.* **61**, 15–20.
17. Tsai, A.P., Inoue, A., and Masumoto, T. (1989). Icosahedral, decagonal and amorphous phases in AlCuM (M = transition metal) systems. *Mater. Trans., JIM* **30**, 666–76.
18. Dubost, B., Colinet, C., and Ansara, I. (1988). Constitution and thermodynamics of quasicrystalline and crystalline AlCuLi alloys. In *ILL/CODEST workshop*, (ed. C. Janot and J.M. Dubois) pp. 39–52. World Scientific, Singapore.

19. Dubost, B., Lang, J.M., Sainfort, P., and Audier, M. (1988). Les monoquasicristaux d'alliage Al6CuLi3, *C. R. Acad. Sci.* (série générale) **5**, 333–46.
20. Cahn, J.W., Shechtman, D., and Gratias, D. (1986). Indexing of icosahedral quasiperiodic crystals. *J. Mat. Res.* **1**, 13–26.
21. Denoyer, F., Heger, G., Lambert, M., Lang, J.-M., and Sainfort, P. (1987). X-ray diffraction study on uniformly oriented quasicrystals. *J. Phys. (Paris)* **48**, 1357–61.
22. Bancel, P.A., Heiney, P.A., Stephens, P.W. Goldman, A.I., and Horn, P.M. (1985). Structure of rapidly quenched Al–Mn. *Phys. Rev. Lett.* **54**, 2422–425.
23. Steurer, W. (1990). The structure of quasicrystals. *Z. Krist.* **190**, 179–234.
24. Todd, J., Merlin, R., Clarke, R., Mohanty, K.M., and Axe, J.D. (1986). Synchrotron X-ray study of a Fibonacci superlattice. *Phys. Rev. Lett.* **57**, 1157–60.
25. Karkut, M.G., Triscone, J.-M., Ariosa, D., and Fischer, O. (1986). Quasiperiodic metallic multilayers: growth and superconductivity. *Phys. Rev.* **B34**, 4390–3.
26. Chattopadhyay, K., Lele, S., Thangaraj, N., and Ranganathan, S. (1987). Vacancy ordered phases and one-dimensional quasiperiodicity. *Acta. Met.* **35**, 727–33.
27. He, L.X., Li, X.Z., Zhang, Z., and Kuo K.H. (1988). One-dimensional-quasicrystal in rapidly solidified alloys. *Phys. Rev. Lett.* **61**, 1116–18.
28. Wang, N., Chen, H., and Kuo, K.H. (1987). Two-dimensional quasicrystal with eightfold rotational symmetry. *Phys. Rev. Lett.* **59**, 1010–13.
29. Cao, W., Ye, H.Q., and Kuo, K.H. (1988). A new octagonal quasicrystal and related crystalline phases in rapidly solidified Mn_4Si. *Phys. Stat. Sol.(a)* **107**, 511–19.
30. Wang, N., Fung, K.K., and Kuo, K.H. (1988). Symmetry study of the Mn–Si–Al octagonal quasicrystal by convergent beam electron diffraction. *Appl. Phys. Lett.* **52**, 2120–1.
31. Wang, Z.M. and Kuo, K.H. (1988). The octagonal quasilattice and electron diffraction patterns of the octagonal phase. *Acta Cryst.* **A44**, 857–63.
32. Ma, L., Wang, R., and Kuo, K.H. (1988). Decagonal quasicrystals and related crystalline phases in rapidly solidified Al–Ir, Al–Pd and Al–Pt. *Scripta Met.* **22**, 1791–6.
33. Kuo, K.H. (1987). Some new icosahedral and decagonal quasicrystals. *Materials Science Forum* **22–4**, 131–40.
34. Bendersky, L. (1985). Quasicrystal with one-dimensional translation symmetry and a ten-fold rotation axis. *Phys. Rev. Lett.* **55**, 1461–3.
35. Shoemaker, C.B., Keszler, D.A., and Shoemaker, D.P. (1989). Structure of *m*-$MnAl_4$ with composition close to that of quasicrystal. *Acta Cryst.* **B45**, 13–20.
36. Fung, K.K., Yang, C.Y., Zhou, Y.Q., Zhao, J.G., Zhan, W.S., and Shen, B.G. (1986). Icosahedrally related decagonal quasicrystal in rapidly cooled Al–14-at.%–Fe alloy. *Phys. Rev. Lett.* **56**, 2060–3.
37. Dong, C., Li, B.G., and Kuo, K.H. (1987). Decagonal phase in rapidly solidified Al–Co alloy. *J. Phys. F: Metal Phys.* **17**, L189–92.
38. Li, X.Z. and Kuo, K.H. (1988). Decagonal quasicrystals with different periodicities along the tenfold axis in rapidly solidified Al–Ni alloys. *Phil. Mag. Lett.* **58**, 167–71.

39. Ma, Y. and Stern, E.A. (1987). Fe and Mn sites in noncrystallographic alloy phases of Al–Mn–Fe and Al–Mn–Fe–Si. *Phys. Rev.* **B35**, 2678–81.
40. He, L.X., Wu, Y.K., and Kuo, K.H. (1988). Decagonal quasicrystals with different periodicities along the tenfold axis in rapidly solidified $Al_{65}Cu_{20}M_{15}$(M = Mn, Fe, Co, or Ni). *J. Mater. Sci.* **7**, 1284–6.
41. Zhang, H. and Kuo, K.H. (1989). The decagonal quasicrystal and its orientation relationship with the vacancy ordered CsCl phase in Al–Cu–Ni alloy. *Scripta Met.* **23**, 355–8.
42. Chen, H., Li, D.X., and Kuo, K.H. (1988). New type of two-dimensional quasicrystal with twelve-fold rotational symmetry. *Phys. Rev. Lett.* **60**, 1645–8.
43. Shechtman, D., Blech, I., Gratias, D., and Cahn, J.W. (1984). Metallic phase with long-range orientational order and no translational symmetry. *Phys. Rev. Lett.* **53**, 1951–3.
44. Zhang, H., Wang, D.H., and Kuo, K.H. (1988). Quasicrystals, crystalline phases and multiple twins in rapidly solidified Al–Cr alloys. *Phys. Rev.* **B37**, 6220–5.
45. Inoue, A., Kimura, H.M., and Masumoto, T. (1987). Formation, thermal stability and electrical resistivity of quasicrystalline phase in rapidly quenched Al–Cr alloys. *J. Mater. Sci.* **22**, 1758–68.
46. Anlage, S.M., Fultz, B., and Krishnan, K.M. (1988). Icosahedral phase formation in rapidly quenched aluminium–ruthenium alloys. *J. Mat. Res.* **3**, 421–5.
47. Bancel, P.A. and Heiney, P.A. (1986). Icosahedral aluminium–transition-metal alloys. *Phys. Rev.* **B33**, 7917–22.
48. Chen, H.S., Phillips, J.C., Villars, P., Kortan, A.R., and Inoue, A. (1987). New quasicrystals of alloys containing s, p and d elements. *Phys. Rev.* **B35**, 9326–9.
49. Schurer, P.J., Werkmann, R.D., and van der Woude, F. (1988). Hume-Rothery rule in quasicrystals. In *Quasicrystalline materials* (ed. C. Janot and J.M. Dubois) pp. 75–82. World Scientific, Singapore.
50. Inoue, A., Kimura, H.M., Masumoto, T., Tsai, A.P., and Bizen, Y. (1987). Al–Ge–(Cr or Mn) and Al–Si–(Cr or Mn) quasicrystals with high metalloid concentration prepared by rapid quenching. *J. Mater. Sci. Lett.* **6**, 771–4.
51. Chen, H.S. and Inoue, A. (1987). Formation and structure of new quasicrystals of $Ga_{16}Mg_{32}Zn_{52}$ and $Al_{60}Si_{20}Cr_{20}$. *Scripta Met.* **21**, 527–30.
52. Janot, C., Pannetier, J., Dubois, J.M., Houin, J.P., and Weinland, P. (1988). Stoichiometry in Al–Mn-based icosahedral phases. *Phil. Mag.* **B58**, 59–67.
53. Gratias, D., Chan, J.W., and Mozer, B. (1988). Six-dimensional Fourier analysis of the icosahedral $Al_{73}Mn_{21}Si_6$ alloy. *Phys. Rev.* **B38**, 1643–6.
54. Inoue, A., Bizen, Y., and Masumot, T. (1988). Quasicrystalline phase in Al–Si–Mn system prepared by annealing of amorphous phase. *Met. Trans.* **A19**, 383–6.
55. Heiney, P.A., Bancel, P.A., Goldman, A.I., and Stephens, P.W. (1986). Extended X-ray absorption fine-structure study of Al–Mn–Ru–Si icosahedral alloys. *Phys. Rev.* **B34**, 6746–51.
56. Ma, Y. and Stern, E.A. (1988). Short-range structure of Al–Mn and Al–Mn–Si aperiodic alloys. *Phys. Rev.* **B38**, 3754–65.
57. Nanao, S., Dmowski, W., Egami, T., Richardson, J.W., and Jorgensen, J.D. (1987). Structure of Al–Mn–Cr–Si quasicrystals studied by pulsed neutron scattering. *Phys. Rev.* **B35**, 435–40.

58. Tsai, A.P., Inoue, A., and Masumoto, T. (1988). New quasicrystals in $Al_{65}Cu_{20}M_{15}$ (M = Cr, Mn or Fe) systems prepared by rapid solidification. *J. Mater. Sci. Lett.* **7**, 322–6.
59. Tsai, A.P., Inoue, A., and Masumoto, T. (1988). Al–Fe–Ta alloy prepared by rapid solidification. *Japan. J. Appl. Phys.* **27**, L5–8.
60. Tsai, A.P., Inoue, A., and Masumoto, T. (1988). New stable quasicrystals in Al–Cu–M (M = V, Cr or Fe) systems. *Trans. JIM* **29**, 521–4.
61. Tsai, A.P., Inoue, A., Yokoyama, Y., and Masumoto, T. (1990). Stable icosahedral AlPdMn and AlPdRe alloys. *Mat. Trans. JIM* **31**, 98–103.
62. Sainfort, P. and Dubost, B. (1986). The T2 compound: a stable quasicrystal in the system Al–Li–Cu–(Mg)?. *J. Phys. (Paris)* **47**, C3-321–30.
63. Mai, Z., Zhang, B., Hui, M., Huang, Z., and Chen, X. (1987). Study of large size quasicrystal in Al_6Li_3Cu alloy. *Mat. Sci. Forum* **22–4**, 591–600.
64. Sastry, G.V.S., Rao, V.V., Ramachandrarao, P., and Anantharaman, T.R. (1986). A new quasicrystalline phase in rapidly solidified Mg_4CuAl_6. *Script. Met.* **20**, 191–3.
65. Shen, Y., Shiflet, G.J., and Poon, S.J. (1988). Stability and formation of Al–Cu–(Li,Mg) icosahedral phases. *Phys. Rev.* **B38**, 5332–7.
66. Mukhopadhyay, N.K., Chattopadhyay, K., and Ranganathan, S. (1989). Synthesis and structural aspects of quasicrystals in Mg–Al–Ag system: Mg_4Al_6Ag. *Met. Trans.* **A20**, 805–12.
67. Henley, C.L. and Elser, V. (1986). Quasicrystal structure of $(Al,Zn)_{49}Mg_{32}$. *Phil. Mag.* **B53**, L59–66.
68. Mukhopadhyay, N.K., Thangaraj, N., Chattopadhyay, K., and Ranganathan, S. (1987). A comparative electron microscopic study of Al-based and Mg-based quasicrystals. *J. Mater. Res.* **2**, 299–304.
69. Kelton, K.F., Gibbons, P.C., and Sabes, P.N. (1988). New icosahedral phases in Ti-transition metal alloys. *Phys. Rev.* **B38**, 7810–13.
70. Zhang, Z., Ye, H.Q., and Kuo, K.H. (1985). A new icosahedral phase with $m\bar{3}\bar{5}$ symmetry. *Phil. Mag. A* **A52**, L49–L52.
71. Yang, Q.B. (1988). Structures of Ti2(Ni,V) in crystalline and quasicrystalline phases. *Phil. Mag. Lett.* **57**, 171–6.
72. Kuo, K.H., Dong, C., Zhou, D.S., Guo, X.Y., Hei, Z.K., and Li, D.X. (1986). A Friauf–Laves (Frank–Kasper) phase related quasicrystal in a rapidly solidified Mn_3Ni_2Si alloy. *Scripta Met.* **20**, 1695–98.
73. Kuo, K.H., Zhou, D.S., and Li, D.X. (1987). Quasicrystalline and Frank–Kasper phases in a rapidly solidified $V_{41}Ni_{36}Si_{23}$ alloy. *Phil. Mag. Lett.* **55**, 33–7.
74. Poon, S.J., Drehmann, A.J., and Lawless, K.R. (1985). Glassy to icosahedral phase transformation in Pd-U-Si alloys. *Phys. Rev. Lett.* **55**, 2324–7.
75. Poon, S.J. (1992). Electronic properties of quasicrystals. An experimental review. *Adv. Phys.* **41**(4), 303–63.
76. Donnadieu, P. and Degand, C. (1993). Evidence of intermediate states between approximant crystal and quasicrystal. *Phil. Mag.* **68**(3), 317–28.
77. Nükura, A., Tsai, A.P., Inoue, A., Masumoto, T., and Yamamoto, A. (1993). Novel face-centered icosahedral phase in AlMgLi System. *Jpn. J. Appl. Phys.* **32**, L1160–3.
78. Cornier-Quiquandon, M., Bellissent, R., Calvayrac, Y., Cahn, J.W., Gratias, D., and Mozer, B. (1993). *J. Non-Cryst. Solids* **153–4**, 10–14.

79. De Boissieu, M., Durand-Charre, M., Bastie, P., Carabellis, A., Boudard, M., Bessière, M., *et al.* (1992). Centimeter-size single grain of the perfect AlPdMn icosahedral phase. *Phil. Mag. Lett.* **65**, 147–53.
80. Boudard, M., de Boissieu, M., Janot, C., Heger, G., Beeli, C., Nissen, H.U., *et al.* (1993). *J. Non-Cryst. Solids* **153–154**, 5–9.
81. Shield, J.E., McCallum, R.W., Goldman, A.I., Gibbons, P.C., and Kelton, K.F. (1993). Phase stabilities in the AlCuRu system. *J. Non-Cryst. Solids* **153–154**, 504–8.
82. Pierce, F.S., Poon, S.J., and Guo, Q. (1993). *Science* **261**, 737-9.
83. Kycia, S.W., Goldman, A.I., Lograsso, T.A., Delaney, D.W., Black, D., Sutton, M., *et al.* (1993). Dynamical X-ray diffraction from an icosahedral quasicrystal. *Phys. Rev. B* **48**(5), 3544–7.
84. Boudard, M. (1993). Structure et propriétés dynamiques de la phase icosaèdrique AlPdMn. *Thesis*, Grenoble.
85. Pauling, L. (1985). Apparent icosahedral symmetry is due to directed multiple twinning of cubic crystals. *Nature* **317**, 512–14.
86. Cahn, J.W., Gratias, D., Shechtman, D., Mackay, A., Bancel, P.A., Heiney, P., Stephens, P.W., Goldman, A.I., and Berezin, A.A. (1986). Pauling's model not universally accepted. *Nature* **319**, 102–4.
87. Audier, M. and Guyot, P. (1990). Rhombohedral to icosahedral solid state transformation in the $Al_{65}Cu_{20}Fe_{15}$ alloy. In *Quasicrystals and incommensurate structures in condensed matter*. Third International Meeting on Quasicrystals, Mexico, May 1989 (ed. M.J. Yacaman, D. Romeu, V. Castano, and A. Gomez) pp. 288–99. World Scientific, Singapore.
88. Audier, M. and Guyot, P. (1990). Microcrystalline AlFeCu phase of pseudo icosahedral symmetry. In *Quasicrystals* (ed. M.V. Jaric and S. Lundqvist) pp. 74–91. World Scientific, Singapore.
89. Bancel, P.A. (1989). Dynamical phasons in a perfect quasicrystal. *Phys. Rev. Lett.* **63**, 2741–4.
90. Janot, C., Audier, M., De Boissieu, M., and Dubois, J.M. (1991). AlCuFe quasicrystals: Low-temperature unstability via a modulation mechanism. *Europhys. Lett.* **14**, 355–60.
91. Audier, M. and Guyot, P. (1988). An approach of the structure of icosahedral $Al_{16}Li_3Cu$ by multiple twinning of a rhombohedral crystal. *Acta Metall.* **36**, 1321–8.
92. Guryan, C.A., Goldman, A.I., Stephens, P.W., Hiraga, K., Tsai, A.P., Inoue, A., and Masumoto, T. (1989). AlCuRu: an icosahedral alloy without phason disorder. *Phys. Rev. Lett.* **62**, 2409–12.
93. Calvayrac, Y., Quivy, A., Bessière, M., Lefebvre, S., Cornier-Quiquandon, M., and Gratias, D. (1990). Icosahedral AlCuFe: towards ideal quasicrystals. *J. Phys. (Paris)* **51**, 417–31.
94. Menguy, N. (1993). Transition de phase reversible icosaèdrique-rhomboèdrique d'un alliage $Al_{63.5}Fe_{12.5}Cu_{24}$. *Thesis*, Grenoble.
95. Menguy, N., Audier, M., Guyot, P., and Vacher, M. (1993). Pentagonal phases as a transient state of the reversible icosahedral-rhombohedral transformation in AlFeCu. *Phil. Mag.* B, 68 n°5, 595–606.
96. Menguy, N., de Boissieu, M., Guyot, P., Audier, M., Elkaim, E., and Lauriat, J.P. (1993). Single crystal X-ray study of a modulated icosahedral AlCuFe phase. *J. Phys. I (France)* **3**, 1953–68.

97. Audier, M., Bréchet, Y., de Boissieu, M., Guyot, P., Janot, C., and Dubois, J.M. (1991). Perfect and modulated quasicrystals in the system AlFeCu. *Phil. Mag. B* **63**, 1375–93.
98. Gratias, D., Calvayrac, Y., Devaud-Rzepski, J., Faudot, F., Harmelin, M., Quivy, A., and Bancel, P.A. (1993). The phase diagrams and structures of the ternary AlCuFe system in the vicinity of the icosahedral region. *J. Non-Cryst. Solids* **153–154**, 482–88.
99. Dong, C., Dubois, J.M., de Boissieu, M., Boudard, M., and Janot, C. (1991). Growth of stable AlPdMn icosahedral phase. *J. Mater. Res.* **6**, 2637–45.
100. Boudard, M., de Boissieu, M., Janot, C., Dubois, J.M., and Dong, C. (1991). The structure of the icosahedral AlPdMn quasicrystal. *Phil. Mag. Lett.* **69**, 197–206.
101. Janot, C., de Boissieu, M., Boudard, M., Vincent, M., Durand, M., Dubois, J.M., and Dong, C. (1992). Single crystal X-ray diffraction study of AlPdMn icosahedral quasicrystal. *J. Non-Cryst. Solids* **150**, 322–26.
102. Tsai, A.P., Yokoyama, Y., Inoue, A., and Masumoto, T. (1992). Chemically driven structural change in quasicrystal AlPdMn alloys. *J. Non-Cryst. Solids* **150**, 327–31.
103. Audier, M., Durand-Charre, M., and de Boissieu, M. (1993). AlPdMn phase diagram in the region of quasicrystalline phases. *Phil. Mag. B* **68**, 607–18.

3

High-dimensional crystallography

You may say I'm a dreamer
but I'm not the only one
John Lennon (*Imagine*)

3.1 Introduction

As illustrated in Chapter 1, 3-dim quasiperiodic structures can be defined and constructed as 3-dim irrational cuts of objects which are periodically distributed in an n-dim space ($n > 3$).

From the experimental point of view (cf. Chapter 2), a quasicrystal is simply an arrangement of atoms which, in diffraction experiments, produces infinitely sharp (δ-function) Bragg peaks in a pattern which exhibits overall 'forbidden' symmetry. These diffraction patterns must be treated as for regular 3-dim crystals in a systematic crystallographic description of the quasicrystals. As such, the crystallography of quasicrystals, or quasicrystallography for short, is simply high-dimensional crystallography with a generalization of the concept of space groups.

Thus, specifying the structure of a quasicrystal should be formally easy, via the indexing of diffraction patterns and modelling of the density distributions to compare Fourier components with the measured intensities of the Bragg peaks.

However, there is a basic difference between regular periodic 3-dim crystals and the high-dimensional periodic structures used to generate 3-dim quasicrystals. In the former, the density distribution is basically a periodic arrangement of point objects, i.e. the averaged atom positions. In the latter, the n-dim periodically distributed objects have $n - 3$ dimensions, i.e. sizes and shapes in addition to positions. Thus, choosing a particular motif in the n-dim unit cell may not be that easy and *a priori* refinement procedures may suffer from too large a number of parameters to be determined.

In this chapter we are going to illustrate that thinking of the quasicrystal structure as a periodic structure in higher dimensions is not merely an amusing mathematical abstraction. For the sake of illuminating advantages and difficulties of the quasicrystallography procedures, the concept and techniques will be analysed in detail with simple examples of 1-dim quasiperiodic structures. The rest of the chapter is mainly devoted to a complete description of high-dimensional crystallography for 3-dim icosahedral quasicrystals.

3.2 The basic principles of quasicrystallography

3.2.1 The general scheme of experimental crystallography

As mentioned in Chapter 1, the structure of condensed matter is mostly investigated via diffraction experiments. A diffraction pattern is a set of Bragg peaks whose intensities are equal to the squared amplitude of the structure factors.

The derivation of a regular crystal structure from diffraction data is equivalent to the reconstruction of the phases of the structure factors $F(\boldsymbol{Q})$. The structure factors are the Fourier components of the density distribution $\rho(\boldsymbol{r})$. They cannot be obtained directly from the observed Bragg peak intensities $I(\boldsymbol{Q})$ which, in turn, are the Fourier components of the so-called Patterson functions (PF). A correspondence scheme can be summarized as follows:

$$\begin{array}{ccccc} I(\mathbf{Q}) \leftrightarrow & F(\boldsymbol{Q})\cdot F(\boldsymbol{Q})^* & \leftarrow & F(\boldsymbol{Q}) & \\ & \updownarrow & & \updownarrow & \\ \rho(\boldsymbol{r})^*\rho(\boldsymbol{r}) = & \mathrm{PF} & \leftarrow & \rho(\boldsymbol{r}) & \leftrightarrow \text{Model} \end{array} \tag{3.1}$$

The $I(\boldsymbol{Q})$ quantities are actually integrated intensities of the measured Bragg peaks. They contain thermal parameters and must be corrected for quite a number of undesired effects: atomic shape factor (for X-rays only), absorption, multiple scattering, polarization effects, etc. Diffuse scattering and weak peaks are usually not included and the subset of observed Bragg peaks is restricted to the available $\boldsymbol{Q}$ range in the given experiment. These are of course rather strong limitations which have to be taken seriously. In particular, termination or truncation effects come from the finite number of measurable Bragg peaks. As a consequence, density maps (or rather PF maps) obtained from a direct Fourier transform of $I(\boldsymbol{Q})$ are not fully reliable, owing to spurious broadening and wavy oscillations of the density features.

Possible ways from $I(\boldsymbol{Q})$ to $\rho(\boldsymbol{r})$ (eqn 3.1), can be listed as follows:

1. *Trial and error*
   A 'reasonable' $\rho(\boldsymbol{r})$ distribution is postulated (from physico-chemichal arguments) with 'free parameters' (e.g. atom coordinates). The Fourier components of this $\rho(\boldsymbol{r})$ model are used to calculate $I_{\mathrm{cal}}(\boldsymbol{Q})$ which finally are fitted to the measured $I(\boldsymbol{Q})$ through refinement of the 'free parameters'. Such a method is certainly difficult to apply to QC because of the excessive number of parameters that would be involved in the specification of the n-dim atomic objects (size and shape).
2. A priori *relationships between $F(\boldsymbol{Q})$ and $I(\boldsymbol{Q})$ (direct methods)*
   Analytical relationships between structure factors can be used for phase reconstruction. Then, a Fourier transform procedure results in density maps, modelling, and further refinements. This is a very popular method

in normal crystallography which does not easily apply to QC, except if the shape of the n-dim atomic objects is known in advance.[1]

3. *Patterson methods*
According to eqn (3.1), it is straightforward to calculate PF from $I(\mathbf{Q})$. These PFs are convolution products of $\rho(\mathbf{r})$. Though usually distorted by the previously mentioned experimental limitations, they may give valuable insight into the actual $\rho(\mathbf{r})$ to be used for modelling and further refinements. This was first applied to quasicrystals by Gratias and co-workers[2–4] and also by Steurer.[5] This approach will be further discussed later in the chapter.

4. *Contrast variation*
This can be used, with either neutron scattering via isotopic substitution or anomalous X-ray diffraction, to separate partial structure factors of a complex structure. For instance, according to eqn (3–1), the diffracted intensity for a binary alloy A–B is

$$I(\mathbf{Q}) = |b_A F_A + b_B F_B|^2, \tag{3.2}$$

with $b_{A,B}$ the neutron scattering length or X-ray atomic scattering factor for A,B atoms and $F_{A,B} = \Sigma_{A,B}\rho_{A,B} \exp(i\mathbf{Q}\cdot\mathbf{r}_{A,B})$ the corresponding partial structure factors, or

$$I(\mathbf{Q}) = |b_A^2 |F_A|^2 + b_B^2| |F_B|^2 + 2b_A b_B |F_A| |F_B| \cos\Delta\phi,$$

with $\Delta\phi$ the phase difference between F_A and F_B. The unknown quantities are $|F_A|$, $|F_B|$, and $\Delta\phi$. They can be determined for each diffraction peak by measuring at least three renormalized independent intensities $I(\mathbf{Q},b_A/b_B)$, b_A/b_B being changed thanks to isotopic (or isomorphous) substitution or anomalous dispersion effects. This has been successfully applied to QC structure and will be illustrated later in the book.[6–9]

3.2.2 *Particular aspects of quasiperiodic structures*

As in Chapter 1, 1-dim examples are going to be used for an easy illustration of the problems arising in quasicrystallography. Let us repeat what was said in Chapter 1 about the Fibonacci sequence as generated by the one-dim cut of periodic square lattice arrangement of segments A_{perp} (Fig. 3.1 is a repetition of Fig. 1.30). The Fourier component of the 1-dim Fibonacci structure is then obtained by projecting the 2-dim Fourier pattern of the 2-dim periodic structure into the 'physical' 1-dim reciprocal space (again, Fig. 3.2 reproduces Fig. 1.31).

The Fourier components of the 1-dim quasicrystal are then expressed as:

$$F(Q_{par}) = \sum_{hh'} \delta(Q_{par} - Q_{par}^{hh'})\, G\,(Q_{perp}^{hh'})/a^2, \tag{3.3}$$

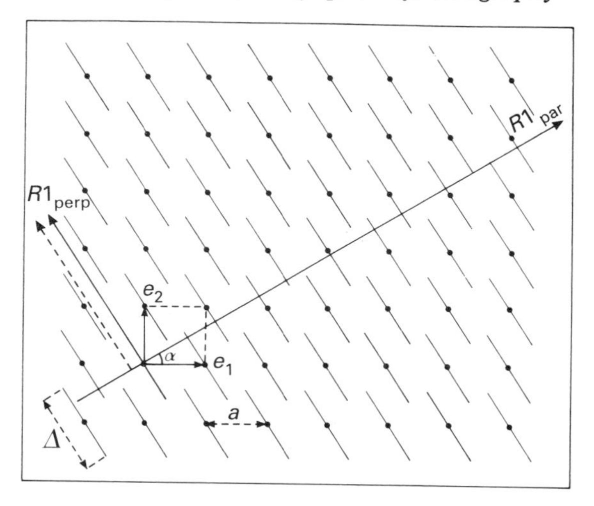

FIG. 3.1 The Fibonacci chain as generated by 1-dim cut of the decorated 2-dim square lattice.

with

$$Q_{\text{par}}^{hh'} = (2\pi/a)\ (h + h'\tau)/(2 + \tau)^{1/2},$$

$$Q_{\text{perp}}^{hh'} = (2\pi/a)\ (h - h'\tau)/(2 + \tau)^{1/2},$$

$$G(Q_{\text{perp}}^{hh'}) = \Delta \left(\sin \frac{Q_{\text{perp}}\Delta}{2}\right) \left(\frac{Q_{\text{perp}}\Delta}{2}\right)^{-1},$$

as demonstrated in Section 1.4.5, using the same notation and definitions.

Thus, assuming that a set of Bragg peaks have been 'experimentally' measured for the 1-dim quasicrystal, the positions of these Bragg peaks, as given by Q_{exp} values, must be compared to the $Q_{\text{par}}^{hh'}$ sequence. If the 2-dim structure is a simple (primitive) square lattice, as exemplified in Fig. 3.1, all values h,h' (integers) are permitted and the Bragg peaks are indexed by the two integers which 'lift' the 1-dim diffraction pattern into the relevant 2-dim space. It is easy to demonstrate that the usual extinction rules also apply to quasicrystallography. For instance, if Bragg peaks only show up for (h,h') indices such that $h + h'$ are even numbers, this means that the 2-dim square structure has atomic objects at the lattice sites *and* at the body centre of the unit cell (see problems).

The first difficulty arising at this stage is due to $Q_{\text{par}}^{hh'}$ being a very dense set of values, even if terms with large values of $Q_{\text{perp}}^{hh'}$ give no significant contributions to the experimental diffraction pattern (see eqn 3.3). The expected corresponding drawbacks may be:

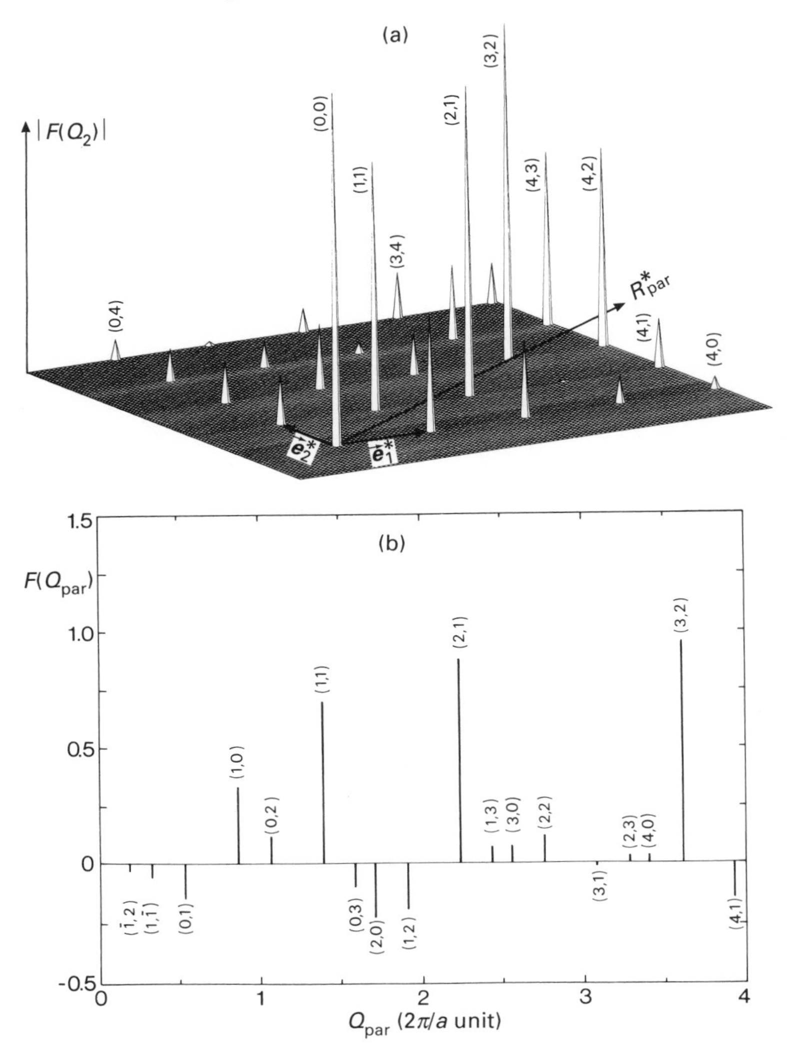

FIG. 3.2 Fourier pattern of a Fibonacci chain as projected from the reciprocal lattice of the decorated square lattice.

(1) some difficulties in accurately accounting for extinction rules;

(2) incorrect indexing;

(3) misattributed intensities to a single set of indices when peaks are not experimentally separated or are degenerate (powder patterns).

Once indexed in n-dim, the diffracted intensities can be used to work out the structure, or calculate $\rho(\boldsymbol{r}_n)$ using the scheme of eqn (3.1). As already stated, PF can readily be obtained using the indexed $I(\boldsymbol{Q}_n)$ as Fourier component or, alternatively a $\rho(\boldsymbol{r}_n)$ density distribution is deduced if phases of structure factors $F(\boldsymbol{Q}_\mathrm{n})$ can be reconstructed. The point to be made here is that only some of the Bragg peaks with small $Q_{\mathrm{perp}}^{hh'}$ are actually measured. Consequences of such an 'experimental' truncation are illustrated in Fig. 3.3 and 3.4 showing simulated density and PF maps for the 2-dim structure of Fig. 3.1. In this simulation, a set of Q_{par} and Q_{perp} values, limited to 8.5 and 1.5 $(2\pi/a)$, respectively, has been used to calculate the 21 corresponding strongest structure factors and diffracted intensities. Then, $\rho(\boldsymbol{r})$ and PF of the 2-dim structures have been calculated via a direct Fourier transform procedure applied to the 21 selected reflections. Density (or PF) maps, along with profile of the density (or PF) features are presented in Fig. 3.3 and 3.4. Both PF and $\rho(\boldsymbol{r})$ maps show unambiguously that the atomic objects in the 2-dim structure are located at the square lattice sites. But the size and shape (a flat plateau) of the actual A_{perp} cannot easily be inferred from the experimental profiles, especially if only FP maps are available. The situation becomes even more intricate when different atomic objects have to be included into the high-dim periodic structure. This will be dealt with further in the next section. Modelling the A_{perp} objects for further refinement of the structure requests additional information such as atomic density and composition of quasicrystal, symmetry constraints, restriction on short atomic distance, etc. This will be illustrated with a real life example in Chapter 4. In any case, it is clear that this modelling/refinement state is less easily achieved in quasicrystallography than in a regular 3-dim structure derivation. In the example of a 1-dim monatomic structure, there is only the 'lattice parameter' to be refined for a periodic sequence, but many parameters for the Fibonacci sequence, i.e. the lattice parameter of the square lattice, one size parameter for the A_{perp} object and an infinite number of shape parameters if the density profile of A_{perp} is completely unknown. In practical cases (see Chapter 4) the problem can reasonably be solved with good and numerous diffraction data . . . and a little imagination!

3.2.3 Further problems . . . and further solutions

So far, our 'high-dimensional' square lattice has been decorated by a single A_{perp} atomic object, located at the lattice sites, except when extinction rules have been mentioned in relation with body centre decoration.

When different A_{perp} atomic objects are considered, the crystallography approach has to achieve the specification not only of their sizes and shapes but also of their relative positions. Equation (3.3) and others of that type have to be rewritten:

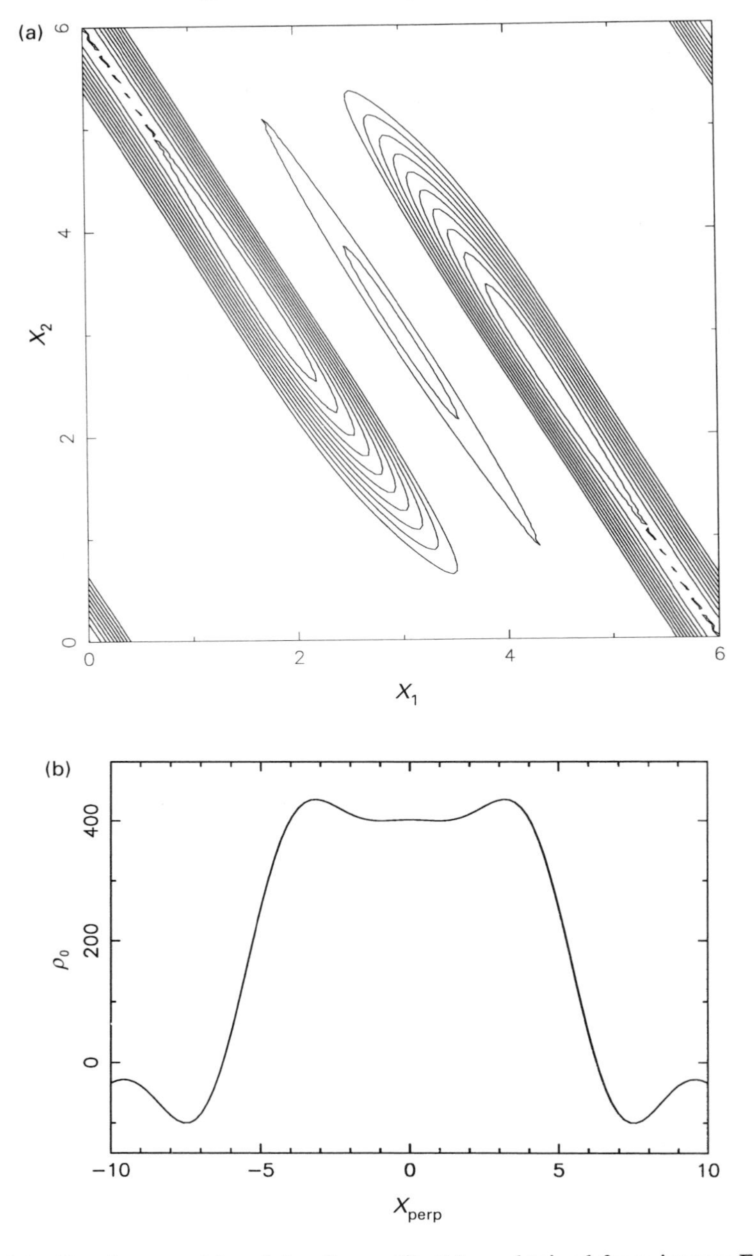

FIG. 3.3 Density map (a) and density profile (b) as obtained from inverse Fourier transform of a set of Fourier component of the Fibonacci chain (Fig. 3.1), restricted to $Q_{par} \leqslant 8.5$ and $Q_{perp} \leqslant 1.5$ (in $2\pi/a$ units). A single unit cell only is represented.

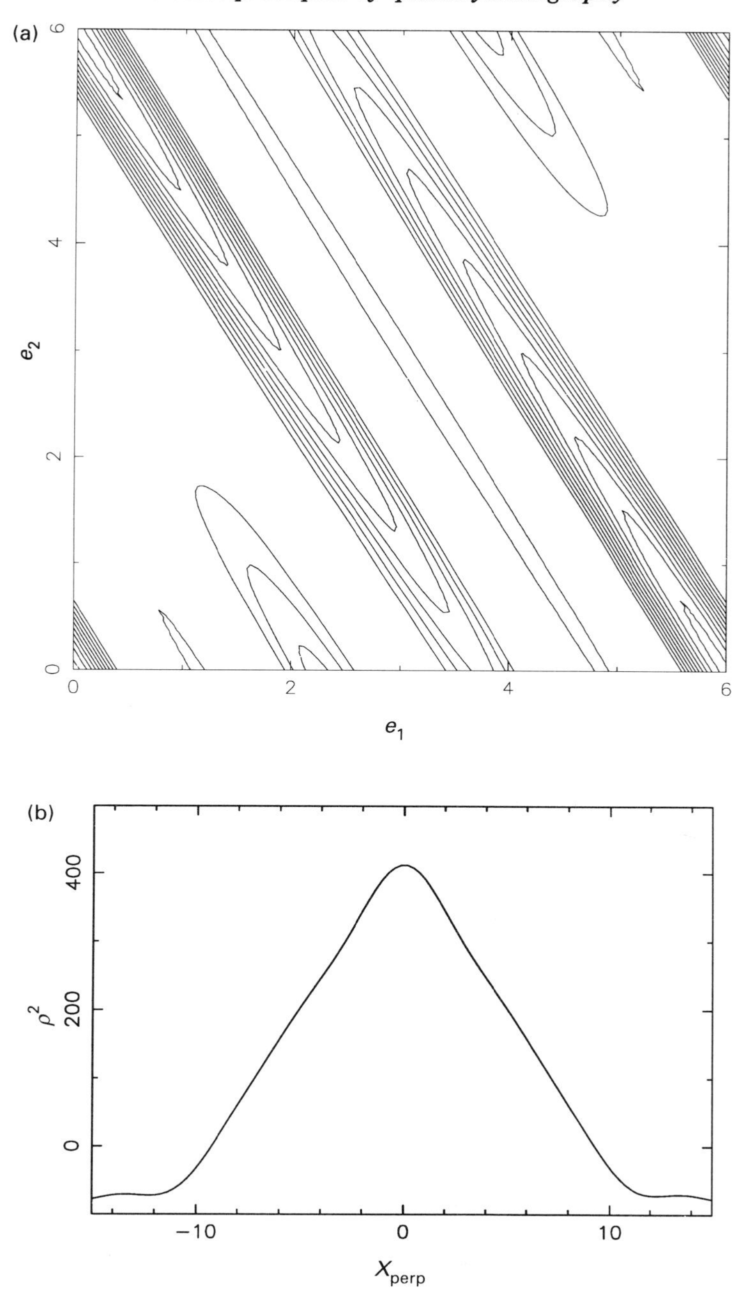

FIG. 3.4 Same as for Fig. 3.3, but for Patterson functions.

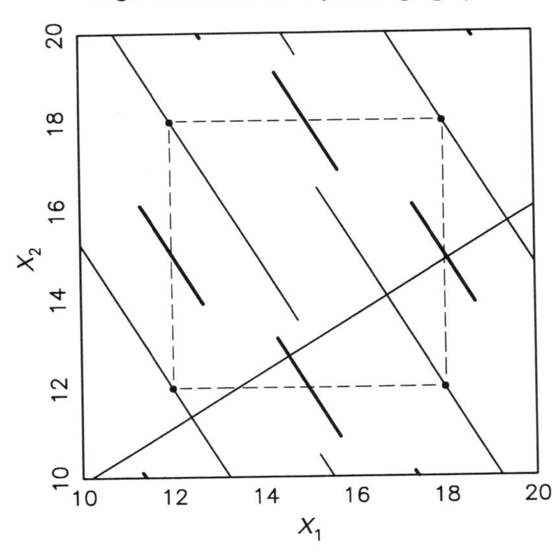

FIG. 3.5 Unit cell of square lattice associated to a 1-dim QC, with two different A_{perp} volumes at origin and mid-edge positions. Compared to the situation of Fig. 3.1, new atom distances are generated in the $R1_{\text{par}}$ physical space.

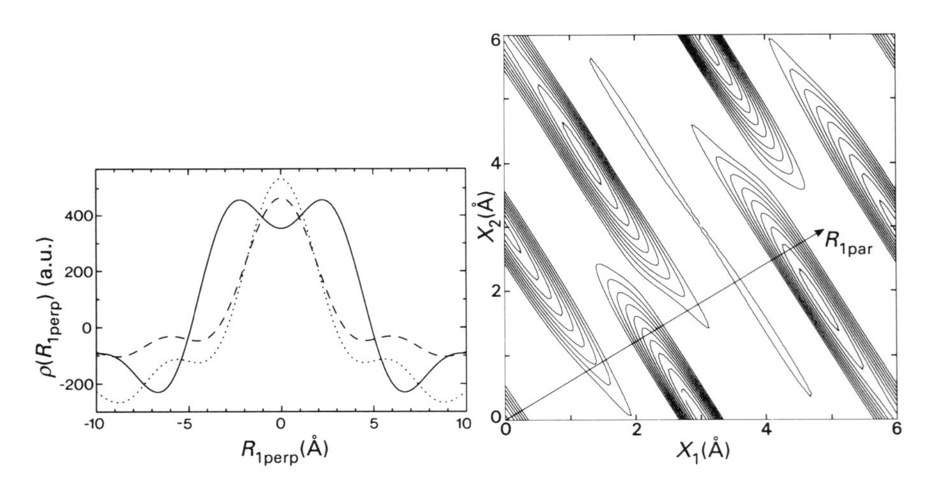

FIG. 3.6 Calculated density map (a) and profile (b), corresponding to the structure of Fig. 3.5, when a restricted set of Fourier components is used. In the simulation, vertex segments have been taken twice as large as the mid-edge ones (9 Å vs. 4.5 Å).

$$F(Q_{\text{par}}) = \frac{1}{a^2}\,\delta(Q_{\text{par}} - Q_{\text{par}}^{\text{lat}}) \sum_j b_j G_j(Q_{\text{perp}}) \exp(\mathrm{i}\boldsymbol{Q}_2 \boldsymbol{X}_j), \qquad (3.4)$$

in which $\boldsymbol{X}_j$ are the positions of the atomic objects in the high-dim unit cell. An example of such a structure is shown in Fig. 3.5. In this example, the square unit cell is decorated by A_{perp} line segments at the lattice sites and by other, shorter, line segments at the mid-edge positions. The interested reader will easily derive the 1-dim quasiperiodic structure which is now obtained by the physical cut, in place of the Fibonacci sequence. Indexing the corresponding diffraction pattern might also be a very fruitful exercise. Again, a simulation of density and PF map reconstruction can be obtained by selecting the 21 strongest Bragg peaks ($Q_{\text{par}} \leqslant 8.5$ and $Q_{\text{perp}} \leqslant 1.5$ in $2\pi/a$ units), as already done in the previous section for the simple Fibonacci case. The resulting $\rho(\boldsymbol{r})$ and PF maps are presented in Figs. 3.6 and 3.7, along with the profiles in complementary space of their density features. The $\rho(\boldsymbol{r})$ maps, as compared with the structure shown in Fig. 3.5, clearly reproduce the right positions for both lattice and mid-edge sites. But, according to their density profiles, two different mid-edge A_{perp} atomic objects are wrongly obtained. Moreover, two spurious atomic positions are generated in the cut 1-dim quasicrystal, due to the wavy background of the reconstructed density features in 2-dim. Still, this density map is a good acceptable starting point for further modelling and refinement of the structure; the profiles of the density features can indeed by easily modelled into a flat plateau to cope with atomic density/composition and to rule out unphysical pair distances in the 2-dim quasicrystal.

Unfortunately, the raw product of a diffraction pattern being intensity of the Bragg peak, modelling has very often to be contented with PF instead of $\rho(\boldsymbol{r})$ map (Fig. 3.7). The PF map being a self-convolution of the $\rho(\boldsymbol{r})$ map makes life a little more difficult. Features are still clearly visible at lattice and mid-edge sites; a simple drawing may convince the reader that features would not be obtained without atomic objects A_{perp} at lattice and mid-edge sites. Now, a body centre feature shows up in the PF map of Fig. 3.7. A very careful examination of the relative integrated intensity of this feature is required before deciding that it comes from an overlapping of mid-edge atomic objects and not from the presence of a true body centre decoration. Obviously enough, further extraction of sizes and shapes of both lattice sites and mid-edge A_{perp} are even more risky if only *total* PF maps are available (see Problem 3.17). This is where contrast variations (with neutron diffraction) enter usefully since, basically, the various measured diffraction patterns allow us to get the scattered signal from each atom family independently, and formally drive back the problem to the simple case of monatomic systems. Simulating the result of such contrast variations experiment, Figs. 3.8 and 3.9 show the *partial* $\rho(\boldsymbol{r})$ and PF maps for the mid-edge objects

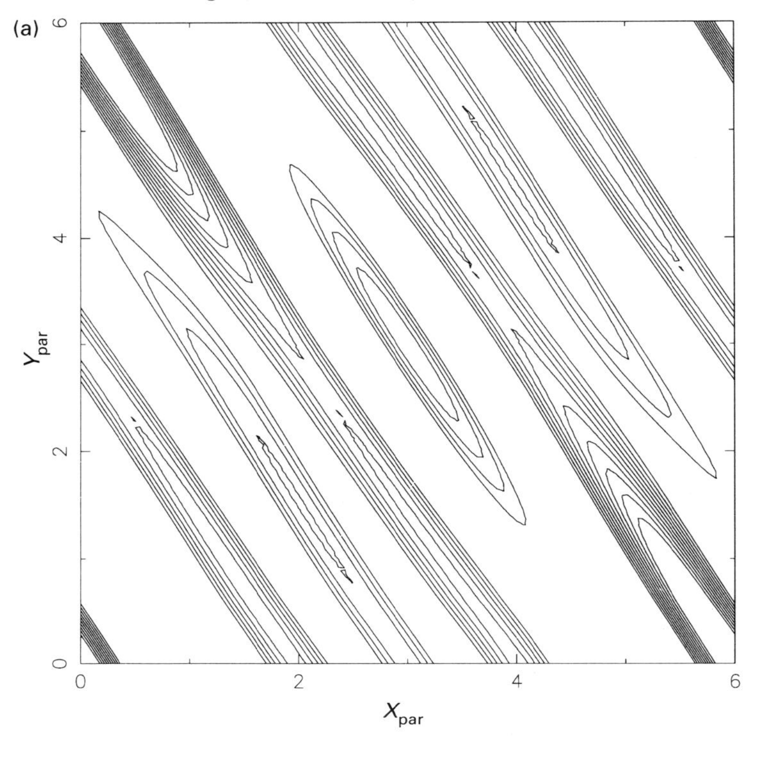

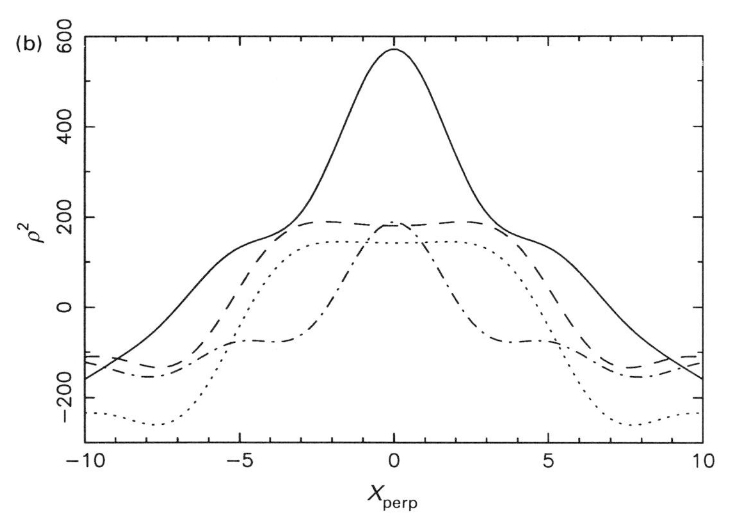

FIG. 3.7 Same as in Fig. 3.6 but for Patterson functions.

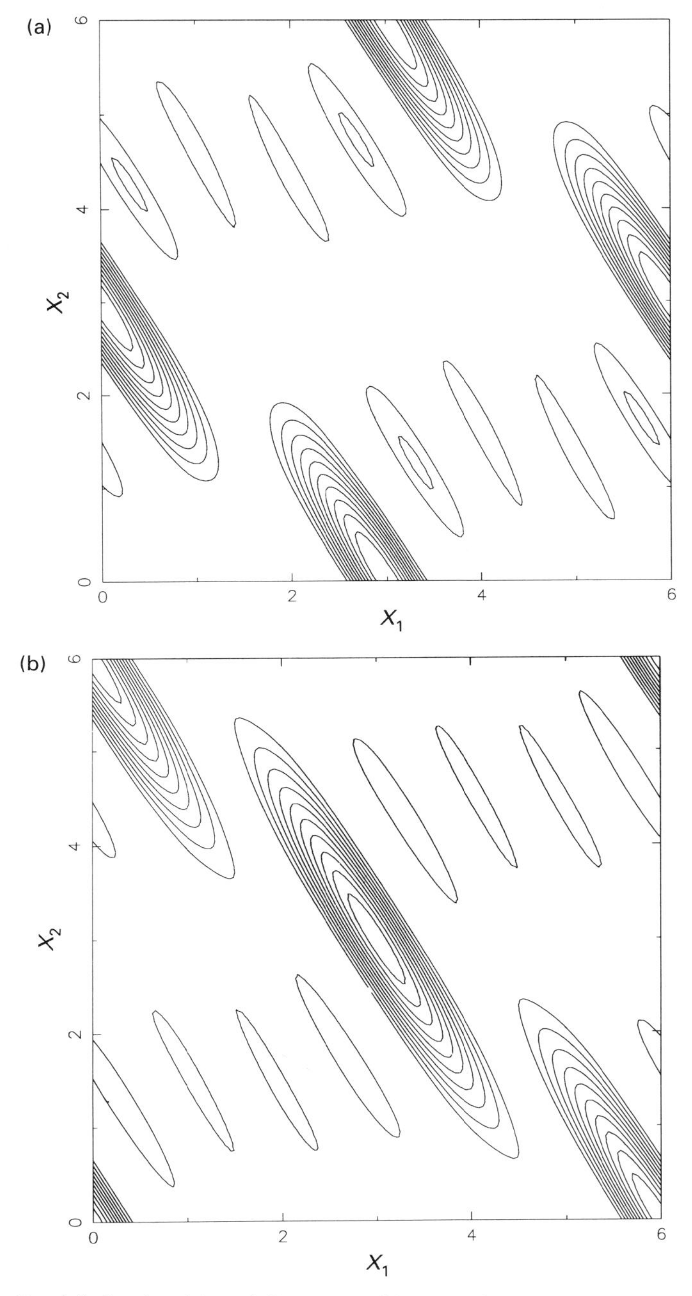

FIG. 3.8 Partial density (a) and Patterson (b) maps for the mid-edge site of the structure shown in Fig. 3.5 as simulated within the conditions of Figs. 3.6 and 3.7.

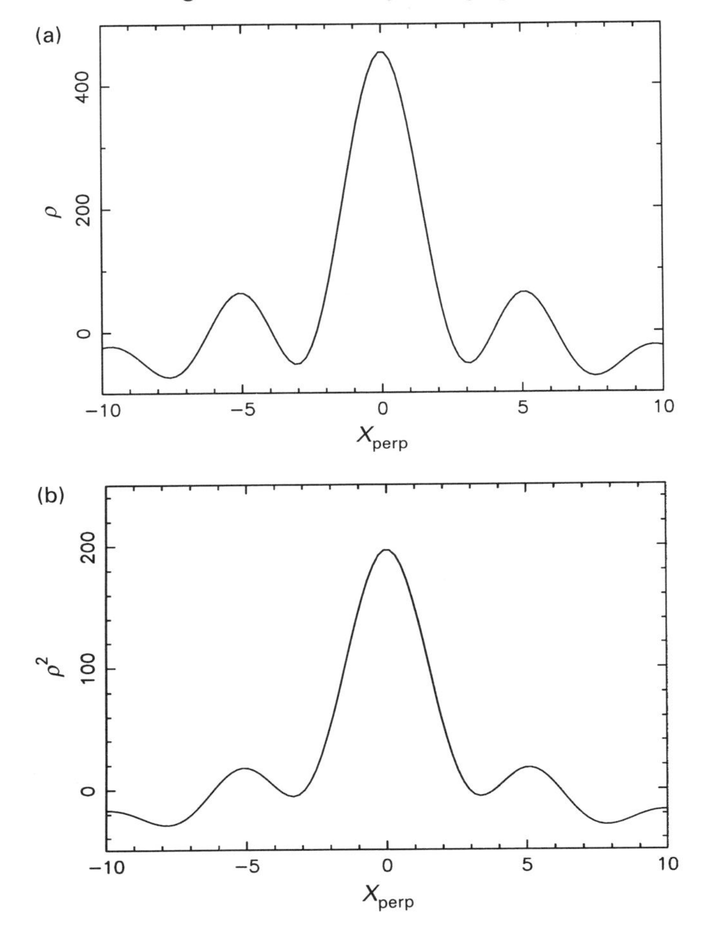

FIG. 3.9 Same as for Fig. 3.8, but for corresponding profiles.

alone. The partial maps for the lattice sites are identical to those shown in Figs. 3.3 and 3.4. The structure problem is now reasonably tractable.

Anyhow, it may certainly be stated that positions and sizes of the atomic surfaces in the n-dim atomic structure related to a QC are safely derived from diffraction data, especially when single crystal data and contrast variation measurements are available. Good insight into the shape of these atomic surfaces may also be expected, but the details escape the crystallographic investigation. As long as these details are related to short-range order features, the crystallography approach may be usefully complemented by the measurement of the pair distribution functions (PDF) or, better, partial pair distribution functions (PPDF). As explained in Chapter 1, the PDF and PPDF can be obtained using methods currently employed for short-range order structural information on non-crystalline materials (liquids, glasses, amorphous alloys, etc). Diffraction patterns are measured over a large Q range and a continuous Fourier transform (FT) of the whole (corrected) scattering signal gives the averaged isotropically regrouped probability of atomic

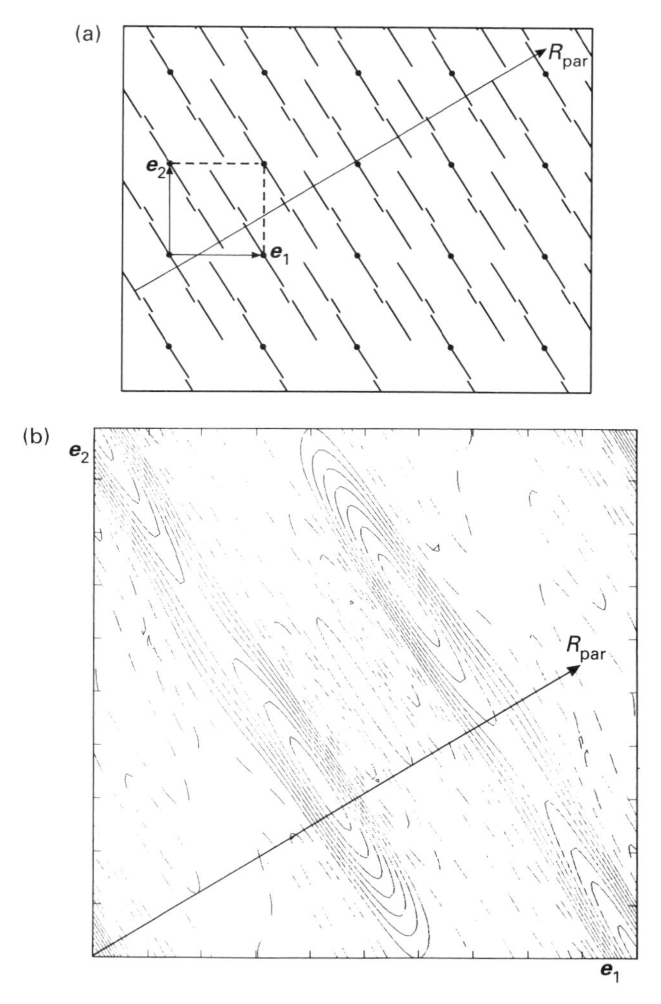

FIG. 3.10 (a) Example of a simple parallel component which maintains the same long-range order but modifies the length of some of the tiles. In (b) is shown the density map calculated with Fourier component that would be available from a diffraction experiment.

pairs as a function of pair distances. Contrast variation with neutron diffraction allows the PPDF to be determined from convenient data sets. The weak point of such a procedure is that angular information is obviously lost. But the whole (diffuse) signal, which is scattered out of the strong peaks, is reintroduced into the FT and thus contributes to the PDF and/or PPDF. One circumstance where these PDFs or PPDFs may be useful is when the atomic surfaces are no longer flat in R_{perp} but have 'parallel components'. One simple example of such a parallel component is shown in Fig. 3.10, along with the corresponding density map that would be derived from a diffraction experiment spanning the same restricted set of (Q_{par}, Q_{perp}) values as defined above in this section. Obviously, it would not be easy to infer the original parallel component from such a density map. In contrast, the curves presented in Fig. 3.11 demonstrate the usefulness of measuring the PDF (or PPDF) in the derivation of possible parallel components. The figure compares a simulated PDF, with artificial Gaussian broadening of the density peaks, for the 1-dim quasicrystal generated in Fig. 3.10 (with a parallel component) and the same structure but without the parallel component. The differences are mainly at the level of splitting or not splitting of the nearest-neighbour distances, as presented in Table 3.1. This point will be demonstrated with studies of real quasicrystals later on in the book (Chapter 4).

The simple procedure of working out n-dim inverse Fourier transforms of the integrated intensities of the measured Bragg peaks, or of the deduced structure factors if phase reconstruction is possible, obviously suffers from at least two major drawbacks: (i) the experimental errors are not accounted for and (ii) all the non-measured structure factors are set to zero, producing dramatic truncation effects.

This is where the maximum entropy method (MEM) enters usefully,[10–12] as the now accepted way to recover a positive distribution or image (here

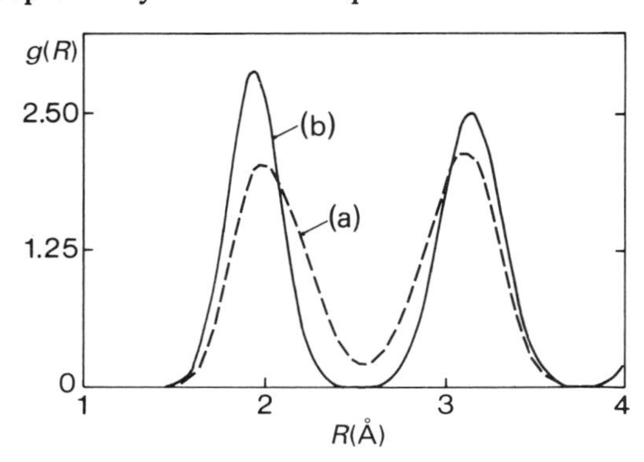

FIG. 3.11 Simulated pair distribution functions for the 1-dim quasicrystal generated in Fig. 3.10(a) with (*a*) and without (*b*) the parallel component in the A_{perp} object (kink).

Table 3.1 Pair distances d and corresponding pair number $n(d)$ calculated for the 1-dim quasicrystal of Fig. 3.10, with and without parallel components

With parallel component		Without parallel component	
d(Å)	$n(d)$	d(Å)	$n(d)$
1.95	0.424	1.95	0.730
2.20	0.308	—	—
2.65	0.030	—	—
2.90	0.262	—	—
3.15	0.930	3.15	1.224
5.10	0.046	5.10	0.046

Patterson or density map) from a given set of diffraction data. Spectacular achievements have been reported, for instance in NMR and astrophysics investigations.[13]

The density $\rho(\boldsymbol{r})$ and the structure $F(\boldsymbol{Q})$ are related by the standard equations

$$F(\boldsymbol{Q}) = \frac{1}{V}\int_r \rho(\boldsymbol{r}) \exp(\mathrm{i}\boldsymbol{Q}\cdot\boldsymbol{r})\mathrm{d}\boldsymbol{r}, \tag{3.5}$$

$$\rho(\boldsymbol{r}) = \sum_{Q} F(\boldsymbol{Q}) \exp(-i\boldsymbol{Q}\cdot\boldsymbol{r}), \tag{3.6}$$

where V is the volume of the n-dim unit cell. For an n-dim structure whose density features are flat atomic objects in R_{perp}, eqn (3.5) becomes

$$F(\boldsymbol{Q}) = \frac{1}{V}\sum_{x_n^i} G^i(Q_{\text{perp}}) \exp(\mathrm{i}\boldsymbol{Q}\cdot\boldsymbol{x}_n^i) \; . \tag{3.7}$$

The general problem to be solved here is to retrieve the 'best' density $\rho(\boldsymbol{r})$, which is called the image, so that (i) the measured Fourier components agree with the computed ones within error bars, (ii) no assumption is made as regards what is not measured, and (iii) the least informative image is chosen. It follows from the latter point that any significant feature in the reconstructed image must be truly present. The solution to the problem has long been known to be the image with maximum entropy,[10–12] subject to the experimental constraints and the available prior knowledge.

The symmetry requirements of the quasicrystal structure must be first incorporated into eqns (3.5)–(3.7) by replacing the exponential term with its average, namely

$$< \exp(\mathrm{i}\boldsymbol{Q}\cdot\boldsymbol{r}) > = \frac{1}{g}\sum_{\hat{R}\in G} \exp(\mathrm{i}\boldsymbol{Q}\cdot\boldsymbol{r}),$$

as taken over all operations $\hat{R}$ of the factor space group G of order g. Then the integral in eqn (3.5) is discretized, yielding M pixels of equal size into the asymmetric unit ($M = M_x M_y$ in the 2-dim example of Fig. 3.10). Equation (3.5) now reads:

$$F(\boldsymbol{Q}) = \frac{1}{M}\sum_{j=1}^{M} \langle\cos \boldsymbol{Q}\cdot\boldsymbol{r}_j\rangle \rho(\boldsymbol{r}_j) \ . \tag{3.8}$$

The exponential is replaced by cos for the simple but usual case of centro-symmetry.

In MEM calculations, two master equations are used:

$$S\{\rho(\boldsymbol{r})\} = -\sum_{j=1}^{M} q_j \log (q_j/p_j), \tag{3.9a}$$

with

$$q_j = \rho(\boldsymbol{r}_j) \Big/ \sum_{j=1}^{M} \rho(\boldsymbol{r}_j)$$

and

$$p_j = \rho_0(\boldsymbol{r}_j) \Big/ \sum_{j=1}^{M} \rho_0(\boldsymbol{r}_j)$$

($\rho_0(\boldsymbol{r}_j)$ quantifies our prior knowledge of the density before the diffraction experiment is run. In particular, p_j must be set equal to zero for those pixels j which are expected to have no density at all, and

$$\chi^2\{\rho(\boldsymbol{r})\} = \sum_{k=1}^{M} \left(\frac{F_{\mathrm{mes}}(\boldsymbol{Q}_k) - F_{\mathrm{cal}}(\boldsymbol{Q}_k)}{\sigma(\boldsymbol{Q}_k)}\right)^2, \tag{3.9b}$$

with $\sigma(\boldsymbol{Q}_k)$ the experimental standard deviation associated with the $F_{\mathrm{mes}}(\boldsymbol{Q}_k)$ structure factor; F_{cal} is calculated using eqn (3.6). The best image $\rho(\boldsymbol{Q})$ must maximize the entropy $S\{\rho(\boldsymbol{r})\}$ subject to $\chi^2 = N$, the number of symmetrically independent data points. The algorithm used,[14-17] developed at Cambridge (UK), is iterative and converges when ∇S is parallel to $\nabla\chi^2$. The solution is unique.

The same simulated data set as the one yielding the Fourier density map presented in Fig. 3.10 has been treated by MEM. No prior knowledge of the density was assumed beside centrosymmetry, and consequently p_j, was set equal to a constant ($= 1/M$). The spectacular result is shown in Fig. 3.12. Spurious noise and truncation waves no longer corrupt the density map.

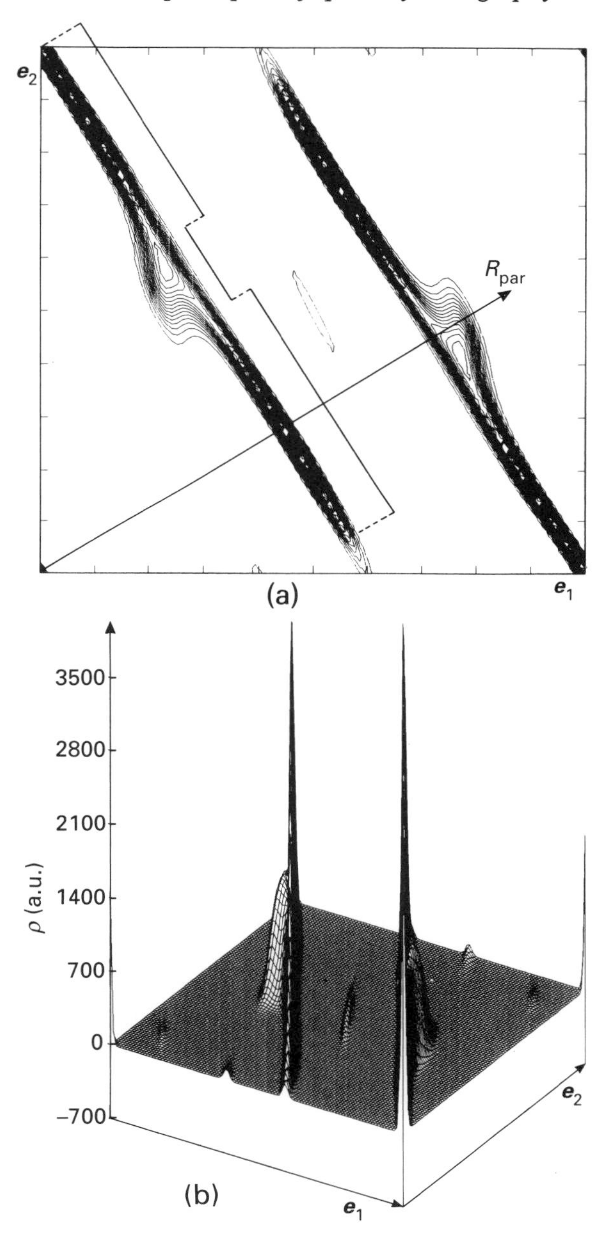

FIG. 3.12 (a) Computer simulation similar to that shown in Fig. 3.10 but resulting from a MEM procedure. The kink related to the parallel component is now clearly visible and reasonably reminiscent of the actual atomic surface (superimposed on the map for the sake of comparison). A pseudo 3-dim representation is also shown (b).

Moreover (and this is of the greatest importance for quasicrystals), details such as the small feature with a parallel component can now be inferred safely from the experimental map. The average density, which in principle is equal to $F(0)$ is also well reproduced by MEM (84.3 to be compared to 82.4, the original true average density). This is another very positive point which may be used to place the measured structure factors on an absolute scale, since the atomic density is a measurable quantity. The two major consequences are: (i) the scale factor currently needed in modelling is now constrained, and (ii) data sets collected independently (i.e. contrast variations data sets, or powder vs. single crystal diffraction patterns) can easily be renormalized to be used altogether (e.g. for calculation of partial structure factors).

Using MEM certainly would not justify a final modelling to be discarded and detailed tailoring of the n-dim atomic surfaces will in any case be the ultimate stage of a quasicrystal structure derivation.[18] However, the MEM procedure obviously makes life easier by providing a lot of details and accurate features beforehand.

3.2.4 'Tailoring' the n-dim atomic objects: final modelling of quasicrystal structure

When modelling and refining the structure of a quasicrystal, the experimentally deduced positions, sizes, and shapes of the atomic object in the related periodic n-dim structure are used as a first approximation. Re-tailoring these atomic objects may be required to eliminate unphysically short atomic distances, and chemically unacceptable bondings, or conversely, to generate desired types of atomic clusters.

Looking at any of the 1-dim examples of quasicrystals presented in this chapter, it can easily be understood that increasing the size of the n-dim atomic objects and/or introducing supplementary types of atomic objects modifies the structure of the quasicrystal. New nearest-neighbour atom distances are generated and, in 2-dim or 3-dim quasicrystals, new types of atomic clusters may be created. Digging holes or adding pieces here and there in the n-dim atomic objects is certainly the way to cope with physical and/or chemical requisites.

The statement is illustrated in Fig. 3.13, again with the example of a 1-dim quasicrystal. The associated 2-dim square lattice is now decorated with line segments at lattice sites and body centre positions. The physical cut generates in R_{par} a 'distance' d in the 1-dim quasiperiodic structure, due to the distance between lattice sites and body centre atomic objects. This distance does exist if the two corresponding A_{perp} objects have a common cross-section when projected on to the complementary space R_{perp}. Assuming that this distance is to be eliminated for physical reason imposes that the perpen-

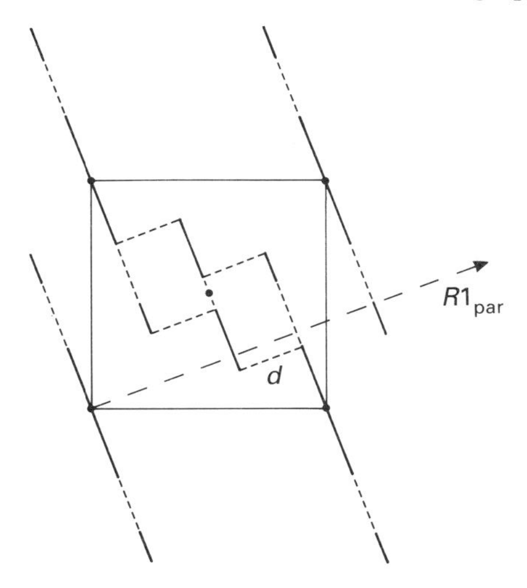

FIG. 3.13 Origin plus body-centre positions may generate unphysical distances 'd' in the 1-dim quasiperiodic structure. For this distance to be avoided, the projection of the body-centre A_{perp} on to the origin A_{perp} (in $R1_{perp}$) must be empty. The solution proposed in the figure is a central hole (dashed line) in the BC volume and an empty 'ring' in the origin volume.

dicular projections of the two A_{perp} concerned have an empty common cross-section. The solution proposed in Fig. 3.13 is to dig a central hole in the body centre object and an empty 'ring' in the lattice site object, with proper sizes. But, obviously, this solution is not unique.

Conversely, one can imagine that additional pair distances, or desired local symmetries for atomic clusters are generated if pieces are conveniently added to the existing A_{perp} object. This again will be exemplified in the next chapter, with the derivation of real quasicrystal structures.[18]

3.2.5 The high-dim representation of some imperfection: phason shift and strain

The n-dim period description of a physical quasiperiodic structure is also very useful when other problems beyond structure are concerned. Hydrodynamics and defects are interesting examples. As stated many times in this book, structure factors related to the intensities of Bragg peaks in diffraction patterns are the Fourier components of the density distribution, i.e.

$$\rho(\boldsymbol{r}) = \sum_{\boldsymbol{Q}} F(\boldsymbol{Q}) \exp(\mathrm{i}\boldsymbol{Q}\cdot\boldsymbol{r}) \tag{3.6}$$

In any case, the structure factor $F(Q)$ is a complex number with amplitude and phase. To be concrete, consider again a linear periodic chain of balls with mass m_a connected by identical springs and constrained to have an average separation a. The equilibrium position of the nth ball is

$$x_{an} = na \ .$$

The mass density of this system is simply

$$\rho(x) = \sum_n m_a \delta(x - x_{an}),$$

and can be expressed by the discrete Fourier transform

$$\rho(x) = \sum_h \rho_h \exp(\mathrm{i}hQ_0 x),$$

where $Q_0 = 2\pi/a$ and h is an integer (Miller index).

Suppose now that the balls are not at their equilibrium positions, but displaced by quantities $u(x)$ depending on their location in the chain. This produces additional phases in the density expansion, which is now written

$$\rho(x) = \sum_h \rho_h \exp[\mathrm{i}\{\mathrm{h}Q_0 x + hQ_0 u(x)\}] \tag{3.10}$$

The $u(x)$ quantities can represent *phonons*, i.e. atomic displacements due to the dynamics (vibrations) of the system. They can also represent any distortion of the structure (defects, strain/stress effects, etc.). Consider now the system composed of two parallel chains, one with mass m_a and lattice parameter a, the other with mass m_b and lattice parameter b. The ratio a/b is constrained to be an irrational number. The balls interact within each chain and between chains and reach equilibrium positions x_{an}, x_{bm}. The total mass density is thus

$$\rho(x) = \sum_{n,m} m_a \delta(x - x_{an}) + m_b \delta(x - x_{bm}),$$

and can be expressed as a double discrete Fourier transform:

$$\rho(x) = \sum_{h,h'} \rho_{hh'} \exp\{\mathrm{i}Q_0(h + h'a/b)x\},$$

depending on two integers, which justifies a 2-dim representation for the density distribution.

Assuming again that the balls are displaced by $u(x)$ with respect to their equilibrium positions, produces additional phases, such as

$$\rho(x) = \sum_{h,h'} \rho_{hh'} \exp\{i(h + h'a/b)Q_0 x + (h + h'a/b)\, Q_0 u(x)\}. \tag{3.11}$$

These additional phases in the density expansion are also expressed in terms of the two integers h,h' and, consequently, can be represented in the n-dim description of the quasicrystal.

A general form for the quasiperiodic density expansion is then

$$\rho(x) = \sum_{h,h'} \rho_{hh'} \exp[i\{\boldsymbol{Q}_{hh'}\boldsymbol{x} + \phi_{hh'}(x)\}], \tag{3.12}$$

where $\boldsymbol{Q}_{hh'}$ are vectors of the n-dim reciprocal space. The phase $\phi_{hh'}(x)$ can be decomposed into two components, either on the basis of the n-dim reciprocal space (as it is in eqn (3.11)) or on the physical (parallel) and complementary (perpendicular) subspaces, i.e.

$$\phi_{hh'}(x) = Q_{\text{par}}\, U(x) + Q_{\text{perp}}\, W(x). \tag{3.13}$$

This is simply the expression of the fact that displacements, such as positions of the atoms in, say, a Fibonacci chain, are related to both parallel and perpendicular components of the displacement/positions of the atomic objects in the associated 2-dim periodic structure. This is schematically illustrated in Fig. 3.14 with an example in which perpendicular displacements of

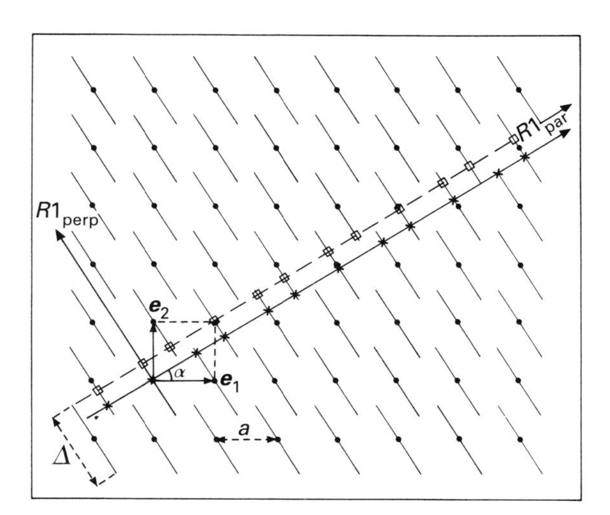

FIG. 3.14 Homogeneous phason shift produced by translation of the A_{perp} objects in the complementary space. For the sake of simplicity in the drawing, it is the physical (R_{perp}) space which has been shifted (dashed line). Atom positions are created and other ones annihilated by the phason shift.

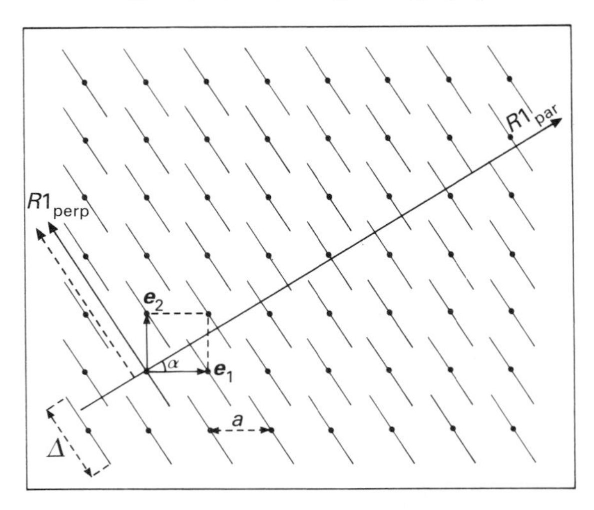

FIG. 3.15 Compared to Fig. 3.14, a homogeneous phonon-like shift produced by translation of the A_{perp} objects in the physical space does not generate any rearrangements of the atoms.

the A_{perp} object make atom positions disappear, to be replaced by other ones.

The vector field $U(x)$ operates in the physical space in a similar way to *phonons* in regular periodic crystals. The vector field $W(x)$ operates in the complementary space and is called a *phason*, by analogy with modulated structures (see Chapter 1).

Particular cases of displacement of the A_{perp} objects are homogeneous translations either in R_{par} or in R_{perp}. These two circumstances are pictured separately in Figs. 3.14 and 3.15. A shift along R_{par} simply represents a uniform translation of the entire system. In contrast, a phase shift along R_{perp} represents some rearrangements of the atoms. However, the total energy of the system remains invariant under this phase shift.

Thus, phonon translations are simply collective continuous motions of the individual atoms with possible distortion of the atomic cells. Conversely, phason translations involve a discontinuous shift of a fraction of the atoms, while the remaining atoms and atomic cells are unaffected. In 3-dim quasicrystals this will correspond to a rearrangement of the tiles. Since this involves significant discontinuous motion, or atomic diffusion, these phason nodes may be pinned, resulting in frozen-in fluctuations or phason disorder. In principle, both phonon-like and phason fluctuations contribute to the position and width of the peaks in the diffraction pattern. From eqns (3.6), (3.12), and (3.13) the structure factor for a given $Q_{hh'}$ is written

$$F_{hh'} = \rho_{hh'} \exp(i\phi_{hh'})$$

or

$$F_{hh'} \simeq \rho_{hh'}(1 + i\phi_{hh'} \ldots)$$

if the vector fields $U(x)$ and $W(x)$ involve only small fluctuations. Intensities of the diffraction peaks are thus Q_{perp} and Q_{par} dependent:

$$\begin{aligned} I(Q_{hh'}) &= F_{hh'}F^*_{hh'} = \rho^*_{hh'}(1 - \phi^2_{hh'} + \ldots) \\ &\simeq \rho^2_{hh'}\exp(-U(x)^2 Q^2_{par} + W(x)^2 Q^2_{perp}) \end{aligned} \tag{3.14}$$

The Debye–Waller factor then has contributions from both phonon and phason fluctuations. While the phonon fluctuations will strongly affect those peaks with a large Q_{par} (physical scattering vector), phason fluctuations most strongly affect those peaks with a sizeable Q_{perp} component. The intensity loss is in principle recovered as diffuse scattering, even at very low temperature.

Broadly speaking, both kinds of strain fields disrupt the positional ordering and lead to changes in the observed diffraction peaks. Depending on the nature of the strain distribution, peaks may be reduced in intensity by a Debye–Waller factor, shifted in position, or broadened. The nature of these diffraction effects will be identical for equivalent phonon and phason fields. However, because the fields each couple to their own conjugate vectors, the diffraction effects will scale with Q_{par} (the physical scattering vector at a Bragg reflection) in the presence of phonon strains, and with Q_{perp} for phason strains. Thus conventional lattice strains produce a peak broadening which increases monotonically with Q_{par}. Analogous phason strains will produce a peak broadening which is monotonic in Q_{perp}. The phonon and phason strain diffraction effects are easy to distinguish because the perpendicular wavevector depends in a complicated way on the reciprocal lattice vector. Phason strain diffraction effects have a puzzling or seemingly random behaviour if plotted against the scattering vector, but show smooth monotonic behaviour versus Q_{perp}. In icosahedral quasicrystals a simple case occurs when one measures a series of peaks of equivalent symmetry with Q_{par} vectors which scale by the golden mean. In a τ-series, Q_{perp} scales inversely with Q_{par} and the phason strain peak broadening *decreases* monotonically with Q_{par}.

Unlike the phonon-like fluctuations, the phason displacements are not restricted to small deviations from some average position. Hence they may give rise not only to a reduction in the intensity of the diffraction peaks (Debye–Waller effect), but also to a broadening of the peaks (breakdown of long-range order), or even a shift in the actual position of the peak if the phason fluctuation has a non-zero average gradient along some direction. Phason fluctuations may, along this line, generate a periodic order in an initial quasiperiodic structure. The resulting crystal is always an approxi-

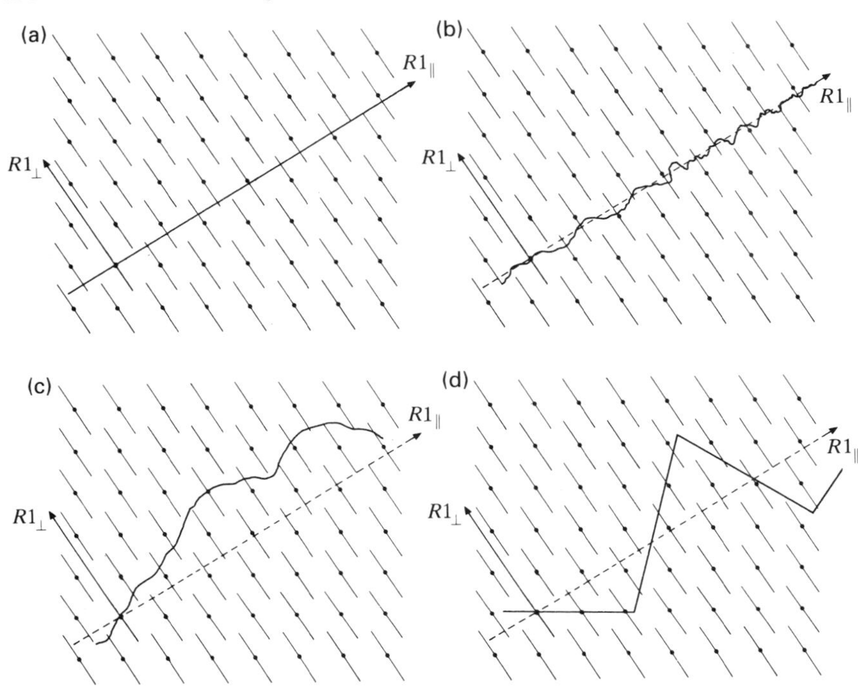

FIG. 3.16 Examples of phason disorder in the 2-dim to 1-dim slicing scheme: (a) perfect slicing (i phase); (b) bounded fluctuations (attenuated Bragg peaks and diffuse scattering); (c) unbounded fluctuations (broad peaks with various lineshapes); (d) 'faceting' of the slice along commensurate directions (microcrystals).

mant for the quasicrystal. An example, with the Fibonacci chain again, corresponds to phason fluctuations that exchange the order of long–short tiles in the sequence. Consider for instance a phason shift with inversion of *LS* pairs into *SL* except when *L* is found between two *S*. This mechanism transforms the Fibonacci sequence: *LSLLSLSLLSLLSLSLLSLSL* . . . into *SLLSLLSL SLLSLLSL SLLSLLSL* . . . which obviously has a period containing the group of tiles *SLLSLLSL*.

Schematic representations of phason fluctuations in the 2-dim to 1-dim slicing are shown in Fig. 3.16. If the slice is randomly undulating (isotropic random fluctuations), its ensemble average will maintain an average orientation, or slope, equivalent to the perfect slice and, on average, the quasicrystal symmetry (say icosahedral) will be preserved. Randomness will be introduced into the cutting of atomic surfaces, inducing position disorder in physical space. This particular disorder is termed **random phason strain**. If the fluctuating slice remains within a bounded distance from the original flat

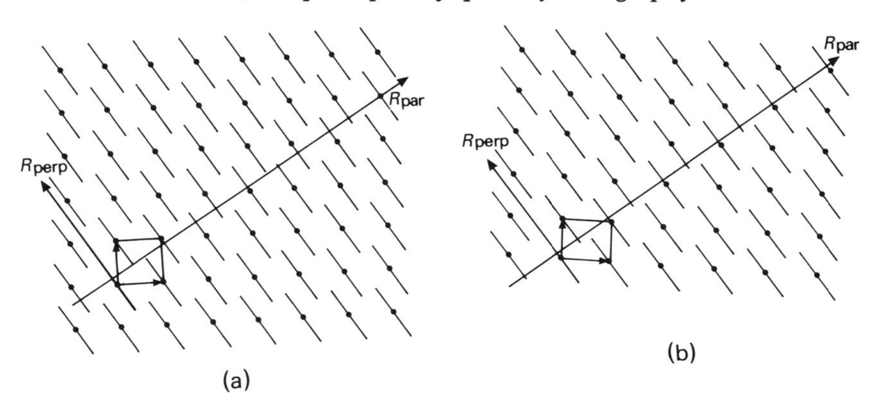

FIG. 3.17 Illustration of a phason strain transformation of the square lattice. The cut structure is (a) quasiperiodic and (b) periodic without and with a uniform phason strain respectively.

cut, then the structure still exhibits long-range translational order but with local violations of matching rules. In this case, diffraction maxima will still occur at icosahedral positions but with intensities decreased by an effective phason Debye–Waller factor. Additional diffuse scattering will appear around the diffraction peaks. Unbounded fluctuations will convert the pattern to broadened peaks with widths that scale monotonically with Q_{perp} and the translational order will be lost. An isotropic icosahedral glass model in which the slice can tear as well as fluctuate also produces peak brodening monotonic with Q_{perp} but with a different power law. Anisotropic fluctuations alter the average orientation of the slice and globally break the icosahedral symmetry. In this case, diffraction maxima will be shifted from their icosahedral positions. An important case is that of a purely linear phason strain which is achieved by simply tilting the slice. Small tilts into rational orientations result in projected structures which are crystalline approximants to the quasicrystal. The shifting of approximant phase Bragg peaks from icosahedral positions is linear in Q_{perp}. This is illustrated in Fig. 3.17, again in the 2-dim to 1-dim slicing scheme. The periodic approximant structure is locally very similar to that of the quasicrystal. One can easily imagine that the quasicrystal structure might be derived from that of the well-known periodic approximant submitted to a suitable 'inverse' uniform phason strain, after lifting its structure in a high-dim space. Phase transitions between quasicrystals and crystals may also occur via phason mechanisms. This will be analysed later in the book and examples have been mentioned in Chapter 2.

3.3 Six-dimensional crystallography for 3-dim icosahedral quasicrystals

3.3.1 Why six dimensions?

As abundantly illustrated in previous sections, non-periodic long-range order can be generated in 3-dim by appropriate cuts of structures which are periodic in n-dim spaces ($n > 3$). If n remains finite, the 3-dim structure is said to be quasiperiodic.

A reciprocal space containing the Fourier patterns of the structures is associated with each direct space. The n-dim representation of a 3-dim quasiperiodic structure leads to the description of its Fourier pattern as the projection of the n-dim periodic Fourier pattern. The quasiperiodic Fourier pattern is dense, defined by vectors whose magnitude ratio is irrational and cannot have a shortest length vector, since if it did, it would have a crystallography symmetry. These statements are perfectly clear from the example of the Fibonacci quasicrystal, as shown in Figs. 3.1 and 3.2: a large number of peaks from the 2-dim periodic Fourier pattern project on to any small segment of the 1-dim reciprocal R^*_{par} space; any $Q^{hh'}_{\text{par}}$ is related to h and $h'\tau$; and smaller and smaller $Q^{hh'}_{\text{par}}$ can be generated by conveniently choosing the 2-dim Fourier component that is projected. Quantitatively, these conclusions are contained in the expression for the possible $Q^{hh'}_{\text{par}}$ values, as given by eqn (3.3):

$$Q^{hh'}_{\text{par}} = (2\pi/a(2+\tau)^{\frac{1}{2}})\,(h + h'\tau)\ .$$

1. The set of values is dense because τ is irrational and h,h' are any integers: a $Q^{h_1h_1'}_{\text{par}}$ value can be as close as wanted to a given $Q^{h_2h_2'}_{\text{par}}$ if values equal to a high-order rational approximant of τ are given to the ratio $\dfrac{(h_1 - h_2)}{(h_1' - h_2')}$.
2. Any $Q^{h_1h_1'}_{\text{par}}$ is a linear combination of the parallel vectors of moduli 1 and τ.
3. $Q^{hh'}_{\text{par}}$ is very small if h/h' is negative and equal to a high-order rational approximant of τ. The consequence here is a scaling ambiguity in the choice of the magnitude of a 'fundamental' wavevector. For a periodic chain, the fundamental wavevector is the shortest-length vector $2\pi/a$ (eqn 1.23); for the Fibonacci chain, eqn (3.3) suggests that a fundamental wavevector could be $2\pi/a\ (2+\tau)^{\frac{1}{2}}$, but it does not correspond to a shortest-length vector. Actually scaling this quantity by τ, or any power of τ, still allows us to describe the full set of $Q^{hh'}_{\text{par}}$ values, since

$$\begin{aligned}\tau(h + h'\tau) &= \tau h + \tau^2 h' \\ &= h' + \tau(h + h') \\ &= H + \tau H'.\end{aligned}$$

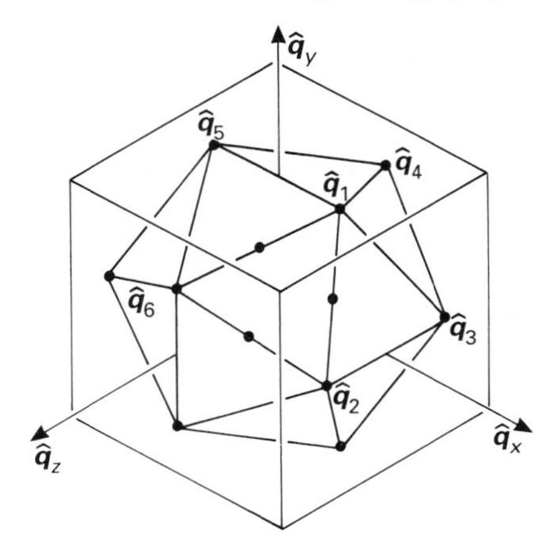

FIG. 3.18 Fundamental vertex reciprocal lattice vectors for a regular icosahedron, as expressed in eqn (3.15). Cubic axes are also shown on which the q_i vectors have the components given in the text.

This is due to the self-similarity of the Fibonacci structure (changing S by $\tau S = L$ and L by $\tau L = L + S$ does not modify the sequence).

In any case, the n-dimensionality of a quasicrystal is clearly defined by the number of independent vectors that must be used to describe its full Fourier pattern. This is just the extension of what is trivial for ordinary 3-dim periodic crystals.

Indeed, the diffraction patterns from crystal structures based on any of the 3-dim space groups may be indexed into a reciprocal lattice with three basis vectors, q_i, which are either orthogonal, or are a rational combination of three orthogonal vectors. Diffraction peaks may be labelled by an integer combination of these reciprocal lattice basis vectors, i.e.

$$Q_{\text{exp}} = \sum_{i=}^{3} n_i q_i .$$

In the course of analyzing diffraction data from quasicrystals of the systems AlMn[2,3,4,19], AlCuFe[20], or AlLiCu,[21] it was found that the diffraction peaks could be successfully indexed with integer n_i by using a 'basis' of six independent vectors pointing to the vertices of an icosahedron (Fig. 3.18), i.e.

$$Q_{\text{exp}} = |Q_0| \sum_{i=1}^{6} n_i q_i \tag{3.15}$$

with $q_1 = (1,\tau,0)$, $q_2 = (\tau,0,1)$, $q_3 = (\tau,0,-1)$, $q_4 = (0,1,-\tau)$, $q_5 = (-1,\tau,0)$, and $q_6 = (0,1,\tau)$ (see Problems).

It is easy to see that the pattern is invariant under inflation or deflation by τ^3, which means that $\tau^3 q_i$ (or $\tau^{-3} q_i$) is a combination of the six q_i as defined in eqn (3.15). Consider for instance $\tau^3 q_1 = (\tau^3, \tau^4, 0)$. If

$$\tau^3 q_1 = \sum_{i=1}^{6} n_i q_i$$

then

$$\tau^3 = n_1 + n_2\tau + n_3\tau - n_5$$
$$\tau^4 = n_1\tau + n_4 + n_5\tau + n_6$$
$$0 = n_2 - n_3 - n_4\tau + n_6\tau$$

or

$$2\tau + 1 = (n_1 - n_5) + \tau(n_2 + n_3)$$
$$3\tau + 2 = (n_4 + n_6) + \tau(n_1 + n_5)$$
$$0 = (n_2 - n_3) + \tau(n_6 - n_4)$$

or

$$n_1 - n_5 = 1; \qquad n_2 + n_3 = 2$$
$$n_4 + n_6 = 2; \qquad n_1 + n_5 = 3$$
$$n_2 - n_3 = 0; \qquad n_6 - n_4 = 0$$

and finally

$$n_1 = 2; \qquad n_2 = n_3 = 1$$
$$n_5 = 1; \qquad n_4 = n_6 = 1$$

In the problem section, the reader is asked to show that τ and τ^2 are not invariance ratios.

Thus, eqn (3.15) describes a Fourier pattern which is dense, has a basis of six vectors, and contains vectors of infinitely small magnitude; the corresponding structure must be quasiperiodic and related to a periodic structure in a 6-dim space. These quasicrystals are called icosahedral quasicrystals and they form, so far, the most common family of quasicrystals.

3.3.2 Possible space group for icosahedral quasicrystals[22,25]

As for any regular periodic structure, the density distributions $\rho(r_6)$ of the 6-dim representation of an icosahedral quasicrystal can be expanded in terms

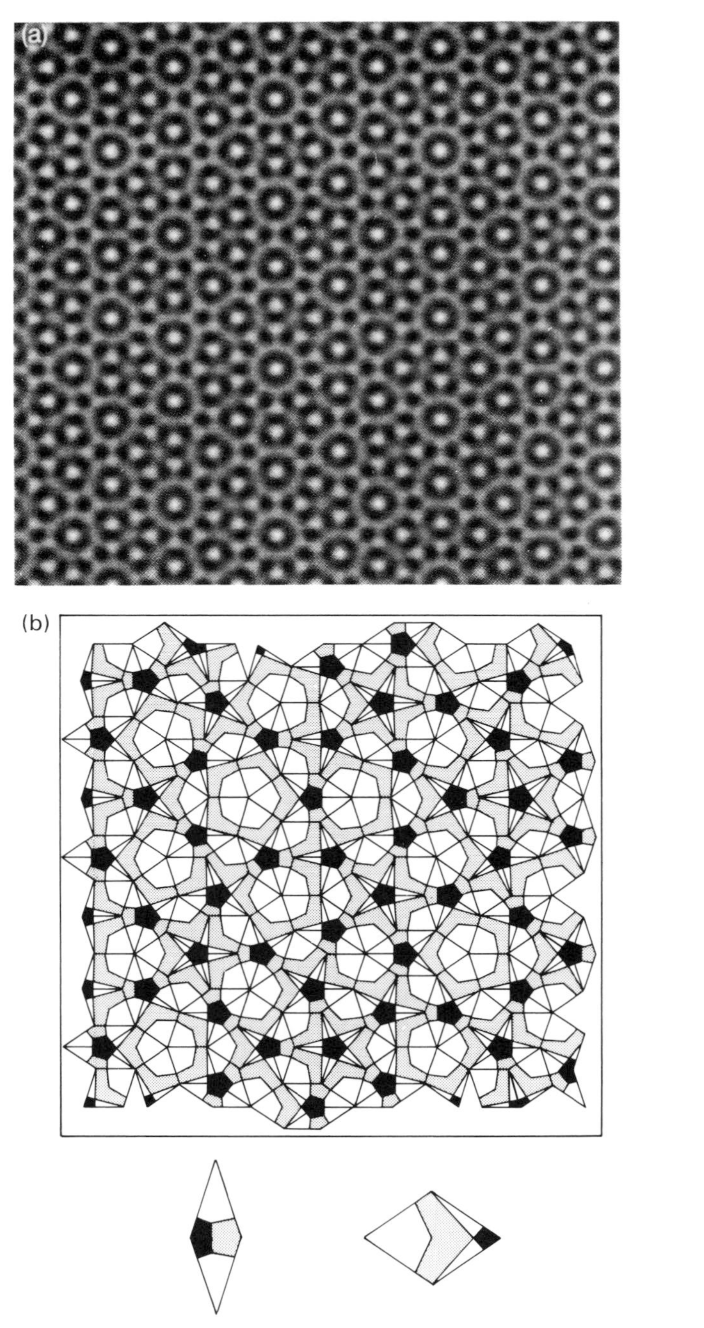

FIG. 3.19 Two-dim Penrose-like density distribution as generated by the superposition of five density waves directed to the vertices of a regular pentagon (a). Compared with a properly decorated true Penrose lattice (b), this illustrates that the symmetry is mainly defined by the lowest order term in the density wave expansion.

of fundamental reciprocal lattice vectors. Rewriting eqn (3.6) in the particular case of icosahedral quasicrystals gives

$$\rho(\boldsymbol{r}_6) = \sum_{\boldsymbol{Q}_6} F(\boldsymbol{Q}_6) \exp(-i\boldsymbol{Q}_6 \cdot \boldsymbol{r}_6). \tag{3.16}$$

The diffraction spots are at $\boldsymbol{Q}_6$ in R_6^* with intensities $|F(\boldsymbol{Q}_6)|^2$, where $F(\boldsymbol{Q}_6)$ is the Fourier transform of the contents of the unit cell and thus depends upon the atomic objects A_{perp} which are associated with the lattice points, i.e.

$$F(\boldsymbol{Q}_6) = \frac{1}{a^6} \sum_j G^j(\boldsymbol{Q}_{\text{perp}}) \exp(-i\boldsymbol{Q}_6 \cdot \boldsymbol{X}_6^j), \tag{3.17}$$

with the same notations as in previous sections. (G^j the Fourier transform of A^j_{perp} and X^j_6 positions A^j_{perp} in the 6-dim unit cell).

The $\boldsymbol{Q}_6$ vectors are linear combinations of the six basis vector of the 6-dim reciprocal space. An equivalent expression for $\rho(\boldsymbol{r}_6)$ is

$$\rho(\boldsymbol{r}_6) = \sum_{\boldsymbol{Q}_6} |F(\boldsymbol{Q}_6)| \cos\{\boldsymbol{Q}_6 \cdot \boldsymbol{r}_6 + \phi(\boldsymbol{Q}_6)\}, \tag{3.18}$$

where $\phi(\boldsymbol{Q}_6)$ describes the relative phases of the density waves. The above function is periodic in its argument with period 2π.

The physical mass-density distribution of the actual icosahedral 3-dim quasicrystal is obtained by the appropriate 3-dim cut of $\rho(\boldsymbol{r}_6)$, i.e. all complementary components of $\boldsymbol{r}_6$ are set to zero and we are left with the physical (parallel) components:

$$\rho(\boldsymbol{r}_{\text{par}}) = \sum_{\boldsymbol{Q}_{par}} |F(\boldsymbol{Q}_6)| \cos\{\boldsymbol{Q}_{\text{par}} \cdot \boldsymbol{r}_{\text{par}} + \phi(\boldsymbol{Q}_6)\}, \tag{3.19}$$

where the $\boldsymbol{Q}_{\text{par}}$ might, for example, correspond to linear combinations of reciprocal basis vectors $\boldsymbol{q}_i$ pointing to the vertices of an icosahedron (Fig. 3.18 and eqn (3.15)); there is a one-to-one correspondence between $\boldsymbol{Q}_6$ and $\boldsymbol{Q}_{\text{par}}$. In Landau's theory,[26–28] the symmetry of a crystal is defined by the lowest order term in the density wave expansion, since higher order terms will generally not lower the symmetry. This is illustrated in Fig. 3.19. We have also mentioned in the previous section of this chapter that the energy, and thus the symmetry, cannot depend upon the phase ϕ since a phase shift simply corresponds to a shift of the origin (Fig. 3.20). An apparent drawback for further discussion is the previously mentioned ambiguity in the choice of fundamental reciprocal lattice vectors for a Fourier pattern which is invariant under inflation or deflation by τ^3. The relation with the 6-dim periodic structure again relieves the difficulty, and the status of 'lowest order

FIG. 3.20 A phonon-like strain, here parallel to the horizontal axis of the figures, has some smearing effects but does not wash out the symmetries.

term' in eqn (3.19) is simply assigned to that in which $\boldsymbol{Q}_{\text{par}}$ corresponds to the fundamental reciprocal lattice vectors in the 6-dim reciprocal space. Still, there is some freedom in choosing the independent $\boldsymbol{q}_i$ in 3-dim since there are different ways to build an icosahedron with elementary vectors. Instead of the six vectors pointing to the vertices of an icosahedron, one could also choose the 10 vectors pointing towards the centre of the triangular faces or the 15 edge vectors. These possible choices result in different kinds of structure.

By choosing the set of basis vectors pointing to the vertices, the 'fundamental' term of the density expansion can be written

$$\rho_1(\boldsymbol{r}_{\text{par}}) = F_1 \cos\left(\sum_{i=1}^{6} \boldsymbol{q}_i \boldsymbol{r}_{\text{par}} + \phi_1\right).$$

In addition to the 6-dim translational symmetries, the density must have the rotational point group symmetries. For instance, the 5-fold axis along, say, $\boldsymbol{q}_1$, requires (see Fig. 3.18) that the density in 3-dim space is invariant under the transformation

$$\boldsymbol{q}_1 \to \boldsymbol{q}_1, \boldsymbol{q}_2 \to \boldsymbol{q}_3 \to \boldsymbol{q}_4 \to \boldsymbol{q}_5 \to \boldsymbol{q}_6 \to \boldsymbol{q}_2,$$

which in turn implies that the density $\rho(\boldsymbol{r}_6)$ in the 6-dim space is invariant under the transformation which leaves invariant the plane spanned by the

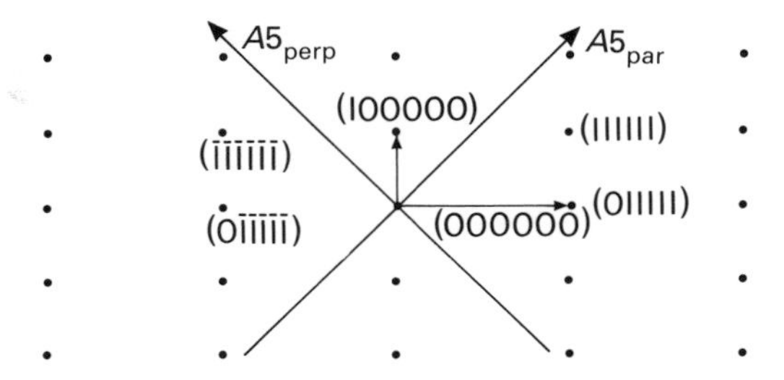

FIG. 3.21 Five-fold plane of a 6-dim crystal, containing one 5-fold axis in the physical space $A5_{\text{par}}$ and one 5-fold axis in the complementary space $A5_{\text{perp}}$.

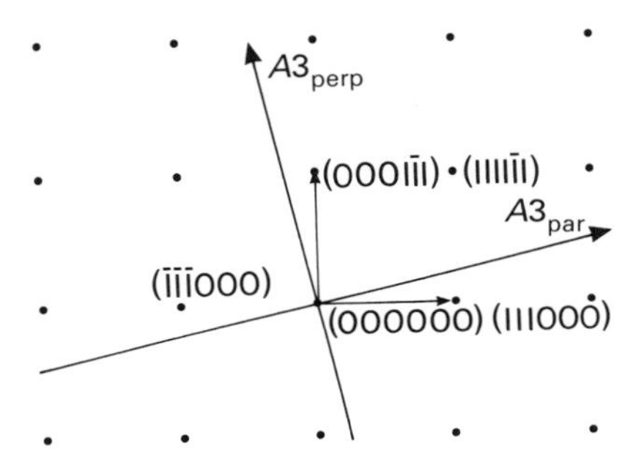

FIG. 3.22 Same as in Fig. 3.19 but for a 3-fold plane.

vectors (1, 0, 0, 0, 0, 0) and (0, 1, 1, 1, 1, 1). Thus, this plane, as shown in Fig. 3.21, is a 5-fold plane and is a plane of symmetry for the 6-dim structure. It contains 5-fold axes both parallel and perpendicular to the real world. Thus, the 6-dim structure has six 5-fold 'mirror' planes. It is easy to demonstrate that there are also ten 3-fold 'mirror' planes spanned by vectors such as (1, 1, 1, 0, 0, 0) and $(0, 0, 0, \bar{1}, 1, 1)$and fifteen 2-fold 'mirror' planes spanned by vectors such as (1, 1, 0, 0, 0, 0) and (0, 0, 1, 1, 1, 1) (Figs. 3.22 and 3.23). In summary, for each rotation which transforms the vectors $\boldsymbol{q}_i$ among themselves, there is a corresponding symmetry operation in 6-dim: the point group of the 6-dim crystal is said to be isomorphous with the icosahedral point group. Vectors along R_{par} and R_{perp}, respectively, transform among themselves under all symmetry operations: physical and complementary subspaces are invariant.

Thus, the 6-dim structure corresponding to the choice of $\boldsymbol{q}_i$ pointing to the vertices of an icosahedron is invariant under integer translations and 2-fold, 3-fold, and 5-fold mirror planes. This defines a 6-dim space group which, by analogy with the usual notations for 3-dim crystals, may be

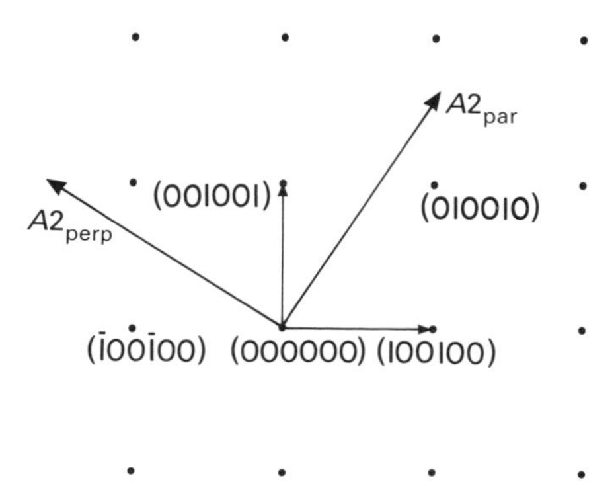

FIG. 3.23 Same as in Fig. 3.19 but for a 2-fold plane.

denoted $P352$, and a structure which may be termed simple or primitive icosahedral (SI or PI). Diffraction spots will be produced at all positions given by eqn (3.15), without systematic extinctions; missing reflections, if any, will be so-called accidental extinction due to the decoration of the 6-dim unit cell.

3.3.3 Body-centred and face-centred icosahedral quasicrystals

Now consider a 6-dim structure with fundamental reciprocal lattice vectors having physical (par) components formed by the 15 pairs of edge vectors $\boldsymbol{e}_{ij}$ of the regular icosahedron. The 'fundamental' term in the 3-dim density expansions is now

$$\rho_1(\boldsymbol{r}_{\text{par}}) = F_1 \cos\left(\sum_{ij=1}^{6} \cos \boldsymbol{e}_{ij}\,\boldsymbol{r}_{\text{par}} + \phi_1\right).$$

Edge vectors of a regular icosahedron are the sum or the difference of two vertex vectors (see Fig. 3.18 again), i.e.

$$\boldsymbol{e}_{ij} = \pm \boldsymbol{q}_i \pm \boldsymbol{q}_j \,.$$

As the $\boldsymbol{q}_i$ vectors are the physical components of 6-dim reciprocal lattice vectors directed towards six of the vertices of the 6-dim unit cell, $\boldsymbol{e}_{ij}$ must be the physical components of 6-dim reciprocal lattice basis vectors directed towards 30 of the face centres of the 6-dim unit cell. Thus, in this case, the reciprocal lattice is *face-centred icosahedral* (FCI). The faces are not equivalent since $\boldsymbol{q}_1 - \boldsymbol{q}_2$ and $\boldsymbol{q}_1 + \boldsymbol{q}_2$ are vectors of different lengths. In the FCI configurations, the acceptable $\boldsymbol{Q}_6$ vector must correspond to $\boldsymbol{Q}_{\text{par}} = \Sigma_i n_i \boldsymbol{q}_i$ restricted to n_i values, allowing the 15 pairwise associations of $\boldsymbol{q}_i$ that generate the $\boldsymbol{q}_{ij}$ edge vectors. The procedure is actually feasible if and only if the 'total number' of involved $\boldsymbol{e}_i$ vectors (i.e. $\Sigma_{i=1}^{6} n_i$) is even. Hence the FIC structure has systematic extinction rules which distinguish it from the PI lattice described in the previous section.

Similarly, if the edge vectors $\boldsymbol{e}_{ij}$ are reexpressed in terms of $\pm \boldsymbol{q}_i \pm \boldsymbol{q}_j$ in the density expansion, the corresponding 6-dim distribution is a periodic function only of the sums or differences of the 6-dim cube integer translations of the form $(\pm\frac{1}{2}, \pm\frac{1}{2}, \pm\frac{1}{2}, \pm\frac{1}{2}, \pm\frac{1}{2}, \pm\frac{1}{2})$ (half-diagonal of the 6-dim cube). So, while the 6-dim reciprocal lattice is FCI, the 6-dim direct space lattice is body-centred icosahedral (BCI). The corresponding space group is naturally denoted *C*235.

A similar analysis can be made to demonstrate that a 6-dim structure, with reciprocal lattice vectors having their Q_{par} components directed towards the triangular face of a regular icosahedron, is an FCI structure with a BCI reciprocal lattice. The corresponding space group is denoted *F*235 (see Problems).

Quasicrystals of the AlMn and AlLiCu families [29,30] have been found to have *P*352 structure while AlFeCu [31] or AlPdMn [32] have been assigned the *C*235 space group.

Based on the above formal analysis, the next section will now address more technical questions about the indexing procedure for the diffraction patterns of icosahedral quasicrystals.

3.3.4 The choice of a coordinate system in 3-dim for the PI space group [33,34]

Taking the basis vector along an important symmetry direction simplifies the crystallographic formulation. For any space group with a unique rotation axis, the z-axis is taken parallel to that axis. For the icosahedral space-group, high symmetry axes are 5-fold (6), 3-fold (10) and 2-fold axes (15). Taking one of the 5-fold axes as the z-axis leaves the other five in a ring 63.43° from this axis, with both acute and obtuse angles between them which are rather difficult to keep track of. The use of a 3-fold axis as z-axis would correspond to rhombohedral bases (two sets of three 5-fold axes at either acute or obtuse angles with each other) or a hexagonal coordinate system (the 3-fold z-axis plus the three 2-fold axes at right angles to it). The simplest alternative is a cubic coordinate system based on a set of three orthogonal 2-fold axes of the icosahedron (Fig. 3.24). (There are five such sets all equivalent to each other via the rotations of a 5-fold axis). The main advantages here are those of orthogonal axes; in particular, the indices of a plane and the direction perpendicular to it are the same (see Section 1.2).

In this cubic coordinate system, the 'Miller indices' of the six 5-fold planes are all of the form $(1\tau 0)$; the ten 3-fold planes (or axes) are (111) for four of them and $(\tau^2 1\ 0)$ for the other six; three 2-fold axes are along the

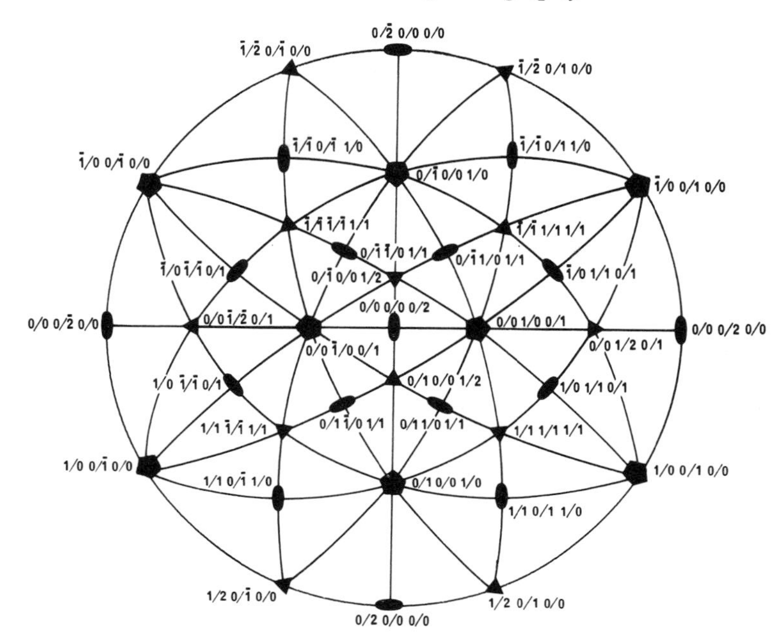

FIG. 3.24 Standard stereographic projection for the icosahedral point group as seen along a 2-fold axis and with the axes of an orthogonal coordinate framework aligned along one of the five sets of mutually perpendicular 2-fold axes. The number indicates the indexing system described in the text.[33]

coordinate axes (0 0 0) and the remaining twelve have the form ($\tau\,\tau - 1$). The general form of these indices is

$$(h + h'\tau \quad k + k'\tau \quad l + l'\tau), \tag{3.20}$$

or (h/h' k/k' l/l') for short in which h, h', k, k', l, l' are all integers, as expected for Miller indices of a 6-dim periodic structure.

Consider now the expression for the $\boldsymbol{Q}_{\text{par}}$ vectors in the physical 3-dim reciprocal space as expressed using the basis of the six 5-fold vectors (eqn 3.15), i.e.

$$\boldsymbol{Q}_{\text{par}} \underline{\propto} \sum_{i=1}^{6} n_i q_i, \tag{3.21}$$

where n_i are any integers; the q_i can now be indexed in the cubic coordinate system as:

$$q_1 = (1/0\,0/1\,0/0)$$
$$q_2 = (0/1\,0/0\,1/0)$$

$$
\begin{aligned}
q_3 &= (0/0\ 1/0\ 0/1) \\
q_4 &= (\bar{1}/0\ 0/1\ 0/0) \\
q_5 &= (0/1\ 0/0\ \bar{1}/0) \\
q_6 &= (0/0\ \bar{1}/0\ 0/1).
\end{aligned}
\tag{3.22}
$$

To express $\boldsymbol{Q}_{\text{par}}$ in terms of the 3-dim cubic coordinates, the q_i given by eqn (3.22) must be substituted in eqn (3.21). After proper regrouping one finds

$$\boldsymbol{Q}_{\text{par}} = ((n_1 - n_4)/(n_2 + n_5),\ (n_3 - n_6)/(n_1 + n_4),\ (n_2 - n_5)/(n_3 + n_6));$$

that is, as expected, an indexing of the form (3.20) with

$$
\begin{aligned}
h &= n_1 - n_4 \qquad & h' &= n_2 + n_5 \\
k &= n_3 - n_6 \qquad & k' &= n_1 + n_4 \\
l &= n_2 - n_5 \qquad & l' &= n_3 + n_6
\end{aligned}
\tag{3.23a}
$$

or vice versa:

$$
\begin{aligned}
2n_1 &= h + k' \qquad & 2n_4 &= -h + k' \\
2n_2 &= l + h' \qquad & 2n_5 &= l + h' \\
2n_3 &= k + l' \qquad & 2n_6 &= -k + l'\ .
\end{aligned}
\tag{3.23b}
$$

This gives expressions for parity rules that the six indices must obey and which correspond to extinction rules. It can be verified that the restriction (3.23*b*) generates four, and only four, kinds of index sets:

(1) h, h', k, k', l, l' are all even;

(2) four indices are even within the sequence (odd/even, even/odd, even/even);

(3) two indices are even (even/odd, odd/even, odd/odd);

(4) all six indices are odd and cyclic permutations.

As an example, let us consider $\boldsymbol{Q}_{\text{par}}$ vectors with indices 0 or 1. To satisfy the parity rules, only the following combinations are acceptable:

(1/0 0/1 0/0) (and cyclic permutations of the sort (0/0 1/0 0/1)
(0/1 1/0 1/1) and permutations
(1/1 1/1 1/1)

and those with changed signs.

3.3.5 *Some useful properties*

Some properties that are evidenced by the six integer indexing described above are listed below. The interested reader is invited to work out the problems of this chapter, where some guides are proposed for the demonstration of these properties.

3.3.5.1 Scaling properties Scaling $\boldsymbol{Q}_{\text{par}}$ by τ satisfies the parity rules only if two of the six Miller indices are even. Scaling by τ^3 preserves the parity rules for all $(h/h'\ k/k'\ l/l')$. This is because icosahedral quasiperiodic structures are self-similar with the inflation/deflation ratio equal to τ^3, as explained earlier.

3.3.5.2 Modulus properties Starting from $\boldsymbol{Q}_{\text{par}} = (h/h'\ k/k'\ l/l')$ one can calculate $|\boldsymbol{Q}_{\text{par}}|^2$ as the sum of the squared components $(h + h'\tau)^2 + (k + k'\tau)^2 + (l + l'\tau)^2$. Using then the special properties of τ^n, one finds

$$|\boldsymbol{Q}_{\text{par}}|^2 \propto N + M\tau,$$

with

$$\begin{aligned} N &= 2\sum_{i=1}^{6} n_i^2 = h^2 + k^2 + l^2 + h'^2 + k'^2 + l'^2 \\ M &= h'^2 + k'^2 + l'^2 + 2(hh' + kk' + ll') . \end{aligned} \tag{3.24}$$

These expressions are very useful, especially when indexing powder diffraction patterns where only $|\boldsymbol{Q}_{\text{par}}|$ are obtained from the experiment. Some sort of parity rules also naturally apply to (N, M) 'indexing'. For instance *N is an even integer*, since it appears in eqn (3.24) as twice the square of the length of the 6-dim vector with components n_i.

If the six indices are all even, eqns (3.24) tell us that N is a multiple of four and so is M; N and M are also both multiples of four if two indices are even. If four indices are even or if all the six indices are odd, N is a multiple of four plus two and $M = 4m + 1$ (with m an integer).

3.3.5.3 The projection matrix that relates $\boldsymbol{Q}_6$ and $\boldsymbol{Q}_{par}$ In 3-dim, the six-vector $\boldsymbol{q}_i$ along the 5-fold axes are not orthogonal, but they are the parallel (physical) component of six vectors which are along the edges of the 6-dim cubic unit cell. The set of the six numbers n_i then represents a position vector $\boldsymbol{Q}_6$ in the 6-dim cubic reciprocal lattice. This $\boldsymbol{Q}_6$ vector projects on to $\boldsymbol{Q}_{\text{par}}$ in the physical reciprocal space with components

$$\begin{aligned} h + \tau h' &= (n_1 - n_4) + \tau(n_2 + n_5) \\ k + \tau k' &= (n_3 - n_6) + \tau(n_1 + n_4) \\ l + \tau l' &= (n_2 - n_5) + \tau(n_3 + n_6). \end{aligned} \tag{3.25}$$

This is only another way to express the parity rules of eqn (3.23*a*), which themselves result from identification of two expressions of Q_{par} in two coordinate frameworks. This is also a way to obtain the (3×6) projection matrix which transforms the 6-dim vector $(n_1, n_2, n_3, n_4, n_5, n_6)$ into the physical 3-dim vector $(h + \tau h', k + \tau k', l + \tau l')$, i.e.

$$Q_{par} = \begin{pmatrix} h + \tau h' \\ k + \tau k' \\ l + \tau l' \end{pmatrix} = \begin{pmatrix} 1 & \tau & 0 & -1 & \tau & 0 \\ \tau & 0 & 1 & \tau & 0 & -1 \\ 0 & 1 & \tau & 0 & -1 & \tau \end{pmatrix} \begin{pmatrix} n_1 \\ n_2 \\ n_3 \\ n_4 \\ n_5 \\ n_6 \end{pmatrix}. \tag{3.26a}$$

The Q_6 vector has also components in the complementary (or perpendicular) 3-dim subspace which can also be scanned by vectors along 2-fold axes perpendicular to the axes of the chosen physical cubic coordinate system, i.e.

$$Q_{perp} = \begin{pmatrix} h' - \tau h \\ k' - \tau k \\ l' - \tau l \end{pmatrix} = \begin{pmatrix} -\tau & 1 & 0 & \tau & 1 & 0 \\ 1 & 0 & -\tau & 1 & 0 & \tau \\ 0 & -\tau & 1 & 0 & \tau & 1 \end{pmatrix} \begin{pmatrix} n_1 \\ n_2 \\ n_3 \\ n_4 \\ n_5 \\ n_6 \end{pmatrix}. \tag{3.26b}$$

Including renormalization, the rotation from basis vectors along the 6-dim cube axes (n_i) to basis vectors along six 2-fold axes (three in each of the physical and complementary subspaces) is then described by

$$R = \frac{1}{\{2(2+\tau)\}^{\frac{1}{2}}} \begin{pmatrix} 1 & \tau & 0 & -1 & \tau & 0 \\ \tau & 0 & 1 & \tau & 0 & -1 \\ 0 & 1 & \tau & 0 & -1 & \tau \\ -\tau & 1 & 0 & \tau & 1 & 0 \\ 1 & 0 & -\tau & 1 & 0 & \tau \\ 0 & -\tau & 1 & 0 & \tau & 1 \end{pmatrix}. \tag{3.26}$$

3.3.5.4 Maximum and minimum M values corresponding to a given N According to eqns (3.24), $|Q_{par}|^2 \propto N + M\tau$, which means that $M > N/\tau$. We also have $N = 2\Sigma_{i=1}^{6} n_i^2$ and the squared modulus of Q_6 can be expressed using eqns (3.26):

$$|Q_6|^2 = 2(2+\tau)\sum_{i=1}^{6} n_i^2$$
$$= (2+\tau)N.$$

As any projection must result in modulus shortening,

$$|Q_{\text{par}}|^2 < |Q_6|^2$$

or

$$N + \tau M < (2+\tau)N;$$

that is

$$M < \tau N.$$

Regrouping all the conclusions related to N and M gives:

(1) N is any even number;

(2) if N is a multiple of four, the associated values of M are also multiples of four, smaller than τN and larger than $-N/\tau$;

(3) if N is not a multiple of four, the associated values of M are of the form $4m + 1$, within the same limits.

To illustrate the statements, let us take a few examples:

(1) $N = 2$ (not a multiple of 4) $\rightarrow$ $-N/\tau < M = 4m + 1 < 2\tau$

$$\underset{-1.236\ldots}{-N/\tau} \qquad \underset{3.236\ldots.}{2\tau}$$

The largest possible value of M corresponds to $m = 0$, so $M_{\text{max}} = M_0 = 1$; for $m = -1$, the obtained M value is -3 which is already below $-N/\tau$. Thus the only possible couple is $(N, M) = (2, 1)$.

(2) $N = 4 \rightarrow -2.472\ldots < M = 4m < 6.472\ldots.$
Possible values of M are 4, 0.

(3) $N = 12 \rightarrow -7.416\ldots < M = 4m < 19.416\ldots.$
Possible values of M are 16, 12, 8, 4, 0, -4. For each couple (N, M) the indices of the corresponding Q_{par} vectors must be derived, according to the expression (3.24) of (N, M) and the extinction rules. For instance, with $(N, M) = (12,16)$ the six indices must be:

(a) *all even*: they must be equal to 2 (three of them) and 0 to satisfy the N value:

(2/2 0/2 0/0) is all right for $M = 16$,
(2/0 2/0 0/2) gives $M = 12$ and is to be rejected;
(0/2 2/2 0/0) is also all right.

(b) *four even indices*. This cannot be accommodated since the two other odd indices would be equal to 1 which leaves 10 to be shared among four even squared indices; this is impossible.

Table 3.2 The $N = 12$ reflections of a PI quasicrystal

M	(n_i)	h/h'	k/k'	l/l'	Multiplicity
16	(210010)	2/2	0/2	0/0	60
	$(11111\bar{1})$	0/2	2/2	0/0	12
12	$(1111\bar{1}1)$	0/2	0/2	0/2	20
	$(20100\bar{1})$	2/0	2/2	0/0	60
8	(201001)	2/0	0/2	0/2	120
4	$(2100\bar{1}0)$	2/0	0/2	2/0	120
0	$(111\bar{1}\bar{1}\bar{1})$	2/0	2/0	2/0	20
−4	$(\bar{1}1111\bar{1})$	$\bar{2}/2$	2/0	0/0	12
	$(20\bar{1}001)$	2/0	$\bar{2}/2$	0/0	60

(c) *two even indices* and all *six odd indices* cannot be accommodated either.

When worked out 'by hand', the selection procedure of the correct indexing, though obvious in its essence, is actually rather tedious. A computerized approach is of course more efficient. The complete listing of the $N = 12$ indexings is given for illustration in the Table 3.2.[33]

3.3.5.5 Reformulation with proper units Up to now, dimensionless vectors have been used. If the $\boldsymbol{Q}$ vectors, in 6-dim or in their 3-dim $\boldsymbol{Q}_{\text{perp}}$ and $\boldsymbol{Q}_{\text{par}}$ components, have to be scaled with proper units, each of the above expressions has to be multiplied by $2\pi/a\ \{2(2+\tau)\}^{\frac{1}{2}}$. For instance, the diffraction peak positions have to be expressed as

$$\boldsymbol{Q}_{\text{exp}} = \boldsymbol{Q}_{\text{par}} = [2\pi/a\{2(2+\tau)\}^{\frac{1}{2}}]\,(h/h',k/k',l/l') \tag{3.27}$$

and

$$|\boldsymbol{Q}_{\text{par}}|^2 = \{4\pi^2/2(2+\tau)a^2\}\,(N + M\tau)$$

$$|\boldsymbol{Q}_{\text{perp}}|^2 = |\boldsymbol{Q}_6|^2 - |\boldsymbol{Q}_{\text{par}}|^2 = \{4\pi^2/2(2+\tau)a^2\}\tau^2(N\tau - M) \text{ etc} \ldots$$

Note that $|\boldsymbol{Q}_{\text{perp}}|$ takes its smallest value for the maximum value of M. Note also that in the expression of $|\boldsymbol{Q}_{\text{par}}|^2$, M can take the value zero if and only if N is a multiple of four. This means that the diffraction features cannot be separated into a subset of so-called 'fundamental' reflections with $M = 0$ and a subset of 'satellite' reflections with $M \neq 0$: the icosahedral crystals are not modulated crystals whatsoever and there is no hidden basis 3-dim periodic structure to which a modulation would be applied. Thus the difference between quasicrystals and modulated structures is made quite clear when looked at in reciprocal space. Remember that, from this point of view, the 1-dim Fibonacci model is *not* a quasicrystal.

3.3.6 *Indexing other structure patterns*

The detailed account, as reported in previous sections of this chapter, deals with the indexing of PI quasicrystals. The same approach can be performed for other space groups, in particular FCI and BCI. When compared to those related to the PI structure the results differ in particular in the corresponding extinctions rules and scaling properties. For instance, scaling by τ preserves all parity rules for the FCI reciprocal lattice while a τ^3 scaling is required for the PI one. These extinction rules are compared in Table 3.3 and the calculated diffraction patterns are shown in Fig. 3.25.

Icosahedral quasicrystals are not the only ones and many other structures were evoked in Chapter 2. Axial quasicrystals are probably the second most popular family of quasiperiodic structures. The principles for indexing their Fourier patterns rely also usefully on high-dimensional crystallographic space groups. For instance, the decagonal quasicrystals[35–37] can crudely be viewed as some sort of perioding piling up of 2-dim pentagonal structures. Thus, each position vector requires four 'quasiperiodic' components in the plane plus one periodic component along the axis. The crystallography of the system is then well described in a proper 5-dim (at least) space. Technical details for further documentation of the indexing problems can be found in some of the published papers.[1,33,34,38,39] Further reading on quasicrystallography is also available.[40,41]

Practical examples of structure determination for real quasicrystals will be developed in the next chapter.

3.3.7 *Direct space description and basic principles for a cut algorithm*

So far, the cut procedure has only been evoked and illustrated, without really going into the practical algorithm that may be used to derive the atom positions in the 3-dim real world from the n-dim periodic representation. The purpose of the present section is to present at least the basic principles that must be used in a cut algorithm.

Let us start to illustrate these principles with the now archetypal example of the Fibonacci chain as obtained from the 1-dim cut of a 2-dim square lattice (Fig. 3.1). Any node of the square lattice has coordinates $(Ha, H'a)$ on the basis $(\boldsymbol{e}_1, \boldsymbol{e}_2)$, H and H' being integers. This node has also $(r_{\text{par}}, r_{\text{perp}})$ components in the physical R_{par} and complementary R_{perp} spaces, respectively. Given the slope $\tan\alpha = 1/\tau$ of R_{par} in the example shown in Fig. 3.1, the two tiles of the Fibonacci chain have lengths

$$L = \tau = a\cos\alpha$$
$$S = 1 = a\sin\alpha.$$

Straightforward calculations lead to the relations

Table 3.3 Extinction rules for icosahedral reciprocal lattices

Indices	PI(a^*)	FCI($2a^*$)	BCI($2a^*$)
n_i	Any integer	$\Sigma n_i = 2n$	All even or all odd
h/h' k/k' l/l'	$\left.\begin{matrix} h + k' \\ k + l' \\ l + h' \end{matrix}\right\}$ even numbers	Same as for P plus $h + k + l$ even; $h' + k' + l'$ even	All even plus $\left.\begin{matrix} h + l + h' + k' \\ h + k + l' + k' \\ l + k + h' + l' \end{matrix}\right\}$ multiple of 4
–	Direct lattice is PI(a)	Direct lattice is BCI(a)	Direct lattice is FCI(a)

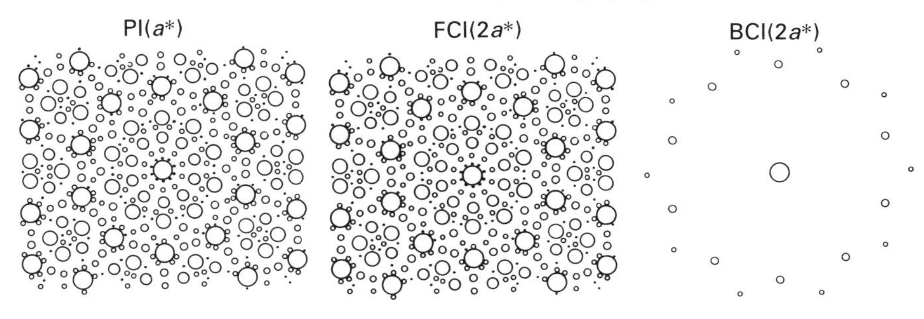

FIG. 3.25 Calculated diffraction patterns for the three icosahedral reciprocal Bravais lattices, as they appear in 5-fold zone axes.[33]

$$
\begin{aligned}
a^2 &= 1 + \tau^2 = 2 + \tau; \\
r_{\text{par}} &= Ha\cos\alpha + H'a\sin\alpha \\
&= H\tau + H' \\
r_{\text{perp}} &= H'a\cos\alpha - Ha\sin\alpha \\
&= H'\tau - H.
\end{aligned}
$$

For convenience, r_{par} and r_{perp} are re-expressed in terms of the lattice parameter a of the square lattice, i.e.

$$
\begin{aligned}
r_{\text{par}} &= \frac{\mathrm{a}}{(2+\tau)^{\frac{1}{2}}}(H\tau + H') \\
r_{\text{perp}} &= \frac{\mathrm{a}}{(2+\tau)^{\frac{1}{2}}}(H'\tau - H).
\end{aligned}
\tag{3.28}
$$

The set of r_{par} values contains the 'atom positions' of the Fibonacci chain, but not all these values correspond to an actual cut of the A_{perp} objects by R_{par}. For r_{par} to be a real cut (or real physical atom position) the corresponding (H, H') 2-dim node must be at a distance r_{perp} from R_{par} of less than half the length of the A_{perp} segment (this is obvious from consideration of Fig. 3.1). Thus, if Δ is taken equal to $a(1+\tau)/(2+\tau)^{\frac{1}{2}}$, the atom positions for the Fibonacci chain are

$$x = a(H\tau + H')/(2+\tau)^{\frac{1}{2}},$$

with the condition on (H, H')

$$|(H'\tau - H)| \leqslant (1+\tau)/2 = 1.309\,016\,994\ldots$$

The simple corresponding cut algorithm is then:

(1) $H = 0$ $\quad H'\tau \leqslant (1+\tau)/2$ imposes $H' = 0$
0 is an atom position.

(2) $H = 1$ $\quad |(H'\tau - 1)| \leqslant (1+\tau)/2$ gives $H' = 0$, $H' = 1$.
Atom positions: $L = \tau$, $L + S = \tau + 1$.

(3) $H = 2$ $\quad |(H'\tau - 2)| \leqslant (1+\tau)/2$ gives $H' = 1$, $H' = 2$.
Atom positions: $2L + S$, $2L + 2S$.

(4) $H = 3$ $\quad |(H'\tau - 3)| \leqslant (1+\tau)/2$ gives $H' = 2$.
Atom positions: $3L + 2S$.

etc.

More generally, H' must be taken equal to $[H/\tau]$ or $[H/\tau] + 1$ in which $[x]$ is the integer part of x.

The series of tiles obtained is: *LSLSLLSLLS* . . ., as expected for the Fibonacci chain. We conclude that the cut procedure is simply a test of r_{par} possible values with respect to an upper limit of r_{perp}. The procedure readily extends to 3-dim quasiperiodic structures as a cut of n-dim periodic distribution of $(n\text{-}3)$-dim atomic objects. Let us consider the example of an icosahedral quasicrystal that would result from the rational 3-dim cut of a 6-dim cubic lattice, decorated with a single spherical atomic object. Call r_0 the radius of this sphere, chosen according to criteria that will be explicitly analysed in Chapter 4 (consistency with diffraction data, existence of a minimum atom pair distance, density and composition of the material, etc.). The formalism introduced in previous sections to explain the six-integer indexing of the reciprocal space for an icosahedral quasicrystal also applies to direct space, as a simple extension of the regular crystallography approach. Thus, formulae such as (3.27) readily translate in direct space into

$$\boldsymbol{r}_{\text{par}} = \frac{a}{\{2(2+\tau)\}^{\frac{1}{2}}}\,(H/H', K/K', L/L'),$$

$$|\boldsymbol{r}_{\text{par}}|^2 = \frac{a^2}{2(2+\tau)}\,(S + \tau T), \tag{3.29}$$

$$|\boldsymbol{r}_{\text{perp}}|^2 = \frac{\mathrm{a}^2}{2(2+\tau)}\,(S\tau - T).$$

H, H', K, K', L, and L' are the 6-dim coordinates of the lattice sites of the 6-dim cubic lattices.

Now, the cut algorithm to generate the atom positions in our 3-dim real world is a simple test of the $\boldsymbol{r}_{\text{par}}$ possible vectors, with respect to the conditions $|r_{\text{perp}}| \leqslant r_0$. As an example of application of this algorithm, let consider the set of $\boldsymbol{r}_{\text{par}}$ vectors that corresponds to $S = 12$, as given in Table 3.4.

If now the radius of the atomic object is taken equal to the $(S = 2, T = 1)$

Table 3.4 The $S = 12$ set of parameters for calculating $\boldsymbol{r}_{\text{par}}$ and $\boldsymbol{r}_{\text{perp}}$ vectors. N_0 is the number of vectors $\boldsymbol{r}_{\text{par}}$ with the same length

T	H/H'	K/K'	L/L'	N_0	Symmetry
16	2/2	0/2	0/0	60	2-fold + inversion
	0/2	2/2	0/0	12	5-fold
12	0/2	0/2	0/2	20	3-fold
	2/0	2/2	0/0	60	2-fold + inversion
8	2/0	0/2	0/2	120	–
4	2/0	0/2	2/0	120	–
0	2/0	2/0	2/0	20	3-fold
−4	$\bar{2}$/0	2/0	0/0	12	5-fold
	2/0	$\bar{2}$/0	0/0	60	2-fold + inversion

$|r_{\text{par}}|$ value, the acceptable atom positions among the ones listed in Table 3.4 must be selected according to

$$S\tau - T \leqslant 2 + \tau = 3.618\,033\,989\ldots;$$

that is

(1) $S = 12, T = 16$
$S\tau - T = 12\tau - 16 = 3.416\ldots$ is acceptable.
This corresponds to 60 vector positions, on 2-fold axes, plus inversion generated positions, and 12 vectors on 5-fold axes, at a distance from the origin given by

$$|r_{\text{par}}|^2 = \frac{a^2}{2(2+\tau)}(12 + 16\tau)$$

or

$$|r_{\text{par}}| = 2.2882\ldots a.$$

(2) $S = 12, T = 12$
$S\tau - T = 12\tau - 12 = 7.416\ldots$: not acceptable.

(3) $S = 12, T = 8$
$S\tau - T = 12\tau - 8 = 11.416\ldots$: not acceptable.

(4) $T = 4, 0$, or -4 also give unacceptable values.

The procedure, in its systematics, is obviously boring, but quite efficient with computer help! See Problem 3.16.

Extension of the method to more complicated 6-dim structures is straightforward in principles. If several different atomic objects rather than a single sphere must be considered, then the test on the $\boldsymbol{r}_{\text{par}}$ values has to be conducted successively within each subset of $\boldsymbol{r}_{\text{par}}$ associated with its corresponding A_{perp}. Spherical shapes of course are a little simplistic and actually rather unphysical. But for more realistic shapes, say a triacontahedral A_{perp}, the procedure remains basically the same, except that the test value on $|r_{perp}|$ now has to be conducted on $\boldsymbol{r}_{perp}$. In other words, the acceptance limit is direction-dependent, due to loss of spherical isotropy. Unfortunately defining these acceptance limits may not be that easy. This point will be addressed in the next section.

3.4 Some further consideration of the atomic objects of the *n*-dim image

3.4.1 A summary of the general properties

As explained so far, describing a quasiperiodic structure via irrational slicing of a higher-dimensional periodic image makes crystallography approaches easier. However, practical problems arise when it is required to describe the high-dim atomic objects (or atomic surfaces A_{perp}) in detail. In particular, finding the exact location of their boundaries in R_{perp} is one of the main problems that is still not completely solved in the structural determination of quasicrystals.

As we have repeatedly stated, no diffraction experiment will ever be sufficiently precise to give the boundaries of the atomic object in R_{perp} by inverse Fourier transform. Some improvements are obtained when contrast variation measurements are feasible, but modelling is always required. Such modelling is necessarily achieved via some *arbitrary choice* of a finite number of parameters that are intended to characterize the atomic volume boundaries. These parameters may be the radius of a sphere in the simplest naive approach, or any other geometrical features of the chosen polyhedron. Then refinement procedures give the best-fit model within the corresponding *class of structure*. There is obviously no guarantee that other classes of structure, where the atomic objects are differently parametrized, would not yield a crystallographic fit that is as good or even better. This suggests that the famous residual parameters may be much less significant in quasicrystallography than in normal crystallography for validating a possible model on the basis of diffraction data alone.

However, before despairing we must realize that we are dealing with the 'skin' of the atomic objects, which is only a rather small part of the whole structure. Let us consider, for example, the trivial case in which a 'true' triacontahedral atomic object is 'mistaken' for a sphere of equal volume. It is easy to calculate that the differences induced in the resulting structures owing to non-equivalent 'skins' of the atomic objects, involve about 5–10 per cent of the atomic positions. Moreover, most of these 'wrong' positions can be eliminated by an *ad hoc* tailoring procedure because they generate unphysically short pair distances. Again, however, this *ad hoc* tailoring does not necessarily have a unique solution. In any case, even if only a few per cent of atoms in the structure have been mistakenly positioned and/or attributed to the wrong chemical species, this may result in dramatic consequences in the evaluation of physical properties (dynamics, conductivity, magnetism, etc.). Finally, it is also frustrating to know that the boundaries of the A_{perp} object *must* be very well defined boundaries but to have to be content with an approximation to them!

Improvements in the state of the art can be expected by imposing the following conditions to the A_{perp} objects.

1. They cannot have a 'thickness' in R_{perp} if they are to generate point-like atoms in the physical space via the slicing procedure. This condition is obviously easy to satisfy.
2. They must have the point group symmetry of the structure (e.g. icosahedral, decagonal, etc.).
3. They cannot intersect nor can they approach each other any closer than a threshold distance related to the physical sizes of the atoms. This is called the **hard core** condition, which can be satisfied in practice via the *ad hoc* tailoring procedure mentioned above.
4. They must allow energy translational invariances of the quasiperiodic structure parallel to both R_{par} and R_{perp} spaces. Condition (1) automatically contains this condition for global translations in R_{par}. The invariance law in R_{perp} means that the A_{perp} surfaces must connect in such a way that there is no 'annihilation–creation' of atoms under any R_{perp} (phason-like) translation. This is called the **closeness condition.**[42,43]

As illustrated in Chapter 1, it is easy to see that the above four conditions are satisfied for a Fibonacci chain with 'atomic line segments' whose length is constrained to the value $\Delta = a(\cos\alpha + \sin\alpha)$ (notation previously defined).

We now briefly describe some alternative approaches to designing the A_{perp} objects corresponding to 3-dim quasicrystals.

3.4.2 From the sphere approximation to faceted objects

Reducing the A_{perp} objects of the high-dim image to their spherical approximation is obviously accepting a low resolution description of the structure. Here, the expression 'low resolution' means that in the Fourier transform $G(Q_{perp})$ of the A_{perp} atomic surfaces the high-order Fourier components (corresponding to large Q_{perp} values) are not really accounted for. Sphere sizes are mostly fixed by density and composition constraints.

One possible method of introducing the high-order Fourier components is to parametrize the atomic surfaces in terms of linear combinations of symmetry-adapted functions associated with their point group symmetry[44]. In the case of an icosahedral quasicrystal, the perpendicular space is three-dimensional. The *spherical harmonics* are then a natural choice for expressing the boundaries of any radial functions $r(\theta, \phi)$. Hence the set of symmetry-adapted orthonormalized functions, invariant for icosahedral point group symmetry, can be chosen according to the decomposition

$$r(\theta, \phi) = \sum_{l,i} a_{li} Z_{li}(\theta, \phi) \tag{3.30}$$

with

$$Z_{li}(\theta, \phi) = \sum_{m} Z_{lm}(i) Y_{lm}(\theta, \phi)$$

in which Y_{lm} are the classical spherical harmonics, Z_{lm} are determined by the point group symmetry of the A_{perp} plus the normalization conditions of Z_{li}, and a_{li} are continuous parameters to be fitted in structural diffraction analysis; the index i allows for the possible existence of several orthogonal invariant functions within the same subspace of functions having a given value of l.

If the point group is large enough, there will be many empty subspaces. For instance, with the icosahedral point groups there is a single invariant function (for l up to 15) only for l values of 0, 6, 10, and 12. Beyond $l = 15$, contributions to eqn (3.30) are expected to be very weak. As an illustration, Fig. 3.26 shows that these four components are sufficient for the reconstruction of an icosahedron certainly beyond experimental resolution.[44] Any physical constraint on the A_{perp}, such as those induced by realistic atomic

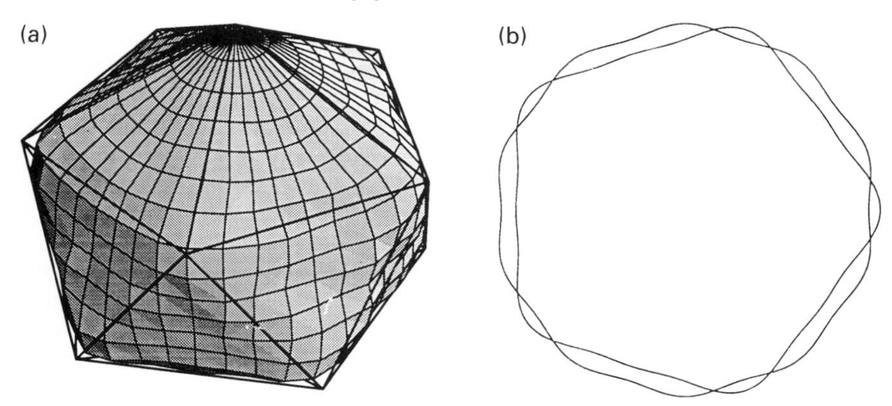

FIG. 3.26 Approximation of an icosahedron by four harmonics in eqn (3.30); (b) two sections of the same surface perpendicular to the z axis, at $z = 0$ and $z = a/\sqrt{5}$.[44]

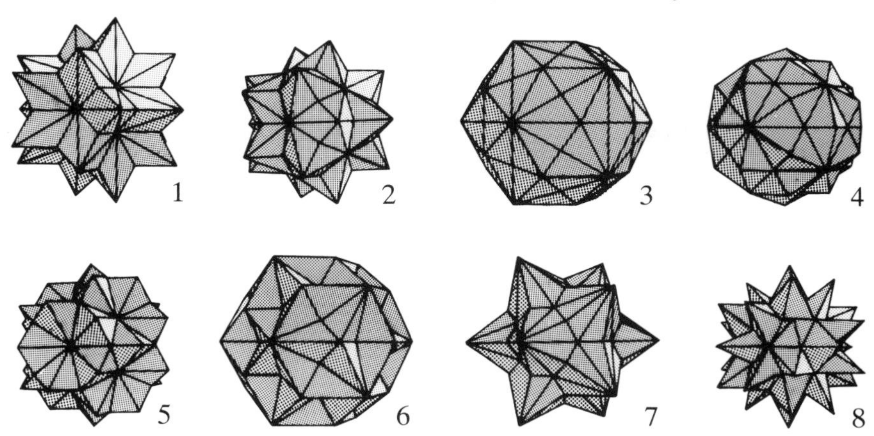

FIG. 3.27 The eight basic polyhedra bounded by 2-fold planes in R_{perp} for the 6-dim structure of the icosahedral phase (redrawn from reference [42, 43]).

distances, density, composition, etc., can be introduced in the refinement via penalty functions.

This is probably a good basis for allowing successful least-squares refinement processes to obtain realistic faceted A_{perp} objects (applications to AlLiCu and AlPdMn quasicrystals were under way at the time of writing[45]).

3.4.3 Formal faceting conditions of the A_{perp} atomic surfaces

The hard-core and closeness conditions mentioned above are satisfied[42,43,46] if the A_{perp} objects are bounded by piecewise connected surfaces, mostly parallel to the complementary space, without overlapping in this space, and globally invariant under point group symmetries. These conditions are satisfied for surface boundaries which are mirror planes of the structures. As a consequence, possible faceted A_{perp} volumes for the 6-dim images of icosahedral quasicrystals would have 2-fold, 3-fold, or 5-fold plane boundaries (see Section 3.3.2). This point has been demonstrated in detail for 2-fold plane boundaries,[42,43,46] and the shapes of the eight corresponding polyhedra are presented in Fig. 3.27.

The acceptable volumes for decorating the 6-dim cube must be one of these polyhedra, or any τ-scaling and/or intersection of them. Obviously, this leaves a number of alternative solutions and the formal faceting conditions, as they stand, have to be considered as a negative test to reject improper solutions.

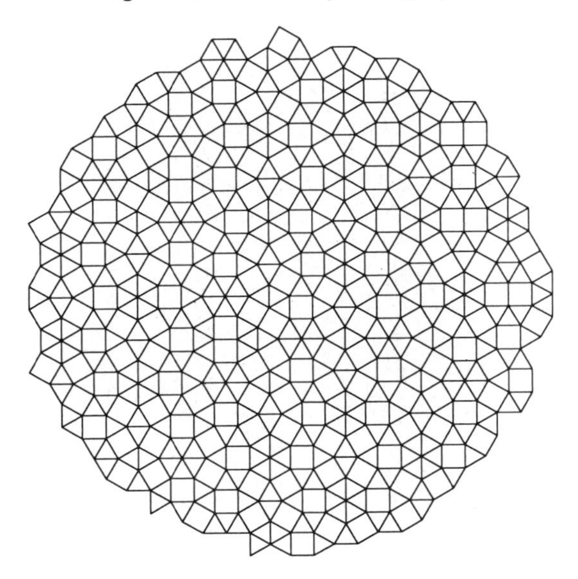

FIG. 3.28 Typical finite portion of a quasiperiodic dodecagonal square–triangle tiling.[47]

3.4.4 *Is it compulsory to have polyhedral A_{perp}?*

So far we have assumed that the A_{perp} atomic objects of the high-dim image are (faceted) polyhedra. This has induced conditions for these atomic objects. It may be of interest to consider whether the polyhedral solution is imposed in every case. There is no general answer to this question, and the point has received very little investigation, with restriction to 1-dim and some 2-dim quasiperiodic structures.

One example has been reported by Baake *et al.*[47] They generated a quasiperiodic dodecagonal tiling of the plane (Fig. 3.28) using squares and regular triangles arranged with simple deflation–inflation symmetries. This 2-dim structure has been 'lifted' (embedded) into a 4-dim periodic lattice and the acceptance domain (or A_{perp} objects) has been iteratively constructed to generate the vertex set of the square–triangle tiling. The result is shown in Fig. 3.29. The procedure leads to a fractally bounded A_{perp}. It can be shown that there is no polyhedral alternative solution if the square–triangle tiling is to be obtained with a single type of A_{perp}.

Recently, Zobetz[48] has demonstrated that 1-dim quasiperiodic chains can be derived from 2-dim periodic arrangements of fractally shaped atomic surfaces. The 1-dim structures of interest are generated via inflation rules,

FIG. 3.29 The acceptance domain filled by the projection of the lift of 32 000 vertices in perpendicular space, as obtained from the structure shown in Fig. 3.28.[47]

as for a Fibonacci chain, but with a generalized scale factor τ^{-n}. For τ^{-1} there are two possible deflation rules for the initial line segments L and S, i.e. (i) $L \to L'S'$, $S \to L'$, and (ii) $L \to S'L'$, $S \to L'$; both combinations yield the Fibonacci chain. For τ^{-2} there are three different ways to deflate L, namely $L'L'S'$, $L'S'L'$, and $S'L'L'$, and two different ways to deflate S, namely $L'S'$ and $S'L'$. For τ^{-3} there are 10 ways to deflate L and three ways to deflate S. Each combination of deflation rules yields a quasiperiodic chain which can be 'lifted' (embedded) into a 2-dim square-lattice structure with convenient A_{perp} segments. The substitution matrix is entirely determined by the scale factor. Structures deduced from a given scale factor only differ by tile (segment) rearrangements.

The results demonstrate that the above quasiperiodic structures have similar A_{perp} objects if they belong to the same local isomorphism class (chains with similar frequencies of inter-point distances). It can be verified that there are 38 possible combinations of the various deflation rules corresponding to the scale factors τ^{-1}, τ^{-2}, and τ^{-3}. These 38 structures belong to nine different local isomorphism classes (labelled from 0 to 8, with the Fibonacci chain in class 0). The nine corresponding types of atomic surfaces are characterized by their profiles in R_{perp}, as shown in Fig. 3.30. Their main properties are as follows.

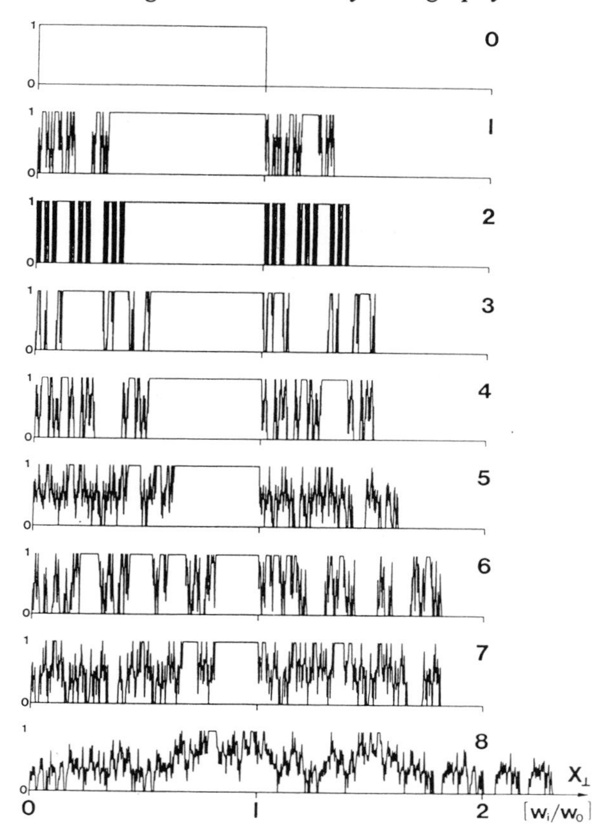

FIG. 3.30 Distribution functions (atomic surfaces) of the lattice points projected onto perpendicular space of the nine different types of quasilattice; w_0 denotes the width of the atomic surface of the original Fibonacci quasilattice (type 0).[48]

1. They extend in R_{perp} in proportion to the periodic strips (length and frequency) that appear in the chain;
2. They are self-similar, i.e. they remain similar to themselves after rescaling;
3. As a consequence of point 2, they are fractal objects, except for A_{perp} associated with class 0. They become increasingly fractal as the deflated strip increases. This is obviously linked to the fact that the 'choice' for types of A_{perp} intercepts with physical space increases; the spiky profiles accommodate fewer simple sequences of tiles and self-similarity preserves the closeness condition.

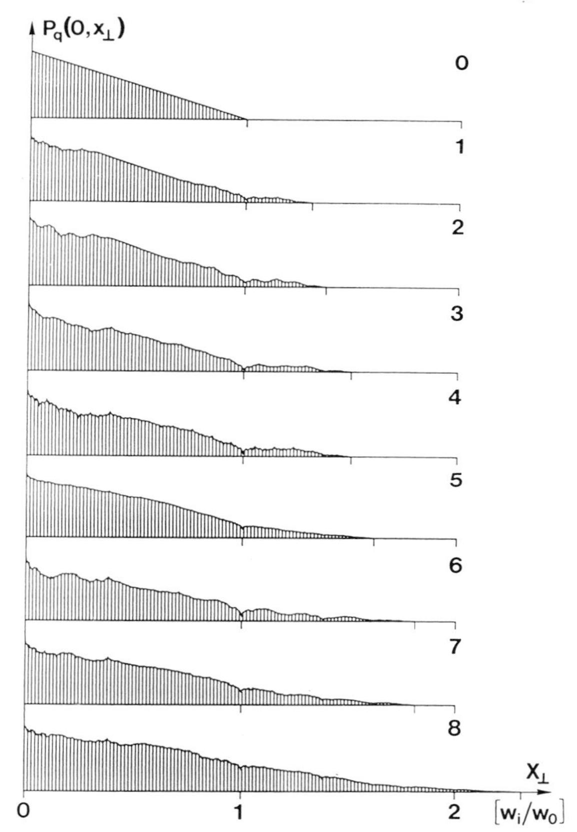

FIG. 3.31 Patterson syntheses of the nine different types of atomic surfaces shown in Fig. 3.30.[48]

In the case of the simple Fibonacci chain, a diffraction experiment would approach reasonably well the width of the A_{perp} segments via the profile of Patterson functions in R_{perp} (see Section 3.2). But it would be a very tedious task to make similar attempts for the other classes! This is clearly visible in Fig. 3.31 which shows the Patterson syntheses of the nine different types of atomic surfaces in the perpendicular space. The self-overlapping procedure that generates the Patterson functions from the density distribution erases all fractal features.

An interesting general question is certainly related to the possible extension of similar conclusions to generic 3-dim self-similar tilings. There is no definitive answer so far, but some recent studies[49] have suggested that quasiperiodic tilings exhibiting simultaneously self-similarity and regular atomic surfaces would be a rarity. The chance of finding such a case increases if the inflation procedure operates on a small number of tile prototypes (two, for instance, as in a Penrose tiling) and with a small scale factor. Unfortunately, simplicity is not always the chosen alternative in real life! Such fractal geometries for the high-dimensional images of quasiperiodic structures are expected to produce Debye–Waller like effects scaling with Q_{perp} in diffraction experiments. This point will be illustrated in the next chapter.

3.5 Problems

3.1 Consider a modification of the structure shown in Fig. 3.1 by adding an atomic object at the body centre of the square unit cell, identical to the one sited at the vertex sites.

(a) Determine the new 1-dim quasiperiodic structure.

(b) Determine the new 2-dim reciprocal lattice, the intensities of the Fourier components and the Fourier pattern of the 1-dim quasicrystal.

(c) Calculate Q_{par} and Q_{perp} with two integer indices (h, h').

(d) Determine the extinction rules.

3.2 Assuming that Bragg peaks can be measured in a diffraction experiment with intensities ranging from 100 to 1, give a list of the Fourier components actually observed for the structure of Fig. 3.1. Plot these intensities as a function of $Q_{\text{perp}}^{hh'}$.

3.3 Using the cut-projection method, show that the L segment positions in a Fibonacci chain are given by $m_i = \lfloor i\tau \rfloor = 1, 3, 4, 6, 8$ ($\lfloor x \rfloor$ is the largest integer in x).

3.4 Generate a 2-dim periodic lattice by a proper cut of a cubic 3-dim structure.

3.5 Work out Problem 1.1 again, but with the structure shown in Fig. 3.5.

3.6 Derive qualitatively the expected Patterson map from the structure of Fig. 3.5 by using a transparent foil to self-overlap the structure.

3.7 Determine the equilibrium positions of the atoms in the two interacting chains described in Section 3.2.5.

3.8 Find the structure resulting from a phason wave deformation of a Fibonacci chain.

3.9 Demonstrate that the components of the six 5-fold vectors of a regular icosahedron are of the form $(1, \tau, 0)$ in the cubic coordinate system

shown in Fig. 3.18. Calculate the angle between one chosen 5-fold axis and the ring of the other five.
Derive the stereographic projection with one of the 5-fold axes as the z-axis. Repeat with one of the 3-fold axes as the z-axis.

3.10 Demonstrate that the basis of six vectors pointing to the vertices of an icosahedron is not invariant under inflation or deflation by τ or τ^2. Investigate the case for the edge vectors of the icosahedron.

3.11 Reasoning as in Section 3.3.2 and 3.3.3, demonstrate that a 6-dim structure, with reciprocal vectors having their Q_{par} components along the 3-fold axes of a regular icosathedron, is FCI with BCI reciprocal lattice.

3.12 In the cubic coordinate system as defined in text, find the indices of the 3-fold and 2-fold directions.

3.13 For the PI structure demonstrate that (a) scaling Q_{par} by τ preserves the parity rules for only some of the Fourier components and (b) scaling by τ^3 preserves all parity rules.
Calculate $|Q_{par}|^2$ and $|Q_{perp}|^2$ as expressed with the h/h' k/k' l/l' indices.
Demonstrate that M in eqn (3.24) is of the form either $4m$ or $4m + 1$, depending on N values, in relation to parity rules effects on h/h' k/k' l/l' indices. Calculate M values associated with $N = 10$ and derive the corresponding h/h' k/k' l/l' indices.

3.14 By repeating the derivation presented in Section 3.3.4 and 3.3.5, determine the indexing, extinction rules, and scaling properties for an FCI reciprocal lattice. Show the differences from a PI reciprocal lattice by listing the expected Fourier components.

3.15 Following the method described in Section 3.3.7, derive atom positions for $S = 2, 4, 6$ in the case of the simple spherical model also proposed in Section 3.3.7. How are the results modified if a second spherical object is added in the body centre position of the 6-dim cube? Try to describe the short-range order of the 3-dim physical structure.

3.16 Use the computer program given below to determine atom positions in a canonical 3-dim Penrose tiling. The 6-dim structure is a primitive cubic with vertex decoration by a triacontahedron. Use any convenient plotting program to visualize the results.

```
c ***************************************************************
c This program generates the coordinates of a 3__D
c Penrose tiling with an edge parameter apr.
c
c Results are in the data file Penros3d.dat.
c Coordinates are expressed in a cubic basis
c following the convention of Cahn, Shechtman, Gratias
c (J of Mat. Res., 1 (1986) 13-26), all in angstroms.
c
c The generated cluster is spherical with radius rmax.
c If rmax is increased, nx should also be increased.
c
c ***************************************************************
      dimension a2tria6d(15, 6), xpe(3), xpa(3)
      dimension a2triaper(15,4)
c
c a2tria6d : 6 dim coordinate of the face center of the
c            standard triacontahedron. (only the 15 positive
c            part is used)
c a2triaper: Projection in perpendicular space
c
      logical test
      fol=30
      nx=3
      rmax=30.0
c
c Penrose edge parameter apr, and corresponding 6D lattice
c  parameter a6d.
c
      apr=4.6
      a6d=4.6*sqrt(2.)
      tau=1.618033989
      co=a6d/sqrt(2.*(2.+tau))
c
c Perpendicular shift of the physical space E//
c
      x0p=0.5
      y0p=1.0
      z0p=0.4
c
c read the 6-dim coordinate of the triacontahedron
c in the data file a2tria6d.dat
c
```

```
      open(27,file='a2tria6d.dat',status='old')
      read(27,*) ((a2tria6d(i, j), j=1, 6), i=1, 15)
      close(27)
c
c projection in perpendicular space of the 6-Dim vector
c
      call projper(15, a2tria6d, a2triaper)
      open(fol,file='penros3d.dat',status='new')
      write(fol, 100)rmax, nx
100   format(2x, 'Atomic coordinates of a 3-D Penrose',/,
     1 2x, 'atoms are inside a cluster with rmax=', f7.2,
     2 /, 2x, 'maximum indice of the 6-D loop is nx=', i2)
      write(fol, 110)
110   format(/,2x,'  x  y  z  r  n1  n2  n3  n4  n5  n6')
c
c Loop on the 6-Dim indices
c
      do n1=-nx, nx
       do n2=-nx, nx
        do n3=-nx, nx
         do n4=-nx, nx
          do n5=-nx, nx
           do n6=-nx, nx
c
c Projection of the 6-Dim vector in perpendicular space

c             xpe(1)=co*((n2+n5)+tau*(n4-n1))+x0p
              xpe(2)=co*((n1+n4)+tau*(n6-n3))+y0p
              xpe(3)=co*((n3+n6)+tau*(n5-n2))+z0p
              rpe=sqrt (xpe(1)*xpe(1)+xpe(2)*xpe(2)+xpe(3)*xpe(3))
c
c Check if the vector fall inside the triacontahedron
c
              call invol(15, a2triaper, xpe, test)
               if (test.eq..true.)then
c
c projection in physical space E parallel
c
                xpa(1)=co*(n1-n4+tau*(n2+n5))
                xpa(2)=co*(n3-n6+tau*(n1+n4))
                xpa(3)=co*(n2-n5+tau*(n3+n6))
                rpa=sqrt(xpa(1)*xpa(1)+xpa(2)*xpa(2)+xpa(3)*xpa(3))
c
c check if atom is inside the cluster with radius rmax
```

```
c
                          if (rpa.le.rmax) then
                           write (fol, 125), xpa (1), xpa(2), xpa(3), rpa, n1, n2
1                         , n3, n4, n5, n6
                          end if
                        end if
                      end do
                    end do
                  end do
                end do
              end do
            end do
125          format (2x, 4(f8.3,x),2x,6(i3, x))
      stop
      end
c
c Check if a vector is inside a volume defined
c by planes.
c
      subroutine invol (n, a, xn, test)
      logical test
      dimension xn(3), a(n, 4)
      test=true.
        do i=1, n
         sc=0
          do j=1, 3
          sc=sc+xn(j)*a(i, j)
          sc1 = -sc
          end do
         if (sc.gt.a(i, 4).or.scl.ge.a(i, 4)) then
          test=.false
          go to 10
         end if
        end do
10    RETURN
      END
c
c projection in perpendicular space of a set
c of 6-Dim vector
c
      subroutine projper(n, a6d, aper)
      dimension a6d(n, 6), aper(n, 4)
```

```
      tau=1.618033989
      xapr=4.6
      xa6d=xapr*sqrt(2.)
      co=xa6d/sqrt(2.*(2.+tau))
      do i=1, n
        aper(i, 1)=co*((a6d(i, 2)+a6d(i, 5))+tau*(a6d(i, 4)-a6d(i, 1)))
        aper(i, 2)=co*((a6d(i, 1)+a6d(i, 4))+tau*(a6d(i, 6)-a6d(i, 3)))
        aper(i, 3)=co*((a6d(i, 3)+a6d(i, 6))+tau*(a6d(i, 5)-a6d(i, 2)))
        aper(i, 4)=0
        do j=1, 3
         aper(i, 4)=aper(i, 4)+aper(i, j)*aper(i, j)
        end do
c
c calculate the perpendicular length of the vector
c
      aper(i, 4)=aper(i, 4)/2.
      end do
      return
      end
```

Data file A2TRIA6D.DAT

```
-1  1  0  1  1  0
 0 -1  1  0  1  1
 1  0 -1  1  0  1
-1  0  1  0  1 -1
 0 -1  1 -1  0 -1
 0  0 -1  1  1  1
 1 -1 -1  0  0  1
 1 -1  1 -1  0  0
-1  0  1  1  1  0
 1 -1  0 -1  0  1
-1  0  0  1  1  1
 1 -1  0  0  1  1
 0 -1  0  1  1  1
 0 -1  1 -1  1  0
-1 -1  1  0  1  0
```

3.17 Consider the figure obtained with a regular triangle whose vertices are decorated by small 'square atoms' and one of the mid-edge positions by a 'circle atom'. Draw the figure corresponding to the 'Patterson function' of the decorated triangle. Now, forget the triangle and try to reconstruct it from its Patterson function. What do you think?

References

1. Steurer, W. (1990). The structure of quasicrystals. *Z. Krist.* **190**, 179.
2. Cahn, J.W., Gratias, D., and Mozer, B. (1988). Six-dimensional Fourier analysis of the icosahedral $Al_{73}Mn_{21}Si_6$. *Phys. Rev.* **B38**, 1643–7.
3. Cahn, J.W., Gratias, D., and Mozer, B. (1988). Patterson Fourier analysis of the icosahedral (Al, Si)—Mn alloys. *Phys. Rev.* **B38**, 1638–42.
4. Cahn, J.W., Gratias, D., and Mozer, B. (1988). A 6-D structural model for the icosahedral (Al, Si)—Mn quasicrystal. *J. Phys.* (*Paris*) **49**, 1225–33.
5. Steurer, W. (1989). Five-dimensional Patterson analysis of the decagonal phase of the system Al–Mn. *Acta Cryst.* **B45**, 534–42.
6. Dubois, J.M., Janot, C., and Pannetier, J. (1986). Preliminary diffraction study of icosahedral quasicrystals using isomorphous substitution. *Phys. Lett.* **A115**, 177–81.
7. Janot, C., de Boissieu, M., Dubois, J.M., and Pannetier, J. (1989a). Icosahedral crystals: neutron diffraction tells you where the atoms are. *J. Phys.: Cond. Matter* **1**, 1029–48.
8. de Boissieu, M., Janot, C., and Dubois, J.M. (1990). Quasicrystal structure: cold water on the Penrose tiling scheme. *J. Phys.: Cond. Matter* **2**, 2499–517.
9. de Boissieu, M., Janot, C., Dubois, J.M., Audier, M., and Dubost, B. (1991). Atomic structure of the icosahedral AlLiCu quasicrystal. *J. Phys.: Cond Matter* **3**, 1–25.
10. Jaynes, E.T. (1983). *Papers on 'probability, statistics and statistical physics.'* (ed. R.D. Rosenkrantz). (Synthese Library, Vol. 158. Reidel, Dordrecht.
11. Tikochinsky, Y., Tishby, N.Z., and Levine, R.D. (1984). Consistent inference of probabilities for reproducible experiments. *Phys. Rev. Lett.* **52**, 1357–60.
12. Livesey, A.K. and Skilling, J. (1985). Maximum entropy theory. *Acta Cryst.* **41**, 113–22.
13. Gull, S.F. and Skilling, J. (1984). *IEE Proceedings* **131(F)**, 646.
14. Papoular, R.J. (1991). *Acta Cryst.* **A47**, 283–95.
15. Skilling, J. and Bryan, R.K. (1984). *Mon. Not. R. Astr. Soc.* **211**, 111.
16. Skilling, J. and Gull, S.F. (1985). In *Maximum entropy and Bayesian method in inverse problems.* (ed. C. Ray Smith and W.T. Grandy Jr) pp. 83–132. D. Reidel, Dordrecht.
17. Livesey, A.K. and Skilling, J. (1985). *Acta Cyst.* **A41**, 113.
18. Duneau, M. and Oguey, C. (1989). Ideal AlMnSi quasicrystal: a structural model with icosahedral clusters. *J. Phys. Paris* **50**, 135–46.
19. Bancel, P.A., Heiney, P.A., Stephens, P.W., Goldman, A.I., and Horn, P.M. (1985). Structure of rapidly quenched Al–Mn. *Phys. Rev. Lett.* **54**, 2422–5.
20. Ebalard, S. and Spaepen, F. (1989). The body-centered-cubic-type icosahedral reciprocal lattice of the Al–Cu–Fe quasi-periodic crystal. *J. Mater. Res.* **4**, 39–43.
21. de Boissieu, M., Janot, C., Dubois, J.M., Audier, M., and Dubost, B. (1989). Partial pair distribution functions in icosahedral Al–Li–Cu quasicrystals. *J. Phys. Paris* **50**, 1689–709.
22. Janssen, T. (1986). Crystallography of quasi-crystals. *Acta Cryst.* **A42**, 261–71.
23. Ostlund, S. and Wright, D.C. (1986). Scale invariance and the group structure of quasicrystals. *Phys. Rev. Lett.* **56**, 2068–71.

24. Rokshar, D.S., Mermin, N.D., and Wright, D.C. (1987). Rudimentary quasicrystallography: the icosahedral and decagonal reciprocal lattices. *Phys. Rev.* **B35**, 5487–95.
25. Gähler, F. (1990). Classification of space groups for quasicrystals with $(2n + 1)$-reducible point groups. In *Quasicrystals and incommensurate structures in condensed matter*, Proceedings of the 3rd international meeting on quasicrystals (ed. H.J. Jacaman, D. Romeu, V. Castaño, and A. Gomez), pp. 69–78. World Scientific, Singapore.
26. Alexander, S. and McTague, J. (1978). Should all crystals be bcc? Landau theory of solidification and crystal nucleation. *Phys. Rev. Lett.* **41**, 702–5.
27. Bak, P. (1985). Phenomenological theory of icosahedral incommensurate (quasiperiodic) order. *Phys. Rev. Lett.* **54**, 1517–19.
28. Sackder, S. and Nelson, D.R. (1985). Order in metallic glasses and icosahedral crystals. *Phys. Rev.* **B32** 4592–606.
29. Shechtman, D., Blech, I., Gratias, D., and Cahn, J.W. (1984). Metallic phase with long-range orientational order and no translational symmetry. *Phys. Rev. Lett.* **53**, 1951-3.
30. Dubost, B., Lang, J.M., Tanaka, M., Sainfort, P., and Audier, M. (1986). Large AlCuLi single quasicrystals with triacontahedral solidification morphology. *Nature* **324**, 48–50.
31. Tsai, A.P., Inoue, A., and Masumoto, T. (1988). New stable icosahedral Al–Cu–Ru and Al–Cu–Os alloys. *Japan. J. Appl. Phys.* **27**, L1587–90.
32. Tsai, A.P., Inoue, A., Yokoyama, Y., and Masumoto, T. (1990). New icosahedral alloys with superlattice order in the AlPdMn system prepared by rapid solidification. *Phil. Mag. Lett.* **61**, 9–14.
33. Cahn, J.W., Shechtman, D., and Gratias, D. (1986). Indexing of icosahedral quasiperiodic crystals. *J. Mat. Res.* **1**, 13–26.
34. Elser, V. (1985). Indexing problems in quasicrystal diffraction. *Phys. Rev.* **B32**, 4892–8.
35. Bendersky, L. (1985). Quasicrystal with one-dimensional translational symmetry and a ten-fold rotation axis. *Phys. Rev. Lett.* **55**, 1461-3.
36. Steurer, W. and Kuo, K.H. (1990). Five-dimensional structure analysis of decagonal $Al_{65}Cu_{20}Co_{15}$. *Acta Cryst.* **B46**, 703–12.
37. He, L.X., Wu, Y.K., and Kuo, K.H. (1988). Decagonal quasicrystals with different periodicities along the tenfold axis in rapidly solidified $Al_{65}Cu_{20}M_{15}$ (M = Mn, Fe, Co, or Ni). *J. Mater. Sci.* **7**, 1284–6.
38. Bak, P. (1985). Symmetry, stability and elastic properties of icosahedral incommensurate crystals. *Phys. Rev.* **B32**, 5764–772.
39. Janssen, T. (1988). Aperiodic crystals: a contradiction in terms. *Physics Reports* **168**, 55–113.
40. Gratias, D. (1988). Introduction à la quasicristallographie. In *Du cristal à l'amorphe.* (ed C. Godrèche) pp. 83–152. Les Editions de Physique, Paris.
41. Mermin, N.D. (1991). Quasicrystallography is better in Fourier space. In *Quasicrystals: the state of the art*, (ed. D.P. DiVincenzo and P.J. Steinhardt), pp. 133–83. World Scientific, Singapore.
42. Cornier-Quiquandon, M., Gratias, D., and Katz, A. (1991). A tentative methodology for structure determination in quasicrystals. In *Methods of structural analysis of modulated structures and quasicrystals.* (ed. J.M. Perez-Mato), pp. 313–32. World Scientific, Singapore.

43. Katz, A. and Gratias, D. (1993). A geometrical approach to chemical ordering in icosahedral structures. *J. Non-Cryst. Solids*, **153–154**, 187–95.
44. El Coro, L., Perez-Mato, J.M., and Madariaga, G. (1993). Systematic structure refinement of quasicrystals using symmetry-adapted parameters. *J. Non-Cryst. Solids*, **153–154**, 155–9.
45. Perez-Mato, J.M. and Elcoro, L. (1994). Private communication.
46. Katz, A. (1990). Hardcore and closeness conditions in quasicrystals. In *Number theory and physics* (ed. J.M. Luck, P. Moussa, M. Waldschmidt, and C. Itzykson), pp. 100–23. Springer, Berlin.
47. Baake, M., Klitzing, R., and Schlottmann, M. (1992). Fractally shaped acceptance domains of quasiperiodic square–triangle tilings with dodecagonal symmetry. *Physica A*, **191**, 554–8.
48. Zobetz, E. (1993). One dimensional quasilattices: fractally shaped atomic surfaces and homometry. *Acta Crystallogr. A*, **49**, 667–76.
49. Godrèche, C., Luck, J.M., Janner, A., and Janssen, T. (1993). Fractal atomic surfaces of self-similar quasiperiodic tilings of the plane. *J. Phys. I (France)*, **3**, 1921–39.

4

Where are the atoms?

When I was a little boy, I thought all rivers were yellow and all nights were orange
Duane Michals

4.1 Introduction

Now that we have discussed the existence of quasiperiodicity and its representation by *n*-dim periodic structures, we must give practical examples of structure determination and try to understand how to think of the quasicrystal in our 3-dim world. We must repeat here the crucial difference between the description of a regular 3-dim crystal in terms of a unit cell plus an atomic basis and quasicrystals. In a regular crystal, one can always describe the structure as a stacking of '*unit cells*', all of which are decorated with atoms in an identical manner. The quasicrystal can be envisioned in the same way, when described in its *n*-dim periodic aspect. That is, we associate a basis with each unit cell in *n*-dim, and repeat this motif for all cells. However, this stacking arrangement in *n*-dim does not translate directly to a tiling in 3-dim. We cannot, in general, reduce the 3-dim crystallographic analysis to a determination of a repeated atomic basis. It would thus seem that any attempt at structural refinement in 3-dim is doomed to failure. The question we are going to address in this chapter is actually 'What is the most appropriate way to describe the atomic motif and how can we achieve this description?'.

The first step is of course to use properly the available diffraction data, as explained in Chapter 3. This will be exemplified in the first section of this chapter, with a special emphasis on phase reconstruction procedures and 6-dim modelling for icosahedral structure. Less generic approaches will also be presented, including giant cell periodic models, decorated quasiperiodic tiling, random tiling, and icosahedral glasses.

4.2 Experimental determination of quasicrystal structures

4.2.1 Data collection and scaling procedures

As for regular crystals, X-rays and neutron diffraction are the natural approach to quasicrystal structure. Beforehand, detailed characterization of the sample must be considered seriously. It is rather easy to obtain quasicrystal-like samples, but it is actually difficult to obtain good quasicrystals. In the early days following their discovery, quasicrystal labels were straightforwardly attributed to materials showing non-crystallographic symmetries in electron diffraction pattern (e.g. five-fold axes). We know now that large cell periodic approximants and/or conveniently oriented grains of periodic crystals can mimic five-fold symmetries. One good example of such a mistaken symmetry has been reported by Audier *et al.*[1] This concerns the AlFeCu system which, depending on composition, preparation history, and/or temperature can form as a perfect icosahedral quasicrystal, a defect modulated quasicrystal, or as rhombohedral periodic crystallites with orientational relationships. Phase transitions between crystalline and quasicrystalline states seem indeed to be more frequently observed than originally expected and this should encourage people to revisit their early quasicrystal investigations.

It is worth noting that diffraction methods (X-rays, neutrons, electrons) are obviously not the only way to collect information about (quasi) crystal structures. Valuable insight has been gained from EXAFS (extended X-ray absorption fine structure),[2, 3] Mossbauer spectroscopy,[4] NMR,[5] field ion microscopy,[6] and tunneling scanning microscopy,[7] not to mention high resolution electron microscopy (see Chapter 2). But quasicrystallography applies mainly to diffraction and is the most direct way to specify atom positions. Neutron and/or X-ray data can be collected on powder or/and single (quasi) crystals. The latter approach is needed if Bragg peaks corresponding to non-equivalent families of indices are to be differentiated. For instance, consider the reflections $(N, M) = (12, 16)$ for an icosahedral quasicrystal (notation as in Chapter 3 and details in Table 3.1). This corresponds to two families of reflections, i.e. the (2/2 0/2 0/0) ones with multiplicity 60 (2-fold directions plus inversion) and the (0/2 2/2 0/0) ones with multiplicity 12 (5-fold directions). These 72 reflections show up at the same $|Q_{\mathrm{par}}|^2 \propto (N + M\tau)$ in a powder diffraction pattern and equal intensities for all peaks must be assumed to go further. Instead, individual peaks are measured in a single crystal (four circles) experiment. As also explained in the previous chapter, contrast variations are highly desirable in order to simplify the structure determination of real polyatomic quasicrystals into monatomic problems. For practical reasons this is usually done on powder samples.

As in regular crystallography, intensities are assigned to peaks once correctly indexed, as described in Chapter 3. This is done by measuring the (integrated) area under the diffraction peaks, disregarding their widths and shapes. The procedure obviously ignores phason disorder, which, with the possible exception of the AlFeCu-like and AlPdMn systems, manifests itself by peak broadenings. But this is a different problem that will be addressed latter on in the book.

Once indexed correctly, which means for instance that the positions of the peaks have to be compared to the sequence expected for icosahedral (or other) quasicrystals (see Chapter 3), the intensities as measured for different diffraction patterns, i.e. different contrast values and/or powder versus single grain, have to be mutually rescaled.[8] As an example, consider the case with diffraction patterns as measured for a binary alloy A–B. The diffracted intensities as expressed by Eqn (3.2) are rewritten below, with the inclusion of a scaling factor δ_c, for each contrast value $c = b_A/b_B$, and of the multiplicity parameter $\mu(Q)$:

$$\delta_c I_o(Q) = \mu(Q)\left(b_A^2|F_A|^2 + b_B^2|F_B|^2 + 2b_A b_B|F_A||F_B|\cos\phi\right). \quad (4.1)$$

Assuming that intensities are measured for n diffraction lines (n values of Q) and p different contrasts (p values of c), we are left with np linearly related eqns (4.1), containing $3n + p$ unknown quantities, i.e. p scaling coefficients δ_c and n sets of the three quantities $|F_A(Q)|$, $|F_B(Q)|$, and $\Delta\phi(Q)$. The number of contrast values is usually four and the system is over-determined as soon as we measure more than four peaks in each given pattern. The unknowns are then obtained using a non-linear square fit procedure, which therefore allows a precise comparison of diffraction intensities coming from different samples without requiring standard calibration.

A special case of interest corresponds to all experimental values of $\Delta\phi$ being equal to 0 or π for icosahedral crystals. The structure factors F_A, F_B are complex quantities, currently expressed by the product of a Bravais lattice contribution F_L and of a motif contribution F_M. In the 6-dim periodic description of the quasicrystal, as for normal 3-dim crystals, each vertex of the Bravais lattice is a centre of symmetry for the whole lattice (centrosymmetry property of a periodic lattice) and, as a consequence F_L must be a real number, with phase equal to 0 or π. The observed phased shift $\Delta\phi$ between F_A and F_B then necessarily comes from the motif contribution F_M. Suppose now that the A atom positions can be assigned to the Bravais lattice sites (vertices). That would imply that ϕ_M^A, the motif phase of F_A, is equal to 0 or π and so would be also ϕ_M^B if $\Delta\phi = 0$ or π. The whole structure would thus be centrosymmetric in its 6-dim periodic representation. The cut-projection procedure that guarantees the quasiperiodic 3-dim structure induces symmetry breaking and the phases of the structure factors are not necessarily preserved. This induces additional basic problems when direct refinement structures are attempted using 3-dim atomic models.

In the following section, the practical case of structure determination for the quasicrystal of the AlLiCu system will be developed.[9] It is not the best quasicrystal but it provides a good example of diffraction studies successfully applied to quasicrystals, with both single crystal samples and isotopic contrast variation effects. The main results of some other structure investigations will also be presented to illustrate particular points.

4.2.2 *Experimentally determined structure of the AlLiCu quasicrystal*

4.2.2.1 Experimental conditions Within the frame of research for light AlLi-base alloys to be used for aerospace purposes, the AlLiCu phase diagram has been revisited carefully along with thermodynamic properties of the phase of interest.[10] The identified phases were the tetragonal θ-Al_2Cu, fcc δ-AlLi, fcc TB-$Al_{7.5}Cu_4Li$, bcc R-Al_5CuLi_3, hex T_1-Al_2CuLi, and icosahedral T_2-Al_6CuLi_3. In particular, it has been shown that only slow cooling rates are required to form the icosahedral T_2-phase which actually behaves like an equilibrium phase, going apparently directly to the liquid state upon heating. The bcc R-phase and the icosahedral T_2-phase have very similar features. Their densities are almost the same (2.46 and 2.47 g cm^{-3}, respectively) and they form within a very narrow composition range: $Al_{5.60}Cu_{1.20}Li_{3.20}$ for the R-phase and $Al_{5.70}Cu_{1.08}Li_{3.22}$ for the T_2-phase (within 3 per cent error bars).

Neutron diffraction is particularly well suited when isotopic contrast variation can be achieved. Lithium has two stable isotopes, ^{6}Li and ^{7}Li, whose respective scattering lengths are $+0.20 \times 10^{-12}$ and -0.222×10^{-12} cm. This allows significant changes in the contrast on the Li sites of the structure when alloys are prepared with different ^{6}Li/^{7}Li mixtures. A 'zero-scatterer' element Li$^{(0)}$ is even easily obtained by mixing about 50/50 per cent of the two isotopes. Copper has also two stable isotopes:^{63}Cu with a scattering length $+0.672 \times 10^{-12}$ cm and ^{65}Cu with a scattering length $+1.102 \times 10^{-12}$ cm. Five samples of the icosahedral phase were produced with natural copper and different $^{6/7}$Li isotopic compositions corresponding to $\langle$Li$\rangle$ scattering length b (Li) = 0.190 (natural Li), and −0.110, 0, +0.102, + 0.20 (pure ^{6}Li isotope) (in 10^{-12} cm), and two more samples with an Li zero-scatterer (b(Li) = 0) and either the ^{63}Cu or ^{65}Cu isotope.

The samples were characterized by powder X-rays and electron diffraction. The shrinkage cavity method[11] was used to produce single (quasi)crystal grains and large pieces of oriented dendrite of the icosahedral phase, for the purpose of single crystal X-rays and neutron diffraction scans. The triacontahedral shape of the single grains obtained is shown in Fig. 4.1.

Using the indexing with six Miller pseudo-cubic indices $(h/h', k/k', l/l')$ (see Chapter 3) and the 6-dim hypercubic lattice constant $a = 7.15$ Å, both powder diffraction (with different contrast parameters) and four-circle diffraction peaks (X-rays and neutrons) were indexed by

$$Q_{\text{exp}} = Q_{\text{par}} = \frac{2\pi}{a\{2(2+\tau)\}^{\frac{1}{2}}}(h + \tau h', k + \tau k', l + \tau l'),$$

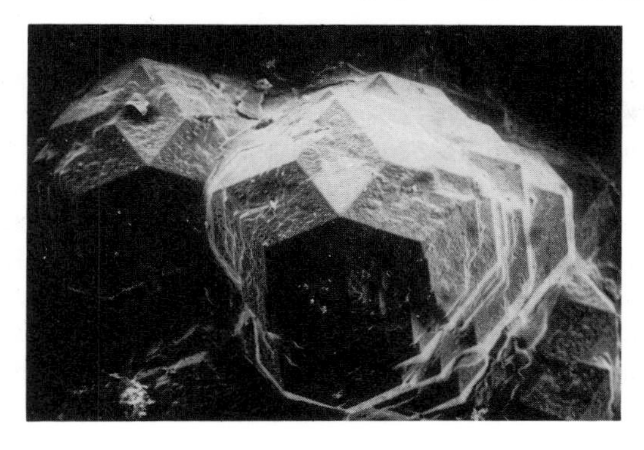

FIG. 4.1 Triacontahedral single grain of the AlLiCu quasicrystal as obtained by the shrinkage cavity method [11] (courtesy of P. Sainford, Pechiney Company).

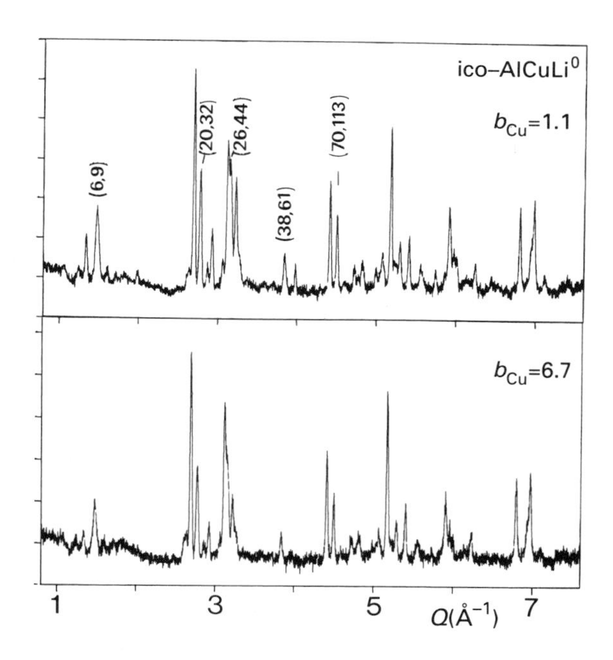

FIG. 4.2 Neutron powder diffraction patterns of the AlLiCu quasicrystal with isotopic contrast (^{63}Cu and ^{65}Cu). The absence of an obvious contrast effect suggests a weak Cu/Al order.

within an error of less than 10^{-3} Å^{-1}. The recorded reflections correspond to $(N, M) \leqslant (228, 368)$. They all belong to a primitive icosahedral Bravais lattice. The AlLiCu quasicrystal is a ternary alloy. Three partial structure factors, with their amplitudes and phases, have to be determined if the structure is to be treated as superimposed monatomic systems. Fortunately enough, the possibility of preparing samples with a 'zero-scatterer' lithium ($Li^{(0)}$) yields some simplifications in as much as any ico-AlCu$Li^{(0)}$ sample actually behaves like a binary compound from the point of view of neutron diffraction. Thus, three such $Li^{(0)}$ bearing samples prepared with natural copper, ^{63}Cu and ^{65}Cu (Fig. 4.2), should lead to the determination of Al/Cu atomic correlations. Actually the powder neutron diffraction patterns corresponding to the extreme contrast obtained with ^{63}Cu and ^{65}Cu look very much the same for both alloys, which means that AlCu order is very weak. Thus the (Al,Cu) atoms can now be treated as a single average species, say atom A, and the ico-phase as a pseudo-binary $A_{68}Li_{32}$ alloy whose partial structure factors may be obtained from Li isotopic substitution (Fig. 4.3). Accordingly, the diffracted intensities at a given scattering vector $Q_{exp} = Q_{par}$ for single crystal experiments can be written

$$\begin{aligned} I(\boldsymbol{Q}_{par}) &= |F(\boldsymbol{Q}_{par})^2| \\ &= |b_A F_A(\boldsymbol{Q}_{par}) + b_{Li} F_{Li}(\boldsymbol{Q}_{par})|^2. \end{aligned} \tag{4.2}$$

where b_A (constant) and b_{Li} (variable) stand for the neutron scattering length of the average A and Li atoms, respectively. The Fs are the corresponding partial structure factors. $I(\boldsymbol{Q}_{par})$ are the integrated intensities of the measured reflections.

In the powder diffraction mode, the amplitudes $|\boldsymbol{Q}_{par}| = Q_{par}$ are the only accessible scattering parameters. The measured intensities are then:

$$I(Q_{par}) = \sum_i \mu_i |F_i(Q_{par})|^2,$$

in which the subscript i scans the different non-equivalent families of reflections showing up at the same Q; μ_i is the multiplicity of equivalent reflections in a given family i. Hereafter, peaks belonging to a single family of equivalent reflections will be referred to as 'simple reflections'. Their powder-diffracted intensities are related to their single crystal-diffracted intensity through their single multiplicity μ. The other reflections will be referred to as 'multiple reflections'. The simple reflections are the only ones which can be used into the rescaling procedure described in the previous section, being the only intensities directly comparable in both powder and single grain modes. Then, X-ray and neutron single grain data being renormalized with respect to each other, 'multiple' reflections can also be treated, relieving all possible degeneracy problems. In the present case the

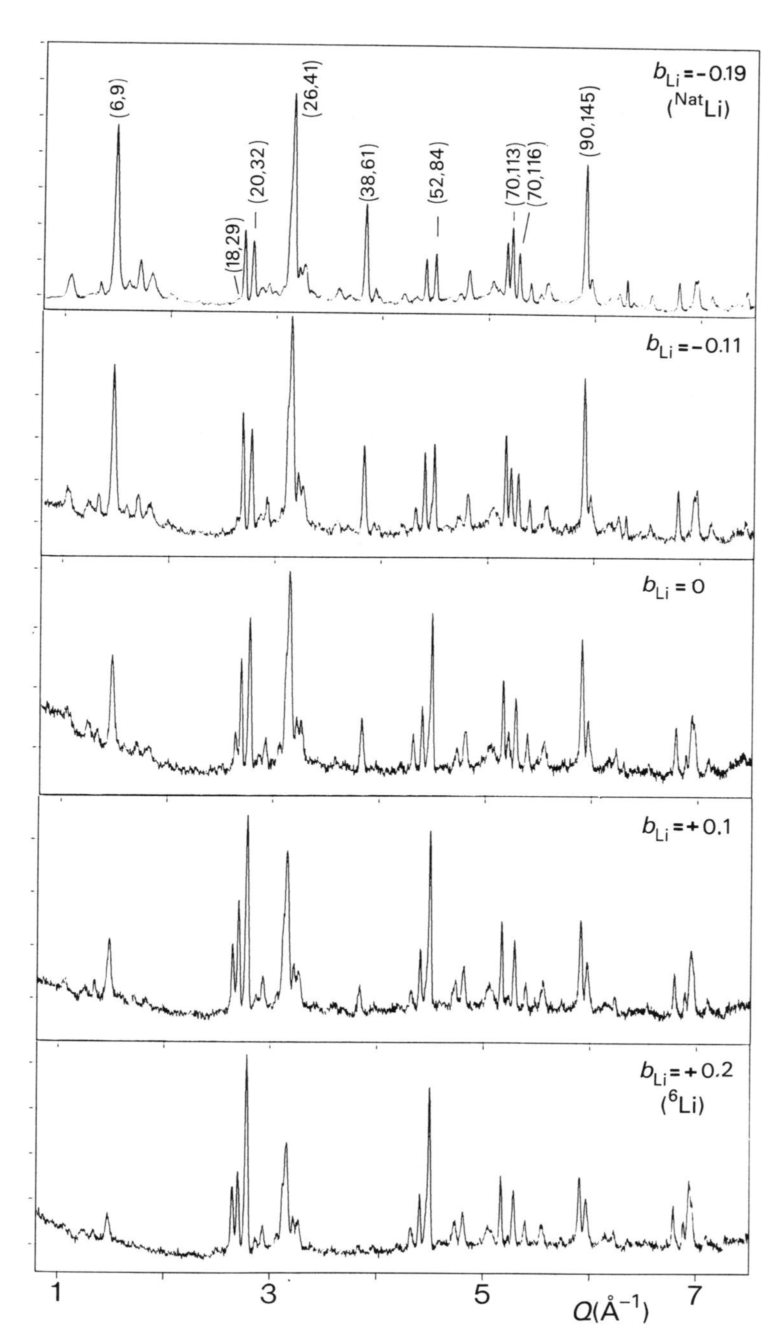

FIG. 4.3 Neutron powder diffraction patterns as measured with samples of the ico-phase containing different ^{6}Li/^{7}Li isotopic mixtures; b_{Li} is the corresponding scattering length. Contrast effects are clearly visible.

phase differences for F_A and F_{Li} are found equal to 0 or π, which is a favourable case for proceeding further in structure specification.

4.2.2.2 Phase reconstruction and 6-dim periodic structure Remembering, from Chapter 1 and 3, that the Q_{perp} dependence of the (partial) structure factors gives the Fourier transforms of the A_{perp} atomic objects in the 6-dim structure, one can expect to get some insight into the A_{perp} shape and/or size by looking at these Q_{perp} dependences. For instance, Fig. 4.4 shows schematically what would be an $F(Q_{perp})$ profile for a spherical A_{perp} object. Actually, the Q_{perp} dependences of the $|F_A|$ and $|F_{Li}|$ values (Fig. 4.5) are just clouds of points which do not suggest a clue to the A_{perp} specification. Consequently we have to work somewhat iteratively through successive steps of approximations. The starting point is of course the independent reflections whose partial structure factors $|F_A|$ and $|F_{Li}|$, with their relative signs, have been obtained from experiment.

A first easy step is to use the six-integer indexing of these reflections of Fourier transform $|F_A|^2, |F_{Li}|^2$, and $|F_A + F_{Li}|^2$ into a 6-dim direct space distribution. The result is partial and total unweighted Patterson functions, using here a procedure first proposed by Gratias and co-workers.[12, 13] Rational cross-sections of these Patterson functions are shown in Fig. 4.6.

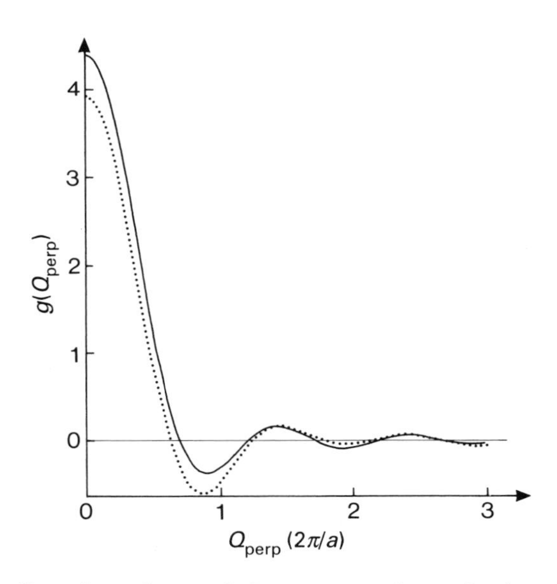

FIG. 4.4 The Q_{perp} dependence of the structure factor is given by the Fourier transform of the A_{perp} atomic objects of the *n*-dim structure. The Q_{perp} dependences for a simple spherical A_{perp} (__) and a spherical shell with an empty core of half its radius (.....) are shown here.

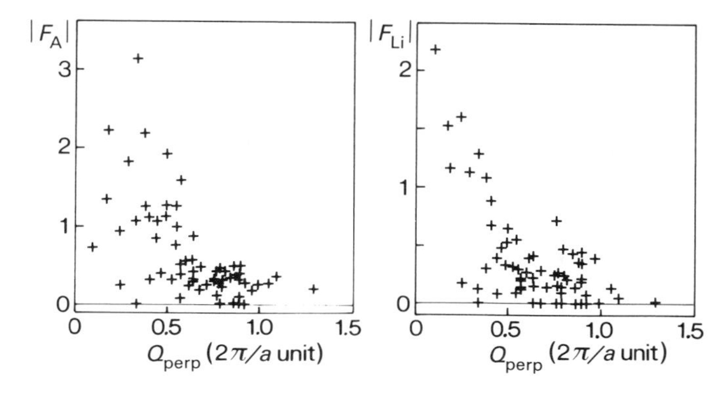

FIG. 4.5 Q_{perp} dependences of the measured amplitudes of the partial structure factors with AlLiCu quasicrystals. There is no evidence for simple behaviour, beyond a rough general decay at 'large' Q_{perp} values.

From the density features visible in the figure, and remembering that Patterson functions illustrate self-overlapping of the structure upon translation, it is possible to conclude that the $A3_{perp}$ volumes in the 6-dim cube are sited at vertex (OR) and mid-edge (ME) positions for the A-atoms and at body-centre (BC) positions for the Li atoms. Thus the partial structure factors must be written:

$$F_A(\boldsymbol{Q}_6) = \frac{1}{V_6}\left(G_{OR}(Q_{perp}) + \sum_{i=1}^{6} G_{ME}^{(i)} \cos(\boldsymbol{Q}_6 \cdot \boldsymbol{r}_i)\right)$$

$$F_{Li}(\boldsymbol{Q}_6) = \frac{1}{V_6} G_{BC}(Q_{perp}) \cos(\boldsymbol{Q}_6 \cdot \boldsymbol{\delta}). \tag{4.4}$$

The G functions are the Fourier transforms of the $A3_{perp}$ volumes, V_6 is the volume of the 6-dim cube, $i = 1, \ldots, 6$ corresponds to the six different mid-edge positions and $\boldsymbol{\delta}$ is the half-diagonal vector of the 6-dim cube.

In the second step, eqns (4.4) are crudely used with spherical $A3_{perp}$ volumes, whose Fourier transforms $G(Q_{perp})$ are easily obtained. The sizes (radii) of the spheres are deduced from both the half-width at half maximum of the Patterson features and the alloy composition and density. Such a sphere would have a radius equal to 8.54 Å for the lithium $A3_{perp}$ volume. Then the $F_{Li}(Q_6)$ are calculated from eqns (4.4) in which the $G_{BC}(Q_{perp})$ term is deduced from the 8.54 Å spherical $A3_{perp}$. The strongest experimental reflections are mainly influenced by size rather than the shape details of $A3_{perp}$. Thus it is reasonable to attribute signs to the strong experimental

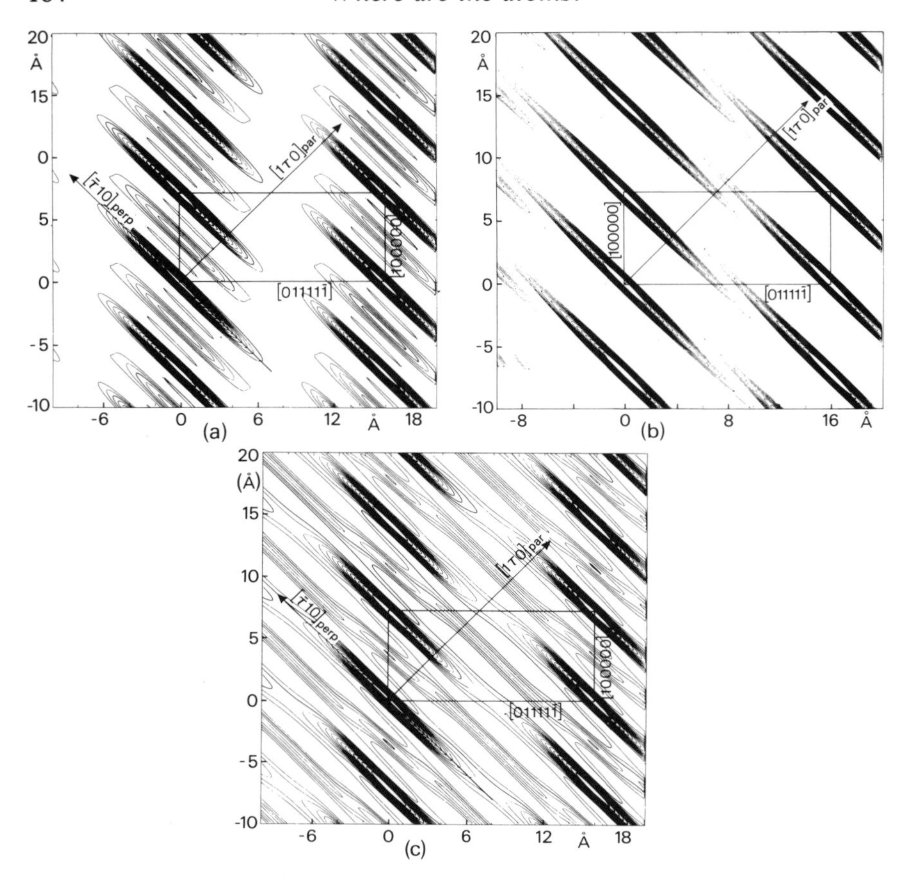

FIG. 4.6 Patterson functions of the 6-dim periodic structure for (a) A, (b) Li, and (c) A + Li atomic sites of a AlLiCu quasicrystal. The figures show a slice of the 6-dim space containing one perpendicular and one parallel 5-fold axis. The map (b) alone cannot tell us whether Li is at BC or origin sites since the PF translates the Li surfaces to the origin. But the map (c) showing A and Li features altogether has a BC density which imposes the existence of a BC surface. This BC surface cannot be due to A atoms when considering the (a) map.

F_{Li} (typically for $Q_{\mathrm{perp}} < 0.5$ in $2\pi/a$ units) identical to those of the spherical approximation. This is actually not contradictory to the decaying rough 'shape' of the $|F_{\mathrm{Li}}|$ behaviour. Now, signs of the corresponding F_{A} can also be derived, since the $F_{\mathrm{A}}/F_{\mathrm{Li}}$ relative signs have been experimentally determined. This step gives about 30 independent reflections with the phases properly reconstructed for both F_{A} and F_{Li} experimental partial structure factors.

In the last step, the above 30 strongest pairs of partial structure factors are Fourier transformed in the 6-dim space. From the deduced partial density distribution, radii equal to 6.5 and 5.6 Å are obtained for the spherical equivalent $A3_{perp}$ volumes of the A atoms (vertex and mid-edge respectively). Using eqns (4.4) again, calculated values of F_A are used to attribute signs to more experimental F_A and then to F_{Li}. One more iteration results in the whole set of measured F_A and F_{Li} having their phases (0 or π) reconstructed unambiguously. The final 6-dim densities are illustrated in Fig. 4.7 and can be used for modelling the 6-dim structure and for refinement procedures. Owing to the good quality of the diffraction data and the resulting rather accurate faceting of the deduced density profiles in 6-dim, the refined model, as described by its A_{perp} atomic object which decorates the 6-dim cubic lattice, must be rather simple (Fig. 4.7). The body-centre lithium volume is basically an 8 Å sphere with 20 added 'bubbles' along the 3-fold axes and 12 holes dug along the 5-fold axes. The vertex A volume is also basically a 6.8 Å deformed sphere with added small volumes along the 5-fold directions. Finally, the mid-edge A volume is more complicated, with 5-fold axial symmetry and cross-sections roughly circular in a plane perpendicular to its 5-fold axis, and roughly elliptical in a plane containing the 5-fold axis and $(1, \tau, 0)_{perp}$ direction. The mid-edge volume has only a $\bar{5}m$ symmetry, the full icosahedral symmetry being recovered when the whole set of mid-edge sites is considered altogether.

4.2.2.3 Modelling further As explained in Chapter 3, the density distribution as obtained from the inverse Fourier transform of the experimental partial structure factors may suffer from some drawbacks, such as truncation effects, presence of parallel components, unphysical short distances, etc. The latter, in particular, imposes an empirical 'tailoring' of the A_{perp} objects for these too short distances to be removed.

Systematically scanning the distances between sites results in a list of what has to be avoided for physical reasons. The strongest constraint corresponds to the distance between the $A3_{perp}$ volumes related to the body-centre Li sites and the mid-edge A sites, which is equal to 0.597 Å along a physical (par) 5-fold axis (Fig. 4.8). This is unacceptably short and is removed by digging holes of proper volume along the twelve 5-fold axes of the BC volume. To restore the lost density, additional small volumes must be added where room is available. Figure 4.8 and the experimental results, as illustrated in Fig. 4.7, show that a clear possibility is around the twenty 3-fold axes.

The same analysis can be done for unphysical distances between ME–ME, OR–ME, and ME–ME pairs of $A3_{perp}$ volumes. The whole procedure results in a set of $A3_{perp}$ models. The vertex A_{OR} volume is a sphere of 6.8 Å radius with an empty central hole of 2.3 Å radius; the mid-edge A_{ME}

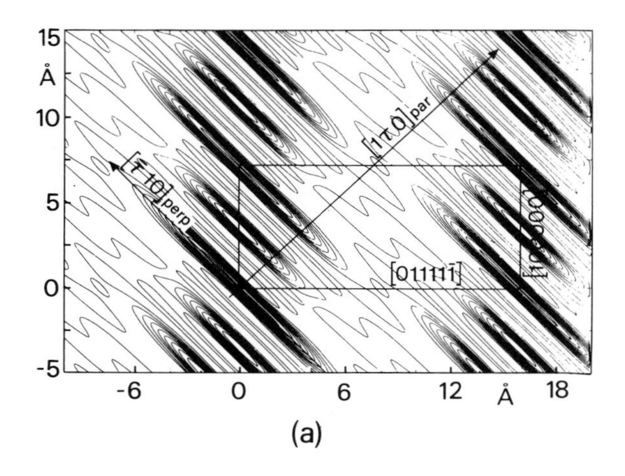

15
Å
10
5
0
-5
-6
0
6
12
Å
18
[1τ0]par
[τ10]perp
[011111̄]
[100000]

(a)

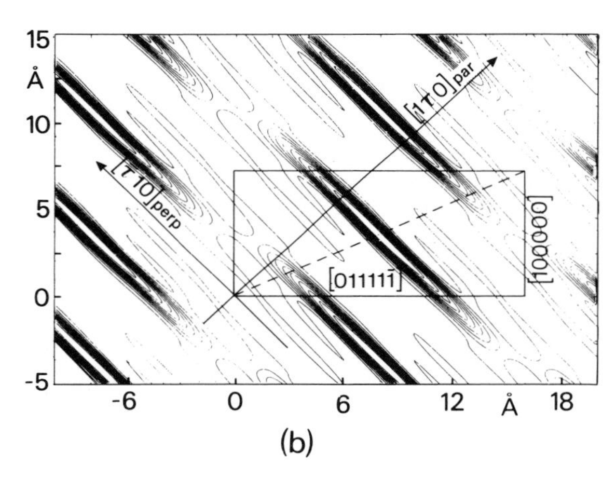

15
Å
10
5
0
-5
-6
0
6
12
Å
18
[1τ0]par
[τ10]perp
[011111̄]
[100000]

(b)

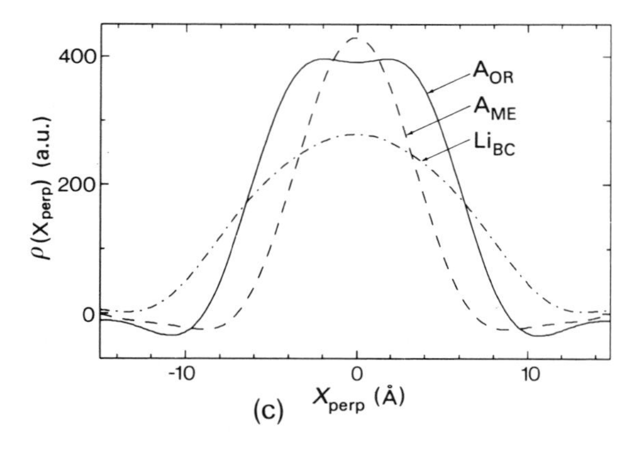

400
200
0
-10
0
10
$\rho(X_{perp})$ (a.u.)
X_{perp} (Å)
A_{OR}
A_{ME}
Li_{BC}

(c)

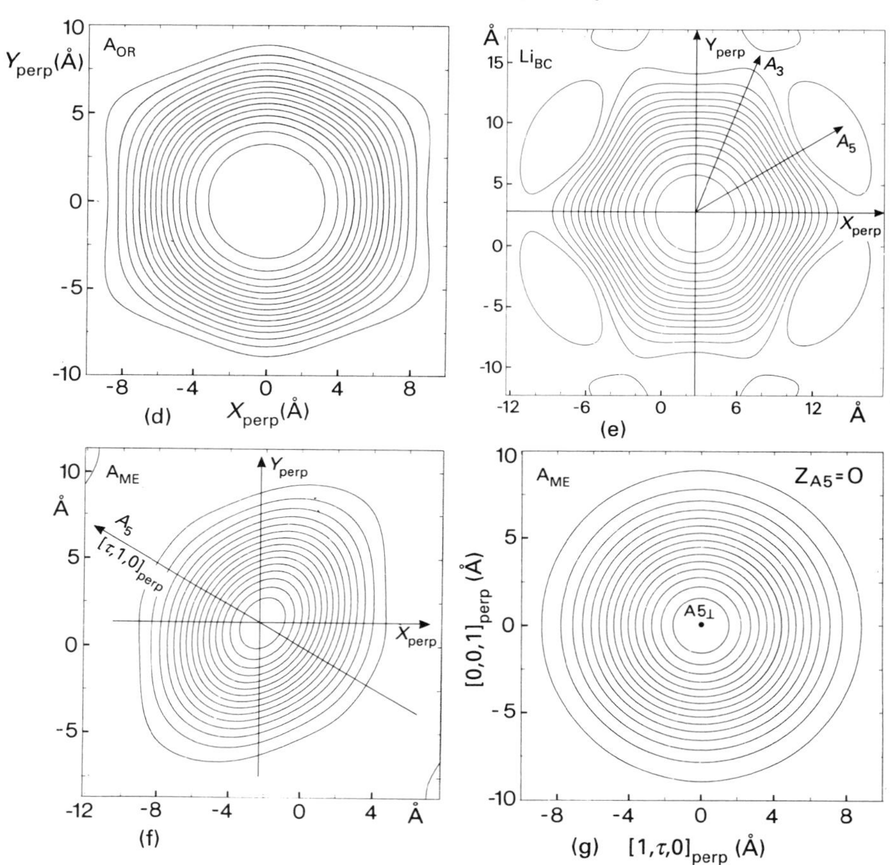

FIG. 4.7 Five-fold axis slices of the 6-dim partial densities of the AlLiCu quasicrystal as obtained from the direct FT of partial structure factors, with their phases reconstructed. (a) Density for A atoms; (b) density for Li atoms; (c) density profile as measured along a 2-fold axis in the complementary space for the $A3_{perp}$ volume corresponding to the A_{OR} (__), A_{ME} (----), and Li_{BC} (....) sites. (d) Density contours of the A_{OR} volume in a plane containing two 2-fold axes of the complementary space (e) Same as in (d) but for the Li_{BC} volume. (f, g) Same as in (d) but for two different cross-sections of the A_{ME} volume.

volume is an axial ellipsoid sited on a five-fold axis with geometrical size given by $a = 4.15$ Å and $b = c = 6.34$ Å; the body-centre Li_{BC} volume is a sphere of 8.5 Å radius with elliptical holes on 5-fold axes (same a, b, c as for the ME site) and additional pieces of small spheres (radius 3.5 Å) on the 3-fold axes (Fig. 4.9).

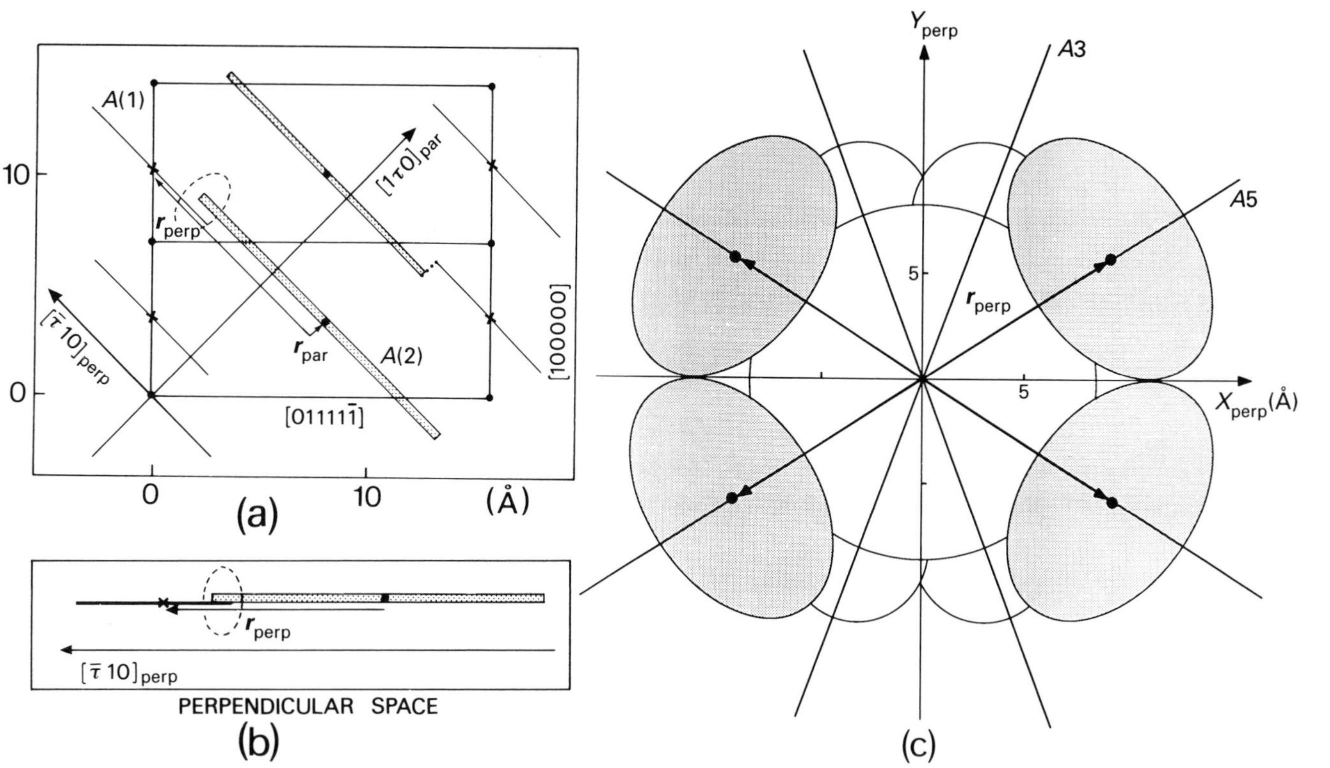

FIG. 4.8 Five-fold axis slice (a) of the 6-dim space (schematic) showing 'overlapping' between BC and ME volumes $A(1)$ and $A(2)$. The short distance r_{par} will show up along a physical (par) 5-fold axis because $A(1)$ and $A(2)$ have overlapping parts (in the dashed loop) when projected on to the perpendicular 5-fold axis (b). When looked at in a 2-fold axis plane of the perpendicular space (c) the $A(1)$ volume is the white 'sphere' and the $A(2)$ volumes are the grey ellipsoids. Clearly, parts of the white 'sphere' have to be dug out if overlapping is to be avoided. Protrusions can be added along the 20 three-fold axes to restore density.

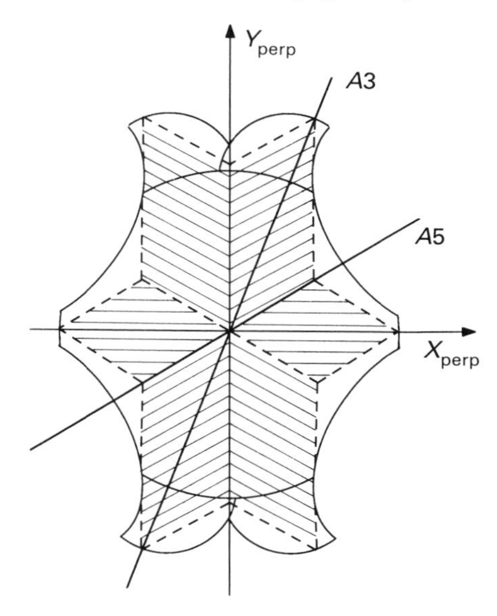

FIG. 4.9 Cross-section through the perpendicular space of the body-centre volume as proposed in the model derived from diffraction data with AlLiCu (solid line). Compare with Fig. 4.7(g). The hatched area within the dashed line shows the stellation that would generate a systematic decoration of the 3-fold axes of a Penrose tiling.

The direct cut of the above 6-dim structure by our physical 3-dim space generates atom positions in a large cluster whose size is only computer time limited. This is of course a somewhat brute force procedure, but quite useful anyhow. We used it to build a spherical cluster and calculate pair distribution functions. These pair distribution functions are shown in Fig. 4.10, where they are compared with the one experimentally measured.[14] The result is quite satisfactory: all atomic distances and weight of the pairs are reasonably reproduced, without spurious unphysical short distances. In particular, the model fits positions and widths of the first distance peaks, which means that, contrary to the ico-AlMn structure,[15] there are no parallel components into the $A3_{perp}$ volumes of the ico-AlCuLi system, at least down to a limit of about 0.05 Å.

Further validation of the model requires calculation of the Fourier components and comparison with the diffraction data (single crystals and powders). The two aluminium $A3_{perp}$ volumes (spheres and ellipsoid) can be Fourier transformed analytically; the BC volume for lithium has been approached by a very dense network of small cubic elements (512 elements

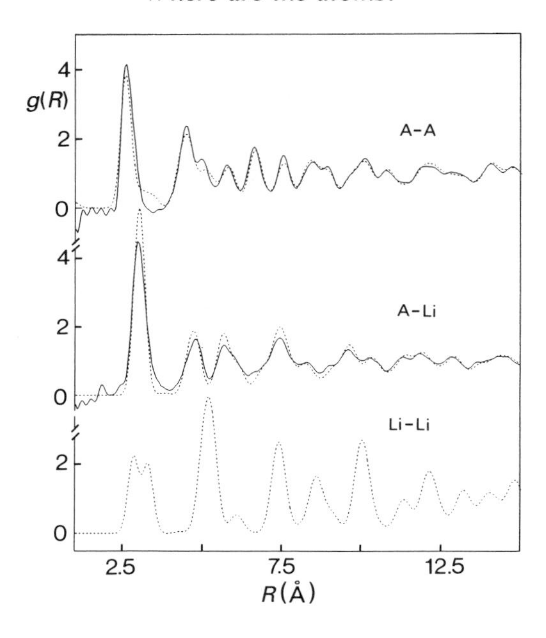

FIG. 4.10 Partial pair distribution functions for the AlLiCu quasicrystal. Model (....) of the present work compared with experimental results from reference [14] (__).

$Å^{-3}$), the Fourier transform then being the sum of contributions from all the cubes.

The adjustment of the model to data was attempted in a way highly reminiscent of classical crystallography with a scaling factor, plus a Debye–Waller factor for each of the three different $A3_{perp}$ volumes as free parameters. The relative Al/Cu compositions were also let free on the two A_{OR} and A_{ME} volumes in order to account for a possible unobserved weak order to be compared with that existing in the crystalline R-phase;[16] as the total concentrations of Al and Cu are fixed, this is only one more parameter (five parameters to be fitted in total). Finally, the residual factors calculated as is currently done in crystallography are less than 0.1, which is acceptable. The Debye–Waller factors are about twice those previously measured in the R-phase[16] but still have reasonable values. The Al/Cu relative concentrations also have little influence on the fit qualities. The retained values obtained from X-ray data are $c(\text{Al}) = 0.879$ for the mid-edge and 0.716 for the vertex sites, instead of the 0.84 value that would correspond to total randomness. Note that perpendicular Debye–Waller factors appeared to have no influence on the fit quality, suggesting that isotropic phason disorder is not a relevant property of the system.

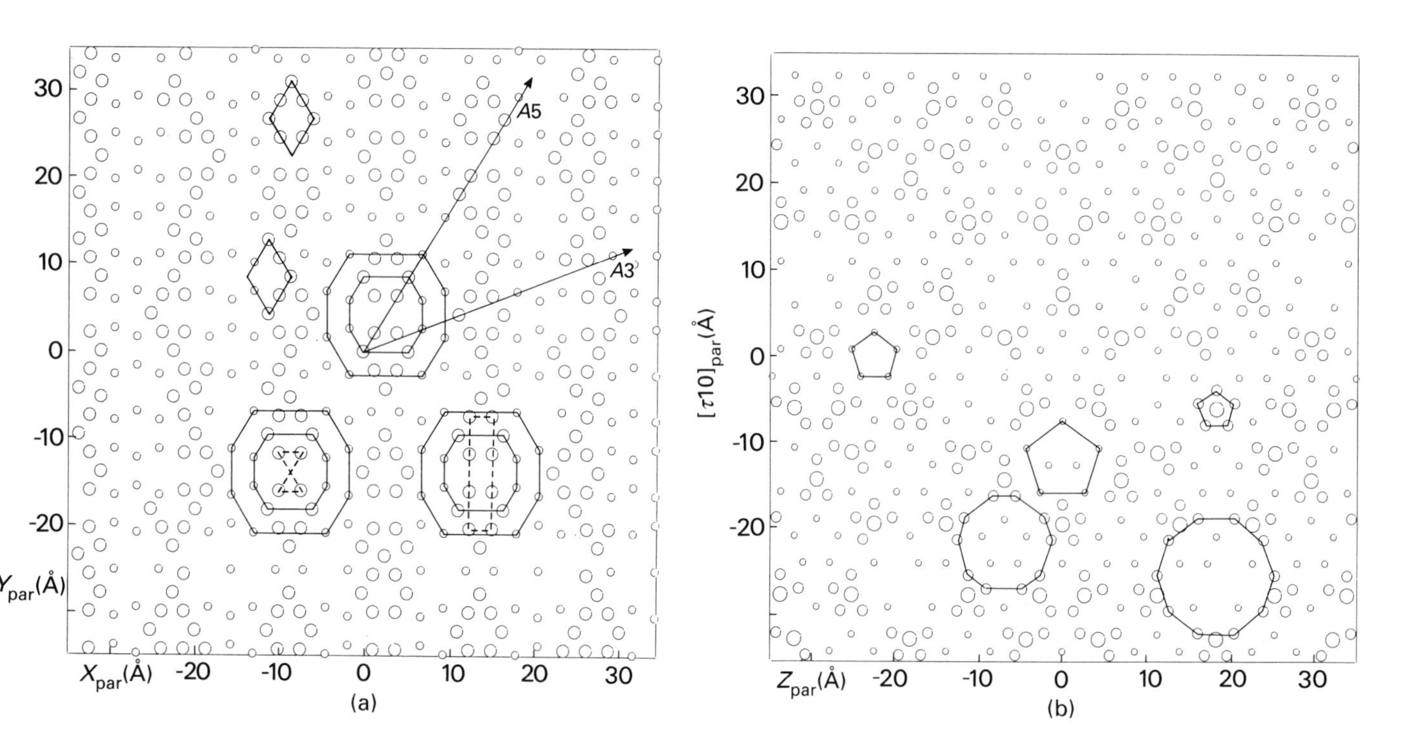

FIG. 4.11 Atomic planes of the 3-dim quasiperiodic structure for AlLiCu as obtained by a cut of the 6-dim model. Large and medium circles correspond to A atoms, and small ones to Li atoms. (a) Two-fold plan. Traces of large and small triacontahedra similar to those of the R-phases are shown. The two lozenges exemplify two different decorations of the faces of the Ammann rhombohedra. (b) Five-fold plan (composed of three different layers).

Tailoring the A_{perp} atomic object obviously implies many parameters (infinity?) to modify the size and shape, beyond the simple procedure just described for removing unphysical short distances. This point will be addressed in a forthcoming section.

4.2.2.4 Atomic structure in the 3-dim physical space Assuming that we are satisfied with the above refined 6-dim periodic structure, the problem is now to go back to the real 3-dim world. As already used in the previous section, the straightforward method of obtaining a list of atomic positions is to generate their three coordinates as the intersection of the 6-dim periodic structure by the physical 3-dim space. This can be visualized through a 2-dim map of the 3-dim density distribution, as exemplified in Fig. 4.11. In this figure, evidence for a certain atomic order is visible, such as polygons or atom rows, which can be attributed to cut profiles in 3-dim triacontahedral and rhombohedral arrangements of the atoms. Lists of atom coordinates and/or 2-dim density maps are very useful when calculations of properties (dynamics, electronic structure, etc.) are concerned.

A geometrical description of the 3-dim atom distribution may be attempted in three different ways:

(1) a comparison with the related crystalline phase (the R-phase in the case of interest) to determine if the same polyhedra occur in both structures;

(2) a comparison with decorated 3-dim Penrose tilings (or other tilings);

(3) a description in term of sequences of atomic planes.

The structure of the R-phase[16] belongs to the *Im*3 (bcc) space group, with a lattice parameter of 13.9056 Å. Atomic arrangements within the cubic cell are represented in Fig. 4.12. The Al, Cu, and Li atoms are distributed over successive shells around the origin (icosahedron, pentagonal dodecahedron, small triacontahedron, truncated icosahedron) which form the so-called 'Samson complex' containing 104 atoms.

The structure of R-Al_5CuLi_3 can then be described as a CsCl-type packing of distorted Samson polyhedra linked in two ways:

(1) along edges of the cubic cell by sharing two aluminium atoms (site 12e);

(2) along the eight body diagonals of the cube by sharing a common hexagonal face of the polyhedra (site 48h). The remaining lithium atoms (site 12e) are found in the interstices formed within the Samson polyhedron packing. They cap the pentagonal faces of a truncated icosahedron. The 12e sites (Li atoms) are located at 24 of the 32 vertices of a 'large' rhombic triacontahedron of radius $r = 8.18$ Å. The eight remaining vertices coincide with the Li in 16f sites already considered in the formation of the underlying dodecahedral shell. The distorted

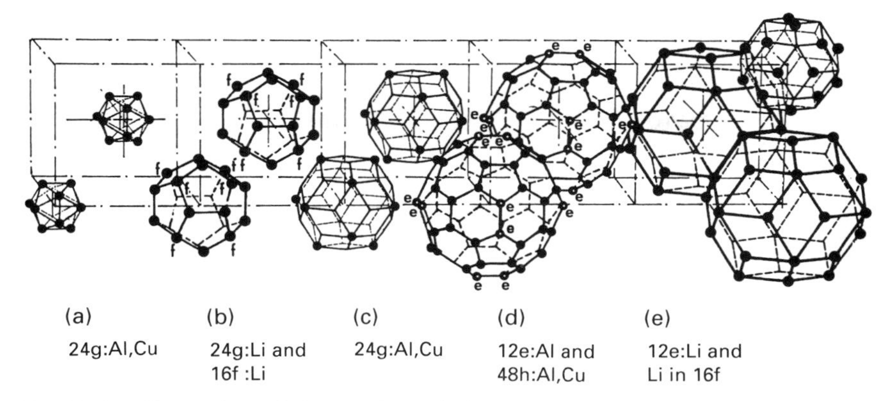

FIG. 4.12 Illustration of the R-Al_5CuLi_3 bcc structure: (a) icosahedra centred at the origin 000 and $\frac{1}{2}\frac{1}{2}\frac{1}{2}$; (b) dodecahedra; (c) the addition of icosahedra to the previous dodecahedra from the small triacontahedra; (d) distorted truncated triacontahedra; and (e) large triacontahedra which are connected either by faces along the ⟨100⟩ directions or through an overlapping volume defining small oblate rhombohedra along the ⟨111⟩ directions.

truncated icosahedra do not have a perfect icosahedral symmetry, which would have forced the atoms in 48h sites to emerge at the surface of the outer triacontahedral atomic shell. There are two triacontahedral shells (so-called 'small' and 'large' heretofore) with diameters in a ratio practically equal to the golden mean τ. All the Al/Cu atoms are in the shells of a 'Soccer ball' (small and large icosahedron plus external shell of the truncated icosahedron), while Li atoms are on the external shell of the large and small triacontahedra.

Owing to the similarities in density and composition, it is natural to wonder whether the structure of the ico-phase can also be described in terms of a network of Samson polyhedra, or of parts of them. The conditions to be fulfilled by the $A3_{perp}$ volumes for generating a given type of atomic cluster in the cut procedure have been analysed by several authors.[17–19] The basic principles are very similar to those used in the procedure of identification and elimination of the too short atomic distances and is schematically pictured in Fig. 4.13). Basically, an atomic distance will be present along a given direction in the 3-dim physical space if and only if A_{perp} atomic objects have a common cross-section when projected on the corresponding direction in the complementary space. A polyhedral arrangement of atoms, or a cluster, is defined by the atomic bonds (distances and directions) linking its centre to its outshell vertices. For instance, the presence of icosahedral clusters in the 3-dim structure will correspond to the existence of families of

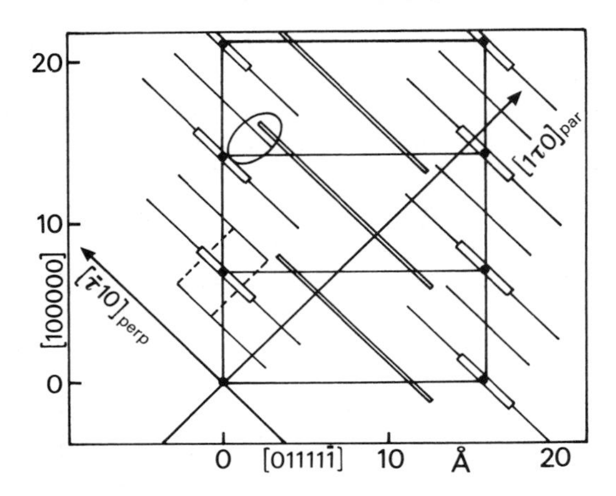

FIG. 4.13 The figure shows a schematic of a 5-fold axis slice of the 6-dim model structure, with cross-sections of the vertex, mid-edge, and body-centre volumes. The loop reproduces the example of too-short distances as detailed in Fig. 4.8. The dashed rectangle shows the acceptance domain for ME–ME distances through the central hole of the vertex volume.

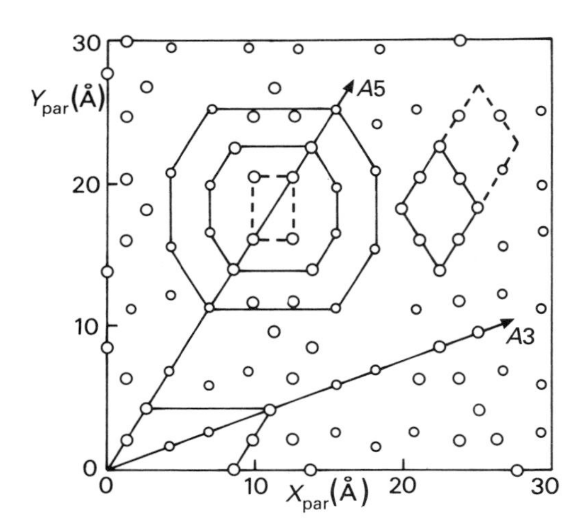

FIG. 4.14 The physical 5-fold axis $[1\tau 0]_{par}$ crosses the $A3_{perp}$ volumes at 3-dim atomic positions. The figure presents a slice of the 3-dim atomic density as deduced from the model (large and small circles are A and Li atoms, respectively). Cross-sections of small (large) triacontahedra and rhombohedral tiles are also shown.

12 equal atomic distances converging along 5-fold axes. In the 6-dim structure, this is equivalent to say that a given $A3_{perp}$ volume has 12 neighbours distributed along pertinent directions and in such a way that the cut procedure generates the proper atomic pairs. In the 6-dim structure model of the AlCuLi quasicrystal, one vertex A_{OR} volume is surrounded by 12 mid-edge A_{ME} volumes. The cut procedure generates A–A distances equal to 2.528 Å if the projected of the A_{OR} and A_{ME} into the perpendicular space have parts of their volumes in common. This is visible in Fig. 4.14. The 12 A_{ME}, having a small common volume projecting on the perpendicular space (Fig. 4.15), roughly a sphere of 1.65 Å radius, indicate that actually small icosahedra of A atoms will be found in the 3-dim atomic structure. This common volume is called the acceptance domain and gives, in particular, the occurrence rate of the corresponding clusters in the structure. As shown in Fig. 4.13, 4.15, and 4.16, the acceptance domains corresponding to the outer shell of the soccer balls are identical to that of a small icosahedron; both correspond to the common volume in perpendicular space of the 12 A_{ME} adjacent to a given A_{OR}. More generally, it can be demonstrated that all the shells present in a soccer ball have the same acceptance domain, though coming from a different association of $A3_{perp}$ volumes.

Thus, all the atomic shells typical of the R-phase structure are also found in the ico-phase, up to the so-called large triacontahedron. The cross-section at $z = 0$ of the 3-dim density distribution presented in Fig. 4.14 shows also very clearly the presence of the two (small and large) triacon-

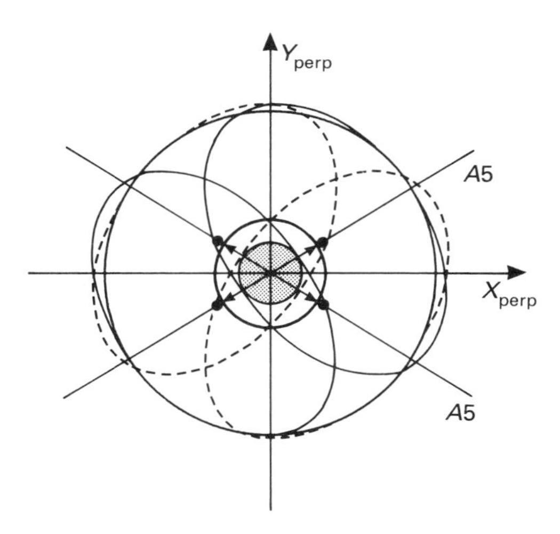

FIG. 4.15 Schematic view of the definition of the acceptance domain for small icosahedra as the common region between a vertex and 12 neighbour mid-edge $A3_{perp}$ when projected into the perpendicular space. This is a 2-dim cross-section.

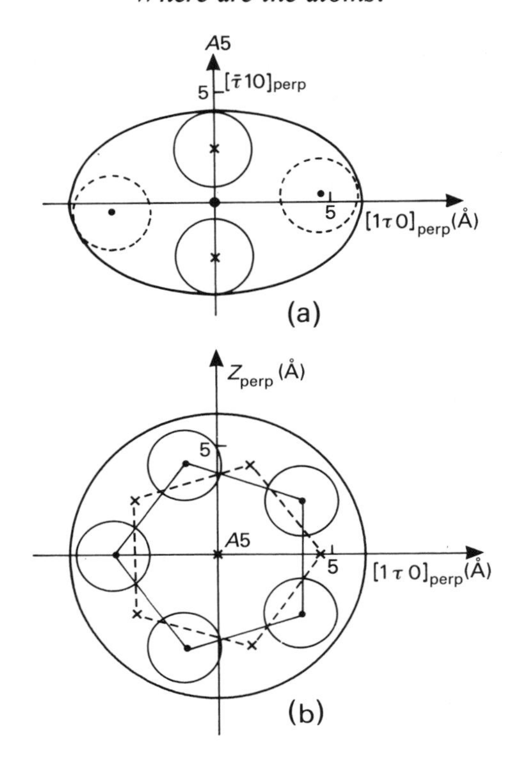

FIG. 4.16 Representation of the acceptance domains for small icosahedra (solid line circles) and for soccer balls (dashed circles) when projected on the mid-edge volume. (a) is a cross-section containing the 5-fold axis of the ellipsoid; (b) is a cross-section perpendicular to the 5-fold axis with two pentagons.

tahedra and the other icosahedral clusters. The acceptance domains may be finely faceted, but the spherical approximation is sufficient to provide at least an estimate of the proportion of Al/Cu atoms within the soccer ball clusters. This proportion is found equal to only 28 per cent, while soccer balls contain all the atoms of the R-phase structure. Only 7 per cent of the Li atoms are also found in these clusters. These soccer balls are weakly connected only along 3-fold axes by having hexagonal faces in common; this is at variance with the (distorted) soccer balls of the R-phase which share additional Al atoms along 2-fold axes. The drawing shown in Fig. 4.16 illustrates that the acceptance domain of the soccer balls cannot be enlarged significantly; the limitation comes from the size of the equatorial circle of the ellipsoidal A_{ME} volume which, in turn, cannot be larger if too short A_{ME}–Li_{BC} distances have to be avoided. On the other hand, the acceptance domain for the small icosahedra and small triacontahedra could be enlarged without too much inconvenience by elongating the ellipsoidal A_{ME} volume in its 5-fold directions. This idea of completing the structure by adding small triacontahedra has been suggested for designing an atomic model in 3-dim and will be described later in this chapter. At this level, the 6-dim structure cannot very easily lead to a more detailed comparison between the ico- and R-phase structures.

Looking now into the cut 3-dim structure (see Fig. 4.11), with the view of a confrontation with a decorated Penrose tiling (PT), is also interesting. In Fig. 4.11, images of rhombohedral tiles are indeed visible. The Al atoms generated by a cut of the 6-dim vertex volume are sited on 3-dim vertices of the 3-dim PT, but owing to the central hole of the A_{OR} volume and its external size, some vertices are unoccupied. The Al atoms generated by a cut of the mid-edge volume are also in mid-edge positions in the 3-dim PT, but again with partial occupancy only. The Li atom positions are generated mostly on the triad axis of the prolate rhombohedra, in a $\tau/1/\tau$ partition, but also occasionally on edge positions at 1.93 Å from unoccupied vertices. Thus, when the structure is assumed to be 3-dim PT-like, the decoration

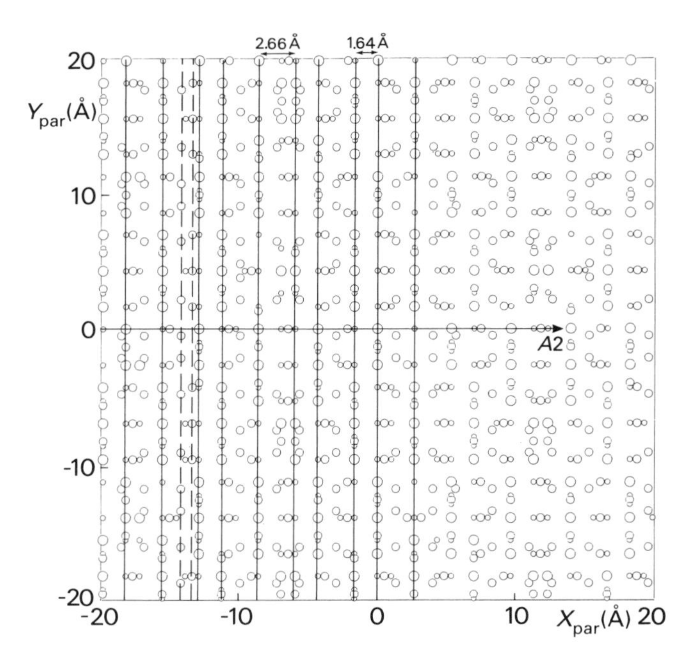

FIG. 4.17 Structure of the 3-dim density as described by families of 2-fold atomic planes as obtained from projections of atom positions contained into a 40 × 40 × 40 Å cube. A_{OR}, A_{ME}, and Li_{BC} atoms are shown as large, medium, and small open circles, respectively. Solid and dashed lines exemplify denser and less dense planes.

of the tiles is not unique. This somewhat reduces the relevance and the usefulness of such a description for the structure of interest.

Finally, there is the alternative way of visualizing the 3-dim structure in terms of 'rough atomic planes'. Such a description can be obtained rather easily from an appropriate physical cut and projection of the 6-dim periodic structure. The point is exemplified in Fig. 4.17, showing a family of atomic 'planes' perpendicular to a 2-fold direction. Two different average repetition distances, namely 1.643 and 2.658 Å in a ratio of $1/\tau$, and arranged into a Fibonacci sequence, are observed. Their average d-spacing corresponds to one of the strongest diffraction peaks, i.e. $(N, M) = (20, 32)$. The 2-fold planes are also the densest ones and more distant from each other (as compared with 3-fold or 5-fold plane families). They are formed by two different types of 'layer': (i) dense 'layers' with A atoms (originating from both vertex and mid-edge volumes) and Li atoms; and (ii) less occupied 'layers' with either A or Li atoms alone. Three- or five-fold planes never contain the three types of atoms simultaneously. Such a description is reminiscent of a property of the simple 3-dim PT in which planes of atoms and columnar structures have been pointed out by Duneau and Katz.[17] This is also related to one of the procedures leading to quasiperiodic order, namely the dual multigrid method which allows us to build a quasiperiodic lattice by intersecting conveniently oriented families of planes in quasi-periodic sequences (see Chapter 1). Figure 4.17 shows clearly that the 'planes' do not have the usual flatness of those found in normal crystals; their fuzziness cannot be better characterized. As they are, these planes allow us to understand somewhat the observed morphology of the single (quasi)crystal AlCuLi (Fig. 4.1) along rules analogous to those of classical crystal growth. Indeed, the triacontahedral single grains which have been obtained have facets perpendicular to 2-fold axes and edges along 5-fold

FIG. 4.18 Scanning electron microscopy image of a dodecahedral single grain of the AlFeCu quasicrystal. The size is roughly 1 mm across (courtesy of M. Audier).

FIG. 4.19 Dodecahedral single grains of the AlCuRu quasicrystal as obtained by slow solidification from the melt (courtesy of Professor H. U. Nissen, ETH, Zürich).

axes, the densest atomic planes and rows. This is what is normally expected for a periodic crystal. Accordingly, it may be conjectured that AlSiMn quasicrystals would like to grow into dodecahedral and/or icosidodecahedral grains because of a different atomic decoration which produces highest density in 5-fold or 3-fold atomic planes.[20, 21] These 5-fold and 3-fold atomic planes should be also the densest ones for AlFeCu[22] and AlPdMn[23] quasicrystals, since their single grains have been observed in the form of regular pentagonal dodecahedra (Figs. 4.18 and 4.19).

4.2.3 An insight into the experimental determination of AlMn-like quasicrystal structures: an example of parallel components in the atomic surfaces

AlMn-like quasicrystals, the first quasicrystalline systems to be discovered,[24] have also been the subject of the first achievements in structural investigations.

Pioneering works by Cahn *et al.*[12, 13] have actually founded the basis of experimental quasicrystallography. They measured X-ray and neutron powder diffraction patterns and produced the first 6-dim Patterson functions of a quasicrystal, demonstrating how simple the QC structure is when described in a proper high-dimensional space. The (total, not partial) 6-dim Patterson functions gave them information on the position and crude extension of the atomic surfaces in the 6-dim cube. The shapes were taken arbitrarily as spheres or spherical shells, whose sizes and main features were determined by comparison with a periodic approximant, conveniently lifted into the high-dim space (see Section 4.3.3).

Contrast variation has also been applied to the AlMn system, specifying more clearly the shapes, sizes, and positions of the atomic objects in the 6-dim cube.[15,20] The main drawback with this system is that, in the absence of a single grain of sufficient size, the structure derivation must rely entirely on *powder* diffraction. This may produce isotropic smearing of the deduced $A3_{\text{perp}}$ volumes and makes difficult direct reasonable reshaping when modelling. Additional constraints must then be introduced, as suggested by Duneau and Oguey,[18] who solved the problem by building $A3_{\text{perp}}$ volumes which (i) discard unphysical short distances and (ii) force the largest possible fraction of the structure to show up into icosahedral clusters. In this approach, the $A3_{\text{perp}}$ volumes have to be generated using three basic elements:

(1) a triacontahedron TR, whose centre to vertex distance along the 5-fold axis is 7.44 Å (if $a = 6.5$ Å for the 6-dim cube). Note that a convenient 3-dim physical cut of TR gives a regular Penrose tiling (see Problem 3.16);

(2) a volume called A obtained by intersecting a τ^3 inflated TR with a regular dodecahedron of 5-fold radius equal to 9.74 Å ($= a\tau^3/2\sqrt{2}$),

(3) a volume sA which is a τ^3 deflated A volume, almost spherical, with a radius of 2.3 Å.

Duneau and Oguey[18] have proposed a model which is not consistent with diffraction data because they placed an $A3_{\text{perp}}$ volume at BC positions for Mn atoms. But the good modifications of their model are obvious:

(1) a TR volume at the origin for T atoms;

(2) an A volume with a TR empty core also at the origin for Al atoms;

(3) an sA volume at the body-centre of the 6-dim cube for Al atoms.

From such a 6-dim structure, an 'average' Penrose tiling can again be described (4.6 Å edge) with Mn atoms at the vertices, Al atoms mainly on the long diagonals of the rhombohedron faces ($1/\tau$ partition) and on the triad axis of the prolate rhombohedra ($1/\tau/1$ partition). But again this is rather meaningless, as the decoration is not unique and changes from tile to tile.

A further weak point of such models is their relatively poor ability to reproduce the measured integrated intensities of the Bragg peaks. This is illustrated in Fig. 4.20, where calculated intensities are compared to the ones measured for the two extreme values of the neutron contrasts used. Mismatching of model to data is also quite visible when comparing the

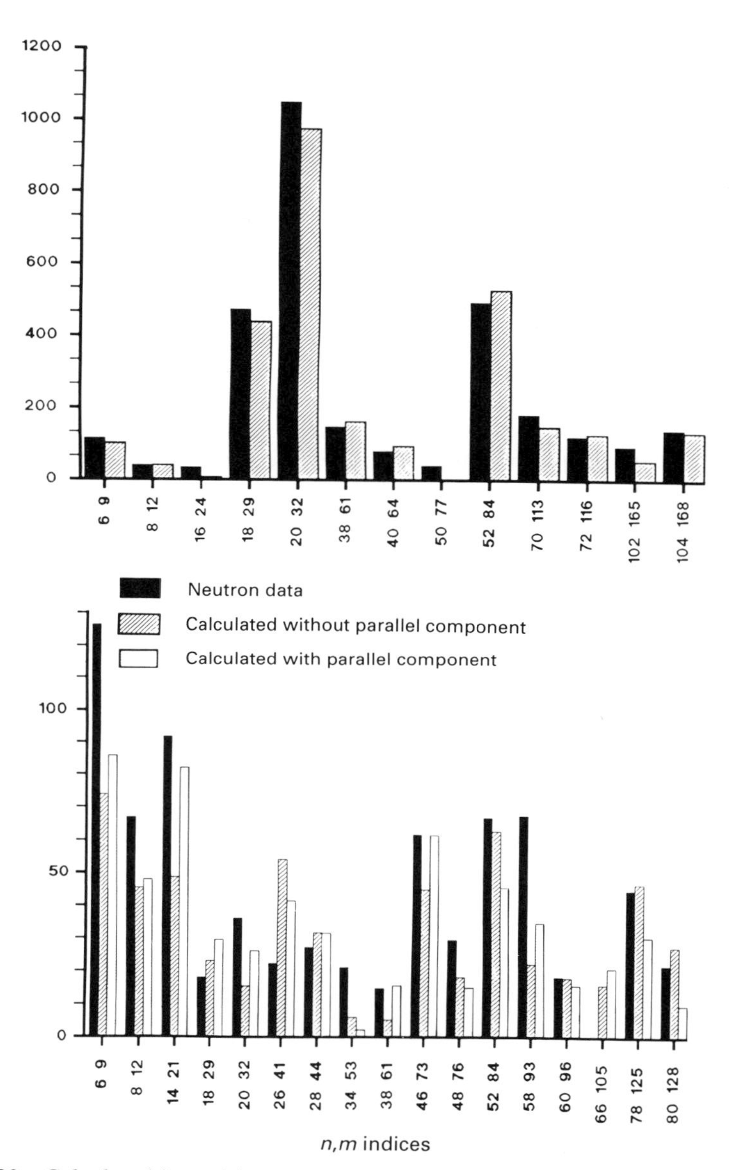

FIG. 4.20 Calculated intensities compared to integrated intensities of neutron diffraction peaks measured with quasi-crystalline alloys: $Al_{74}Si_5(FeCr)_{21}$ ($b_T = +0.658 \times 10^{-12}$ cm, similar to X-ray contrast with $Al_{74}Si_5Mn_{21}$) and $Al_{74}Si_5Mn_{21}$ ($b_T = -0.373 \times 10^{-12}$ cm) (top and bottom parts, respectively). The calculated intensities have been obtained within a spherical approximation with or without parallel components, as explained in text. In the top part of the diagram measured and calculated intensities compare reasonably and are not very sensitive to the parallel component. In the bottom part, the agreement is globally poorer but improves when parallel components are introduced.[15]

pair distribution functions, as shown in Fig. 4.21. Any attempted models using spherical or non-spherical $A3_{perp}$ volumes, with a variety of permitted slight modifications, have reached a comparable degree of failure when compared with neutron diffraction data for the AlMn system. Thus, a fundamentally different ingredient seems to be necessary if improvements are to be expected. This ingredient is actually related to the $A3_{perp}$ not being flat in the complementary space (see Chapter 3). In particular, if the physical cut of the TR gives a regular Penrose tiling, digging holes here and there in the TR will not split the vertex to vertex distances as suggested by experimental data (Fig. 4.21). By contrast, even very simplistic changes in the $A3_{perp}$ volumes have deep influences, as long as parts of these volumes are shifted in the physical space. For instance, let us take the TR(Mn) volume in which 12 sA small volumes are identified by the positions of their centres, namely $[a/\{2(2+\tau)\}^{\frac{1}{2}}]\{(\tau,1,0)_{perp}+(1,\tau,0)_{par}\}$ and equivalent vectors, with respect to the origin. Let us now displace these sA volumes along their 5-fold axis in the physical space such that their centres are now at vectors $(\tau,1,0)_{perp}+(1+\alpha)(1,\tau,0)_{par}$, without modifying the rest of the TR. The $A(Al_0)$ empty shell can be modified similarly by shifting 30 small spheres of volume sA along the 2-fold axes in the 3-dim space. The obvious effect of such parallel components is to expand ($\alpha > 0$) or contract ($\alpha < 0$) the tile size, without altering the symmetry or destroying quasiperiodicity. Adjusting α to fit the data now gives a better agreement between model and experiment (see Fig. 4.20 and 4.21). As explained in Chapter 3, such parallel components are very difficult to specify in detail straight from the diffraction data. They may even be completely washed out into background and truncation effects. The prospect of using maximum entropy methods may improve the situation. Globally, AlMn-like quasicrystals are probably poorly ordered materials, and quasicrystallography methods may be relevant only to their average structure.

4.2.4 Structures of the perfect quasicrystals of the AlFeCu and AlPdMn families

The most promising quasicrystals for the purpose of diffraction studies are certainly those of the families AlCuM (with M = transition metals)[26, 27] or AlPdMn[23] because they can be grown as perfect large single grains (Fig. 4.18 and 4.19). According to reported data,[28–30] the structure of AlFeCu in 6-dim space is again very simple, but now belongs to the $Fm\bar{3}\bar{5}$ space group, at variance with the AlCuLi and AlMn cases. It can be described as a superstructure of a primitive lattice, with three atomic objects located at special points with full icosahedral symmetry of the F-lattices. These points are the two non-equivalent nodes of the underlying primitive lattice plus one of the two non-equivalent body centres, the other one

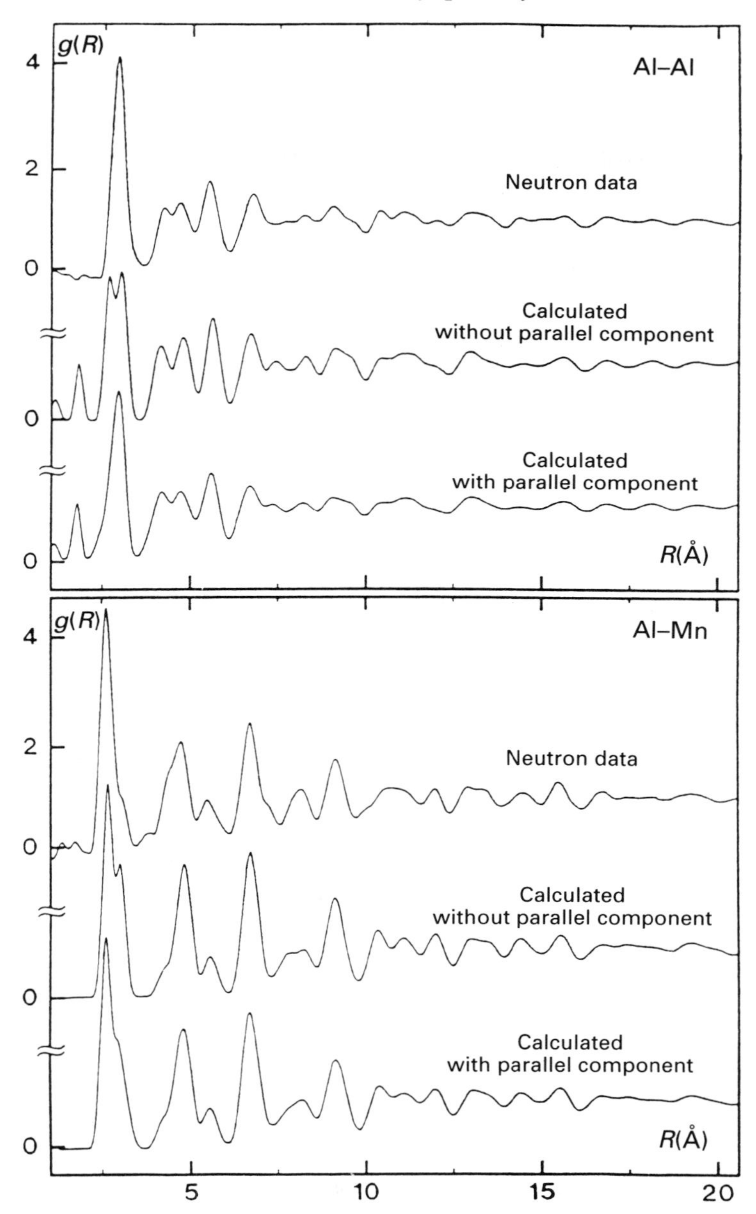

FIG. 4.21 Partial pair distribution functions of the icosahedral $Al_{74}Si_5Mn_{21}$ phase measured directly and calculated within a spherical approximation for the $A3_{perp}$ volumes (see Section 3.3 for details). Agreement between data and simulation is obviously improved by a parallel component effect, which especially reduces spurious splitting of the pair distances around 2.5–2.8 Å.[15]

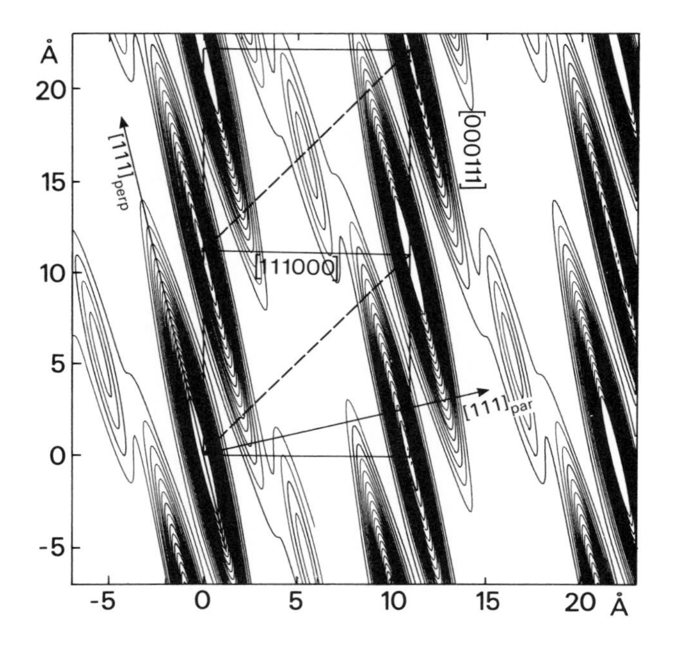

FIG. 4.22 Patterson function of the $A3_{perp}$ volumes obtained from neutron diffraction with an AlFeCu quasicrystal.[31]

being empty. This is clearly visible on experimentally determined Patterson functions[30] (Fig. 4.22) and the A_{perp} density profiles (Fig. 4.23).

Neutron diffraction data with double-isotopic substitution using two isotopes each for Fe and Cu have been measured with six samples of $Al_{62}Cu_{25.5}Fe_{12.5}$. Chemical order into the 6-dim periodic image has been specified.[31–33] If the two non-equivalent sites of the underlying 6-dim simple cubic lattice are denoted n_1 and n_2 and the occupied body centre is denoted bc_1, this chemical order results in a distribution of Al, Fe, and Cu into 'concentric shells' around n_1, n_2, and bc_1. Globally, the structural study of the AlCuFe icosahedral quasicrystal is an example of modelling the data using faceted polyhedra as atomic surfaces along the lines explained in Chapter 3. Suggested atomic surfaces[29,33] are among those shown in Fig. 3.27 (Chapter 3) and reproduced in Fig. 4.24.[33] The small volume at bc_1 is entirely due to Cu. Fe is concentrated in the interior part of the large volumes at n_1 and n_2, while Al is mostly distributed at the periphery of these large volumes and as a small fraction mixed with Fe near the centre of n_2. The rest of the Cu is mainly on n_2 and with a small amount on n_1 mixed with Al or in separated shells.

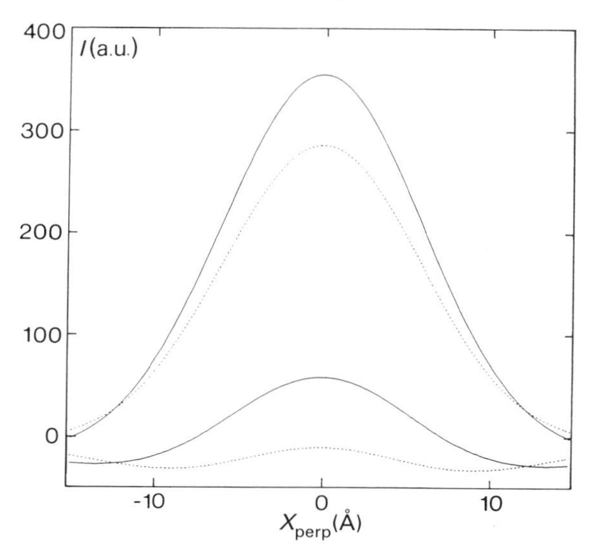

FIG. 4.23 Density profiles corresponding to the Patterson function of Fig. 4.22. Intense profiles correspond to lattice sites, weak profiles to body centre. Solid curves are for even sites, dashed curves for odd sites.

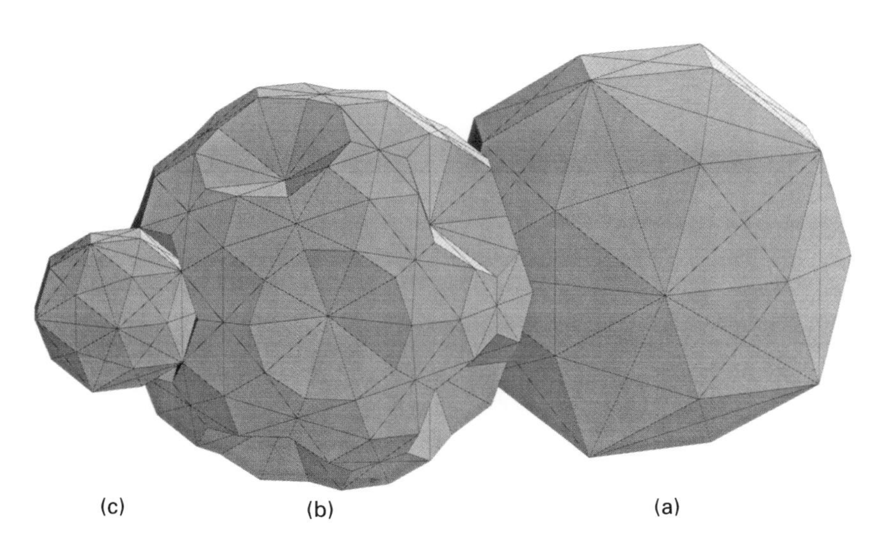

FIG. 4.24 Suggested atomic surfaces for the icosahedral AlFeCu system at sites (*a*) n_1, (*b*) m_2, and (*c*) bc_1[31–33] (courtesy of M. Quiquandon).

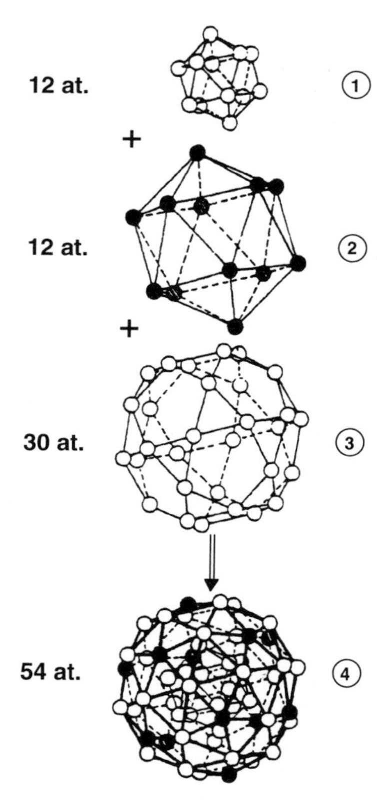

FIG. 4.25 Schematic representation of a Mackay icosahedron (MI) showing the successive 'shells' of atoms: a small internal icosahedron (1), a large icosahedron (2), and an external icosidodecahedron (3) combine to form the MI (4).

The contrast variations obtained via isotopic substitution are not very strong in AlFeCu, and this has imposed an obvious limit on the accuracy of structure specifications. It has not yet been possible to deduce a detailed 3-dim structure from an irrational slice of the above 6-dim image. However, it is clear that a large fraction of the structure is concentrated into packing of atomic clusters whose features appear to be reminiscent of the Mackay icosahedron (Fig. 4.25). Since the atomic surface at bc_1 is a small volume, it is a very favourable case for deduction of the structure from convoluted Patterson images.

The second family of perfect quasicrystals belongs to the AlPdMn system.[23] Its true composition is $Al_{70}Pd_{22}Mn_8$ for a density of 0.069 atoms $Å^{-3}$. We have seen in Chapter 2 that large single quasicrystals, in the centimetre range, can be grown at equilibrium using classical Bridgmad or

Czochralski procedures. Contrast variation can be produced on the Mn site by substituting the σ-FeCr phase and on Pd sites via X-ray anomalous diffraction.[34–36] The structure of the 6-dim image of the AlPdMn icosahedral quasicrystal also belongs to the $Fm\,\overline{3}\overline{5}$ space group, which can be viewed as a primitive cubic lattice with a superstructure induced by chemical order and atomic surfaces at n_1, n_2, bc_1, and bc_2 (labelling as for AlFeCu). Mn contributes to the interior part of the atomic surfaces at n_1 and n_2, with the occupied volume at n_2 being about 1.6 times as large as that at n_1. Pd contributes to the small atomic surface at bc_1 and to the atomic surface at n_2 as a shell around the manganese core. Al may contribute to a very small atomic surface at bc_2 and participate in the atomic surfaces at n_1 and n_2 as external shells. The chemical order is mainly due to the location of Pd at atomic surfaces sited on only half the occupied sites and, to a significantly lesser extent, to an uneven distribution of Mn and Al over these occupied sites.

The shape details of the atomic volumes have not been finalized so far, but features of the low resolution 3-dim structure have been deduced from the spherical approximation via the slicing procedure.

The existence of dense atomic planes in the 3-dim physical structure has been deduced from planar projection of the structure. The densest planes, characterized by large spacings, have been observed perpendicular to the 5-fold axis. These planes are slightly corrugated (rough) and arranged in quasiperiodic (Fibonacci-like) sequences. A systematic search of the very dense 5-fold plane showed that only a finite number of different types of atomic planes has to be considered. These different types of planes are shown in Fig. 4.26. Each type is characterized by a resonably well-defined local order and average chemical composition. Careful examination of magnified images of these 5-fold planes (Fig. 4.27) suggests interesting aspects of the structure in terms of the hierarchical packing of clusters. Rings of 10 atoms are clearly visible in Fig. 4.27. It can also be seen that these rings seem to combine nicely into similar inflated large rings of 10 elementary rings, with a scale factor of τ^3. A larger figure would show that this inflation repeats itself *at infinitum*. Examination of the whole 3-dim structure shows that the 10-atom rings are actually equatorial sections of atomic clusters called pseudo-Mackay icosahedra (PMI)[36] (the exact Mackay icosahedron is shown in Fig. 4.25). The elementary PMI combine into inflated PMI, which are τ^3 times as large, and so *ad infinitum* in such a way that, at any inflation step, the centres of the PMI are arranged in exactly the same fashion as the centres of the previous PMI generation, but scaled up by the τ^3 inflation factor. The elementary PMI are composed of an external shell of 42 atoms (12 vertices of an icosahedron plus 30 vertices of an icosidodecahedron) and an inner shell of eight or nine atoms which is a fragment of a dodecahedron. The external shell is very well defined from diffraction data, while the inner shell is only poorly specified.[36]

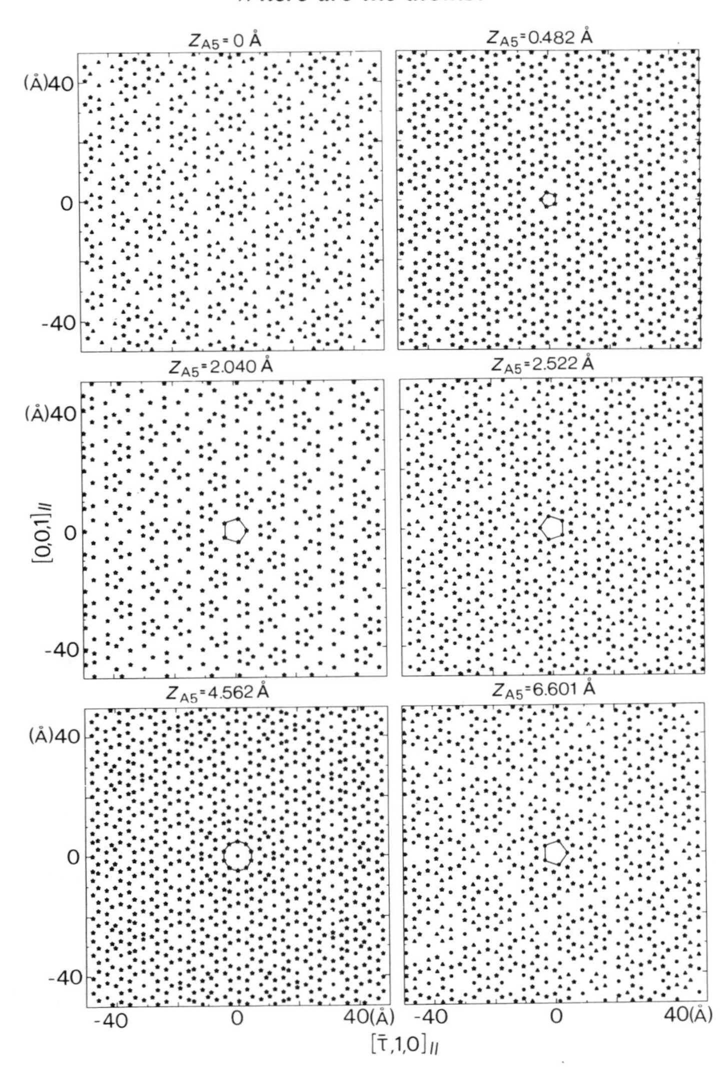

FIG. 4.26 General display of a series of six successive dense atomic planes perpendicular to a 5-fold axis in the structure of the icosahedral AlPdMn quasicrystal. Z_{A5} gives the intercept of the planes with the 5-fold axis. Atomic species are Al (*), Pd (▲), and Mn (●).

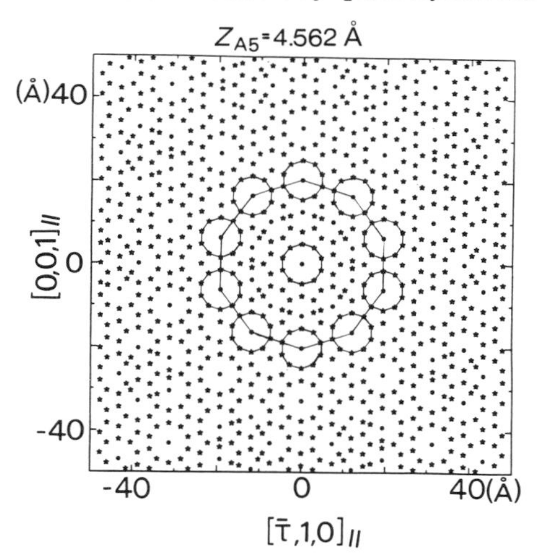

FIG. 4.27 One of the atomic planes shown in Fig. 4.26. Rings of 10 atoms and τ^3-inflated rings are outlined.

Two types of PMI have actually been found in the experimentally determined structure of the icosahedral quasicrystal AlPdMn. One is basically a regular Mackay icosahedron (except for the inner shell) with Mn atoms on the external icosahedron vertices and Al atoms elsewhere, except for some substitution of a few Mn atoms by Pd atoms. The other contains about 20 Pd atoms, and the remainder are Al. According to diffraction data, the atoms and subsequently successive PMI generations are connected along 2-fold and 3-fold bondings. Another experimental fact is the existence of 'connecting units', which are interpenetrating pieces of PMI. They are arranged in shells which have the same density as the PMI, obey inflation growth, and can be regarded as 'interfaces' between the PMI. These interfaces are quite thick and contain about 34 per cent of the total volume ($\sim 1\text{–}50/\tau^9$) to be compared with the 30 per cent of atoms of the clusters which lie on their surfaces, accounting for all inflation levels.

These experimentally determined structures have the following drawbacks: (1) the exact composition of each PMI is not known; (2) two types of PMI are either too many or not enough—a single type would have been compatible with inflated packing of 'average atoms' while three would have allowed the inflated arrangements of Al, Pd, and Mn atoms to be reproduced strictly. The third PMI has not been found either because it does not exist or because it has escaped investigation. The former explanation has been assumed, which means that, from the second generation of inflation onwards, only one kind of packing is available and hence the main feature of the structure of AlPdMn is a skeleton composed of τ^3-inflated PMI packing. This description is rather simplistic, but cannot be improved at present. Elementary PMI have a diameter of about 10 Å and the distance between PMI centres is approximately 20 Å. Obviously, a better description of the atomic surfaces in the 6-dim image of the structures would yield detailed modifications of the inferred 3-dim physical structure. However, the gross features which appear in the hierarchical packing of inflated clusters would certainly be retained.

A diffraction study of this AlPdMn icosahedral quasicrystal using high resolution anomalous X-ray diffraction close to the Pd edge strongly suggests that defining the atomic surfaces as faceted polyhedra may be not fully justified. Measurements with anomalous effects close to the Pd edge have allowed determination of the partial structure factor for the arrangement of the Pd atoms as if they were alone. The intriguing result of this measurement was a partial Pd atomic surface that was about twice as large as that expected from both the spherical approximation and the macroscopic constraints due to composition and density. In other words, the partial Pd shell could not be fitted between the Mn core and the Al external shell of the n_2 atomic surfaces as described above.

To account for this, it is necessary to inject some chemical disorder either produced by partial occupancy on the Pd sites or generated by phason fluctuations. The former is self-explanatory, but the latter needs further analysis. In the ideal model, the atomic surfaces in the 6-dim image are perfectly centred on lattice site. If random bounded position fluctuations in the perp space are introduced, long-range order of the structure is not affected, i.e. Bragg peaks are still observed but their intensities are affected by a decaying Debye–Waller-like term DW_{perp}. Thus the observed structure factors become

$$|F_{obs}| = |F_p| \exp(-B_{perp} Q^2_{perp})$$

where F_p stands for the perfect structure, without phasons, and corresponds roughly to the spherical model. Now, a plot of ln (F_{obs}/F_p) versus Q^2_{perp} should show a linear decay. This is actually observed, and thus demonstrates a very clear Debye–Waller-like trend with $B_{perp} = 1.2$ Å^{-2}. In effect, $B_{perp} = 1.2$ Å^{-2} is also obtained by fitting the F_{Pd} data to a modified spherical model when an overall DW_{perp} is introduced as a free parameter, with the important result that the usual weighted R factor improves strongly. A large DW_{perp} has also been directly measured[37] using high-resolution quasielastic neutron scattering.

The B_{perp} value obtained here corresponds to average fluctuations in perpendicular space of the order of 1 Å, which is small compared with the radius of the atomic surfaces (~10 Å). However, in the case of Pd n_2 atomic surface such a 'skin' contains as many as about 40 per cent of the atoms.

Such a large proportion of displaced atoms has dramatic consequences as far as the structural specification of quasicrystals is concerned. In particular, it is not clear whether a description in terms of a perfect quasicrystal convoluted with phason fluctuation is adequate, and the validity of atomic surfaces as faceted 3-dim objects may be questionable. As an example of non-faceted atomic surfaces, we have reported in Chapter 3 the cases of inflation-generated 1-dim and 2-dim quasiperiodic structures whose atomic surfaces have fractally shaped borders in their high-dim images. One may wonder here whether fractal atomic objects are likely to be revealed as a generic aspect of inflated structures. In any case, further experiments are needed to establish the exact origin of such large phason fluctuations. Indeed, phason fluctuations can be static or dynamic. Static phason fluctuation could result from surfaces being fractal objects or made of small disconnected pieces. Another possible source is random tiling structures (to be described later in this book) which are not likely to occur in a material whose structural perfection is revealed by the observation of dynamic scattering (see Chapter 2). The former does not induce diffuse scattering, whereas the latter does. Dynamic phasons are temperature dependent, while static phasons are not. The alternative to phason fluctuations, i.e. mere chemical disorder, has been ruled out by demonstrating that the atomic surface at the n_1 site is also affected by a DW_{perp} factor even though it is forced into perfect chemical order as it is mainly occupied by Al atoms (plus a very small amount of Mn atoms). This was achieved by measuring the total structure factor just below the Pd edge to increase the relative contribution of Al.

In conclusion, icosahedral quasicrystals show basically two types of periodic 6-dim images: primitive cubic or NaCl-like cubic. Quasicrystals whose structure is not perfectly ordered, e.g. AlLiCu or AlMn-like systems, belong to the former group. Thus the quasicrystallographic approach gives their average structures and the description of the sliced 3-dim physical structure remains rather vague. Perfect quasicrystals, such as the AlFeCu or AlPdMn systems, belong to the latter group. The NaCl structure then results from a superstructure induced into the primitive lattice by chemical-order effects. The low-resolution structures of these perfect icosahedral quasicrystals are well understood and can certainly be described as hierarchies of atomic clusters. Questions about the details, with respect to the exact shapes of atomic surfaces in the 6-dim images, are still outstanding. However, enough is known to allow structures and properties to be modelled, at least in principle, at the level of their basis aspects.

4.2.5 Structures of decagonal quasicrystals

Apart from the icosahedral phases, very few quantitative structural investigations had been carried out on other quasicrystalline materials at the time of writing. Exceptions are the decagonal phases.[38–41] These phases are very interesting crystallographically because extinction in the electron diffraction pattern suggests that a 10_5-screw axis and a *c*-glide plane exist, giving a quite general high-dimensional space group. The decagonal phase presents periodicity in one direction and quasiperiodicity with a 10-fold symmetry in the orthogonal plane. Full periodicity is recovered in (5 + 1)- or (4 + 1)-dimensional spaces giving equivalent descriptions.

From the physical point of view the decagonal phases are also of interest because they concentrate the double aspect of periodicity versus quasiperiodicity which may facilitate comparison. Indeed, physical properties in any material are always strongly influenced by structure and chemical perfection, or rather by defects with respect to perfection. For instance, comparison of an icosahedral phase with one of its crystalline modifications, may result in false conclusions since an attempt is being made to relate two different samples. Conversely, a single sample of a decagonal phase offers a unique opportunity to measure and compare properly the crystalline property parallel to the 10-fold axis and the quasicrystalline properties perpendicular to this axis. For instance, the ratio of the electrical resistivities in these two directions in an $Al_{70}Ni_{15}Co_{15}$ decagonal phase is about 30 to 1,[42] and this is necessarily an intrinsic observation.

In addition to many metastable (disordered) decagonal phases such as $Al_{78}Mn_{22}$, an increasing number of stable, almost perfect phases have been obtained in various chemical systems: AlM(1)M(2) (M(1) = Cu or Ni, M(2) = Co or Rh)[26, 43–45]; AlPdM (M = Mn, Fe, Ru or Co).[46, 47] Single-crystal X-ray structural analyses of decagonal phases have been reported for $Al_{78}Mn_{22}$,[48] $Al_{65}Cu_{20}Co_{15}$,[39, 40] and $Al_{70}Ni_{15}Co_{15}$.[40, 49]

Most of the quantitative structure analysis of the decagonal phases have been carried out via 5-dim Patterson analysis followed by least-squares refinement of single-grain X-ray diffraction data, sometimes assisted by electron diffraction and high-resolution transmission electron microscopy. The perpendicular space is two-dimensional and the atomic surfaces appear as pentagons at the vertices of a 4-dim cubic lattice (Fig. 4.28). Generally, the physical 3-dim structures are quasiperiodic atomic layers stacked according to different sequences. For instance, the $Al_{70}Ni_{15}Co_{15}$ decagonal phase is composed of a single type of quasiperiodic planar arrangement of atoms (Fig. 4.29(a)), while the $Al_{70.5}Mn_{16.5}Pd_{13}$ structure shows an ABA stacking of two 'layers', one flat (Fig. 4.29(b)) and the other slightly puckered (Fig. 4.29(c)).

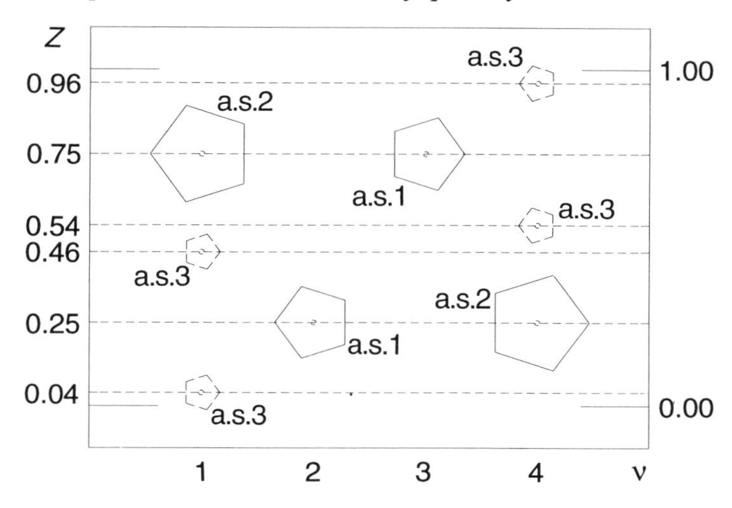

FIG. 4.28 Schematic presentation of the high-dim unit cell for the $Al_{65}Cu_{20}Co_{15}$ decagonal structure[41] (z is the periodic direction; ν determines the Wickoff positions). Eight atomic surfaces are visible; they belong to three different types (a.S.1, a.S.2, a.S.3) and are deduced from Patterson analysis of diffraction data (courtesy of P. Launois).

The description in terms of stacked layers may be inappropriate for two main reasons. The first is related to the growth morphology of the decagonal phases, which often show decaprismatic needle-like domains. Second, the interatomic distances are identical for atomic bonds within layers and between adjacent layers. Thus it may be better to describe the decagonal structure in terms of columnar clusters. These columns are parallel to the 10-fold axis and are formed by channels around a central channel. Their cross-section of about 30 Å corresponds to the diagrams in Fig. 4.29. The central channel is formed by the vertices of stacked pentagonal antiprisms. All triangle faces of the antiprisms are tessellated by slightly distorted tetrahedra that are face-connected pairwise. The central channel is surrounded by 10 more pentagonal antiprismatic channels with half their faces tessellated by terahedra. These peripheral channels are linked via edge-sharing tetrahedra. The centres of the columns occupy the nodes of a Fibonacci pentagrid with short and long distances d and τd respectively ($d = 30$ Å is the column diameter). This guarantees a self-similar hierarchical tiling of the space. An alternative description of the 3-dim decagonal structure has been proposed[50–52] in terms of binary tiling with interpenetrating clusters. There are two plane layers per vertical period in this binary tiling model, (but different sequences may be envisaged). Each layer has 5-fold symmetry, but the layer

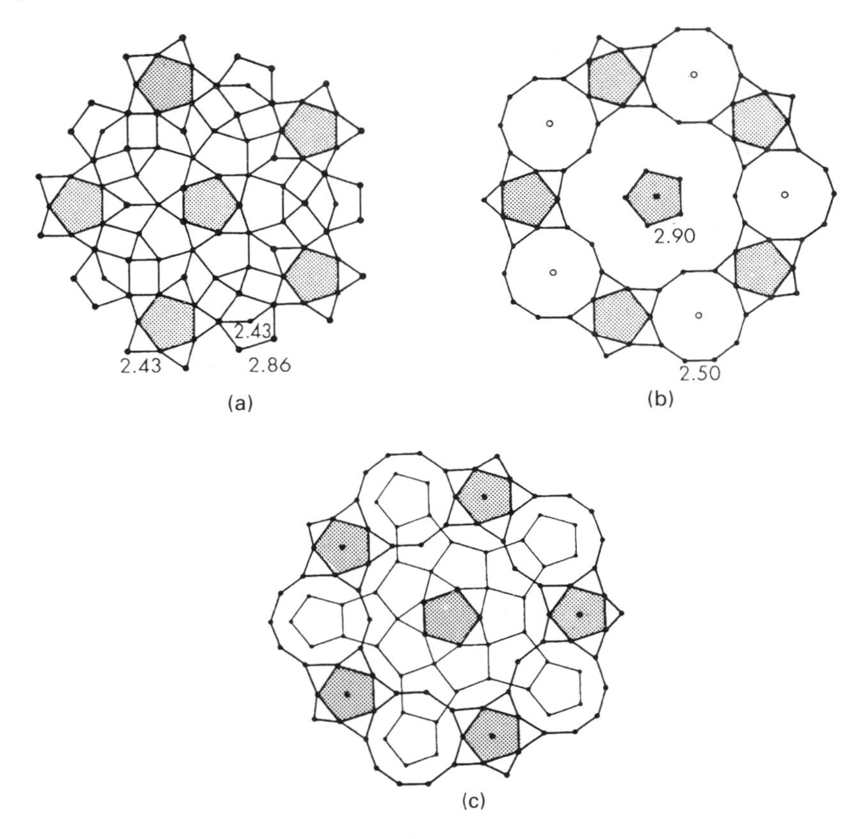

FIG. 4.29 Sections of a 30 Å diameter column in decagonal structures: (a) for $Al_{70}Ni_{15}Co_{15}$ at $x_3 = 1/4$; for $Al_{70.5}Mn_{16.5}Pd_{13}$ at (b) $x_3 = 1/4$ and at (c) x_3 from 0.065 to 0.115 (puckered plane). Distances indicated are in ångströms.[40, 48]

at $z = c/2$ is rotated by 36° with respect to the layer at $z = 0$, generating an overall decagonal symmetry. Each layer tiling is achieved using *one* building block, which is a decorated decagonal cluster of 10.3 Å radius, and *two* allowed overlapping connections, without either atomic displacements of any sort or 'glue' atoms. This is shown in Fig. 4.30 which also illustrates that such a structure is equivalent to a vertex decoration of two Penrose rhombi. The high-dim periodic image has been obtained by lifting the tiling into a 5-dim space. The atomic surfaces which are then obtained are not as simple as those deduced from diffraction data (Fig. 4.28). However, they preserve an overall pentagonal pancake aspect with subdivision faceting. They are presented in Fig. 4.31, and they occupy four different sites in the high-dim cubic image.

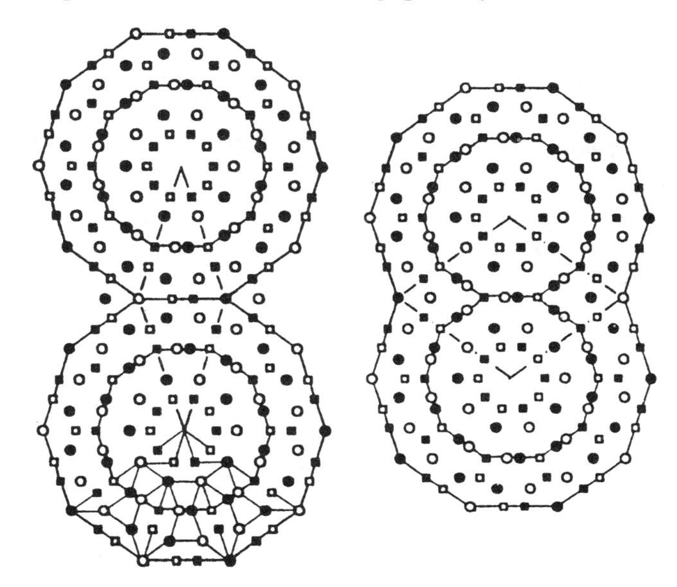

FIG. 4.30 Two allowed overlappings of decagonal clusters and corresponding tiles. A portion of the decagonal network hosting the atoms is shown: circles, Al; squares, transition metal; open symbols, $z = 0$; solid symbols, $z = c/2$ layer.[50–52]

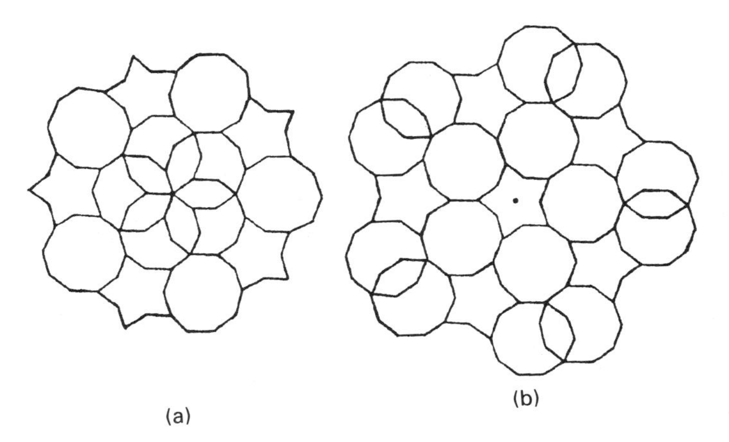

FIG. 4.31 Atomic surfaces of 'ideal' structure (no phason disorder): (a) transition metal; (b) Al. Decomposition to smaller polygons classifies local environments.[50]

Decagonal phases have many crystalline counterparts which frequently form microcrystalline twinned arrangements.[50–52] In principle, extinctions of the reflections $h_1\ h_2\ \bar{h}_2\ \bar{h}_1\ h_5$ with $h_5 = 2n + 1$ are typical of decagonal quasicrystalline order. They cannot occur with five or ten incoherently scattering crystalline twinned individuals with overall 10-fold symmetry except if the approximant structure also contains a *c*-glide plane.[41] The ability to differentiate a quasicrystal from a high-order approximant is also resolution limited since the diffraction resolution in reciprocal space ΔQ defines the sampling size in direct space ($\Delta R \approx 2\pi/\Delta Q$). For instance, measuring diffraction patterns with 0.001 Å^{-1} resolution allows a quasicrystal to be differentiated from its periodic approximants with lattice parameters less than about 6 000 Å. The main problem with the decagonal phase systems is doubt about their quasicrystalline perfection, in view of the observation of quite large amounts of diffuse scattering. They do not reach the quality of icosahedral quasicrystals, and samples are very often twinned microcrystals[41] or phason-perturbed quasicrystals.[44]

4.2.6 *Another way of solving the phase problem*

As described previously in this chapter, diffraction experiments give the intensities of the Fourier components of the structure. The determination of the structure by inverse Fourier transform requires solving the phase problem and it is not a trivial task. When contrast variations can be achieved this is of great interest and helps considerably.

For other cases, an astute technique has been developed by Jaric *et al.*[53] Since most of the quasicrystalline phases, like incommensurate structures, may have closely related 'parent' crystalline phases, one can try to reconstruct the missing phases of the quasicrystal via an optimization algorithm from the corresponding phases of the parent crystal. The correspondence between the two compounds can be pictured in their *n*-dim representation. The crystalline phase, or approximant structure, is obtained by applying a uniform (phason) strain in the perpendicular space to the *n*-dim lattice which is designed for generating a quasicrystal in the cut procedure. This has been exemplified by the one-dimensional Fibonacci structure in Chapter 3 (Fig. 3.17). The 2-dim periodic lattice is deformed (strained) in such a way that now the physical space R_{par} always intersects the atomic segments A_{perp} at the same two levels; R_{par} has also a rational slope with respect to the new rows of the 2-dim lattice. The tiles of the 1-dim structure are still of two sizes, but they are distributed periodically (*LS LS LS*. . .) instead of quasiperiodically.

In reciprocal space, the effect of a phason strain is to displace the spots of the quasicrystal and put them in coincidence with those of the crystal. Generally, several spots of the icosahedral phase match the position of a single spot in the approximant crystal, due to the lowering in symmetries. But if both amplitudes and phases of the 'distorted' quasicrystal are constrained to be equal to those measured for the approximant crystal, then the phases can be reconstructed unambiguously. The method has been applied by Jaric *et al.*[53, 54] to the case of the AlLiCu icosahedral quasicrystal with its *R*-phase. Using the same experimental data (single grain X-ray and neutron diffraction measurements) they obtain atomic objects decorating the 6-dim cube which are very similar to the ones directly deduced without the *a priori* assumption of crystal similarities from the inverse Fourier iterative procedure (Section 4.2.2).[9]

We conclude that the association of contrast variation effects with single crystal-like investigations allows good experimental determinations of partial structure factors, with their amplitudes and phases. The procedure for data analysis, though being worked out in *n*-dim space, is reminiscent of the early age of standard crystallography: Patterson functions are used to suggest a density model which is further refined up to convergence with diffraction data.

Three-dimensional structures are then derived, using the cut procedure. The straightforward approach is to build a 3-dim cluster of large size (say more than 100 × 100 × 100 atoms) and to list the corresponding coordinates for further use (electronic structure calculations, magnetic properties, atomic vibrations, etc.).

Alternative methods are the description in terms of atomic planes or cluster statistics. This kind of approach demonstrates that polyhedra typical of the related crystal structures are also basic elements for the quasicrystals. But these polyhedra alone are not sufficient to specify the whole quasicrystal structure, whose detail features are still to be analysed further. In particular it is not well understood so far why these polyhedra are quasiperiodically stacked. The description in terms of Penrose tiling or other space tiling schemes is not necessarily relevant. Its most dangerous aspects are a false simplicity and the false feeling that one can rely on an underlying (quasi) lattice. Assuming that the A_{perp} volumes of the *n*-dim structure can be designed properly, the whole information about atomic positions is actually contained in this higher space periodic description, within the usual very condensed aspect of the decorated unit cell. Unfortunately, we still do not know how to project this *whole* information (if possible!) into physical space. In the cut procedure, part of the structural information is unavoidably lost and remains hidden in 6-dim, resulting in limitations of size and type of clusters or atomic planes, unsatisfactory space tiling, etc.

It is interesting to note that all the structures of quasicrystals presently known are 'simple' structures where the *n*-dim atomic objects are located at high symmetry special points of the *n*-dim structure. This may suggest that further simplification would be gained by describing the structure in a space with even higher dimensionality (e.g. 10 or 12 instead of 6 for the icosahedral structures). The result might be that the A_{perp} atomic objects would merge into a single one, sited at the lattice node of the 10- or 12-dim unit cell. One possible advantage would be also to gain centrosymmetry (phase equal to 0 or π) when this is not achievable at lower dimensions. One good illustration of the above statement lies in the property of a canonical 2-dim Penrose tiling that can be equivalently generated either from a 4-dim space with five atomic objects located at special points of the 4-dim cubic lattice or from a 5-dim space with a single A_{perp} at the lattice node.

In a way, experimental quasicrystallography may be considered as less conclusive than normal crystallography. Actually, refinement procedures cannot apply to the many parameters that would be needed to specify all details of the high-dim atomic volumes. However, the approach to improvement is well identified, and what is known already gives grounds for optimism about the structure and properties of quasicrystals.

Less generic approaches to quasicrystal structure have tried to describe their atomic packing directly in the physical 3-dim space. This is the purpose of the next section.

4.3 Three-dimensional atomic models

4.3.1 General statements about the 3-dim approach

The description of quasicrystal structures by their n-dim periodic counterparts is obviously generic and global. Long-range order easily propagates by a copy at each lattice site of an n-dim unit cell decorated by $(n - 3)$-dim dimensional atomic objects. Local atomic environments can be forced to be, for instance, close to that of related crystals, by properly tailoring the n-3 dim atomic objects. This also avoids unphysical arrangements, e.g. short atomic distances. In this procedure, quasiperiodicity is guaranteed. As another great advantage, the Fourier transform is easy, since it is derived directly from a periodic geometry. This makes the understanding of reciprocal space and the adjustment of models to diffraction data simple. The main disadvantage is the difficulty of describing the structure properly and completely in its 3-dim physical aspect.

This is why different specific approaches have been developed, all based on direct 3-dim space filling with atomic models. Some of these atomic models try anyway to maintain consistency with quasiperiodicity. They start with a 3-dim Penrose-like tiling and assume that elementary rhombohedra of the same geometrical kind are decorated the same way, irrespective of the way these rhombohedra are packed with respect to each other. It is thus possible to ensure local atomic environments close to those of related crystals and the 3-dim atomic structure is perfectly understood in direct space. Long-range order can be propagated by τ^n inflation ($n = 1, 3$), thus coping with this aspect of diffraction data for icosahedral quasicrystals. The interatomic distances are handled well in the model and density/ composition parameters are good criteria for testing the resulting structure. The main disadvantages are that quasiperiodicity is not guaranteed and the difficulty of Fourier transforming the model makes it difficult to fit diffraction data. At the two extreme ends of such 3-dim atomic models, the particular cases of periodic twinned structures and icosahedral glass have been proposed.

A number of experiments have ruled out proposals based on twinned cubic crystals with several hundred atoms in the unit cell.[55, 56] Actually, any set of lines in a diffraction pattern can be indexed within experimental accuracy if a sufficiently large cubic unit cell is used, especially when close packed structures are concerned. It is, however, instructive to consider the minimum size of the cubic cell required for consistency with the scale of the experimental patterns. For instance, the two lowest-order diffraction spots observed along 2-fold direction in the single crystal diffraction pattern of AlLiCu, measured with neutrons,[9] are located at 1.06 Å^{-1} and 1.71 Å^{-1}, that is, in a ratio 1.613. The best approximant is given by $21/13 = 1.615$. This implies a minimum edge length for the cubic cell of over 77 Å, containing of the order of 20 000 atoms. Twinning on this scale should certainly show up in converging beam electron diffraction measurements and/or in high resolution electron microscopy images. Anyhow, this is a very traditional crystallography view and can be directly studied with the usual background in the original papers.[55] The rest of this chapter will thus be devoted only to quasiperiodic tiling and 'random' tiling of the 3-dim space.

4.3.2 *Classes of 'quasilattice' and decorations of tiles*

Building a 3-dim atomic model for a quasicrystal structure requires two major ingredients: the choice of an atomic decoration for the elementary tiles and an algorithm to pack these tiles together. Let us consider again the 1-dim illustration of this statement. To build a chain, we must start by selecting unit cells, say two segments with length L and S ($L/S = \tau$) and atomic decoration of these cells, say one atom per unit cell. Now, we must choose an algorithm to generate the chain. Possible schemes are the following:

(1) a periodic algorithm, in which the sequence LS is repeated, resulting in the one-dim crystal

$$LSLSLSLSLS\ldots;$$

(2) a quasiperiodic algorithm in which we start with L and successive strings are built by deflating L into LS and S into L, i.e.:

$$LLSLSLLSLLSLSLLSLSL\ldots;$$

(3) a 'glass' algorithm in which a chain is grown by adding L and S tiles chosen at random from a pool;

(4) various 'random tiling' algorithms, in which some of the Fibonacci ingredients are preserved, thus generating some order in the 'glass'. For instance, it can be imposed that the number of L tiles is τ times as large as the number of S tiles; local rules can also be forced, ensuring for instance that never two S and/or never three L would occur in sequence.

Thus there are two basic classes of packing that may be considered.[57]

Class A: icosahedral glasses, or more generally a 'random' tiling. This is a network generated by putting together clusters of atoms, or structural units, having icosahedral symmetry and possibly atomic bonds or simply atoms in common (vertex, edge, or face connections are equally possible). No other constraint except preservation of bond orientational order is imposed. Early attempts to interpret the observed electron diffraction patterns of AlMn compounds were based on this scheme. A demonstration that enforcement of orientational order between icosahedra suffices to promote sharp diffraction peaks matching experimental widths has been proposed by Stephens and Goldman.[58, 59] They actually used numerical simulations. Fig. 4.32 shows an example of a 2-dim pentagonal glass. Missing bonds simply correspond to links that do not belong to any pentagon, but do not necessarily mark regions of weaker density. However, 3-dim models may exhibit loose packing and too large an amount of disorder, compared with experimental data. 'Random' Penrose tiling can be generated in a way that makes density and disorder acceptable. This is simply obtained by using the same two tiles as for a perfect Penrose lattice (Fig. 4.33), but with 'weak' matching rules instead of the very strict ones.[60] This kind of random tiling, which generates configuration entropy in the structure, has been invoked to interpret high temperature stability of the AlFeCu quasicrystal and to understand the growth process. This point will be addressed in more detail in Chapter 6.[61–63]

Actually, Penrose-like random tiling can be also generated using the n-dim representation of a 3-dim quasilattice. Any kind of disorder in 3-dim is related to defects in and/or displacements of the $(n-3)$-dim atomic objects in the n-dim periodic structure. Rough A_{perp} objects instead of smooth ones generate phason-like disorder and allow 'controlled randomness' in the cut 3-dim quasicrystal. A 1-dim example is pictured in Fig. 4.34, showing that the sequence of LS tiles can be disturbed by changing the length of some A_{perp} segments. Such a phason-like disorder can be accounted for by some sort of static Debye–Waller factor in the perpendicular

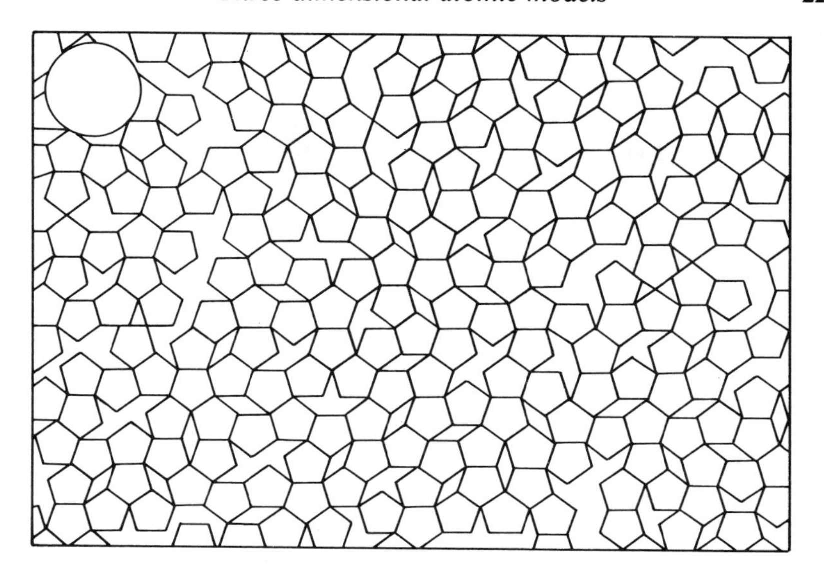

FIG. 4.32 Random tiling of a plane with regular pentagons and forced edge orientational order.

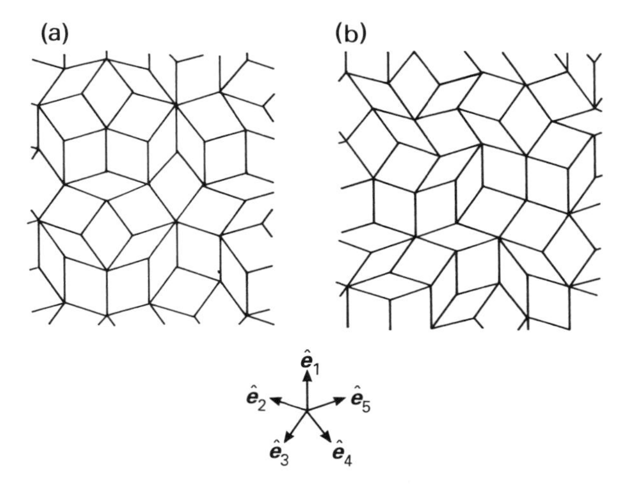

FIG. 4.33 (a) Perfect 2-dim Penrose tiling by rhombuses with 36° and 72° acute angles and (b) random tiling by the same tiles.

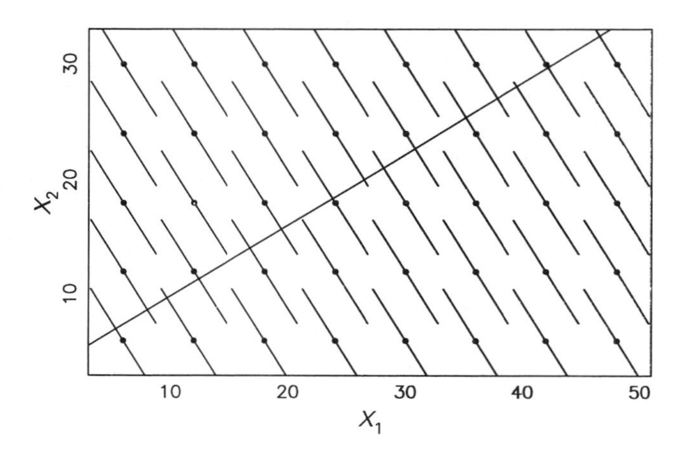

FIG. 4.34 The Fibonacci sequence of L,S tiles is perturbed when the 2-dim atomic objects are too large: phason-like translation does not preserve the atomic density.

space:[62] the sharpness of the Bragg peaks will basically be preserved but they will show 'wings' of diffuse scattering.[64, 65]

Class B instead must be coined 'perfect quasicrystals'. The network is now completely deterministic in its construction (strict matching rules, for instance, or deflation procedures). Nevertheless, the decoration of each specific tile may be context-dependent, in the sense that proximity rules may forbid the simultaneous occupation of two sites at too short a distance from each other. Most of the atomic decoration models proposed so far have been defined in this framework.

Of course, real quasicrystals are not simply tilings, nor even necessarily simply related to tilings. However, diffraction data have shown that almost all of them conform to the basic properties of ideal tilings. In other words, order in real quasicrystals can be considered as the major geometric properties. The presence of defects can be regarded as a perturbation of the ideal structure, in a way that is highly reminiscent of that which is well accepted for normal periodic 3-dim crystals.

Now, the decoration of the tile is always based on observed similarities or relations of the quasicrystal with some crystalline compounds of the same chemical system. The basic idea is that local structures or clusters must be kept the same, but reconnected in another way, with possible strain effects. The most popular experimental way to identify crystalline counterparts of a quasicrystal is electron diffraction measurement with the support of high resolution images.[66, 67] This was first used to demonstrate that AlMnSi quasicrystals grow in a coherent orientation relationship with the lattice of the cubic α-AlMnSi modification[68] (Fig. 4.35). An alternative approach is to measure the partial pair correlation function of the quasicrystal, when possible, using isotopic substitution for contrast variation in neutron diffraction, as is currently carried out for amorphous structures (see Chapter 1). Then a comparison with the same patterns, but for crystalline modification, gives direct insight into the expected similarities. In Fig. 4.36[14] are shown partial pair distribution functions for the ico- and R-phases of the AlCuLi system, giving obvious evidence for equivalent short-range atomic order.

Again the n-dim description gives further justification to this kind of approach. The so-called crystalline related phase of quasicrystals can indeed be viewed as a periodic approximant. This will be advocated further in the next section, along with a more quantitative definition of the periodic approximants.

4.3.3 The periodic approximants of a quasicrystal structure: basic definitions

Repeating here what has already been stated in previous sections, quasicrystals and crystals can be viewed as irrational and rational 3-dim cuts through an n-dim periodic structure respectively (see Section 1.4). This is illustrated in Fig. 4.37 in the simple case of 1-dim structures which are cut through a 2-dim square lattice. If the 1-dim space R_{par} passes by one and only one node of the square lattice, its slope is irrational and the 1-dim structure is quasiperiodic. If now several sites of the square lattice are intercepted by R_{par}, then the 1-dim structure is periodic with a periodicity given by the distance between nearest square lattice sites on R_{par}. For instance, if R_{par} has been drawn to joint the site (0, 0) and (n, 1), the period is obviously equal to $a(n + 1)^{\frac{1}{2}}$, a being the lattice parameter of the 2-dim structure.

Let us consider the particular case pictured in Fig. 4.37, in which R_{par} has a slope $1/\tau$. The quasiperiodic structure is the famous Fibonacci sequence, with two tiles (segments) of length $L = \tau = a\cos\alpha$ and $S = 1 = a\sin\alpha$. If now periodic structures are generated with proper cuts having slopes as close as possible to $1/\tau$, they are qualitatively reminiscent of the quasiperiodic structure. As shown in Fig. 4.37, this is the case for cuts passing by the sites (1, 0) or (1, 1) or (2, 1) or (3, 2) or (5, 3), etc. Interestingly enough, the slopes of the corresponding rational cut are the successive orders of the approximant number to $1/\tau$ (or τ equivalently). This is the origin of the 'periodic approximant' terminology. It is clear that the higher the order, the better the periodic approximant mimics the quasicrystal structure.

The successive approximants of a Fibonacci chain can also be generated

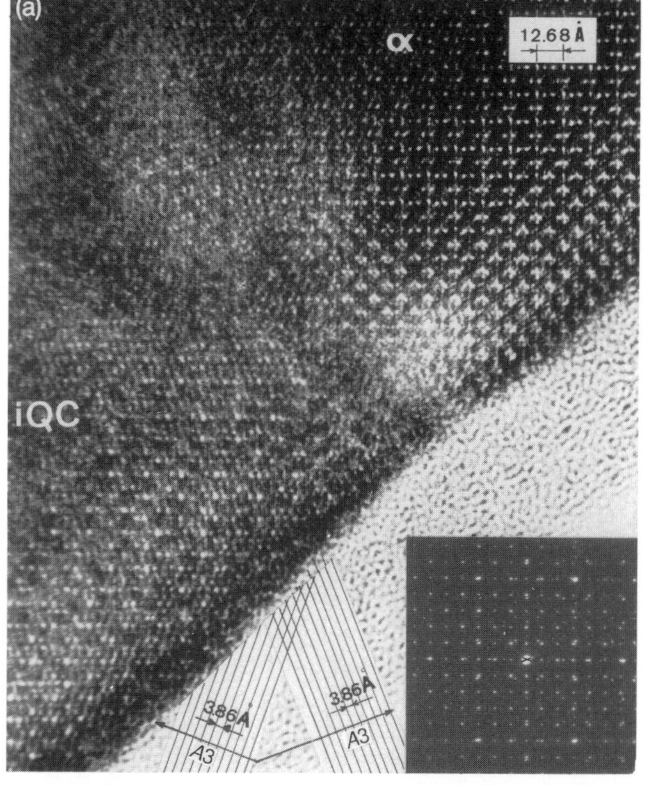

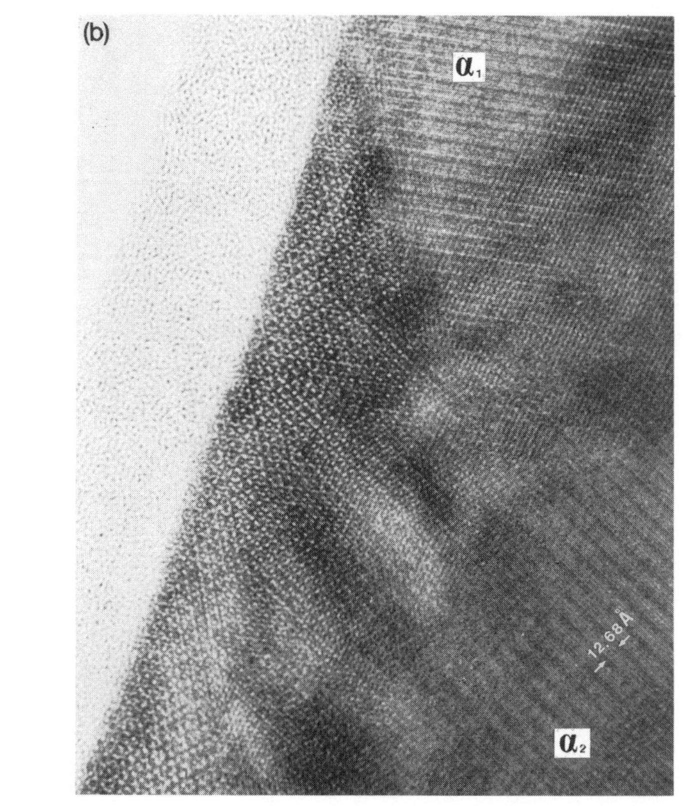

FIG. 4.35 High resolution electron microscopy micrograph of the α- and i-phases coexisting in as-quenched AlMnSi ribbons. (a) and (b) show orientational relationships $[100]_\alpha \| [A2]_i$ and $[503]_\alpha \| [A5]_i$, respectively[55] (courtesy of M. Audier).

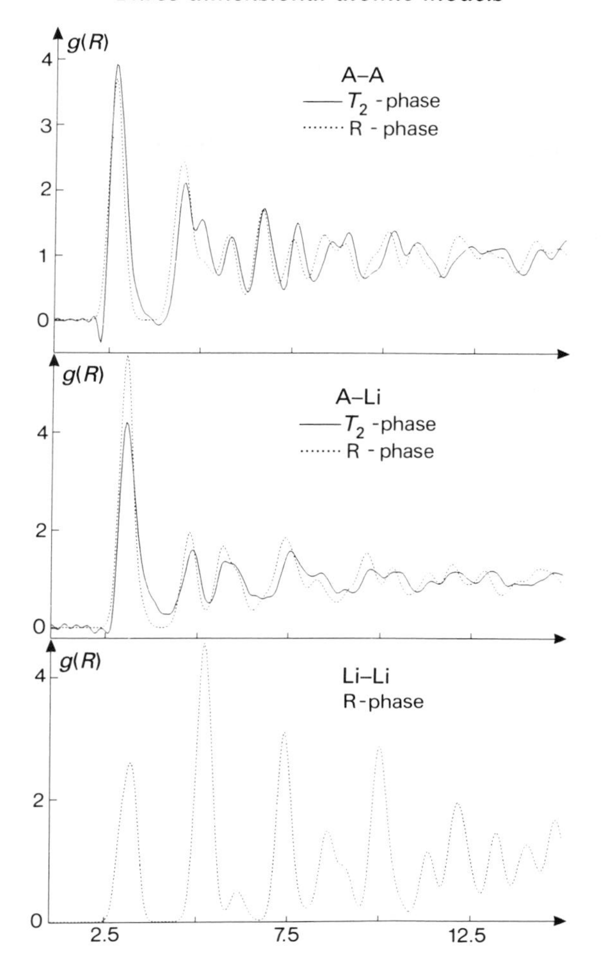

FIG. 4.36 Experimentally determined partial pair distribution functions of the icosahedral AlCuLi phase (—) compared with those deduced from the R-phase crystalline structure (····). The Li–Li correlations in the ico-phase have not been measured.

directly in the physical space. The starting point is the periodic chain, made of S segments:

$$SSSSSSSSS\ldots.$$

Then the 'deflation rule $S \rightarrow L$ is applied. The result is another periodic chain, made of L segments:

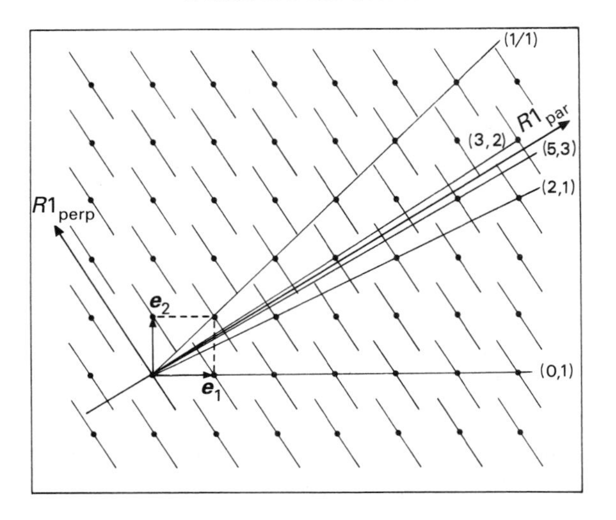

FIG. 4.37 A cut of a 2-dim square lattice with slope $1/\tau$ gives a Fibonacci chain (quasiperiodic structure). Cuts with slope 0/1, 1/1, 2/1, 3/2, 5/3 . . . give periodic structures with repetition lengths equal to a, $a\sqrt{2}$, $a\sqrt{3}$, $a\sqrt{5}$, $a\sqrt{8}$, etc.

$$LLLLLLLLL \ldots .$$

Now, the replacement rule $L \rightarrow LS$ generates the periodic approximant:

$$LSLSLSLSLSLSLS \ldots .$$

with period LS.

By applying both $S \rightarrow L$ and $L \rightarrow LS$ rules iteratively, successive periodic approximants with increasing periods are produced, e.g.

$$LSLLSLLSLLSLLSL \ldots$$

with period LSL, or

$$LSLLSLSLLSLSLLSLSLLS \ldots$$

with period $LSLLS$, etc.

As the period increases, the number of L segments with respect to that of the S segments also increases. The ratio of these two numbers follows the series 1/1, 2/1, 3/2, 5/3, 8/5, . . ., which is the series of rational approximants to the golden mean τ.

In the scheme pictured in Fig. 4.37, the 2-dim structure is kept the same while the 1-dim space is conveniently modified for generating the quasicrystal and its periodic approximants. This is somewhat unphysical, since R_{par} is supposed to be the real world and has been defined once for

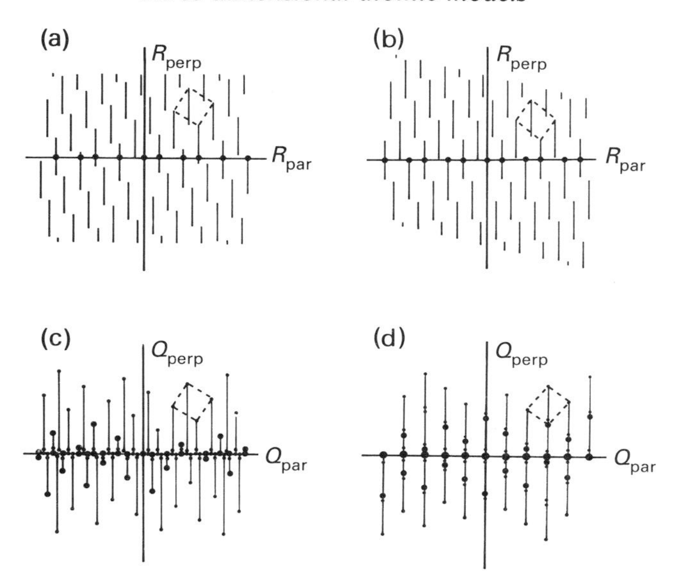

FIG. 4.38 (a) When cut by a straight line of irrational slope, the perfect 2-dim square lattice allows a Fibonacci chain to be generated. (b) A uniform shear strain brings two opposite corners of the 2-dim unit cell on to lines parallel to the physical space. The cut structure is still a packing of the same two tiles L and S but the sequence is periodic. Magnitudes of the associated 2-dim Fourier components are shown in (c) and (d) in the corresponding reciprocal space. As indicated, the 1-dim Fourier components are obtained by projection (redrawn from reference[54]).

all! The alternative physical scheme is pictured in Fig. 4.38. In this scheme, R_{par} is kept the same for the complete procedure, but the 2-dim periodic structure is typical of each of the generated 1-dim structures. The perfect square lattice that permits us to generate the Fibonacci chain (Fig. 4.38(*a*) is submitted to a uniform shear strain so that its site (H, H') is dragged parallel to R_{perp} (to coincide with a point in R_{par}) (Fig. 4.38(*b*). Now the resulting cut structure is periodic but has kept the same two tiles as the Fibonacci chain. Moreover, the period is obviously equal to the r_{par} coordinate of the node (H, H') in the perfect square lattice; that is

$$a_{rat} = H\tau + H'$$

This kind of uniform shear in the perpendicular space is a good example of 'phason strain', which generates 'absolute periodicity defects' in the quasicrystal structure.

To avoid unit problems, the period a_{rat} can be re-expressed as:

$$a_{\text{rat}} = \frac{a}{(2+\tau)^{\frac{1}{2}}}(H\tau + H'). \tag{4.5}$$

This comes from $\tau = a\cos\alpha$ and $1 = a\sin\alpha$ and thus $\tau^2 + 1 = a^2$ (with $\tau^2 = \tau + 1$). The periods of the successive approximants of the Fibonacci chain are:

$$\frac{a}{(2+\tau)^{\frac{1}{2}}}\tau, \qquad \frac{a}{(2+\tau)^{\frac{1}{2}}}(\tau+1) = \frac{a}{(2+\tau)^{\frac{1}{2}}}\tau^2,$$

$$\frac{a}{(2+\tau)^{\frac{1}{2}}}(2\tau+1) = \frac{a\tau^3}{(2+\tau)^{\frac{1}{2}}}, \ldots, \frac{a\tau^n}{(2+\tau)^{\frac{1}{2}}}, \ldots.$$

They correspond to successive τ-inflated periodic structures.

Similarly, periodic approximants of the icosahedral structures can be generated when convenient uniform phason shears are imposed on the 6-dim cubic structure. Calculation of the corresponding periods is a straightforward extension of the 1-dim case (see problems) and one obtains

$$a_{\text{rat}} = \frac{2a}{\{2(2+\tau)\}^{\frac{1}{2}}}(T\tau + S), \tag{4.6}$$

T/S being again the successive orders of approximant numbers to τ. This can also be expressed in terms of the successive powder of τ, i.e.

$$a_{\text{rat}}(n) = 2a\tau^n/\{2(2+\tau)\}^{\frac{1}{2}}. \tag{4.7}$$

If this formula is applied to the 1/1 approximant ($S = T = 1$) of the AlLiCu quasicrystal (lattice parameter a of the 6-dim cubic lattice equal to 7.15 Å, according to diffraction data) one finds $a_{\text{rat}} = 13.9174$ Å, to be compared to 13.9056, the measured lattice parameter of the crystalline R-phase.[16] All these various considerations are justifications for proposing 3-dim atomic models of quasicrystal structures based on the comparison with their crystalline periodic approximants. The procedure, though rather tedious, will be described below.

The method proposed by Jaric *et al.*[53] for reconstructing phases of the structures factors as deduced from diffraction data (see Section 4.2.4) is also a consequence of the above relation between a quasicrystal and its periodic approximants. In Jaric's method the comparison is made in reciprocal spaces, as pictured in the parts (*c*) and (*d*) of Fig. 4.38: the 'phason' shear allows us to shift the quasicrystal spots (Fig. 4.38(*c*) to coincide with the 1/1 approximant spots (Fig. 4.38(*d*)). Phases of the structure factors are then identified (the 1-dim Fourier components are obtained by projection).

Approximant periodic structures have also been used to constrain the atomic objects in the 6-dim structure of the quasicrystal.[69]

4.3.4 Examples of the 3-dim tiling model for icosahedral quasicrystals

The interested reader can find details about various quasicrystal models based on a comparison with the periodic approximant in the review paper by Audier and Guyot.[70] In this section we will restrict ourselves to the presentation of some atomic models for the AlLiCu quasicrystal.

In all cases, the starting point is the structure of the (bcc)-R-phase (Fig. 4.12) as summarized in Section 4.2.2.3 and which is an excellent 1/1 approximant of the quasicrystal. Thus the basic idea is to decorate the prolate and the oblate rhombohedra of a 3-dim Penrose tiling in such a way that the local atomic arrangements of the R-phase are preserved to a great extent. In this structure, both the Al and the Cu atoms are along 5-fold axes, basically distributed within icosahedral and triacontahedral shells. The Li atoms are along 3-fold directions, distributed on a dodecahedral shell and on a large triacontahedron. This can be somewhat preserved if the Al and Cu atoms are on vertices plus edge positions while Li atoms are on the triad axes (possible only for the prolate rhombohedron) of the elementary rhombohedra in a Penrose tiling. Such an atomic distribution is actually the basis of the Elser and Henley model.[71, 72] However, they proposed a slightly modified version of a 3-dim Penrose lattice in which the elementary tiles are the prolate rhombohedron plus a rhombic dodecahedron built up from two prolate and two oblate rhombohedra (Fig. 4.39). The advantage with respect to a canonical Penrose tiling is a better decomposition of the R-phase unit cell. Up to now, this model has not really matched the experimental diffraction data. The simple version, i.e. canonical Penrose tiling with the above decorated prolate and oblate rhombohedra, has indeed been compared to diffraction data.[73, 74] A rather good agreement has been obtained, as long as only X-ray data (single grain) are concerned and with the condition of accepting an unrealistic AlCu/Li chemical randomness. Such a randomness, in particular, has proved to be incompatible with the measured (neutron diffraction) partial structure factors. It seems clear now that it is not possible to fit the diffraction data with any decoration of the elementary Penrose tiles if this decoration is forced to be unique, as for the unit cell of a crystal.

Audier and Guyot[75] have tried to take advantage of the inflation properties of a Penrose lattice to push away the problem of a single decoration scheme for the tiles. Instead of considering rhombohedra with 5.05 Å edges (as deduced from diffraction data), they have proposed a decoration scheme for rhombohedra with $5.05 \times \tau^3 = 21.4$ Å edges. These rhombohedra are now large enough to accept, in place of single atoms, clusters of atoms such as the small and large triacontahedra of the R-phase (Fig. 4.12). The large triacontahedra are sited on the vertices and the small

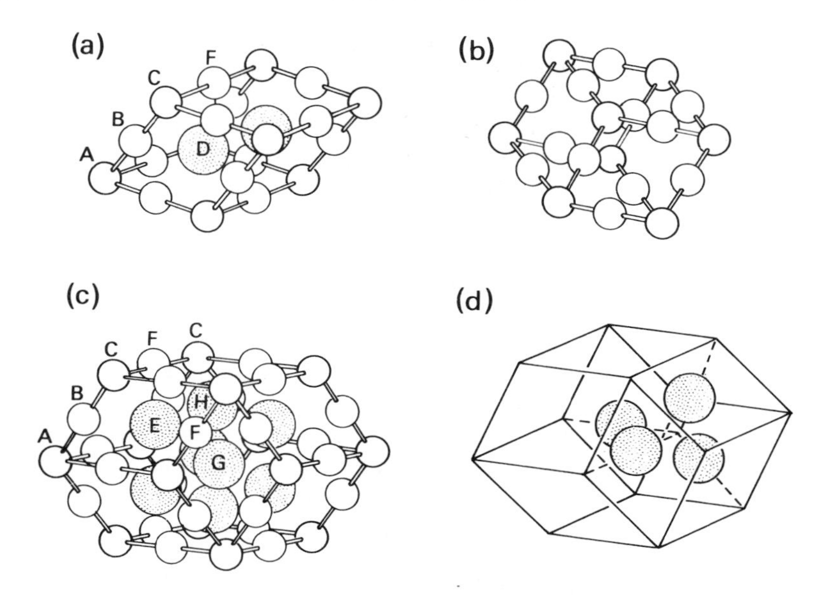

FIG. 4.39 The Henley and Elser model [72]—atomic decoration of the prolate (a) and oblate (b) rhombohedra. Al/Cu atoms (small open circles) are sited on vertices and mid-edge points; Li atoms (large shaded circles) lie on the triad axis of the prolate. The decorated rhombic dodecahedron is shown in (c) and in (d) (partially stripped to illustrate the internal unoccupied vertex with its four surrounding Li atoms).

ones on the faces (Fig. 4.40); there are actually 10 small triacontahedra in a prolate tile, positioned in 3-fold symmetry, four of which are internally tangent to the faces while the remaining six have their centres in the faces. The prolate rhombohedra contain the equivalent of one full large triacontahedron plus seven small triacontahedra. In the oblate tile, there are six small triacontahedra (equivalent to three in total) sited on three faces and tangent to the other three. A very interesting consequence of such a decoration scheme is the non-equivalence of the tile faces, which induces some sort of natural matching rules for generating the inflated Penrose lattice. The density of the model is consistent with measured values (between 2.45 and 2.47 to be compared with 2.45 $g\,cm^{-3}$), as long as Al and Cu atoms are added, in an *ad hoc* manner, between the large triacontahedra and their 30 nearest neighbours' small triacontahedra. This has the additional advantage of generating some icosidodecahedral aggregate, also present in the R-phase structure. Consistency of the model with diffraction data is acceptable (residual factors of 0.16 and 0.13 for single crystal data, X-rays, and

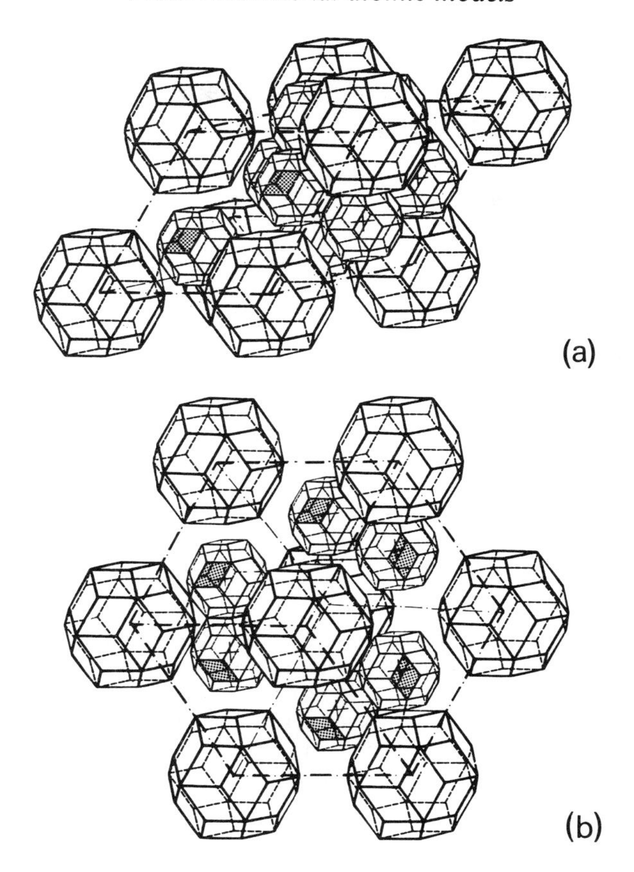

FIG. 4.40 Decoration of the τ^3 inflated rhombohedra[75] (prolate (a) and oblate (b)). Atom sites are not represented. Shadowed faces of the small triacontahedra are tangential to the rhombic faces of the tiles. The other small triacontahedra have centres on tile faces (courtesy of M. Audier).

neutrons respectively). The distributions of atom pair distances also compare quite well with measured patterns (Fig. 4.41 and 4.42).[76] However, it must be realized that in such a large model there are 459 atoms in the prolate tiles and 278 in the oblate ones. This may induce at least two drawbacks. Firstly, a procedure which proposes bigger and bigger golden units might be a blind alley in the sense that it does not allow for a clear distinction of quasicrystals from crystalline approximants. Secondly, we are facing here again the problem of giant unit cells that were discarded, reproachfully, for the periodic model as involving unphysically long-range interactions, and

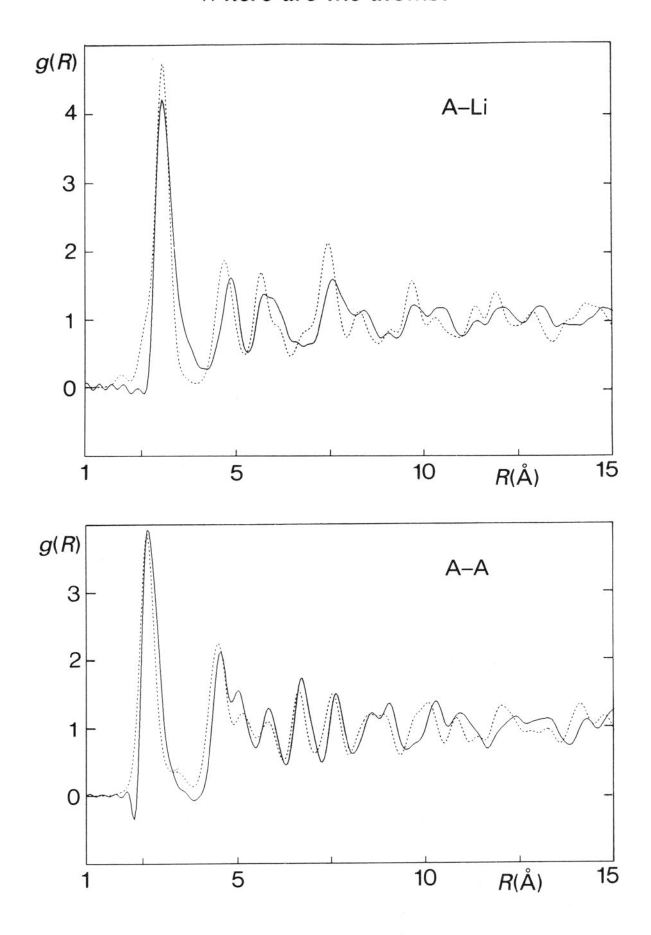

FIG. 4.41 Measured (—) and calculated (····) pair distribution functions of the AlLiCu quasicrystal.[76]

being capable of fitting diffraction data anyway! However, atomic models of this type can be used successfully to predict properties.[77]

4.4 Conclusion

When good diffraction data are available, with in particular the possibility of deriving the partial structure factors and of reconstructing their phases,

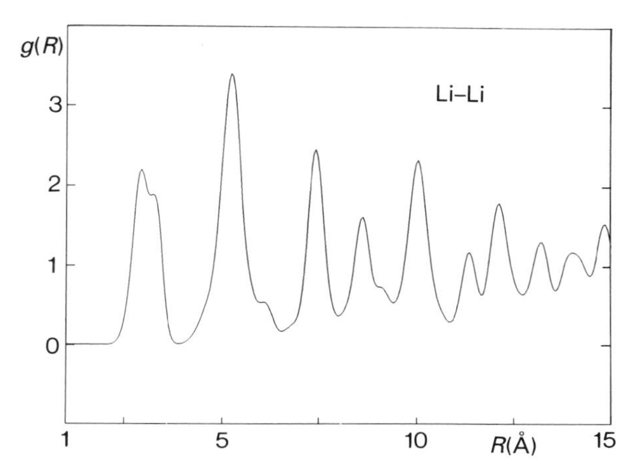

FIG. 4.42 Calculated Li–Li pair distribution function[76] for the AlLiCu quasicrystal. The accuracy of the data does not allow us to measure the partial pair distribution function.

then most of the structural atomic organization of quasicrystalline material can be reasonably specified.

Building directly 3-dim atomic models is certainly less generic and less promising than working out the structure in its n-dim periodic representation.

Good n-dim density maps, as deduced from an inverse Fourier transform of the experimentally determined (partial) structure factors, is certainly the obligatory path for further and more complete modelling. However, this is difficult since a large (infinite!) number of parameters are required when 3-dim atomic objects have to be positioned, sized, and shaped. Obviously, uniqueness of the solution cannot be guaranteed without the addition of other physical constraints (density, composition, shortest possible atom distances, forced existence of atomic clusters, etc.).

As material properties are meaningful in the real world only, the derivation of some 3-dim descriptions is a requisite. The best approximations are either atom positions in a large cluster, as deduced from the n-dim structure by a cut algorithm or periodic approximants with the largest acceptable unit.

In any case, attempts to specify where the atoms are must start by considering diffraction data quantitatively and routinely, as is usually performed in normal crystallography.

Additional problems may be caused by the presence of phason-like disorder when quasicrystals are not perfect. This does not preclude determining the 'average structure' as is current practice in crystallography. Indeed, structural models based on 'random tiling' algorithms have attempted to use these phason-like defects as justifications for entropy stabilization and/ or ease of growth. However, with increasing perfection of quasicrystal samples, the random tiling approach becomes more difficult to accept. Recently, Kycia *et al.*[78] have observed dynamic X-ray diffraction (the Bormann effect) in AlPdMn icosahedral quasicrystals (see Chapter 2). This cannot occur unless perfect long-range order is achieved, and random tiling would not be compatible with such dynamic effects. The existence of thermodynamically stable decagonal quasicrystals also denies the foundations of random tiling considerations. Indeed, entropy of an ensemble applies to the volume, and random tiling would affect the 3-dim arrangement of the decagonal structure. Some randomness would show up in the decagonal planes and along the periodic 10-fold direction. This would result in diffuse scattering, including along the 10-fold axis, and the physical properties would reveal the disorder. This is not found experimentally. Diffuse scattering is always confined in planes perpendicular to the 10-fold axis and the properties along this axis are those of an ordinary periodic metallic phase.

Thus experimental observations continue to converge on the perfect quasicrystal description as deduced from high-dim periodic images, possibly with a variable concentration of phason-like defects.

4.5 Problems

4.1 Consider a 2-dim square lattice decorated by two-segment atomic objects, one at the origin, the other at the body centre. The lengths of these two segments are equal to $a(1 + \tau)$ where a is the lattice parameter of the 2-dim structure. A 1-dim quasiperiodic structure is cut by a 1-dim space R_{par} having a slope $1/\tau$ with respect to horizontal rows of the square lattice. Calculate the 10 strongest Fourier components of the structure in the following cases:

(a) origin and BC sites having equal scattering factors;

(b) the scattering factors have the same values but opposite signs;

(c) the BC sites scatter twice as much the origin site does.

4.2 Consider a 6-dim cubic lattice with decoration by spherical atomic objects at the origin and either at the BC sites or the mid-edge positions. Give qualitative representations of partial and total Patterson maps in both cases.

4.3 Calculate the 10 strongest Fourier components of a Fibonacci chain and of its 0/1, 1/1, 2/1, and 3/2 periodic approximants. Find correspondences between them and derive the phase relations (Jaric method).

4.4 Take the 10 strongest Fourier components of a Fibonacci chain as obtained from the 2-dim representation. Ignore their phases and use their intensities (squared modulus) to simulate the phase reconstruction procedure (a standard fast Fourier transform computer program may be used).

4.5 Using the direct space description presented in Section 3.3.7, calculate the lattice parameter of the periodic approximants of the simple icosahedral quasicrystal (6-dim cubic lattice decorated with a single spherical atomic object sited at the origin). Determine the Fourier components of the 1/1 approximant and find correspondence rules with those of the quasicrystal.

References

1. Audier, M., Bréchet, Y., de Boissieu, M., Guyot, P., Janot, C., and Dubois, J.M. (1991). Perfect and modulated quasicrystals in the system AlFeCu. *Phil. Mag.* **B63**, 1375–93.
2. Sadoc, A. (1989). Local order description of an icosahedral AlCuFe quasicrystal. *Phil. Mag. Lett.* **60**, 195–200.
3. Ma, Y. and Stern, E.A. (1988). Short-range structure of Al–Mn and Al–Mn–Si aperiodic alloys. *Phys. Rev.* **B38**, 3754–65.
4. Dunlap, R.A., McHenry, M.E., O'Handley, R.C., Srinivas, V., and Bahadur, D. (1989). Mössbauer indications for high site symmetry in quasicrystalline $Ti_{56}Ni_{23}Fe_5Si_{16}$. *Phys. Rev.* **B39**, 1942–5.
5. Lee, C., White, D., Suito, B.H., Bancel, P.A., and Heiney, P.A. (1988). NMR study of Li in Al–Li–Cu icosahedral alloys. *Phys. Rev.* **B37**, 9053–6.
6. Elswijk, H.B., de Hosson, J.T.M., van Smaalen, S., and de Boer, J.L. (1988). Determination of the crystal structure of icosahedral Al–Cu–Li. *Phys. Rev.* **B38**, 1681–5.
7. Kortan, A.R., Becker, R.S., Thiel, F.A., and Chen, H.S. (1990). Real space atomic structure of a 2-dim decagonal quasicrystal. *Phys. Rev. Lett.* **64**, 200–3.
8. Roth, M., Lewit-Bentley, A., and Bentley, G.A. (1984). Scaling and phase difference determination in contrast variation experiments. *J. Appl. Cryst.* **17**, 77–84.
9. de Boissieu, M., Janot, C., Dubois, J.M., Audier, M., and Dubost, B. (1991). Atomic structure of the icosahedral Al–Li–Cu quasicrystal. *J. Phys.: Cond. Matter* **3**, 1–25.
10. Dubost, B., Collinet, C., and Ansara, I. (1988). Constitution and thermodynamics of quasicrystalline and crystalline AlCuLi alloys. In *Quasicrystalline materials* (ed. C. Janot and J.M. Dubois) pp. 39–52. World Scientific, Singapore.
11. Dubost, B., Lang, J.M., Tanaka, M., Sainfort, P., and Audier, M. (1986). Large

AlCuLi single quasicrystals with triacontahedral solidification morphology. *Nature* **324**, 48–50.

12. Cahn, J.W., Gratias, D., and Mozer, B. (1988). Patterson Fourier analysis of the icosahedral (Al,Si)–Mn alloys. *Phys. Rev.* **B38**, 1638–42.
13. Cahn, J.W., Gratias, D., and Mozer, B. (1988). Six dimensional Fourier-analysis of the icosahedral $Al_{73}Mn_{21}Si_6$ alloy. *Phys. Rev.* **B38**, 1643–7.
14. de Boissieu, M., Janot C., Dubois, J.M., Audier, M., and Dubost, B. (1989). Partial pair distribution functions in icosahedral Al–Li–Cu quasicrystals. *J. Phys. (Paris)* **50**, 1689–709.
15. de Boissieu, M., Janot, C., and Dubois, J.M. (1990). Quasicrystal structure: cold water on the Penrose tiling scheme. *J. Phys.: Cond. Matter* **2**, 2499–517.
16. Audier, M., Pannetier, J., Leblanc, M., Janot, C., Lang, J.M., and Dubost, B. (1988). An approach to the structure of quasicrystals: a single crystal X-ray and neutron diffraction study of the R-Al_5CuLi_3 phase. *Physica* **B153**, 136–42.
17. Duneau, M. and Katz, A. (1985). Quasiperiodic patterns. *Phys. Rev. Lett.* **54**, 2688–91.
18. Duneau, M. and Oguey, C. (1989). Ideal AlMnSi quasicrystal: a structural model with icosahedral clusters. *J. Phys. (Paris)* **50**, 135–46.
19. Henley, C.L. (1986). Sphere packings and local environments in Penrose tilings. *Phys. Rev.* **B34**, 797–816.
20. Janot, C., de Boissieu, M., Dubois, J.M., and Pannetier, J. (1989). Icosahedral crystals: neutron diffraction tells you where the atoms are. *J. Phys.: Cond. Matter* **1**, 1029–48.
21. Gratias, D. (1988). Some aspects of the structural determination of the quasicrystalline phases. In *Quasicrystalline materials* (ed. C. Janot and J.M. Dubois) pp. 83–96. World Scientific, Singapore.
22. Audier, M., Guyot, P., and Brechet, Y. (1990). High temperature stability and faceting of the icosahedral AlFeCu phase. *Phil. Mag.* **61**, 55–62.
23. Tsai, A., Inoue, A., Yokoyama, Y., and Masumoto, T. (1990). New icosahedral alloys with superlattice order in the AlPdMn system prepared by rapid solidification. *Phil. Mag. Lett.* **61(1)**, 9–14.
24. Stechtman, D., Blech, I., Gratias, D., and Cahn, J.W. (1984). Metallic phase with long-range orientational order and no translational symmetry. *Phys. Rev. Lett.* **53**, 1951–3.
25. Dubois, J.M. and Janot, C. (1989). Partial pair distribution functions in icosahedral Al–Mn studied by contrast variation with neutron diffraction. *J. Phys. (Paris)* **48**, 1981–9.
26. Tsai, A.P., Inoue, A., and Masumoto, T. (1989). Icosahedral, decagonal and amorphous phases in AlCuM (M = transition metal) systems. *Mat. Trans. JIM* **30**, 666–76.
27. Guryan, C.A., Goldman, A.I., Stephens, P.W., Hiraga, K.H., Tsai, A.P., Inoue, A., and Masumoto, T. (1989). AlCuRu: an icosahedral alloy without phason disorder. *Phys. Rev. Lett.* **62**, 2409–12.
28. Calvayrac, Y., Quivy, A., Bessière, M., Lefebvre, S., Cornier-Quiquandon, M., and Gratias, D. (1990). Icosahedral AlCuFe: towards ideal quasicrystals. *J. Phys. (Paris)* **51**, 417–31.
29. Cornier-Quiquandon, M., Quivy, A., Lefebvre, S., Elkain, E., Heger, G., Katz, A., and Gratias, D. (1991) Neutron diffraction study of icosahedral AlCuFe single quasicrystal. *Phys. Rev.* **B44**, 2071–2084.

30. Janot, C., Dubois, J.M., and de Boissieu, M. (1990). The structure of quasicrystals: from diffraction patterns to atom positions. In *Proceedings of a NATO advanced research workshop on Common problems of quasicrystals, liquid crystals and incommensurate insulators*, Preveza, September 1989 (ed. J.C. Toledano) pp. 9–24. Plenum, New York.
31. Cornier-Quiquandon, M., Bellissent, R., Calvayrac, Y., Cahn, J.W., Gratias, D., and Mozer, B. (1993). Neutron scattering structural study of AlFeCu quasicrystals using double isotopic substitution. *J. Non-Cryst. Solids* **153–154**, 10–14.
32. Katz, A. and Gratias, D. (1993). A geometric approach to chemical ordering in icosahedral structures. *J. Non-Cryst. Solids* **153–154**, 187–95.
33. Cornier-Quiquandon, M., Gratias, D., and Katz, A. (1991). A tentative methodology for structure determination in quasicrystals. In *Methods of structural analysis of modulated structures and quasicrystals* (ed. J.M. Perez-Mato), pp. 313–32. World Scientific, Singapore.
34. Boudard, M., de Boissieu, M., Janot, C., Dubois, J.M., and Dong, C. (1991). The structure of the icosahedral AlPdMn quasicrystal. *Phil. Mag. Lett.* **64**, 197–206.
35. Janot, C., Boudard, M., de Boissieu, M., Durand, M., Vincent, H., Dubois, J.M., and Dong, C. (1991). Single crystal X-ray diffraction of AlMnPd quasicrystals. *J. Non Cryst. Sol.* (in press).
36. Boundard, M., de Boissieu, M., Janot, C., Heger, G., Beeli, C., Nissen, H.U., *et al.* (1992). Neutron and X-ray single-crystal study of the AlPdMn icosahedral phase. *J. Phys.: Conds. Matter* **4**, 10149–68.
37. Janot, C., Magerl, A., Frick, B., and de Boissieu, M. (1993). Localized vibrations from clusters in quasicrystals. *Phys. Rev. Lett.* **71**, 871–4.
38. Steurer, W. (1989). Five-dimensional Patterson analysis of the decagonal phase of the system Al–Mn. *Acta Cryst.* **B45**, 534–42.
39. Steurer, W. and Kuo, K.H. (1990). Five-dimensional structure analysis of decagonal $Al_{65}Cu_{20}Co_{15}$. *Acta Cryst.* **B46**, 703–12.
40. Steurer, W. (1993). Comparative structure analysis of several decagonal phases. *J. Non-Cryst. Solids* **153–154**, 92–7.
41. Fettweis, N., Launois, P., Dénoyer, F., Reich, R., and Lambert, M. (1993). Decagonal quasicrystalline or microcrystalline structures: the specific case of Al-Cu-Co(-Si). submitted to *Phys. Rev. B*.
42. Shibuya, T., Hashimoto, T., and Takeuchi, S. (1990). Anisotropic conductivity in a decagonal quasicrystal $Al_{70}Ni_{15}Co_{15}$. *J. Phys. Soc. Jpn* **59**, 1917–20.
43. Grushko, B. (1993). Study of the decagonal phase stability in the AlCuCo(Si) alloy system. *J. Non-Cryst. Solids* **153–154**, 489–93.
44. Zhang, Z., Li, H.L., and Kuo, K.H. (1993). Stable decagonal Al–Cu–Co–Si phase: a phason-perturbed quasiperiodic crystal. *Phys. Rev. B* **48**, 6949–6951.
45. Inoue, A., Tsai, A.P., and Masumoto, T. (1990). Icosahedral and decagonal quasicrystals in Al-Ni-M and Al-Cu-M (M = transition metal) systems. In *Quasicrystals* (ed T. Fujiwara and T. Ogawa), pp. 80–90. Springer Verlag, Berlin.
46. Beeli, C., Nissen, H.V., and Robaday, J. (1991). Stable AlMnPd quasicrystals. *Phil. Mag. Lett.* **63**. 87–95.
47. Tsai, A.P., Inoue, A., and Masumoto, T. (1991) Stable decagonal quasicrystals with a periodicity of 1.6 mm in AlPd–(Fe, Ru or Cs) alloys. *Phil. Mag. Lett.* **64**, 163–8.

48. Steurer, W. and Kuo, K.H. (1990) Five-dimensional structure analysis of decagonal $Al_{65}Cu_{20}Co_{15}$. *Acta Crystallogr.* **B46**, 703–12.
49. Steurer, W. (1991). Five-dimensional structure refinement of decagonal $Al_{78}Mn_{82}$. *J. Phys.: Cond. Matter* **3**, 3397–410.
50. Burkov, S.E. (1991). Structure model of the AlCuCo decagonal quasicrystal. *Phys. Rev. Lett.* **67**, 614–17.
51. Burkov, S.E. (1992). Modeling decagonal quasicrystals: random assembly of interpenetrating decagonal clusters. *J. Phys. (Paris)* **2**, 695–706.
52. Burkov, S.E. (1993). Enforcement of matching rules by chemical ordering in decagonal quasicrystals. *Phys. Rev. B* **47**, 12325–36.
53. Jaric, M.V. and Quiu, S.Y. (1990). From crystal approximants to quasicrystals. In *Quasicrystals*, (ed. T. Fujiwara and T. Ogawa) pp. 48–56. Springer, Berlin.
54. Qiu, S.Y. and Jaric, M.V. (1989). Quasicrystal structure determination AlCuLi. In *Proceedings of the anniversary Adriatic research conference on quasicrystals*, Trieste 1989 (ed. M.V. Jaric and S. Lundqvist) pp. 19–33. World Scientific, Singapore.
55. Pauling, L. (1989). Icosahedral and decagonal quasicrystals as multiple twins of cubic crystals. In *Extended icosahedral structures* (ed. M.V. Jaric and D. Gratias) pp. 137–62. Academic, New York.
56. Cahn, J.W. (1986). Quasiperiodic crystals: a revolution in crystallography. *MRS Bulletin* March–April, pp. 9–14.
57. Henley, C.L. (1987). Quasicrystal order, its origins and its consequences: a survey of current models. In *Comments on condensed matter Physics* Vol. 8, No. 2, pp. 59–117.
58. Stephens, P.W. and Goldman, A.I. (1986). Sharp diffraction maxima from an icosahedral glass. *Phys. Rev. Lett.* **56**, 1168–71.
59. Stephens, P.W. and Goldman, A.I. (1986). *Phys. Rev. Lett.* **57**, 2331.
60. Henley, C.L. (1991). Random tiling models. In *Quasicrystals: the state of the art* (ed. D.P. DiVincenzo and P.J. Steinhardt) pp. 429–518. World Scientific, Singapore.
61. Henley, C.L. (1988). Short range order in aged Al–Mn quasicrystals. *Phil. Mag. Lett.* **58**, 87–9.
62. Elser, V. (1988). The growth of icosahedral phase. In *Aperiodic crystals*, vol. 3 (ed. M.V. Jaric) pp. 105–36. Academic, New York.
63. Strandburg, K.J., Tang, L.H., and Jaric, M.V. (1989). Phason elasticity in entropic quasicrystals. *Phys. Rev. Lett.* **63**, 314–17.
64. Janot, C., Audier, M., de Boissieu, M., and Dubois, J.M. (1991). AlCuFe quasicrystals: low-temperature unstability via a modulation mechanism. *Europhys. Lett.* **14**, 355–60.
65. Tang, L.H. (1990). Random tiling quasicrystals in three-dimensions. *Phys. Rev. Lett.* **64**, 2390–3.
66. Guyot, P. and Audier, M. (1985). A quasicrystal structure model for Al–Mn. *Phil. Mag.* **B52**, L15–19.
67. Loiseau, A. and Lapasset, G. (1987). Relations between quasicrystals and crystalline phases in AlLiCuMg alloys: a new class of approximant structures. *Phil. Mag. Lett.* **56**, 165–71.
68. Cooper, M. and Robinson, K. (1966). The crystal structure of the ternary alloy α-AlMnSi. *Acta Cryst.* **20**, 614–22.
69. Li, F.H. (1991). Quasicrystal structure determination based on high-dimensional

crystal construction. In *Methods of structural analysis of modulated structures and quasicrystals* (ed. J.M. Perez-Mato), pp. 492–502. World Scientific, Singapore.

70. Audier, M. and Guyot, P. (1989). Quasicrystal structure models related to crystalline structures. In *Extended icosahedral structures* (eds M.V. Jaric and D. Gratias) pp. 1–36. Academic, New York.
71. Elser, V. and Henley, C.L. (1985). Crystal and quasicrystal structures in Al–Mn–Si alloys. *Phys. Rev. Lett.* **55**, 2883–6.
72. Henley, C.L. and Elser, V. Quasicrystal structure of $(Al,Zn)_{49}Mg_{32}$. *Phil. Mag.* **B53**, L59–66.
73. Van Smaalen, S. (1989). Three-dimensional Patterson function for the Al_6CuLi_3 quasicrystal. *Phys. Rev.* **B39**, 5850–6.
74. Elswijk, H.B., Bronsveld, P.M., and de Hosson, J.T.M. (1988). Field-ion microscopy contradiction of the quasicrystal model based on twinning of a cubic crystal. *Phys. Rev.* **B37**, 4261–4.
75. Audier, M. and Guyot, P. (1988). The structure of the icosahedral phase. Atomic decoration of the two basis cells. In *Quasicrystalline materials* (eds C. Janot and J.M. Dubois) pp. 181–94. World Scientific, Singapore.
76. de Boissieu, M., Janot, C., Dubois, J.M., Audier, M., Jaric, M.V., and Dubost, B. (1990). About atomic structure of the AlLiCu icosahedral phase. In *Quasicrystals: proceedings of the anniversary Adriatic research conference* (eds M.V. Jaric and S. Lundqvist pp. 74–91. World Scientific, Singapore.
77. Windisch, M., Hafner, J., Krajci, M., and Mihalkovic, M. (1993). Structure and lattice dynamics of icosahedral AlCuLi. *Phys. Rev. B* (in press).
78. Kycia, S.W., Goldman, A.I., Lograsso, T.A., Delaney, D.W., Black, D., Sutton, M., *et al.* (1993). Dynamical X-ray diffraction from an icosahedral quasicrystal. *Phys. Rev. B* **48**, 3544–7.

5

Phonons, phasons, and dislocations in quasicrystals

Truth is never pure, and rarely simple
Oscar Wilde

5.1 Introduction

In studying quasicrystal structures in the last two chapters, it has mostly been assumed that atoms were at rest at their lattice sites and that there was no moving or static defect within the atomic arrangements. Atoms, however, as in normal crystals, are not quite stationary but oscillate around their equilibrium positions as a result of thermal energy, and experience occasional jumps from site to site. These dynamics and defect aspects have a fundamental influence on the thermal, acoustic, optical, and mechanical properties of the materials.

Materials with icosahedral or other quasiperiodic symmetries represent actually a new phase of matter with possibly unique physical properties, which one would like to identify and understand. Are quasicrystals stable or only metastable? How do they respond to external forces? What is the nature of their low-energy excitations? What are their transport properties?

There are well-established methods for addressing these and other related questions about periodic crystals. First, an ideal crystal structure is specified by using diffraction data. Then elementary excitations from the ground state (phonons) and the electronic states (band structure) are determined and used to calculate thermodynamic and transport properties. This process, already quite difficult for complex periodic structures, cannot yet be applied to quasicrystals because the precise positions of atoms and details of the structure remain unknown. Some other approach to the study of their physical properties must be sought. The distinguishing feature of quasicrystals is their unique translational and rotational symmetry. It is natural to ask what properties follow from this symmetry alone and not from the details of the atomic position and interatomic interactions.

Phenomenological elastic and hydrodynamic theories use only the invariance implied by the symmetries to infer the response to slowly varying external forces (elasticity) and the nature of long-wavelength, low-frequency excitations (hydrodynamics). The most interesting hydrodynamic modes are those associated with broken translational invariance of the quasilattice. In the n-dim periodic description of quasicrystal structures these modes may operate in the physical space and/or in the complementary space. We have already mentioned that the former are propagating (phonon-like) and the latter are diffusive (called phason mode) and we will see that cross terms between the two species generate dislocations.

In this chapter, toy models (especially 1-dim spring models) will be used with the aim of following the evolution, from crystal to quasicrystal via modulated structure, of dynamic modes and defects.

The correspondence between phason modes and atomic rearrangements is also relevant in understanding the growth of quasicrystals. Even if the icosahedral phase can be described by a packing of unit cells with effective matching rules, mismatches may be quenched in during solidification. The fact that phasons correspond to rearrangements of unit cells means that relaxation requires diffusive motion of atoms among the unit cells. Deviations from ideal quasicrystal structure may thus come from phason strain. These points will be stressed further in this chapter.

We will start with some basic aspects of dynamics and defects in regular crystals. Modulation will be introduced to give insight into the peculiar effects due to periodicity breaking. The last section will be devoted to phonons, phasons, and dislocations in quasicrystals, presented in a very introductory way and without too much mathematical development.

5.2 Basic knowledge about lattice dynamics and defects in periodic strutures

5.2.1 Elastic waves in solids

In lattice dynamics, the atomic nature may be disregarded when the wavelength is very long and the solid can be treated as a continuous medium. The corresponding vibrations are referred to as **elastic waves**.

Suppose that the wave is longitudinal and denote the elastic displacement at the point x by $u(x)$. The **strain** ε, defined as the change of length per unit length, is given by

$$\varepsilon = \frac{\partial u}{\partial x}.$$

The **stress** σ, defined as the force per unit area, is also a function of ε and is proportional to the strain (Hooke's law) in the linear (elastic) approximation, i.e.

$$\sigma = Y\varepsilon.$$

Y is known as Young's modulus. Let us consider an arbitrary segment of solid, with length dx, sandwiched between positions x and $x + \mathrm{d}x$, with cross-sectional area A and mass density ρ.

The motion of this segment obeys the classical equation

$$(\rho A'\mathrm{d}x)\frac{\partial^2 u}{\partial t^2} = \{\sigma(x + \mathrm{d}x) - \sigma(x)\}A'$$
$$= A'\frac{\partial \sigma}{\partial x}\mathrm{d}x.$$

Using the expression $\sigma = Y\varepsilon = Y(\partial u/\partial x)\mathrm{d}x$ gives finally

$$\frac{\partial^2 u}{\partial x^2} - \frac{\rho}{Y}\frac{\partial^2 u}{\partial t^2} = 0,$$

which is the well-known wave equation in one dimension. Solutions in the form of propagating plane waves, such as

$$u = U_0 \exp\{i(\omega t - kx)\}$$

gives the **dispersion law**

$$\omega = v_s k, \tag{5.1}$$

with $v_s = (Y/\rho)^{\frac{1}{2}}$ via a simple mathematical substitution. The corresponding curve, a simple straight line, is shown in Fig. 5.1. Sound waves in liquid and glasses, optical waves in a vacuum, etc. satisfy this type of dispersion relation. Small wave vectors k, or equivalently long wavelengths λ ($k = 2\pi/\lambda$) correspond to overall translations of the system and do not involve energy transfer ($\Delta E = \hbar\omega$). Increasing the k vector induces more and more local strains in which more and more energy is dissipated. This is the qualitative meaning of ω being an increasing function of k.

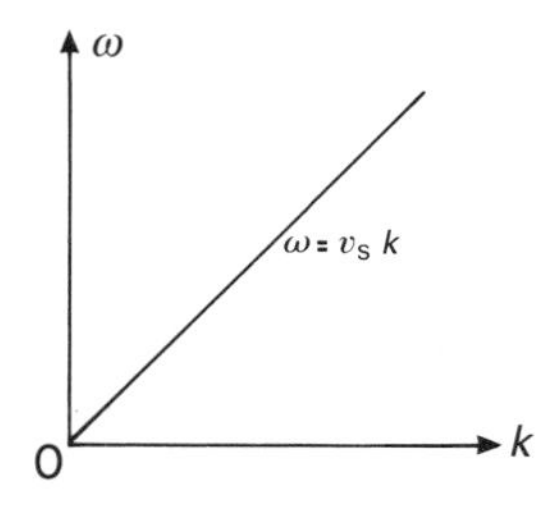

FIG. 5.1 Dispersion curve of an elastic wave.

Typical values of the parameters in eqn (5.1) are $v_s = 5 \times 10^3\,\mathrm{ms^{-1}}$ and, $\rho = 5 \times 10^3\,\mathrm{kgm^{-3}}$, which give $Y = 1.25 \times 10^{11}\,\mathrm{kgm^{-1}s^2}$.

Deviation from the linear relationship (that is, true **dispersion**) is often observed. In particular, dispersion is expected when the wavelength is short with respect to interatomic distances. In the example above, the elastic wave has been supposed longitudinal, i.e. the displacement $u(x)$ is parallel to the propagation vector k. Transverse (or shear) waves must be also considered, with propagating displacement perpendicular to the direction of propagation. This introduces a shear elastic constant, the shear modulus μ, analogous to Young's modulus; the velocity of the shear wave is also related to it by an equation similar to eqn (5.1). The two elastic constants can then be used to describe the propagation of an arbitrary elastic wave in the solid.

Thus far it has been tacitly assumed that the solid is isotropic. Crystals actually are anisotropic and this introduces many more elastic constants than the two needed for the isotropic solid. Their number is fortunately decreased in a substantial manner by symmetry considerations. For instance, in the case of a cubic crystal there are only three independent elastic constants, denoted C_{11}, C_{12}, and C_{44}; C_{11} relates compression stress and strain along the [100] direction, while C_{44} involves shear stress and strain in the same direction; C_{12} relates the compression stress in any given direction to the strain in a perpendicular direction. These constants can be determined by measuring the sound velocities: $(C_{11}/\rho)^{\frac{1}{2}}$ and $(C_{44}/\rho)^{\frac{1}{2}}$ for longitudinal and shear waves, respectively, along the [100] direction, and $\{(C_{11} + 2C_{12} + 4C_{144})/\rho^{\frac{1}{2}}$ for a longitudinal wave along [111].

Boundary conditions, due for instance to space extension of the solid, impose that k takes only specific values. It is easy to demonstrate that if, say, $u(x=0) \equiv u(x=L)$ at any time, the allowed values for k are $k = n2\pi/L$, where n is any positive or negative integer. The resulting discreteness in k induces the same kind of discretness in ω and one has to introduce a new concept, namely the density of states $g(\omega)$ defined in such a way that $g(\omega)\mathrm{d}\omega$ gives the number of allowed propagating modes in the frequency range $\mathrm{d}\omega$ lying between ω and $\omega + \mathrm{d}\omega$. It is easy to calculate $g(\omega)$ for a solid of volume V and with sound velocity v_1 and v_s for longitudinal and shear waves, respectively, i.e.

$$g(\omega) = \frac{V\omega^2}{2\pi^2}\left(\frac{1}{v_1^3} + \frac{2}{v_s^3}\right). \tag{5.2}$$

In the Debye model, the atomic discreteness of the solid is reintroduced at this level, somewhat artificially, by stating that a solid sample containing N atoms has a maximum of $3N$ degrees of freedom and cannot exhibit more than $3N$ propagating modes. Summing $g(\omega)$ to infinity is then disallowed and a maximum value ω_D must be introduced such that:

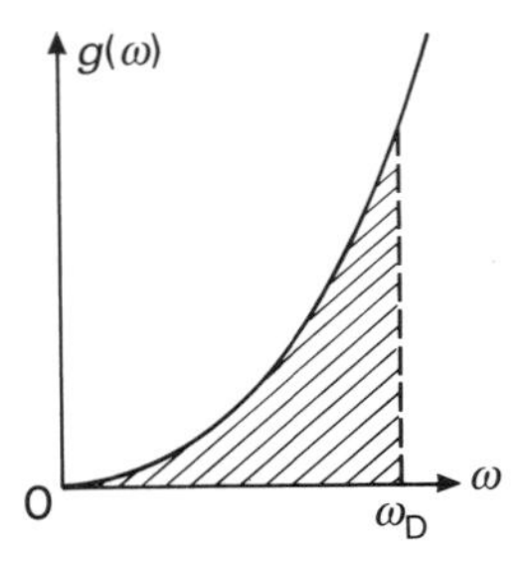

FIG. 5.2 The Debye cut-off procedure. The shaded area is equal to $3N$, the number of degrees of freedom of the system.

$$\int_0^{\omega_D} g(\omega)\mathrm{d}\omega = 3N,$$

or, using eqn (5.2): $\omega_D = \{6\pi^2(N/V)\}^{\frac{1}{3}} v$ (if $v_l = v_s = v$ is assumed for simplicity). The Debye cut-off procedure is illustrated in Fig. 5.2. Physical properties, such as the specific heat, are reasonably well interpreted by the simple Debye model. This model also shows that the energy of propagating modes is quantized, the unit of quantum energy being $\hbar\omega$. It is then justified to introduce a particle-like entity, called a **phonon**, with the energy $\hbar\omega$ and a momentum $\boldsymbol{p} = \hbar\boldsymbol{k}$.

5.2.2 Lattice waves and Brillouin zones

The oversimplification contained in the continuous medium approximation has to be somewhat relaxed for real crystals. In the long wavelength limit, a large number of atoms are moving with approximately the same phase, which means that most of them have no significant relative displacements with respect to each other. Then the linear relation $\omega = v_s k$ still holds. However, as the wavelength decreases and k increases, the atoms start to scatter the wave, because neighbour atoms are now moving with respect to each other. The net result of this scattering is that the propagation is somehow impeded and the velocity of the wave decreases. Then, the dispersion curve is bent downwards, as indicated in Fig. 5.3, as a result of ω being now smaller, for a given k, than the corresponding continuous medium value.

Further qualitative features of the dispersion law can also be derived from the nature of a crystalline solid.

1. Propagating modes with wavevectors $\boldsymbol{k}$ and $-\boldsymbol{k}$ have the same wavelength but travel to the right and to the left parts of the crystal, respectively. Right and left are irrelevant notions for the crystal and both waves

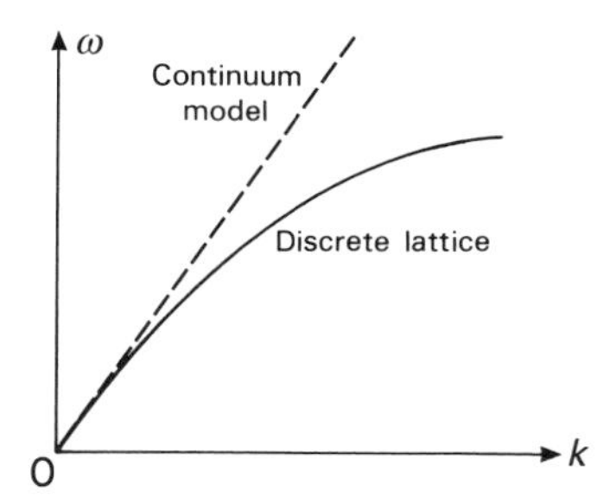

FIG. 5.3 Dispersion curve of a discrete lattice. The dashed line corresponds to the continuum model approximation. The two curves coincide at $k = 0$.

transport the same energy since the relative displacements of the atoms are the same. Then $\omega(\boldsymbol{k}) = \omega(-\boldsymbol{k})$ and the dispersion curves for elastic waves in a crystal are symmetric with respect to reflection around the origin in the $\boldsymbol{k}$ space.

2. Propagating modes with certain wavevectors may correspond to Bragg-like reflections and are, indeed, reflected. Thus, they are not allowed to propagate. Then, the dispersion curve $\omega(\boldsymbol{k})$ may show frequency gaps or forbidden energy ranges.
3. If $\boldsymbol{k}$ is increased by a vector of the reciprocal lattice, say $\boldsymbol{G}$, the mode $\boldsymbol{k}' = \boldsymbol{k} + \boldsymbol{G}$ must correspond to the same atomic displacements if translation invariance in both direct and reciprocal spaces is to be respected. This point is illustrated in Fig. 5.4 for the particular simple case of a 1-dim monatomic chain with periodicity a. Two propagating modes are shown, with $\lambda = 4a$ and $\lambda' = 4a/5$, respectively. These two modes correspond to the same atomic displacements, and thus propagate the same elastic enegy. Their wavevectors $k = 2\pi/4a$ and $k' = 2\pi/(4a/5) = 2\pi/4a + 2\pi/a$ differ by $2\pi/a$, which is the repetition length of the 1-dim reciprocal space. Thus the dispersion curves have the translational

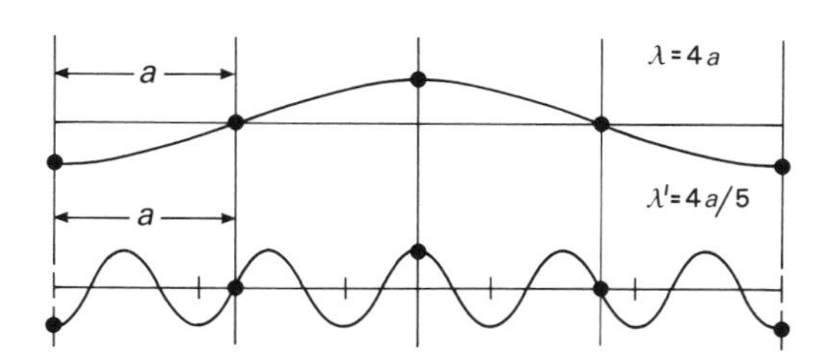

FIG. 5.4 Transverse waves corresponding to $k = \pi/2a$ and $k' = k + 2\pi/a$, or $\lambda = 4a$ and $\lambda' = 4a/5$, respectively.

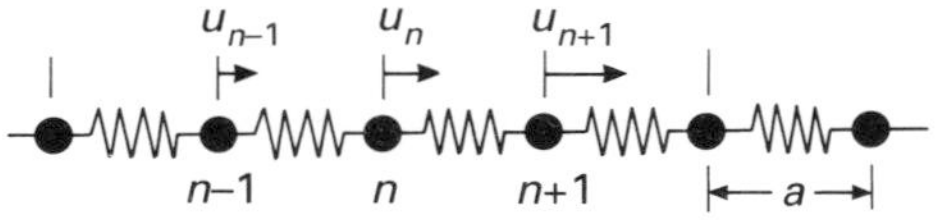

FIG. 5.5 Part of a one-dimensional lattice. The arrows represent atomic displacements from equilibrium positions (exaggerated for illustrative purposes). Springs simulate elastic forces between atoms.

symmetries of the reciprocal space. The wavevectors $\boldsymbol{k}$ have then to be defined in only one unit cell of the reciprocal space; usually the cell surrounding the origin is chosen for convenience and is called the **first Brillouin zone**. For $\boldsymbol{k}$ equal to any vector $\boldsymbol{G}$ of the reciprocal lattice, the propagating modes are those corresponding to $k = 0$ with, in particular, the existence of modes with zero energy ($\omega = 0$).

Let us illustrate quantitatively these important conclusions, with the simple example of a toy system made of identical balls bonded by identical springs in a 1-dim space (Fig. 5.5). When this lattice is at equilibrium, each atom is positioned exactly at the lattice sites of a periodic chain with repetition distance a. As the atoms interact with each other via the spring bonds, any motion of one atom tends to propagate through the entire chain. Consider an atom at position n and its small displacement u_n with respect to its equilibrium position. In an elastic, or so-called harmonic, approximation, with interactions limited to the nearest-neighbour atoms, the equations of motion will be

$$m\frac{\mathrm{d}^2 u_n}{\mathrm{d}t^2} = +\alpha(u_{n+1} - u_n) + \alpha(u_{n-1} - u_n)$$

$$= -\alpha(2u_n - u_{n+1} - u_{n-1}), \tag{5.3}$$

where m is the atom mass. There are as many equations as atoms in the chain. The search for plane wave solutions of the type

$$u_n = U_0 \exp\{\mathrm{i}(nka - \omega t)\}$$

results in the dispersion relation after trivial mathematical manipulation, namely (Fig. 5.6)

$$\omega = \left(\frac{4\alpha}{m}\right)^{\frac{1}{2}} \left|\sin\frac{ka}{2}\right|. \tag{5.4}$$

We can immediately check that $\omega(-k) = \omega(k)$, and that $k' = k + 2\pi/a$ and k give the same q value. Thus k values can be limited to the interval $[-\pi/a, +\pi/a]$ and $\omega = 0$ for $k = 0, 2\pi/a, 4\pi/a, \ldots$.

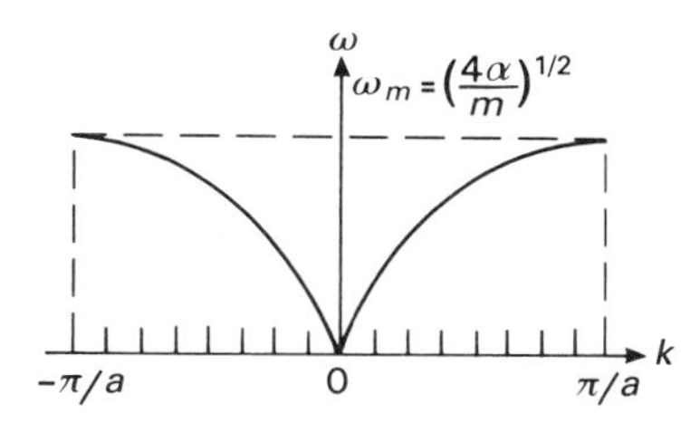

FIG. 5.6 Dispersion curve of the one-dimensional monatomic lattice.

There is no gap in the frequency range, but the chain acts as a low-pass mechanical filter with the maximum frequency for free propagation $\omega_m = (4\alpha/m)^{\frac{1}{2}}$. In the long-wavelength limit $(k \to 0)$, the dispersion relation (5.4) may be approximated by

$$\omega \approx \frac{\omega_m a}{2} k.$$

Thus the sound velocity is equal to $\omega_m a/2$ and can be obtained by spectroscopic measurement of ω_m; this is also a way to determine the elastic constants. It is easy to realize that neighbouring atoms of the chain vibrate in phase for modes in the centre of the Brillouin zone $(k = 0)$ while they are out of phase at the borders of the Brillouin zone $(k = \pi/a)$.

It is also useful to know that the Brillouin zone limits correspond to modes which do not propagate at all, since the group velocity $v_g = \partial\omega/\partial k$ is found equal to zero. This is physically due to the fact that waves scattered by successive atoms in the chain interfere constructively and combine with the incident wave to form a standing wave, which of course has a vanishing group velocity. Frequencies larger than ω_m are not transmitted by the lattice, but they can give rise to localized vibration modes when the oscillations are forced. For $k = \pi/a$ the situation is similar to the Bragg condition, but applied here to elastic waves (Fig. 5.7). Assuming that eigen modes of individual atoms localize would imply, according to eqn (5.3), in which u_{n+1} and u_{n-1} are set to zero, that $\omega_L = (2\alpha/m)^{\frac{1}{2}}$ is out of the energy band. This is obviously not the case.

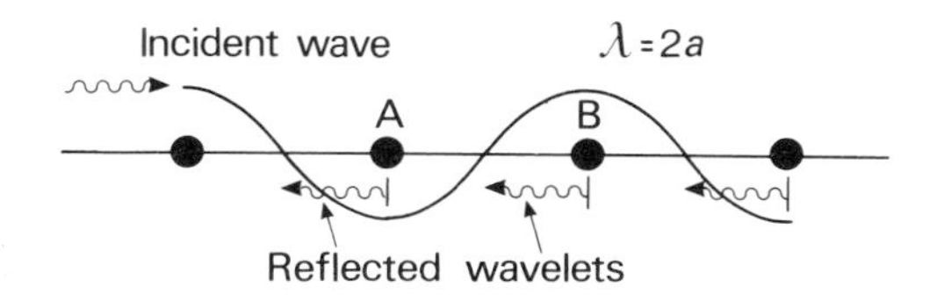

FIG. 5.7 Bragg-like reflection of elastic waves.

Finally, within the band of allowed frequencies the only allowed values of the k vectors are fixed by the boundary conditions, as explained for the continuous medium approximation.

5.2.3 Superstructure effects and energy gaps

Let us consider now a chain of atoms (balls and springs) forming a 1-dim diatomic lattice, with 'white' and 'black' atoms (see Fig. 5.8). The masses are denoted by m_1, and m_2. The distance between one white and one black atom is equal to a again. The dispersion curve for a travelling elastic wave can be qualitatively deduced from that obtained for the monatomic chain.

To start with, the diatomic chain can be averaged to a monatomic chain

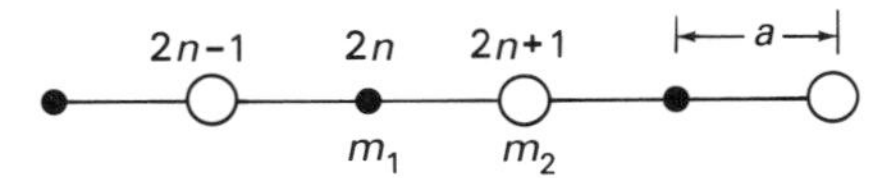

FIG. 5.8 A one-dimensional diatomic lattice. The unit cell has a length $2a$.

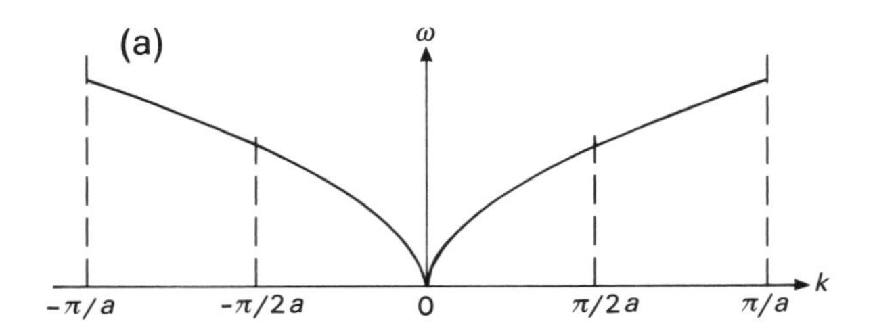

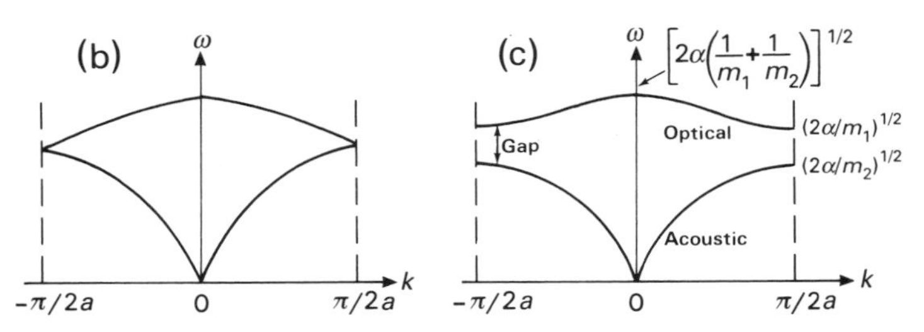

FIG. 5.9 Dispersion curves of (a) the 'averaged' diatomic chain (extended zone scheme); (b) the 'averaged' diatomic chain, but with repetition distance $2a$ accounted for (reduced zone scheme); and (c) the real diatomic chain.

with 'average' atoms of masses m given by $1/m = \frac{1}{2}(1/m_1 + 1/m_2)$. This would result in a dispersion curve, as illustrated in Fig. 5.9(a), directly reproduced from Fig. 5.6. Now, the repetition length in the true diatomic chain is $2a$, not a. This means that the extension of the first Brillouin zone is $2\pi/2a$, not $2\pi/a$, and that the k vector must be taken over the range $[-\pi/2a, +\pi/2a]$, not $[-\pi/a, +\pi/a]$. Thus the pseudo-Brillouin zone of Fig. 5.9(a), also called the extended zone, must be folded upon itself so that the dispersion curve now has two branches, with two different ω values for any k vectors between $-\pi/2a$ and $+\pi/2a$ (Fig. 5.9b) (reduced zone scheme). Finally, it appears (as illustrated in Fig. 5.10) that in the modes corresponding to $k = \pm\pi/2a$ one species of atom is at rest (relatively) while the second species reproduces the standing wave situation as described for the monatomic chain at the Brillouin zone limit $k = \pm\pi/a$). Thus the modes $k = \pm\pi/2a$ do not propagate, which means that there are no ω values defined for this k vector and that an energy or frequency gap must be open in the dispersion curve (Fig. 5.9c). When approaching the gap, the diatomic chain behaves as if only one type of atom were able to move, the other one having a very large effective mass. This suggests ω values for the gap lips as defined by the averaged inertial parameter $1/m = \frac{1}{2}(1/m_1 + 1/m_2)$ with either m_1 or m_2 being infinitely large, i.e. $\omega_1 = (2\alpha/m)^{\frac{1}{2}}$ and $\omega_2 = (2\alpha/m_2)^{\frac{1}{2}}$. For completeness, the largest frequency, corresponding to the $k = 0$ mode of the upper branch of the dispersion curve, is equal to $\{(2\alpha(1/m_1 + 1/m_2)^{\frac{1}{2}}$.

All these results can be obtained quantitatively by writing the equations of motion. By analogy with eqn (5.3), we have

$$m_1 \frac{\mathrm{d}^2 u_{2n}}{\mathrm{d}t^2} = -\alpha(2u_{2n} - u_{2n-1} - u_{2n+1})$$

$$m_2 \frac{\mathrm{d}^2 u_{2n+1}}{\mathrm{d}t^2} = -\alpha(2u_{2n+1} - u_{2n} - u_{2n+2}) \qquad (5.5)$$

If N is the number of unit cell in the chain, there are $2N$ coupled equations that have to be solved simultaneously. Solutions in the form of propagating plane waves lead to the so-called **secular equation**:

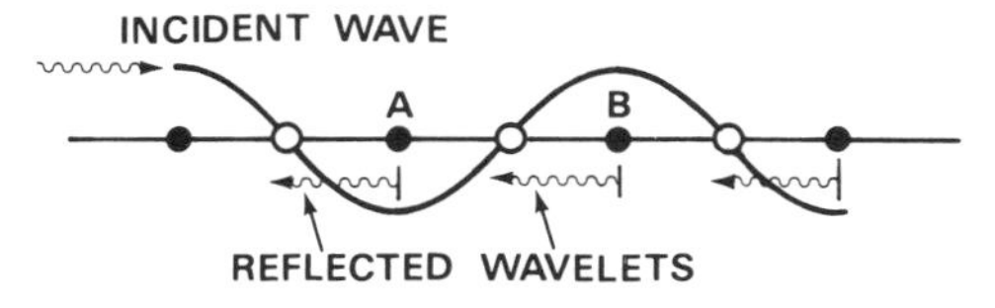

FIG. 5.10 Vibration mode corresponding to $k = \pm\pi/2a$ in the diatomic lattice.

$$\begin{vmatrix} 2\alpha - m_1\omega^2 & -2\alpha\cos(ka) \\ -2\alpha\cos(ka) & 2\alpha - m_2\omega^2 \end{vmatrix} = 0,$$

or

$$\omega^2 = \alpha\left(\frac{1}{m_1} + \frac{1}{m_2}\right) \pm \alpha\left\{\left(\frac{1}{m_1} + \frac{1}{m_2}\right)^2 - \frac{4\sin^2 ka}{m_1 m_2}\right\}^{\frac{1}{2}} \tag{5.6}$$

which gives exactly the two branch dispersion curve produced via physical arguments (Fig. 5.9(*c*)). The lower branch is called the **acoustic branch,** while the almost flat upper one is the **optical branch**. By inserting plane wave solutions and either $\omega = 0$ or $\omega = \{2\alpha(1/m_1 + 1/m_2)^{\frac{1}{2}}$ into the equations of motion (5.5), one finds that the equations are satified at $k = 0$ only if

$U_1 = U_2$ for the acoustic branch, and

$m_1 U_1 + m_2 U_2 = 0$ for the optical branch,

where U_1 and U_2 are the amplitudes of the wave at each atom species.

Thus, for the acoustic branch and in the long wavelength limit, the two atoms in the cell have the same amplitude and also the same phase: the whole lattice oscillates as a rigid body. For the other branch, conversely, the two atoms move π out of phase with each other. As k increases, ω_{opt} decreases, but not that much because the atoms continue to oscillate approximately π out of phase and the main part of the energy involved comes from the two sublattices moving relative to each other. Conversely, ω_{ac} increases significantly with k, as already observed with the monatomic chains, and for the same reasons.

The reason for referring to the lower branch as acoustic is obviously that it does correspond to acoustic modes and sound waves in the long wavelength limit. The optical branch corresponds to ω of the order of $3 \times 10^{13}\,\text{s}^{-1}$, which effectively lies in the infrared region. Furthermore, if the solid is an ionic lattice, the cell carries a strong electric dipole moment, which can be excited by electromagnetic waves.

As for the monatomic chain, the number of allowed k values is N if there are N cells in the chain. Thus the total number of modes inside the first Brillouin zone is $2N$, equal to the number of degrees of freedom in the lattice. The diatomic chain can be viewed as a monatomic chain with a simple periodic mass modulation. It is interesting to observe that the modulation effect is to open gaps in the dispersion curve at half the wave vectors of the modulation (Brillouin zone borders).

5.2.4 Lattice waves in three-dimensional lattices

Without going into mathematical detail, the 3-dim lattice dynamics follow smoothly and logically from the 1-dim case. Consider a non-Bravais 3-dim lattice, with a unit cell containing m atoms (subscripts $j = 1, \ldots, m$). We must now write the equation of motion for each atom in the cell and for the three coordinates, i.e. $3m$ equations or a single matrix equation of order $3m \times 3m$. Normal mode solutions of this system are of the form

$$\boldsymbol{u}_{nj} = \boldsymbol{U}_j \exp\{\mathrm{i}(\omega t - \boldsymbol{k}\cdot\boldsymbol{r})\}.$$

The vector $\boldsymbol{U}_j$ specifies the amplitude as well as the direction of vibration of the atoms, i.e. the **polarization** of the wave. The polarization tells us whether the wave is longitudinal ($\boldsymbol{U}_j$ parallel to $\boldsymbol{k}$) or transverse ($\boldsymbol{U}_j$ perpendicular to $\boldsymbol{k}$).

Substituting $\boldsymbol{u}_n$, into the motion equation gives the secular equation, which is of degree $3m$ in ω^2 and has, therefore, $3m$ roots leading to $3m$ branches. Three of these roots must correspond to the three free translations of the lattice in the long wavelength limit ($\boldsymbol{k} = 0$), i.e. they vanish at $\boldsymbol{k} = 0$ and result in three acoustic branches, classified as TA_1, TA_2, and LA, for obvious reasons. The $3m$-3 remaining roots belong to optical branches that may be classified as TO and LO, with the proper subscripts. The dispersion relations are not necessarily isotropic in the $\boldsymbol{k}$-space. In the example of Fig. 5.11, the experimental dispersion curves are different in germanium for the [100] and [111] directions. The transverse branches may appear to be degenerate, i.e. TA_1 and TA_2, for instance, may coincide in a single branch.

Each frequency branch must satisfy the symmetries of the space group to which the crystal belongs. Beyond the inversion symmetry $\{\omega_j(-\boldsymbol{k}) = \omega_j(\boldsymbol{k})\}$, periodicity in the reciprocal lattice is imposed, i.e.

$$\omega_j(\boldsymbol{k} + \boldsymbol{G}) = \omega_j(\boldsymbol{k}).$$

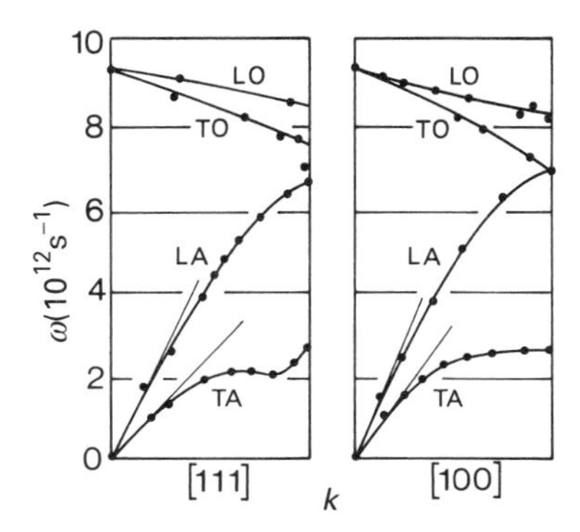

FIG. 5.11 Dispersion curves as measured for germanium in the [100] and [111] directions.

This means that we may confine our attention to the first Brillouin zone (BZ) only. Moreover, all symmetry operations of the point group must be satisfied. For instance, the cubic rotational group contains 48 symmetry operations. This means that only 1/48th of the first BZ has to be scanned to determined the dispersion curves. There are always three zero energy modes for $\boldsymbol{k}$ vectors equal to any reciprocal lattice vector $\boldsymbol{G}$. This is a generic conclusion.

The **density of states** $g(\omega)$ is defined, as before, such that $g(\omega)\mathrm{d}\omega$ gives the number of modes in the frequency range $[\omega, \omega + \mathrm{d}\omega]$. For a given branch $\omega_j(\boldsymbol{k})$, the density of states $g_j(\omega)$ is basically determined by plotting frequency contour in the BZ for $\omega_j(\boldsymbol{k}) = \omega$ and $\omega_j(\boldsymbol{k}) = \omega + \mathrm{d}\omega$ and then counting the number of modes enclosed between these surfaces. At low frequencies, $g_j(\omega)$ usually increases as ω^2, because long wavelength acoustic modes obey the Debye, or continuous medium, approximation. As ω_j increases further, however, $g_j(\omega)$ exhibits some structure, then begins to decrease, and eventually vanishes completely (Fig. 5.12), which corresponds to $(\omega, \omega + \mathrm{d}\omega)$ shells lying progressively outside of the first BZ. The total density of state is simply defined as the sum of the individual densities of all the branches, i.e.

$$g(\omega) = \sum_j g_j(\omega).$$

Because of the anisotropic aspects of the dispersion curves, the energy gaps may be smeared out in the density of states.

The dispersion curves of phonons in crystals and the corresponding density of states are determined by inelastic scattering of neutrons from these materials. By measuring both energy and momentum of the neutrons before and after scattering by the sample and using conservation laws, the wavevectors $\boldsymbol{k}$ and frequency ω of the phonons are deduced.

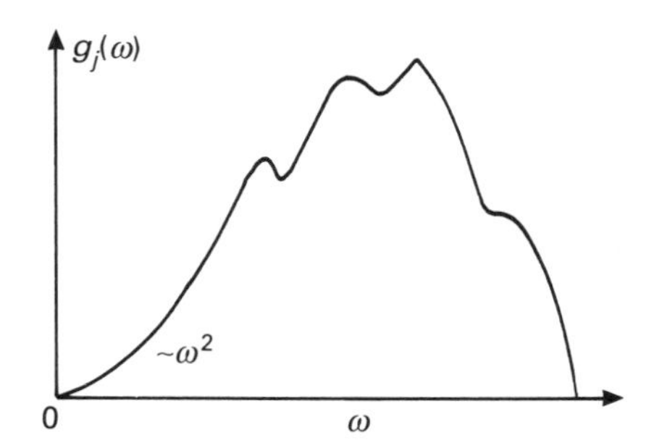

FIG. 5.12 A typical density of state curve.

5.2.5 Strain/stress distribution in periodic lattices due to structure defects

The concept of a perfect crystal is an extremely useful and appealing one. But real crystals are not perfect. The example of the elastic waves as developed in the previous section is actually a good illustration of the 'imperfection' of a crystal with respect to its periodic description. Besides this 'propagating' field of strain that perturbs the average lattice, 'diffusive' strain/stress fields can be associated with so-called structure defects. The most important types of such defects are point defects, line defects, and surface defects. We will not mention any further surface defects, which include the surface of the sample itself and grain boundaries in a polycrystalline sample.

Point defects are just what they say they are: irregularities in the crystal structure, localized to a lattice site or to an equivalent piece of space. Examples are substitutional or interstitial impurities which occupy unduly a site position or a position between the host atoms. Self-interstitial defects can also exist. The region surrounding an impurity is strained, the extent of the strain depending on the kind of impurity atom and its location.

Vacancies are also point defects. They are empty lattice sites from which the usual atom has been removed. They are created, at equilibrium, by thermal excitation, because large (high-temperature) amplitude vibration allows some atoms to leave their sites completely. The region surrounding a vacancy is distorted because the lattice relaxes. Vacancies are involved in many atomic transport mechanism, i.e. atomic diffusion: atoms move by elementary jumps into close vacancies. The mechanism is essentially efficient at high temperature when the vacancy concentration is high enough (about 10^{-4} per atom in a regular metal, near the melting point).

A line defect, also called a **dislocation**, is a linear array of misplaced (dislocated) atoms extending over a considerable distance inside the lattice. This type of defect is primarily responsible for the softness and ductility of some solids, such as metals. If the plastic deformation of a perfect crystal were initiated by parallel slip displacement of two atomic planes, the necessary force would be of the order of the shear modulus μ, whereas in experiments, forces smaller by a factor of 10^{-4} cause plastic flow, thanks to the existence of dislocations. The simplest type of dislocation is an edge dislocation, which can be viewed as an extra half atomic plane being inserted into the crystal lattice (Fig. 5.13). The defect itself is the line L corresponding to the border of this extra half-plane. In the line L, or core of the dislocation, the lattice is structurally changed, whereas outside L it is only distorted, without losing the topology of the perfect lattice. The motion (diffusive again) of the edge dislocation is illustrated in Fig. 5.14; only a single jump

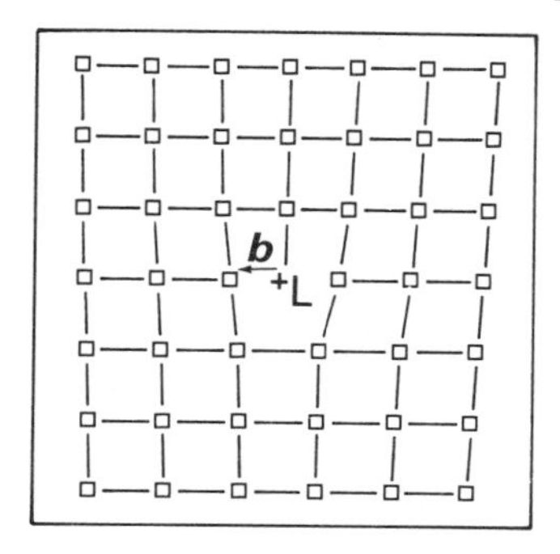

FIG. 5.13 Edge dislocation in a square lattice. An additional row of atoms stops at the defect core L.

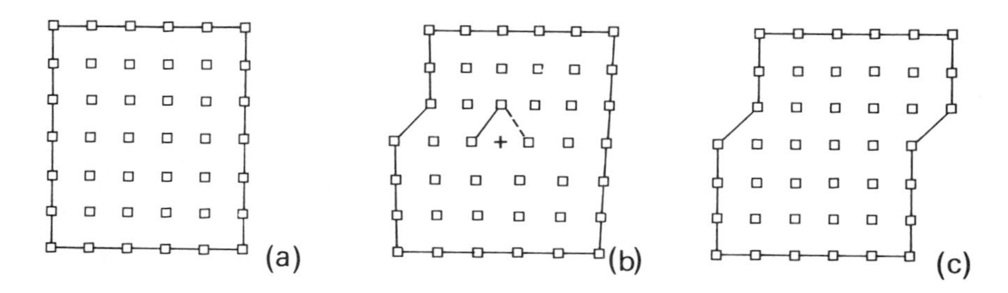

FIG. 5.14 Motion of an edge dislocation: (a) initial state without dislocation; (b) moving the dislocation by a jump of an atom close to the core; (c) the final state, resulting in an equivalent rigid shear.

of one atom near the defect core is required to move the dislocation by one lattice constant. Successive such atomic jumps allow an easy slip motion, ending with the two halves of the crystal in positions equivalent to the result of a rigid shear. The dislocation is characterized by its Burgers vector $\boldsymbol{b}$ which is the missing lattice vector to make a closed loop equivalent to zero translation around the dislocation line in the atomic plane perpendicular to the line (Fig. 5.13).

A more general construction method for dislocation is the Volterra process. It can also be applied for other types of line defect (disclinations for instance). The Volterra process is illustrated in Fig. 5.15 and consists of the following successive steps:

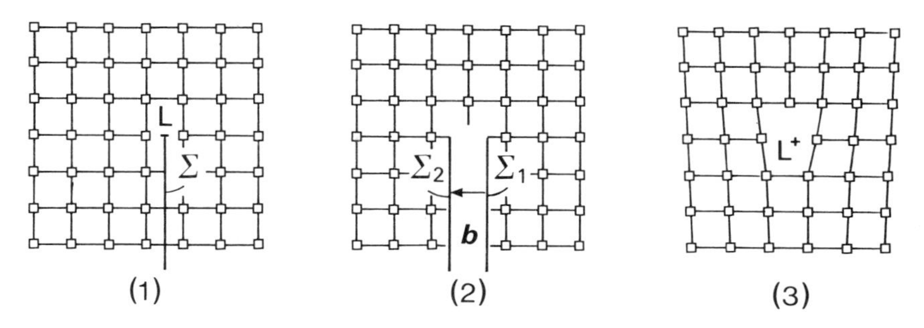

FIG. 5.15 Volterra process for the construction of a dislocation in a square lattice (see text).

(1) cut the crystal along an arbitrary surface which is terminated by a line L;
(2) separate the two lips of the cut by a vector ***b***, the Burgers vector, which is a lattice vector;
(3) fill the space between lips with extra matter and glue it to the lips. The lattice is restored everywhere outside the line L, except for a continuous strain field.

A disclination is also constructed by the Volterra process, but the two lips are related by a lattice rotation instead of a translation (Fig. 5.16). The formation energy for a dislocation, i.e. the associated elastic strain energy is about 10 eV per atomic length, which is considerable when compared with thermal energies. Consequently, dislocations are not thermally created but result from mechanical processes.

In an edge dislocation, the Burgers vector ***b*** is perpendicular to the

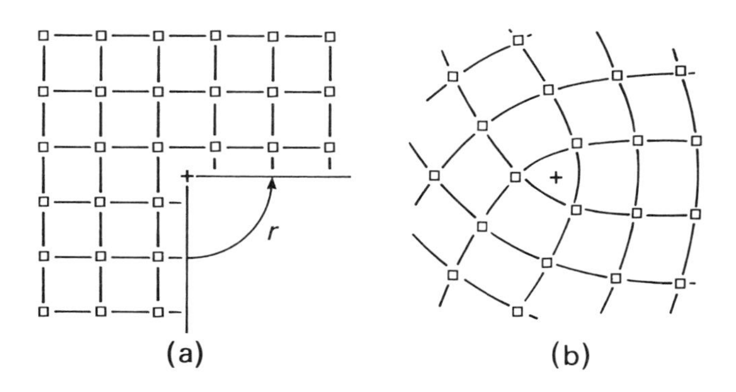

FIG. 5.16 Construction of a $\pi/2$-disclination in a square lattice. A $\pi/2$-sector is removed and lips are glued together.

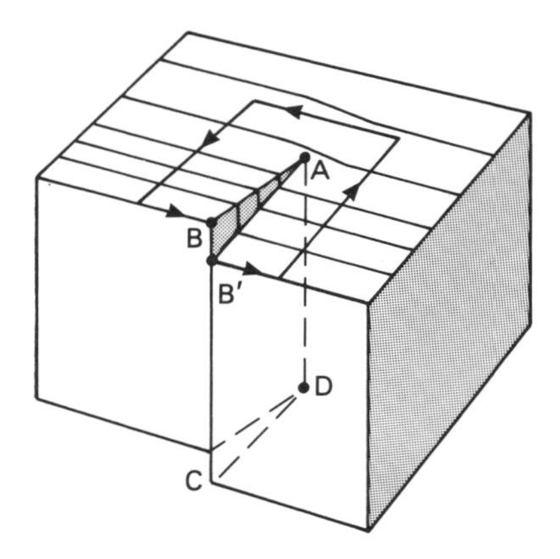

FIG. 5.17 A screw dislocation. The dislocation is represented by the line AD. Lines on top represent traces of vertical atomic planes. The shaded area ABB′ indicates the region of slippage. (Points B and B′ were coincident before the dislocation was created).

dislocation line. There are also so-called **screw dislocations**, with Burgers vector $\boldsymbol{b}$ parallel to the dislocation line. They are created by a planar cut made in the crystal and the subsequent slip up of one side past the other side (Fig. 5.17). If one moves in the atomic planes, around the dislocation, one finds that the plane actually spirals. Changes in volume around a screw dislocation line are less than for an edge dislocation, which involves considerable expansion.

5.2.6 Generalized continuum elasticity and influence of fluctuating strain fields on diffraction patterns

This section will propose a generalized expression of density fluctuation arising from a fluctuating strain field in periodic crystals. The consequences for diffraction will be also derived in a general way, with a view to an easy extension to the case of quasiperiodic structures.

Order parameters distinguishing periodic solids from fluids can be defined in terms of Fourier components of the mass density. Let $\rho(\boldsymbol{r})$ be the mass density at position $\boldsymbol{r}$ in a d-dimensional space ($d = 3$ for regular crystals and $d > 3$ for quasicrystals), for a simple system with a single species of atoms of mass m located at position $\boldsymbol{r}^{\alpha}$. This density can be expressed as

$$\rho(\boldsymbol{r}) = \sum_{\alpha} m\delta(\boldsymbol{r} - \boldsymbol{r}^{\alpha}). \tag{5.7}$$

Its average $\langle \rho(\boldsymbol{r}) \rangle$ over an equilibrium ensemble is a periodic function of $\boldsymbol{r}$ satisfying $\langle \rho(\boldsymbol{r}+\boldsymbol{R}) \rangle = \langle \rho(\boldsymbol{r}) \rangle$ for every vector $\boldsymbol{R}$ in the periodic lattice. Periodicity in coordinate space implies that $\langle \rho(\boldsymbol{r}) \rangle$ can be expressed as a discrete Fourier sum:

$$\langle \rho(\boldsymbol{r}) \rangle = \sum_{\boldsymbol{G}} \langle \rho_{\boldsymbol{G}}(\boldsymbol{r}) \rangle \exp(\mathrm{i}\boldsymbol{G}\cdot\boldsymbol{r}), \tag{5.8}$$

where $\boldsymbol{G}$ is a vector in a d-dim periodic lattice, reciprocal to the lattice of vector $\boldsymbol{R}$. The order parameters $\langle \rho_{\boldsymbol{G}}(\boldsymbol{r}) \rangle \equiv \langle \rho_{\boldsymbol{G}} \rangle$ are mass density wave amplitudes that distinguish the solid from the fluid phase and are independent of $\boldsymbol{r}$ in an equilibrium ensemble in the absence of applied stresses. For a fluid all $\langle \rho_{\boldsymbol{G}} \rangle$ are zero except for $|\boldsymbol{G}| = 0$. Applied stresses will distort the lattice and generate a position dependence of $\langle \rho_{\boldsymbol{G}}(\boldsymbol{r}) \rangle$ which are nothing else than the structure factors at reciprocal lattice vector $\boldsymbol{G}$; the scattering intensity into Bragg peaks is proportional to $|\langle \rho_{\boldsymbol{G}} \rangle|^2$. These order parameters $\langle \rho_{\boldsymbol{G}}(\boldsymbol{r}) \rangle$ are actually complex numbers with an amplitude and a phase, i.e.

$$\langle \rho_{\boldsymbol{G}}(\boldsymbol{r}) \rangle = |\langle \rho_{\boldsymbol{G}} \rangle| \exp\{\mathrm{i}\phi_{\boldsymbol{G}}(\boldsymbol{r})\}. \tag{5.9}$$

As the density must be a real function, one has $\phi_{\boldsymbol{G}}(\boldsymbol{r}) = -\phi_{-\boldsymbol{G}}(\boldsymbol{r})$. Under a rigid translation $\boldsymbol{u}$ of all particles, the average density changes from $\langle \rho(\boldsymbol{r}) \rangle$ to $\langle \rho(\boldsymbol{r}+\boldsymbol{u}) \rangle$, which can be expressed using eqn (5.8) as

$$\langle \rho(\boldsymbol{r}) \rangle = \sum_{\boldsymbol{G}} \langle \rho_{\boldsymbol{G}}(\boldsymbol{r}) \rangle \exp\{\mathrm{i}(\boldsymbol{G}\cdot\boldsymbol{r} + \boldsymbol{G}\cdot\boldsymbol{u})\}. \tag{5.10}$$

This is equivalent to writing that the phases $\phi_{\boldsymbol{G}}$ have been changed into $\phi_{\boldsymbol{G}} = \phi_{\boldsymbol{G}}^0 + \boldsymbol{G}\cdot\boldsymbol{u}$. The free energy F of any periodic system does not depend on the spatially uniform vector field $\boldsymbol{u}$, but spatially non-uniform distortions of the displacement vector defined via eqn (5.10) do, however, increase the free energy. An analytic expansion of the free energy in powers of strain field components yields the elastic free energy F_{el} which can be used to express the partition function of the system, and then to calculate the fluctuations of any function of the strain field, such as the Debye–Waller factor.

Diffraction measurements probe the Fourier transform of the density–density correlation function. The elastic scattering intensity from a sample of volume V is

$$I(\boldsymbol{Q}) = \int d^3\boldsymbol{r} \int d^3\boldsymbol{r}' \exp\{\mathrm{i}\boldsymbol{Q}(\boldsymbol{r}-\boldsymbol{r}')\} \langle \rho(\boldsymbol{r})\rho(\boldsymbol{r}') \rangle. \tag{5.11}$$

Substituting eqn (5.8) into eqn (5.11) gives the expression of the elastic scattering intensity in terms of the mass density wave amplitudes, i.e.

$$I(Q) = V^2\left(\sum_G |\langle\rho_G\rangle|^2 \delta_{Q,G} + (\langle\rho_Q\rho_{-Q}\rangle - \langle\rho_Q\rangle\langle\rho_{-Q}\rangle)\right). \quad (5.12)$$

The first term is the coherent scattering intensity, and gives the intensity at Bragg peaks at reciprocal vector $\boldsymbol{G}$. This intensity is reduced by the Debye–Waller factor, when fluctuations arise in the applied strain field, with respect to the expected 'perfect' unstrained crystal. The Debye–Waller factor is given by the expression:

$$DW = \exp(-2W_G) = |\langle\exp\{i\boldsymbol{G}\cdot\boldsymbol{u}(\boldsymbol{r})\rangle|^2 = \exp\left(-\sum G_i G_j \langle u_i(\boldsymbol{r}) u_j(\boldsymbol{r})\rangle\right). \quad (5.13)$$

An elastic strain field, say for instance a uniform isotropic compression or expansion, also results in a position shift of the Bragg peaks. As the lattice spacing d is related to the Bragg peak positions $\boldsymbol{G}$ via the equation

$$|G| = \frac{2\pi}{d},$$

it straightforwardly follows that

$$|\Delta \boldsymbol{G}| = \frac{\Delta d}{d}|\boldsymbol{G}|.$$

Consequently, strain fields have the strongest influence at large $|\boldsymbol{Q}|$ values. Fluctuating strain fields will result in peak broadening effects.

The last two terms in eqn (5.12) are the contributions arising from the fluctuation in the density wave amplitudes and produce an incoherent background or diffuse scattering, which compensates for the Debye–Waller loss in the Bragg peaks.

To finish with this section we would like to emphasize the important physical concept expressed by eqn (5.10), i.e. a strain field $\boldsymbol{u}$ as applied to a periodic crystal induces a phase shift $\boldsymbol{G}\cdot\boldsymbol{u}$ in the mass density wave amplitudes $\langle\rho_G\rangle$. This will be extended to the problem of strain fields in quasicrystals.

Before this, we would like to illustrate, with simple toy models, the expected influence on lattice dynamics of periodicity breaking.

5.3 Phonons in disordered materials

5.3.1 Vibration modes in an atomic chain with mass defects

Let us consider again the toy system made of balls connected by identical springs in 1-dim space (Fig. 5.5). The balls are still distributed periodically in the chain but their masses can differ. The equations of motion are similar to eqn (5.3), but with the mass parameter bearing the position index:

$$m_n \frac{d^2 u_n}{dt^2} = \alpha(u_{n+1} + u_{n-1} - 2u_n). \tag{5.14}$$

The labels are as defined in Section 5.2.2. If the mass distribution is assumed to obey the proper conditions, m_n can be developed in terms of 'mass waves', i.e.

$$m_n = \sum_k M_k \exp(ikna)$$

or

$$M_k = \frac{1}{N}\sum_n m_n \exp(-ikna) \tag{5.15}$$

with $k = h(2\pi/Na)$ (h integer) if m_n reproduces periodically after a translation of N atomic spacings. Now, solutions of the equation of motion (5.14) are no longer necessarily simple plane waves. They are Bloch-like solutions of the form

$$u_n = a_n \exp(i\omega t) \quad \text{with} \quad a_n = \sum_k A_k \exp(ikna). \tag{5.16}$$

Substituting m_n (eqn (5.15)) and u_n (eqn (5.16)) into eqn (5.14) gives

$$\omega^2 \sum_h M_{p-h} A_h - 4\alpha A_p \sin^2 \frac{p\pi}{N} = 0. \tag{5.17}$$

As a particular example, let us take all $m_n = m$, except for the ball at site n_0 which has a mass $m(1-\epsilon)$ with $\epsilon > -1$. This is the simple impurity problem. Then the above equations transform easily into

$$\begin{gathered} M_h = m\left[\delta_{0h} - \frac{\epsilon}{N}\exp\left(-i\,\frac{2\pi h}{N}\,n_0\right)\right] \\ \left[m\omega^2 - 4\alpha\sin^2\left(\frac{p\pi}{N}\right)\right] A_p \exp\left(i\,\frac{2\pi h}{N}\,n_0\right) = \frac{\epsilon}{N}\, m\omega^2 a_{n_0} \\ a_{n_0} = \sum_h A_h \exp\left(i\,\frac{2\pi h}{N}\,n_0\right). \end{gathered} \tag{5.18}$$

The only true propagating plane wave occurs at $\omega = 0$ with $a_n = A_0$; this is the translation mode at the limit of the acoustic regime. The other modes show up as two families of standing waves.

The first family with $a_{n_0} = 0$ corresponds to modes with a vibration node on the impurity atom. The dispersion law and vibration amplitudes are then given by

$$m\omega_p^2 = 4\,\alpha \sin^2\left(\frac{p\pi}{N}\right)$$

$$a_n = A_p \exp\left(\mathrm{i}\,\frac{2\pi p}{N}\,n_0\right)\left\{\exp\left[\mathrm{i}\,\frac{2\pi p}{N}(n - n_0)\right] - \exp\left[-\mathrm{i}\,\frac{2\pi p}{N}\,(n - n_0)\right]\right\}.$$

The vibration frequencies are those of the perfect lattice.

The second family corresponds to standing waves with a vibration maximum on the impurity atom. Approximate solutions for ω^2 and a_n are

$$m\omega^2 = \frac{4\,\alpha}{1 - 2\epsilon/N}\sin^2\left(\frac{p\pi}{N}\right)$$

$$a_n = A_p \exp\left(\mathrm{i}\,\frac{2\pi p}{N}\,n_0\right)\left\{\exp\left[\mathrm{i}\,\frac{2\pi p}{N}(n - n_0)\right] + \exp\left[-\mathrm{i}\,\frac{2\pi p}{N}\,(n - n_0)\right]\right\}.$$

Values of ω are above or below those of the perfect lattice depending on whether the impurity mass is lighter or heavier respectively, than the normal mass. For a light impurity a localized mode occurs whose features can be calculated exactly. Adding the second eqn (5.18) for all p values gives

$$1 - \frac{\epsilon}{N}\,m\omega^2 \sum_p \frac{1}{m\omega^2 - 4\,\alpha \sin^2(p\pi/N)} = 0.$$

For $m\omega^2$ values larger than 4α, the Σ_p term can be calculated as an integral, i.e.

$$m\omega^2 = \frac{4\alpha}{1 - \epsilon^2}\,.$$

Substituting this result in eqn (5.18) gives the amplitudes

$$a_n = \frac{\epsilon}{N}\frac{1}{1 - \epsilon^2}\,a_{n0} \sum_p \frac{\exp[\mathrm{i}(2\pi p/N)\,(n - n_0)]}{1/(1 - \epsilon^2) - \sin^2(p\pi/N)}\,.$$

The Σ_p term can be calculated as an integral in the complex space, i.e.

$$a_n = (-1)^{|n - n_0|}\left[\frac{1 - \epsilon}{1 + \epsilon}\right]^{|n - n_0|} a_{n0}\,. \tag{5.19}$$

The amplitude is a maximum at the impurity sites and shows an exponential decay on both sides. Atom displacements are localized around the impurity and localization strengthens as the impurity lightens.

5.3.2 *Vibration modes in 'amorphous solids'*

Of course, a mass defect is a rather weak disorder. However, it can be demonstrated that a more general defect, including spring constant changes would produce similar corruption of the dispersion law, i.e. appearance of localized modes.

When the number of 'point defects' is increased in the chain, there are more and more localized modes (for light impurities) and the 'standing states' also tend to localize or at least to become quasilocal excitations. One way of achieving this result is to consider the vibration modes that would have a node of amplitudes on all the impurity atoms. Qualitatively, this would force the amplitudes a_n, as defined by eqn (5.16), to have a large number of non-zero Fourier coefficients. This is equivalent to saying that the same frequency is carried by several superimposed plane waves with different wave vectors. Thus the crude effect of disorder is to expand the vibrational modes in reciprocal space, and thus to confine them in real physical space (so-called weak localization). The modes corresponding to $\omega = 0$ still propagate and, by continuity solution, the low-energy range (pseudo-acoustic regime) also propagates to some extent as damped plane waves. Thus the absence of long-range order prevents the vibrational modes in amorphous solids from being considered as plane waves with well-defined wavevectors. Raman spectroscopy and inelastic neutron scattering experiments suggest that the measured densities of states are related to tunnelling modes which weakly couple some rigid structural units. Strictly speaking, there are no analogues of acoustic and optical phonon modes in amorphous solids. However, the measured density of states and the results of numerical diagonalization of the large dynamic matrix show features which are reminiscent of TA, TO, LA, and LO branches, even if most details are washed out apparently because of the disorder present in the structure. This is illustrated in Fig. 5.18.

To summarize, the presence of disorder in a solid affects the vibrational excitations in a variety of ways. The lack of periodicity prevents proper definition of a reciprocal lattice. Therefore any excitation of the solid cannot be described simply with respect to the underlying lattice by means of plane waves or Bloch formalism, since there is no proper space to define a wavevector as a good quantum number. Furthermore, the description of vibrational excitations in terms of the dispersion law $\omega = f(q)$ becomes inappropriate. Finally, disorder scattering reduces the lifetime of excitations which may become spatially localized rather than extended, either in the strict sense of

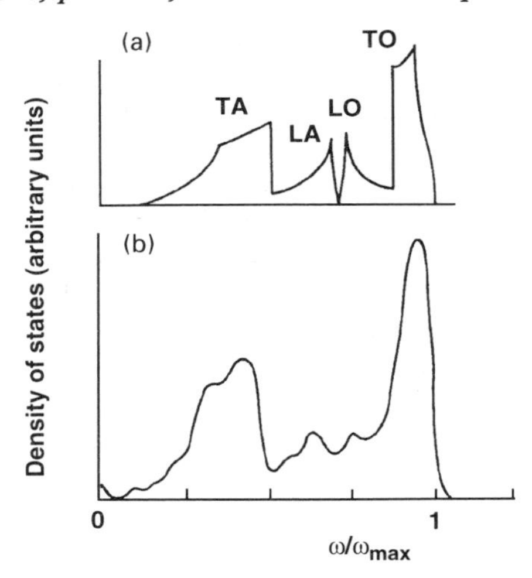

FIG. 5.18 Calculated density of states for (a) a diamond cubic structure and (b) a continuous random network model.

the word or in the form of a complex combination of standing states (weak localization). *Densities of states* which are observed are essentially *dominated* by *short-range order* or cluster-like effects, without the 'Van Hove singularities' which, for a crystal, are discontinuities in the slope of $g(\omega)$ due to Brillouin zone border effects ($d\omega/dq = 0$ in the $\omega - q$ dispersion curves).

One further problem arises in glasses or amorphous material as a result of the relaxation effect of the structure. Consequently, the dynamics observed, for example via density of states measurements, is a combination of both vibration and relaxation modes, i.e. a rather complex association of oscillations around fixed positions and some atomic jumps between these positions. This is reminiscent of the so-called soft modes which are observed in crystalline material when phase transitions are about to take place: prior to atomic rearrangement, interactions soften to facilitate the process. In the case of amorphous material, such 'soft modes' due to vibration–relaxation coupling show up as a bump in the low energy range in the measured density of states. This is known as the Boson peak.

5.3.3 Fractal structures and fractons

Some aspects of the structure and properties of a disordered solid can be described in terms of the concept of fractal structures. One class of fractal structures is related to self-similar geometry which, as already explained earlier in the book, means that details of the object considered are indistinguishable upon length rescaling operations. Such a geometry induces special density properties. If M is the mass of solid inside a sphere of radius R, we have $M = \alpha R^d$; $d = 3$ is known as the Euclidean dimension and α is a dimensionless constant parameter (e.g. $4\pi/3$). For a self-similar fractal object $M = \beta R^{\bar{d}}$; $\bar{d} \leqslant d$ is called the 'fractal dimension' and β is no longer a constant but varies with the true density of the solid ('lacunarity'). Fractal geometry is usually achieved within a space range between length scale limits, i.e. $a \leqslant r \leqslant \xi$. Beyond $r = \xi$, the structure transforms from fractal to Euclidean geometry. Euclidean spaces obey translational invariance (e.g. crystals), while fractal geometry exhibits 'dilatation' symmetry, i.e. with scale invariance. Therefore fractal structures can be considered as intermediate concepts between periodic crystals and disordered (random) materials.

Vibrational modes of a fractal network are called 'fractons' or, incorrectly, fractal phonons. Qualitatively, one can understand that there are two vibrational regimes. For short wavelengths $\lambda < \xi$, fractons are 'localized' within details of the structure scaling with λ. Low-energy modes ($\lambda > \xi$) are propagating phonon-like modes with some stationarity and/or damping aspects. The cross-over frequency ω_c between the two regimes is expressed as

$$\omega_c \propto \xi^{-\bar{d}/\bar{\bar{d}}} \tag{5.20}$$

where $\bar{\bar{d}}$ is called the 'fracton dimension'. Below ω_c, the phonon dispersion relation is Debye like, i.e. $\omega_p(k) \propto k$ or $\omega_p(\lambda^{-1}) \propto \lambda^{-1}$. Above ω_c, the fracton dispersion relation is

$$\omega_f(\lambda^{-1}) \propto \lambda^{-\bar{d}/\bar{\bar{d}}}. \tag{5.21}$$

The corresponding densities of states are

$$\begin{aligned} g_p(\omega) &\propto \omega^{d-1} \qquad (\omega^2 \text{ in 3-dim}) \\ g_f(\omega) &\propto \omega^{\bar{\bar{d}}-1} \end{aligned} \tag{5.22}$$

with some sort of discontinuity at ω_c. It is easy to show that eqns (5.21) and (5.22) are compatible within self-similarity. Assume for instance that b is the scale factor of the self-similar structure. Fracton eigenmodes must have 'steady-state'-like behaviour with wavelength scaling as the geometry, i.e.

$$\lambda(\xi/b) = \frac{1}{b}\lambda(\xi) \tag{5.23}$$

which also gives

$$[\lambda(\xi/b)]^{-\bar{d}/\bar{\bar{d}}} = b^{\bar{d}/\bar{\bar{d}}}[\lambda(\xi)]^{-\bar{d}/\bar{\bar{d}}}$$

or, according to eqn (5.21),

$$\omega(\xi/b) = b^{\bar{d}/\bar{\bar{d}}}\omega(\xi). \tag{5.24}$$

Then, taking the ($\bar{\bar{d}} - 1$) power of eqn (5.24) and accepting eqn (5.22) gives

$$g_f[\omega(\xi/b)] = b^{\bar{d} - \bar{d}/\bar{\bar{d}}} g_f[\omega(\xi)]$$

or

$$g_f[b^{\bar{d}/\bar{\bar{d}}}\omega(\xi)] = b^{\bar{d} - \bar{d}/\bar{\bar{d}}} g_f[\omega(\xi)]$$

which is indeed satisfied within the eqn (21) dispersion law. Thus the dispersion law of the form of eqn (5.21) and the density of states (5.22) are, at least, self-consistent. Typical values for $\bar{d}$ and $\bar{\bar{d}}$ are 2 and 4/3 respectively in the case of site percolation networks of fractal nature. These would give

$$\begin{aligned} \omega_f(k) &\propto k^{3/2} \\ g_f(\omega_f) &\propto \omega^{1/3}. \end{aligned} \tag{5.25}$$

These results are applicable (to some extent) to any self-similar structure, even if the range of self-similarity extends up to infinity ($\xi = \infty$) and if the structure is not fractal. Then the fracton-like dispersion law and density of states may apply down to low frequency (with $\bar{d} = d = 3$ and $\bar{\bar{d}} < d$ depending on the scale factor b). Such eigenmodes would not participate in sound velocity ($d\omega/dq = 0$ for $q = 0$) and contribute poorly to thermal conductivity. Compared with a Debye-like solid, the low-energy density of states would be reinforced at the expense of high-frequency modes.

5.4 Modulation and quasiperiodicity effects on lattice dynamics

5.4.1 A qualitative approach to modulation effects

In Section 5.2.3, we have seen that a diatomic chain made of two atoms with different masses has a dispersion curve exhibiting two branches. A gap is open around frequencies corresponding to a wavevector equal to half the wavevector of the superstructure. We mentioned also that a diatomic chain could be viewed as the simplest *mass modulation* of a monatomic chain. This strongly suggests that modulation and gap openings in the energy bands should be related phenomena.

Beyond masses, atomic interactions are the other ingredients that naturally participate in system dynamics. These atomic interactions are very often mimicked by spring constants in the usual toy model. Modulation of the spring constants, rather than mass modulation, is certainly a more realistic approach to real modulated systems. Actually, a system consisting of a linear chain with a unit cell containing masses $m_1, m_2, \ldots, m_N$ and spring constants $\alpha_1, \alpha_2, \ldots, \alpha_N$ has a dual system, in the sense that

$\alpha_n^* = c^2/m_n$, $m_n^* = c^2/\alpha_n$ (c^2 is an arbitrary constant) constitute a system which has the same vibration spectrum as the original one. Thus, everything that holds true for a chain with a unit cell of N different springs holds for a chain with N different masses, and vice versa. In particular, a modulated mass distribution is equivalent to a modulated spring model.

The above statement can be illustrated by the calculation of the vibration spectrum for a linear chain made of equispaced masses m but connected via two type of springs with constants α_1, α_2 regularly distributed along the chain. This gives two types of atoms, labelled (1) and (2), bonded to types (2) and (1), respectively, by spring constants (α_2, α_1) and (α_1, α_2) in that order. The equations of motion are then also of two types, i.e.

$$m\frac{\mathrm{d}^2 u_n^{(1)}}{\mathrm{d}t^2} = -\alpha_1(u_n^{(1)} - u_n^{(2)}) - \alpha_2(u_n^{(1)} - u_{n-1}^{(2)})$$

$$m\frac{\mathrm{d}^2 u_n^{(2)}}{\mathrm{d}t^2} = -\alpha_2(u_n^{(2)} - u_{n+1}^{(1)}) - \alpha_1(u_n^{(2)} - u_n^{(1)}),$$

with the standard notation. Plane wave solutions for the atomic displacements $u_n^{(i)}$ are of the form

$$u_n^{(1)} = U_0^{(1)}\exp\{\mathrm{i}(2nak - \omega t)\}$$

$$u_n^{(2)} = U_0^{(2)}\exp\{\mathrm{i}(2nak + ak - \omega t)\}.$$

When substituted into the equations of motion, these plane waves can propagate with frequencies given by the secular equation:

$$\begin{vmatrix} (\alpha_2 + \alpha_2) - m\omega^2 & -\alpha_1\exp(\mathrm{i}ak) - \alpha_2\exp(-\mathrm{i}ak) \\ -\alpha_1\exp(-\mathrm{i}ak) - \alpha_2\exp(\mathrm{i}ak) & (\alpha_1 + \alpha_2) - m\omega^2 \end{vmatrix} = 0.$$

Now, the 'dual' tranformation can be made, with

$$\alpha_i = c^2/m_i^* \qquad \text{and} \qquad m = c^2/\alpha^*.$$

Simple derivations result in the dispersion relation

$$\omega^2 = \alpha^*\left(\frac{1}{m_1^*} + \frac{1}{m_2^*}\right) \pm \alpha^*\left\{\left(\frac{1}{m_1^*} + \frac{1}{m_2^*}\right)^2 - \frac{4\sin^2 ka}{m_1^* m_2^*}\right\}^{\frac{1}{2}},$$

which is formally identical to eqn (5.6), i.e. the two branch dispersion curve of a regular diatomic chain.

Thus, according to circumstances, the modulation effects can be analysed using toy models with either mass modulation or spring constant modulation.

Now we may try to generalize the main conclusions by extending the model. Let us consider a 1-dim chain of identical particles with nearest-

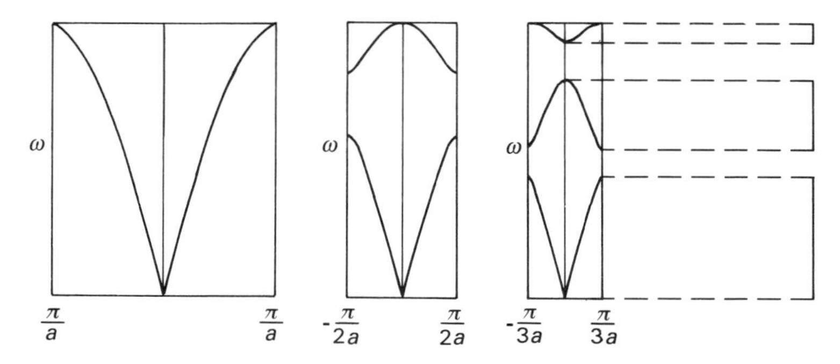

FIG. 5.19 Dispersion curves for monatomic chains with (a) no modulation, (b) modulation with a period $2a$, and (c) $3a$. For the last situation, the three branches are projected on a vertical axis, giving the spectrum of the allowed energy ranges.

neighbour interaction similar to that described in Section 5.2.2. If an external periodic strain field of some sort is imposed on the chain, the system will reach a new equilibrium configuration depending on the applied strain field. It is rather easy to understand qualitatively what would happen if the applied strain field has a space periodicity equal to Na, where a is the lattice constant of the unstrained chain. The strained chain has now gained a superstructure with a repetition distance equal to Na and with N atoms in the superstructure cell. The wavevector of the modulation is then $2\pi/Na$, which means that the dispersion curve will repeat in the k-space with a periodicity equal to $2\pi/Na$ which has to be considered as the extension of the new reduced Brillouin zone. In the extended zone scheme, $N-1$ gaps are expected to open at k vectors equal to π/Na, $2\pi/Na$, ..., $(N-1)\pi/Na$. In other words, the dispersion curve must have N branches in the reduced zone (from $-\pi/Na$ *to* $+\pi/Na$), one being of the acoustic type (all atoms moving with the same phase as a rigid body at the large wavelength limit) the $N-1$ others being optical-like (atoms moving π out of phase with each other; see Fig. 5.19). Details of the dispersion curves are obviously going to be influenced by the true nature and the typical parameters of the applied modulation strain field. In particular, one may expect dramatic effects when the modulation period is no longer commensurate with the lattice parameter.

Let us try to illustrate the point with a periodic chain of balls whose masses are modulated according to

$$m_n = m_0 + \Delta m \cos(qna).$$

Using the calculation method presented in Section 5.3.1, we can write the atomic displacements as (see eqn. 5.16)

$$u_n = a_n \exp(\mathrm{i}\omega t)$$

where

$$a_n = \sum_k A_k \exp(\mathrm{i}kna),$$

and the equations of motions are expressed as

$$\omega^2 \sum_k M_{k'-k} A_k - 4\alpha A_{k'} \sin^2\left(\frac{k'a}{2}\right) = 0$$

where (see eqns (5.15) and (5.17))

$$M_p = \frac{1}{N} \sum_n m_n \exp(-\mathrm{i}pna).$$

Substituting m_n as given above allows calculation of the Fourier coefficient M_p:

$$M_p = m_0 \delta_{0p} + \frac{\Delta m}{N} \sum_n \cos(qna) \exp(\mathrm{i}kna). \tag{5.26}$$

As a particular example, let us consider the simple case in which the mass modulation period is twice the repetition distance of the chain, i.e. $q = 2\pi/2a = \pi/a$. Equation (5.26) can easily be simplified to

$$M_p = m_0 \delta_{0p} + \Delta m\, \delta_{k,\pi/a}$$

which reduces the equations of motions to two different situations as follows:

$$\left[m_0\omega^2 - 4\alpha \sin^2\left(\frac{ka}{2}\right)\right] A_k + \Delta m \omega^2 A_{k-\pi/a} = 0$$

$$\Delta m\, \omega^2 A_k + \left[m_0\omega^2 - 4\alpha \cos^2\left(\frac{ka}{2}\right)\right] A_{k-\pi/a} = 0.$$

Solutions in A_k and $A_{k-\pi/a}$ are not trivial if the secular equation is satisfied, i.e.

$$\omega^2 = \frac{2\alpha m_0 \pm 2\alpha[m_0^2 - (m_0^2 - \Delta m^2)\sin^2(ka)]}{m_0^2 - \Delta m^2}. \tag{5.27}$$

The dispersion law becomes periodic with period π/a instead of $2\pi/a$, two branches are generated, and a gap opens at $k = \pi/a$. It should be noted that eqn (5.27) is strictly equivalent to eqn (5.6), with $m_1 = m_0 - \Delta m$ and $m_2 = m_0 + \Delta m$.

If the mass modulation period is ha instead of $2a$, the Fourier coefficients

M_p (eqn (5.26)) will exhibit h components each, and the equations of motion will form a linear system of h equations with h amplitude coefficients A_k. Consequently, the dispersion law will become periodic with period $2\pi/ha$, h branches, and $h-1$ gaps opening in proportion to the corresponding structure factors.

If the mass modulation period is not a rational expansion of the repetition length, all Fourier coefficients M_p of the mass function have non-zero values. Thus, we are left with as many coefficients M_p as there are m_n variables and the number of independent equations of motion cannot be reduced by the method. In the first Brillouin zone, after folding the whole dispersion law, gaps form a dense set at positions $k_i = \nu q/2$ (modulo $2\pi/a$) where ν is any integer. Low-order gaps (only small values of ν) correspond to significant openings, with the gap width being related to the structure factor of the satellites of the diffraction patterns (see Chapter 1). If the modulation amplitude is weak (Δm small) $M_{\mathrm{p}} \approx m_0 \delta_{0\pi}$ and the dispersion law remains almost unperturbed.

5.4.2 The notions of phason and amplitudon modes

Let us consider again a monatomic chain submitted to a modulation strain field. Fluctuations in the amplitude and/or the phase of this applied field may excite special modes of dynamics in the modulated structure. This is illustrated in Fig. 5.20 showing chains of atoms subjected to displacement strain fields having the same space periodicity (here $N = 3$) but with either different amplitudes (Fig. 5.20(*b*) and Fig. 5.20(*c*)) or different phases (Fig. 5.20(*b*) and Fig. 5.20(*d*)) with respect to the origin of the unconstrained chain. (Fig. 5.20(*a*)) When the amplitude changes, the atoms

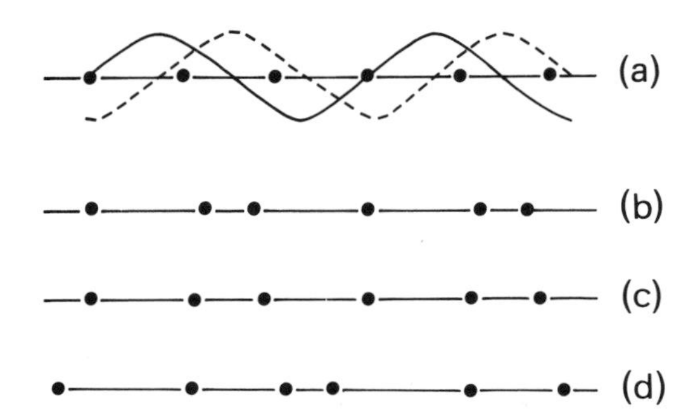

FIG. 5.20 A periodic monatomic chain of atoms (a) is submitted to a modulation strain schematized by the solid wavy line; the new distributions of atom positions are shown in (b) or (c) depending on the amplitude of the modulation; in (d) is shown the atomic arrangement corresponding to a phase shifted modulation (dashed line in (a) with the same amplitude as in (b)).

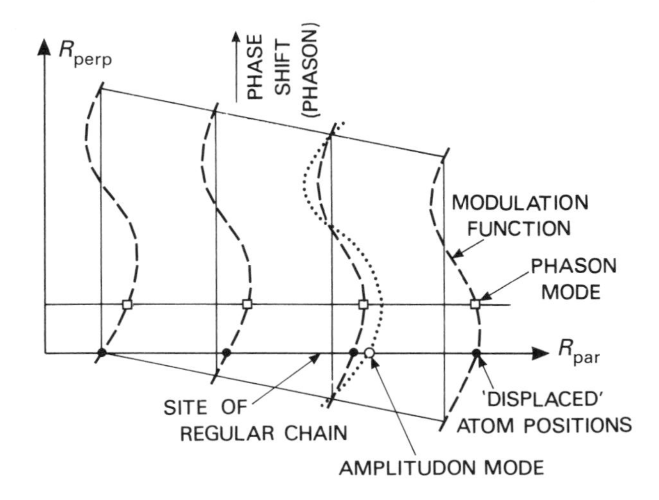

FIG. 5.21 Illustration of phason and amplitudon modes with the high-dim cut scheme of a modulated structure. Modulation of the position drives atoms to new sites (•) displaced from the initial regular lattice site. Phase shifts (□) or amplitude changes (○) in the modulation function modify further these sites.

are more or less shifted to the right-hand side, in between the two fixed atoms limiting the Na superstructure cell. When the phase changes, the displacement field as a whole entity is shifted to the left-hand side. Assuming that these two effects are time-dependent would generate 'amplitude' and 'phase' modes, called **amplitudons** and **phasons**.

As a further example to illustrate the point, the regular diatomic chain can be described as a modulated monatomic chain according to

$$m_{1,2} = m + \Delta m \cos(\pi n + \phi).$$

Odd n values generate the 'light atom' with mass $m_1 = m - \Delta m$, while even n values generate the heavy one with mass $m_2 = m + \Delta m$, as long as the phase shift ϕ of the modulation with respect to the lattice is kept equal to zero. Amplitudons would correspond to Δm variation, while changes in ϕ may generate phasons.

A better view of these special modes can be gained by looking at the modulated structure within its high-dimensional representation. As explained in Chapter 1, all the possible configurations of a modulated structure, corresponding to phase shift of the modulation function, can be 'piled up' in a complementary space, with the advantage of offering a simultaneous view of all the equivalent configurations. Figure 5.21 is a copy of Fig. 1.41, but with the addition of a second position of the physical space. In this figure, amplitudons would correspond to excited changes of the amplitude of the modulation function (dashed wavy lines). Such an amplitude change (here an increase) is exemplified in the figure. It produces additional shifts of the atoms, different in magnitude but all in the same direction if the amplitudon fluctuation is global (rigid oscillations of all the atoms). It is interesting to observe that, even in the large wavelength limit, amplitudons have non-zero energy because the final state of atom distribution is not identical to the initial state.

The phason modes are pictured in Fig. 5.21 by the atom position shifts resulting from a perpendicular translation of the physical space with respect to the modulation function. In such a displacement, some pair distances are increased while some others are reduced. A detailed examination of the position changes would demonstrate that the uniform phason (rigid perpendicular translation) does not imply energy transfer. Thus, phason oscillations must have a frequency equal to zero at the large wavelength limit. This **'large wavelength limit'** actually corresponds to atoms moving as if they were 'stuck' to the modulation function. Consequently, their displacements with respect to the initial unmodulated lattice have a wavevector equal to the wavevector $\boldsymbol{k}_i$ of the modulation function. Accordingly, the dispersion of the phason branch is linear around $\boldsymbol{k}_i$ in a similar way that acoustic phonon branches are linear around the zero wavevector. There is, however, an important difference between phasons and acoustic phonons. For the latter, the damping goes to zero when the wavevector tends to zero and the modes remain well defined. The reason is that for long wavelength acoustic modes, all particles in a neighbourhood move in phase and their mutual distances do not change much during a vibration period. The zero frequency for $\boldsymbol{k} = \boldsymbol{k}_i$ of the phason mode is also a consequence of an energy degeneracy: the total potential energy of the configuration does not depend on the phase but only on the amplitude of the applied modulation field. A change of the phase therefore leads from an atomic configuration to another with the same energy. But in this motion the relative position of the atoms are modified, which gives rise to dissipation effects. Thus, the phason modes always become over-damped if their wavevector difference from $\boldsymbol{k}_i$ is small enough. If this were not the case, phasons would modify the measured phonon dispersion curves by producing significant dips at $\boldsymbol{k} = \boldsymbol{k}_i$ (soft modes). This has indeed been observed for a few real modulated systems.

Thermally excited phonon and phason-like elastic field may be observed simultaneously. Let us call u_a and u_b the displacements of the unstrained chain and of the applied modulation field respectively. The phason modes are related to the relative displacement of the chain with respect to its modulation field, i.e. they are described by a coordinate w expressed as

$$w = \beta_1 (u_b - u_a).$$

Similarly, the phonon modes are related to a coordinate u describing displacements of the chain and the applied modulation field together; u is of the form

$$u = \beta_2(u_b + u_a)$$

The scale factors β_1, and β_2 are arbitrary to some extent. Inversion of the expressions for u and w is straightforward and gives

$$\begin{aligned} u_a &= (\beta_1 u - \beta_2 w)/2\beta_1\beta_2 \\ u_b &= (\beta_1 u + \beta_2 w)/2\beta_1\beta_2. \end{aligned} \tag{5.28}$$

These expressions will prove very useful when incommensurate modulations and quasicrystals are concerned.

To finish describing the general aspects of the problem, it is interesting to note that phason modes, when they do exist, do not form new degrees of freedom of the crystal. The total number remains three times the number of particles. This statement can be illustrated, again with the example of a simple sinusoidal modulation where the equilibrium positions of the atoms are

$$\boldsymbol{u}_n^0 = A\sin(\boldsymbol{k}_i \cdot \boldsymbol{n} + \phi).$$

The motion of the atoms in one phason mode is then given by

$$\begin{aligned} \boldsymbol{u}_n(t) &\simeq A\sin\{\boldsymbol{k}_i \cdot \boldsymbol{n} + \phi_k \sin(\boldsymbol{k}\cdot\boldsymbol{n} - \omega t)\} \\ &\simeq \boldsymbol{u}_n^0 + \tfrac{1}{2}A\phi_k\{\sin\{(\boldsymbol{k} + \boldsymbol{k}_i)\cdot\boldsymbol{n} - \omega t\} \\ &\quad + \sin\{(\boldsymbol{k} - \boldsymbol{k}_i)\cdot\boldsymbol{n} - \omega t\}\} \end{aligned} \tag{5.29}$$

The phason with wavevector $\boldsymbol{k}$ appears here simply as a mixture of phonons with wavevectors $\boldsymbol{k} + \boldsymbol{k}_i$ and $\boldsymbol{k} - \boldsymbol{k}_i$. For a more general modulation scheme, this will turn out to be a mixture of phonons with wavevectors $\boldsymbol{k} \pm m\boldsymbol{k}_i$, m being any integer.

Before moving to incommensurate modulation we would like to describe more quantitatively some of the points developed in this section. This will be addressed via a specific example of a commensurate modulated structure.

5.4.3 The modulated spring model[1]

The model is a modification of that presented in Section 5.2.2, i.e. a 1-dim chain of identical atoms with nearest-neighbour interactions, but with the spring constants varying periodically along the chain. Then the equations of motion (5.3) are given by

$$m\frac{d^2u_n}{dt^2} = -\alpha_{n+1}(u_n - u_{n+1}) - \alpha_n(u_n - u_{n-1}), \tag{5.30}$$

with

$$\alpha_n = \alpha\left\{1 - \varepsilon\cos\left(\frac{2\pi}{b}na + \varphi\right)\right\}$$

and a and b being integers; φ is the phase of the modulation with respect to the chain framework.

We consider now the normal mode vibrations of the chain by substituting

$$u_n^j = U_0^j \exp\{i(nka - \omega t)\}$$

into the equation of motion (5.30); j takes the values 1,2, . . ., N if the modulation period b is fixed at Na. The problem is considered for a given value of the modulation amplitude ε. The system for equations (5.30) is now

$$(\alpha_n + \alpha_{n+1} - m\omega^2)U_0^j - \alpha_n \exp(-ika)U_0^{j-1} - \alpha_{n+1}\exp(ika)U_0^{j+1} = 0. \tag{5.31}$$

There are N equations of this form. The problem then reduces to solving the secular equation given by the tridiagonal determinant, plus extra non-zero elements at the upper right and bottom left. The α_n are given by the second eqn (5.30) and they take N different values between the two extremes $\alpha[1 - \varepsilon\cos\{(2\pi N) + \varphi\}]$ and $\alpha(1 - \varepsilon\cos\varphi)$, incrementing by $2\pi/N$ the argument of the cosine. Note that $\alpha_{n+N} = \alpha_n$ (in particular $\alpha_0 = \alpha_N$). Abbreviating eqn (5.31)

$$\gamma_n^* U_0^{j-1} + \beta_n U_0^j + \gamma_{n+1}U_0^{j+1} = 0 \tag{5.32}$$

(with obvious definitions for β_n, γ_n), the $(N \times N)$ secular equation is now written

$$0 = \begin{vmatrix} \beta_1 & \gamma_2 & 0 & 0 & & \cdots & 0 & 0 & \gamma_N^* \\ \gamma_2^* & \beta_2 & \gamma_3 & 0 & & \cdots & 0 & 0 & 0 \\ 0 & \gamma_3^* & \beta_3 & \gamma_4 & 0 & \cdots & 0 & 0 & 0 \\ 0 & 0 & \gamma_4^* & \beta_4 & \gamma_5 & \cdots & 0 & 0 & 0 \\ \vdots & & & & & & & & \vdots \\ & & & & & & & & 0 \\ \gamma_1 & 0 & 0 & & & \cdots & 0 & \gamma_N^* & \beta_N \end{vmatrix} \tag{5.33}$$

The secular equation is of degree N in β_1, that is of degree N in ω^2, and the solutions are N branches of frequency with $N - 1$ gaps in between. As an example, let us write eqn (5.33) for the simple case $b = 2a$ $(N = 2)$, i.e.

$$\begin{vmatrix} \beta_1 & \gamma_2 + \gamma_1^* \\ \gamma_2 + \gamma_1^* & \beta_2 \end{vmatrix} = 0,$$

with

$$\beta_{1,2} = -m\omega^2 + 2\alpha$$

$$\gamma_2 = \alpha(1 - \varepsilon\cos\varphi)\exp(ika)$$

$$\gamma_1 = \alpha(1 + \varepsilon\cos\varphi)\exp(ika).$$

Then

$$m^2\omega^4 - 4\alpha m\omega^2 + 4\alpha^2(1 - \varepsilon^2\cos^2\varphi)\sin^2 ka = 0.$$

The two branches are given by

$$m\omega_\pm^2 = 2\alpha[1 \pm \{1 - (1 - \varepsilon^2\cos^2\varphi)\sin^2 ka\}^{\frac{1}{2}}].$$

This expression is formally similar to eqn (5.6) giving the two branch dispersion curve of a diatomic chain. Actually, we know that modulating the atom mass or the spring constant gives two systems which are exactly equivalent to each other.

At the Brillouin zone boundary ($k = \pi/2a$), the frequencies are given by

$$m\omega_\pm^2(Bz) = 2\alpha(1 \pm \varepsilon\cos\varphi).$$

The gap is a maximum for $\varphi = 0$, with the approximate value $\Delta\omega(Bz) \simeq \varepsilon$, and vanishes for $\varphi = \pi/2$. More generally, for other modulation periods maximal gaps are still obtained with $\varphi = 0$ but non-zero minimal gaps correspond to $\varphi = \pi/N$, though the difference between these two extreme situations is significant only for small values of N. The description of the structure of the gap distribution when the modulation period is changed can be restricted to b varied over the range $[a, 2a]$, by rational increments of a. The range $[2a, \infty]$ gives a symmetrical structure, as easily deduced from the expression for α_n(eqn 5.16). This convention seems to introduce unphysical situations, with N values ($b = Na$) which are not integers. The problem is easily ruled out by replacing the **rational** fractional values of N by the equivalent ratio of the smallest possible integer. For instance $N = 1.5$ is $N = 3/2$ and changing $2\pi \times 2/3 = 2\pi - 2\pi/3$ by 2π does not modify the definition of α_n; thus, $N = 1.5$ and $N = 3$ gives the same distribution of gaps (two gaps in fact). One may of course anticipate that at the limits of the range, especially close to $b = a(N = 1)$ the distribution of gaps changes quite abruptly. For b strictly equal to a, we have the simple monatomic periodic chain and there is no gap (eqn 5.4). If we now take $N = 1.02$ for instance, we can write: $2\pi/1.02 = 2\pi \times 100/102 = 2\pi(1 - 1/51)$ and the situation is equivalent to $N = 51$, i.e. 51 branches

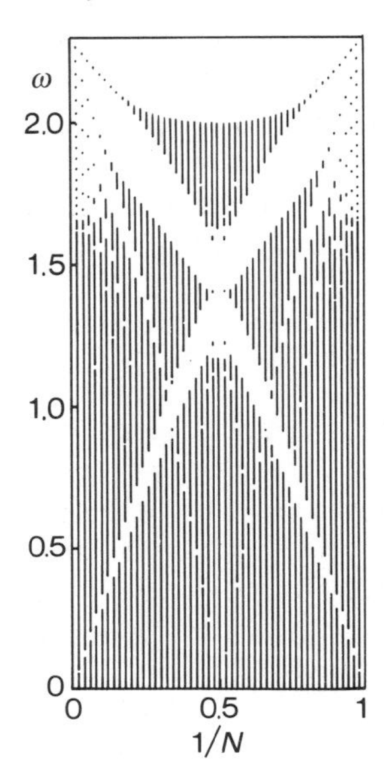

FIG. 5.22 The spectrum of a modulated spring model as a function of the modulation period (see text for details).[1]

of frequencies with 50 gaps. The full calculation of the gap distribution is straightforward in principle, though rather tedious without a computer. Figure 5.22, due to de Lange and Janssen,[1] shows the spectra for $\varphi = 0$ and for a number of discrete values of $1/N$ between 0 and 1. The ratio α/m was arbitrarily taken to be of unit magnitude and $\varepsilon = 10^{-\frac{1}{2}}$ for convenience. As expected, the number of gaps varies widely with the values of $1/N$, but the distribution of the bigger gaps is seen to be very smooth. Two 'diagonal branches' of 'large' gaps cross at $1/N = 0.5$, where there is only one gap. Two other 'branches' of smaller but still easily visible gaps run from bottom to top from $1/N = 0.5$ to $1/N = 0$ and 1 respectively. A third pair of gap branches, though less clear, starts at $1/N = \frac{1}{3}$ and $\frac{2}{3}$, etc. For situations equivalent to large N values, a substantially smaller number of gaps among the $N - 1$ ones are larger than a given small δ value. The very small gaps show up at the upper part of the ω range (upper corners of Fig. 5.22). The relatively large frequency 'bands' which are bounded by equally relatively large gaps are fully propagating modes, with either acoustic or optical characters. The narrow 'bands', or very narrow bands, sandwiching the numerous narrow gaps are necessarily of a peculiar type. Being so narrow they must correspond to completely flat frequency branches, the wavevector value within the reduced Brillouin zone having almost no influence on the frequency. These modes can also be viewed as isolated modes within a large band of forbidden energy. Both analyses suggest that these modes have special characters, midway between the propagating and the localized types. They are sometimes called **critical** modes; they can appear, everywhere in space but with a modulated amplitude due to complex steady state characters.

5.4.4 Excitation in incommensurate phases

The modulation becomes incommensurate when the repetition distance of the chain (in the above example) and the modulation period are not rational. In this case, using the notation of the previous section, the situation is roughly equivalent to $N \to \infty$. The lack of lattice periodicity for the whole system means that the usual treatment of lattice vibrations breaks down. There is no Brillouin zone and therefore no labelling by vectors in this zone or gaps on the boundaries.

We have observed that centres of Brillouin zones are defined by the reciprocal vectors of the Fourier components of the structure. This is the source of the gap openings and zone reduction mechanism resulting from commensurate modulation or superstructure formation. Zone centres correspond also to zero energy acoustic phonons, possibly confused with phason modes when they correspond to the modulation wavevectors.

If now these conclusions have to be extended to incommensurate modulations, one has to consider also some sort of pseudo-zone centres, defined again by the reciprocal vectors of the structure. For a simple static distortion of wavevector $\boldsymbol{k}_i$ as applied to a regular three-dimensional crystal, these vectors were expressed in Chapter 1 by a formula of the sort

$$\boldsymbol{Q} = h\boldsymbol{a}^* + k\boldsymbol{b}^* + l\boldsymbol{c}^* + m\boldsymbol{k}_i,$$

where $\boldsymbol{a}^*$, and $\boldsymbol{b}^*$, $\boldsymbol{c}^*$ are the three reciprocal basis vectors of the undistorted lattice and h, k, l, and m are integers. Thus, it may be considered that the satellite spots with $m \neq 0$ define additional, secondary, pseudo-Brillouin zone centres, with, in particular, 'satellite' acoustic branches going to zero energy at these points. Depending on the satellite spot intensities, the corresponding contributions to excitations may not be observable.

Let us try to illustrate this rather chaotic situation, again using the modulated spring model, but with a modulation of the spring distribution which is incommensurate with the lattice constant. Formally, the equations of motion remain similar to eqn (5.30) except that b and a in the α_n expression are not in a rational ratio and that there are an infinite set of coupled linear equations. One can of course say that the solutions are plane waves of the form

$$u_n = U(2\pi\, na/b)\exp(inka), \tag{5.34}$$

with

$$U(\tau) = U(\tau + 2\pi).$$

This property, however, does not lead to further simplifications. The system remains an infinite set. Since each irrational number may be arbitrarily approximated by a rational one, one may try to draw conclusions if a continuous behaviour is observed when the successive periodic approximants are used. The procedure is actually similar to that has been developed in the previous section, since periodic modulations are again concerned, with growing periodicity when the rational approximant becomes closer and closer to the irrational ratio. For instance, if the modulation period in a modulated spring model is $b = \tau a$, with τ the golden mean, the successive rational approximants of τ are the fractions 1/1, 2/1, 3/2, 5/3, 8/5, 13/8, . . ., i.e. F_n/F_{n+1} with $F_{n+1} = F_n + F_{n-1}$. In the scheme as developed in the previous section, these periodic modulated spring models would correspond to situations equivalent to large unit cells with 1,2,3,5, . . . atoms. The generation of more and more frequency gaps would follow the same scheme as developed for regular commensurate modulation. The spectral bands for the first approximants of $1/\tau$, the $N = F_n$ spectral bands, are shown in Fig. 5.23.[2] From the calculation, but also directly from the diagram, it appears that the spectrum becomes self-similar: the same sequences of frequency bands occur all over the diagram on different scale. When N becomes very large, the total width of the bands tends to zero and the spectrum is a pure point pattern. Numerical calculations show that the gap structure is completely determined by the continued fraction expansion of the irrational ratio. For instance, if the irrational ratio of the two periods (lattice and modulation) is the golden mean τ, then the continuous fraction expansion is:

$$\tau = 1 + \cfrac{1}{1 + \cfrac{1}{1 + \cfrac{1}{1 + \cfrac{1}{1 + \cfrac{1}{1 + \cdots \ddots 1 + \cfrac{1}{1 + \cdots}}}}}}. \tag{5.35}$$

Each term or fraction added to a given degree of expansion generates the division of one frequency band into two sub-bands of width proportional

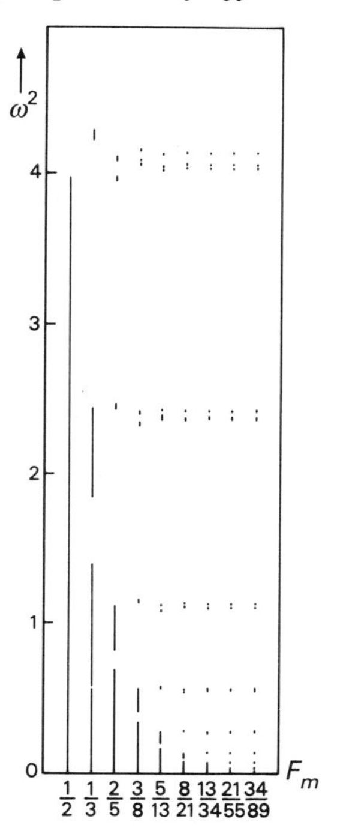

FIG. 5.23 Spectra of a modulated spring model for successive approximants of the Fibonacci chain.[2] Approximant values of $1/\tau$ are indicated on the horizontal axis.

to the corresponding approximants. The subdivision into two sub-bands is clearly visible in Fig. 5.23, particularly for the successive splits of the lowest frequency band.

Now, it can be easily understood that the gap openings depend also on the 'strength' of the modulation, that is on its amplitude. The weaker this amplitude, the closer to that of the 'normal' crystal is the dispersion curve of the incommensurate modulated structure, the differences lying in the appearance of the more pronounced occasional gaps. If the modulation becomes very strong, the whole chain may be 'pinned' by the external modulation potential and uniform translations, i.e. long wavelength acoustic phonons can no longer be excited at zero energy level: the $k = 0$ propagating modes correspond to ω not equal to zero. Examples of numerical calculation of the dispersion curves of a linear chain with increasing modulation potential are shown in Fig. 5.24,[3] which illustrates the effect of modulation amplitude changes for a given incommensurate period of modulation.

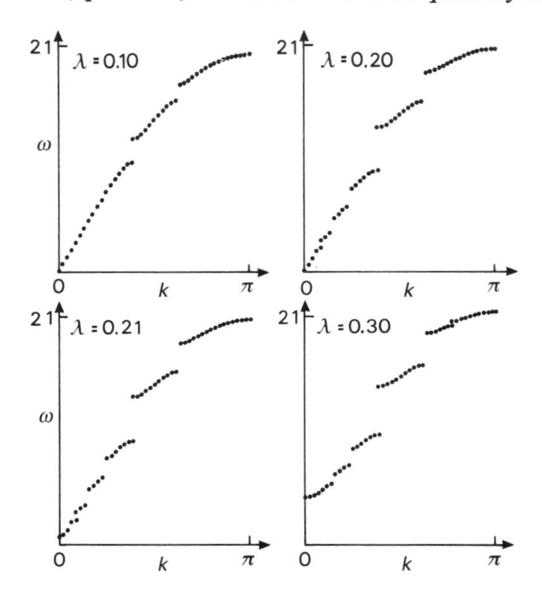

FIG. 5.24 Phonon spectrum of a modulated monatomic chain [3] for several amplitudes of the modulation potential. The λ parameter is a measure of this potential.

5.4.5 *Numerical results for a Fibonacci chain*

If we consider the Fibonacci chain presented in Chapter 1 (Sections 1.4.1 and 1.4.2), the atom positions along the chain are given by eqn (1.28), namely:

$$x_n/S = n(1 + 1/\tau^2) + 1/\tau^2 - (1/\tau)\, F\{(n+1)/\tau\},$$

in which S is a scaling factor and $F\{y\}$ represents the fractional (non-integer) part of y; $F\{y\} \in (0 - 1)$.

The distance a_n between two nearest-neighbour 'atom positions' is then given by

$$a_n = x_{n+1} - x_n = S(1 + 1/\tau^2) - (S/\tau)\,[F\{(n+2)/\tau\} - F\{(n+1)/\tau\}].$$

The average spacing $\bar{a}$ over the chain is calculated easily via the formula

$$\bar{a} = \lim_{N\to\infty} \frac{1}{N} \sum_{n=0}^{N} a_n .$$

In the second term of a_n the difference of the two $F\{\ \}$ quantities is obviously smaller than unity. Summing N terms of the sort results in a number smaller than N which, divided by N, goes to zero when N goes to infinity. Thus, an average spacing can indeed be defined and has the value

$$\bar{a} = S(1 + 1/\tau^2). \tag{5.36}$$

If the structure has an averaged spacing, an average periodic lattice should exist. This suggests that the Fibonacci chain may be viewed as a regular periodic chain with an incommensurate modulation of the interatomic spacing. Rewriting eqn (1.27) here:

$$x_n = S\{n + (1/\tau) \lfloor (n + 1)/\tau \rfloor\},$$

which is an equivalent expression of the atom positions in the chain, the distance between two atoms is

$$\begin{aligned} a_n &= x_{n+1} - x_n \\ &= S\{1 + (1/\tau)(\lfloor (n+2)/\tau \rfloor - \lfloor (n+1)/\tau \rfloor)\} \end{aligned}$$

($\lfloor y \rfloor$ means the largest integer contained in y).

The function in the second term has a period equal to τ, as demonstrated below.

$$\begin{aligned} f(x) &= \lfloor (x + 2)/\tau \rfloor - \lfloor (x + 1)/\tau \rfloor \\ f(x + \tau) &= \lfloor (x + 2)/\tau + 1 \rfloor - \lfloor (x + 1)/\tau + 1 \rfloor \\ &= (\lfloor (x + 2)/\tau \rfloor + 1) - (\lfloor (x + 1)/\tau \rfloor + 1) \\ &= f(x). \end{aligned}$$

This periodic function takes only two different values, namely 0 and 1. Indeed, as $(n + 2)/\tau = (n + 1)/\tau + 1/\tau$, the integer part of $(n + 2)/\tau$ is equal to or larger by one than that of $(n + 1)/\tau$, depending whether the fractional part of $(n + 1)/\tau$ is smaller or larger than $1 - 1/\tau$.

We conclude that the Fibonacci chain is equivalent to a modulated bond (or spring) structure; the basic periodic structure has a repetition distance S and the modulation function takes the values 0, 1 with a periodicity τ, i.e.

$$a_n = S(1 + f(n)/\tau), \tag{5.37}$$

with

$$f(x + \tau) = f(x) = 0 \text{ or } 1.$$

If phason modes are excited, this corresponds to the chain sliding with respect to the modulation function, with the consequence that some bond lengths have to turn from 1 to τ (times S) and vice versa. Thus, in the present case, the phason modes are easily interpreted as dynamical processes that generate collective permutation of occasional (L, S) bond pairs.

As suggested in the previous section, the vibration spectrum of any incommensurate (quasiperiodic) structure is expected to be very chaotic, or at least to have very chaotic components. However, part of this chaos may not show up in experimental approaches, just because most of the phonon modes have scattering intensities that are too weak. This is the case in particular when the phonons are observed in Brillouin zones around weak Bragg reflection.

If a Fibonacci structure is considered as a chain of atoms connected by harmonic springs with nearest-neighbour force constants, its equations of motion are given by

$$m_n \omega^2 u_n = \alpha_{n+1,n}(u_{n+1} - u_n) + \alpha_{n-1,n}(u_{n-1} - u_n)$$

for all n values corresponding to atoms in the sequence. Even if m_n takes a single value, analytical solutions cannot be found for the eigenmode $u_n \exp(\mathrm{i}\omega t)$ because nearest neighbours of the nth atom do not repeat regularly. The equation of motion above can be formally written as

$$u_{n-1} = c_n u_n + d_n u_{n+1}$$

or

$$\mathbf{w}_n = \begin{bmatrix} u_{n-1} \\ u_n \end{bmatrix} = \begin{bmatrix} c_n & d_n \\ 0 & 0 \end{bmatrix} \begin{bmatrix} u_n \\ u_{n+1} \end{bmatrix} = T_n \mathbf{w}_{n+1} \tag{5.38}$$

where

$$c_n = 1 + \frac{\alpha_{n+1,n}}{\alpha_{n-1,n}} - \frac{m_n \omega^2}{\alpha_{n-1,n}}$$

$$d_n = -\frac{\alpha_{n+1,n}}{\alpha_{n-1,n}}.$$

T_n is called a 'transfer matrix'. When applied recursively, transfer matrices relate the u_n to each other. From eqn (5.38) we can deduce that

$$\mathbf{w}_1 = T_1 T_2 T_3 \ldots T_{N-1} T_N \mathbf{w}_{N+1} = Q \mathbf{w}_{N+1}. \tag{5.39}$$

For fixed boundary condition $u_0 = u_{N+1} = 0$ and eqn (5.39) becomes

$$\begin{bmatrix} 0 \\ u_1 \end{bmatrix} = \begin{bmatrix} Q_{11} & Q_{12} \\ Q_{21} & Q_{22} \end{bmatrix} \begin{bmatrix} u_N \\ 0 \end{bmatrix}.$$

Because of the self-similar properties of the Fibonacci sequence it is easy to verify that $T_n = T_{n-2} T_{n-1}$, which simplifies the calculation of the Q matrix.

For a non-trivial solution, the vibrational frequency ω must satisfy the eigenvalue equation:

$$Q_{11}(\omega) = 0.$$

This equation must be solved numerically. The vibrational density of states is then given in terms of the statistics of the eigenfrequencies:

$$g(\omega) = \sum_k \delta(\omega - \omega_k)$$

The character of the eigenstates is well described by the participation ratio

$$P(\omega_k) = \frac{\Sigma_n |\boldsymbol{e}_n(\omega_k)|^2}{N\Sigma_n |\boldsymbol{e}_n(\omega_k)|^4} \tag{5.40}$$

in which $\boldsymbol{e}_n(\omega_k)$ is the eigenvector (unit vector) defining the polarization of the atomic displacement for the mode of frequency ω_k on the atom at position $\boldsymbol{n}$. The eigenvectors obey the orthogonality relation:

$$\sum_n |\boldsymbol{e}_n(\omega_k)|^2 = 1.$$

If the mode is extended all $|\boldsymbol{e}_n(\omega_k)|^2$ are equal and of the order of $1/N$ to satisfy the orthogonality relation; as a consequence $|\boldsymbol{e}_n(\omega_k)|^4$ is equal to $1/N^2$ whatever n and $\Sigma_n |\boldsymbol{e}_n(\omega_k)|^4 = 1/N$. This means that the participation ratio is $P(\omega_k) \approx 1$ for extended modes. If the mode is strictly localized, $|\boldsymbol{e}_n(\omega_k)|^2 = 0$ for all n except one; as a result $\Sigma_n |\boldsymbol{e}_n(\omega_k)|^2 = \Sigma_n |\boldsymbol{e}_n(\omega_k)|^4 = 1$ and $P(\omega_k) \approx 1/N$, i.e. almost zero. Between these two cases, the space distribution of the eigenvector $\boldsymbol{e}(\omega_k)$ is such that most of them may have non-zero but different values. The participation ratio can thus take a value between $1/N$ and 1, which is a measurement of the degree of localization. The three cases are illustrated in Fig. 5.25.

From a more physical point of view, critical vibrational modes can be described as exponentially localized wave packets which tunnel from one cluster to its copies in the structure, if such identical clusters appear in the structure. An equivalent way of stating the same thing is to consider the squared vibrational amplitude $|U_n|^2$ on the atom at site n and to count the number $N(\epsilon)$ of sites where $|U_n|^2$ is larger than any small threshold ϵ, as a function of the system size L. For extended modes, $|U_n|^2$ is the same at all sites and $N(\epsilon) \sim L$. For a localized state, $|U_n|^2 = 0$ almost everywhere, except for a few sites, and $N(\epsilon)$ is bounded to a constant value which does not depend on L. For a critical state $N(\epsilon)$ will scale as L^β with $0 < \beta < 1$.

As explained in the previous sections, incommensurate modulation of a periodic structure and quasiperiodicity force the eigenvectors to be described by superimposition of numerous Fourier components whose relative contributions tend to become uniform as the mode frequency increases. This induces situations similar to those shown in Fig. 5.25(*b*) and 5.25(*c*), with modes that are extended only at very low frequency, and become progressively more confined (critical) to finish as fully localized modes in the

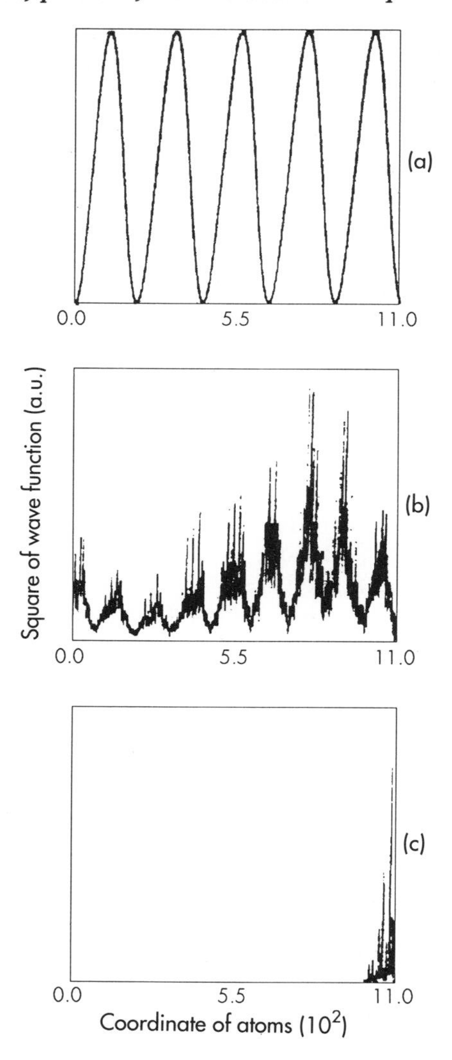

FIG. 5.25 Schematic presentation of the space variation of $|\boldsymbol{e}_n(\omega_k)|^2$ corresponding to (a) extended, (b) 'critical', and (c) localized eigenmodes.

high-frequency range. It should be noted that the number of Fourier components of the eigenvectors also increases for periodic modulation when the modulation period is expanded. This suggests that the participation ratio should decrease, particularly at high frequencies, when the successive orders of periodic approximants of a quasicrystal are considered. We shall return to this point when we discuss calculations for 3-dim models.

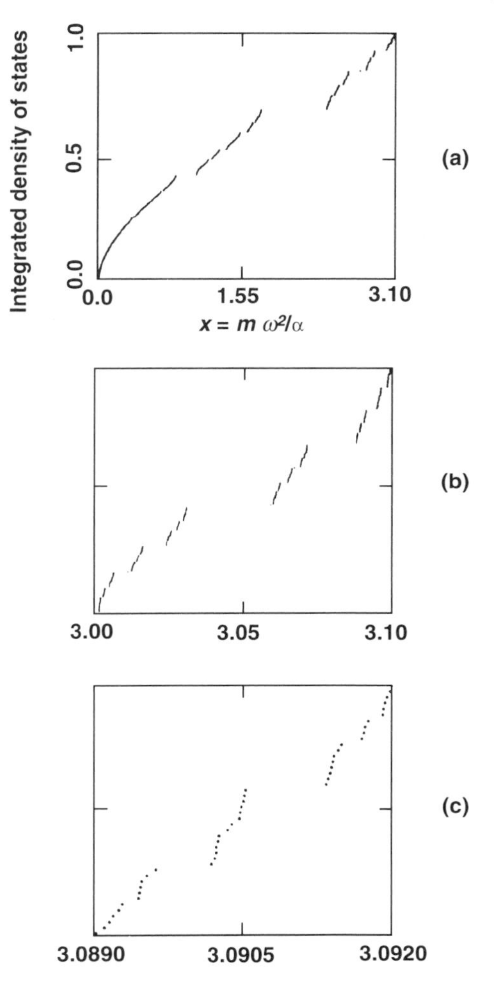

FIG. 5.26 (a) Integrated density of states of the Fibonacci chain deduced from the transfer matrix method. The scaled frequency x is $m\omega^2/\alpha$. (b) and (c) are enlargements of (a) to show the self-similarity behaviour. At $x \to 0$, the curve has the $\sqrt{x}$ behaviour of the continuous-medium approximation.

Indeed, when applied to a Fibonacci chain with $m_n = m$ and two force constants α and $\alpha\tau$, the transfer matrix calculation leads to the following conclusions.

1. In the low-frequency range ($m\omega^2/\alpha \ll 0.1$) the density of states is almost identical to that of a periodic chain and the corresponding modes are extended ($P \sim 1$).

2. In the high-frequency range, the density of states is self-similar and exhibits a dense set of gaps. The participation ratio decreases as ω increases, demonstrating localization effects at the upper part of the frequency range.

Features of the integrated density of states, defined as $g_{\text{int}}(\omega) = \int_0^{\omega} g(\omega)\,\mathrm{d}\omega$ are shown in Fig. 5.26.

Experimentally, phonons can be observed by measuring changes in the energy and momentum of neutrons after scattering by the sample. The incoming momentum is chosen to match one of the vectors of the reciprocal lattice, so that phonons are scanned in the $(\boldsymbol{k}, \omega)$ space around the corresponding Bragg peak. In other words, the experiment probes the excitations (fluctuations) of one of the mass density waves as introduced in Section 5.2.6.

If the structure is modulated or quasiperiodic, scanning the $(\boldsymbol{k}, \omega)$ space around a strong Bragg peak will simultaneously probe the numerous medium and weak mass density waves which densely fill the reciprocal space. The risk of observing nothing but a smooth messy noise has to be tested. Numerical calculations have been performed by Benoit *et al.*[4] in order to investigate whether pseudo-Brillouin zones and corresponding pseudo-dispersion curves can be defined for a quasiperiodic structure despite the dense character of the reciprocal space.

Benoit *et al.*[4] have considered a Fibonacci chain with atoms corresponding roughly to aluminium (mass and one of the force constants) and an average atom spacing equal to 3.6327 Å^{-1}. Positions and intensities of the most intense Bragg peaks have been calculated, using formulae like eqns (1.29) and (1.30). The results are given in Table 5.1 and compared to those obtained with a perfect periodic chain made of repetition of the two same bonds as for the Fibonacci chain. Disordered Fibonacci chains have also been considered, with 1 per cent, 10 per cent, and 20 per cent of (L, S) bond inversions; the consequences for Bragg peaks intensities are shown in Table 5.2.

Then, the inelastic intensity $S(k, \omega)$ has been calculated for points in the (ω, k) space around Bragg peaks and the corresponding $\omega(k)$ dispersion curves have been deduced from the maxima of $S(k, \omega)$. Details about the numerical calculation method may be found in reference [4].

Some of the results are illustrated in Fig. 5.27. Interesting conclusions can be summarized as follows:

1. Acoustic modes are clearly visible, with the same frequency for a given wavevector whatever the chosen Bragg peak, the intensity of the phonon being in proportion to the Bragg peak intensity. This is the same as what is observed for normal crystals.

2. When the (ω, k) space is scanned between two strong Bragg peaks, one

Table 5.1 Intense Bragg peaks ($I/I_0 > 0.05$) for a perfect periodic chain and a Fibonacci chain[4]

			Fibonacci		Periodic	
i	h	h'	Bragg peaks G_i(Å^{-1})	Intensity $I(G_i)$	Bragg peaks G_i(Å^{-1})	Intensity $I(G_i)$
1	0	0	0	1	0	1
2	1	0	1.0689	0.1127	0.9130	0.1313
3	0	1	1.7296	0.4930	1.8260	0.5437
4	1	1	2.7986	0.7726	2.7390	0.8042
5	1	2	4.5281	0.9076	3.6520	0.0076
6	2	2	5.5971	0.3225	4.5650	0.9218
7	1	3	6.2578	0.2304	5.4780	0.3702
8	2	3	7.3267	0.9638	6.3910	0.2695
9	2	4	9.0563	0.6780	7.3040	0.9696
10	3	4	10.1253	0.6032	8.2170	0.0378
11	2	5	10.7859	0.0617	9.1300	0.7119
12	3	5	11.8549	0.9860	10.0430	0.6496
13	4	5	12.9238	0.1805	10.9560	0.0674

Table 5.2 Changes in Bragg peak intensities according to (L, S) bond inversion in a Fibonacci chain[4]

		I/I_0 for defect concentrations C		
i	Bragg peaks G_i(Å^{-1})	$C = 1\%$	$C = 10\%$	$C = 20\%$
1	0	1	1	1
2	1.0689	0.57	0.27	0.15
3	1.7296	0.49	0.21	0.09
4	2.7986	0.58	0.28	0.15
5	4.5281	0.83	0.43	0.22
6	5.5971	0.49	0.20	0.08
7	6.2578	0.47	0.20	0.15
8	7.3267	0.66	0.53	0.45
9	9.0563	0.57	0.26	0.14
10	10.1253	0.47	0.20	0.15
11	10.7859	0.34	0.36	0.16
12	11.8549	0.60	0.43	0.42
13	12.9338	0.57	0.25	0.13

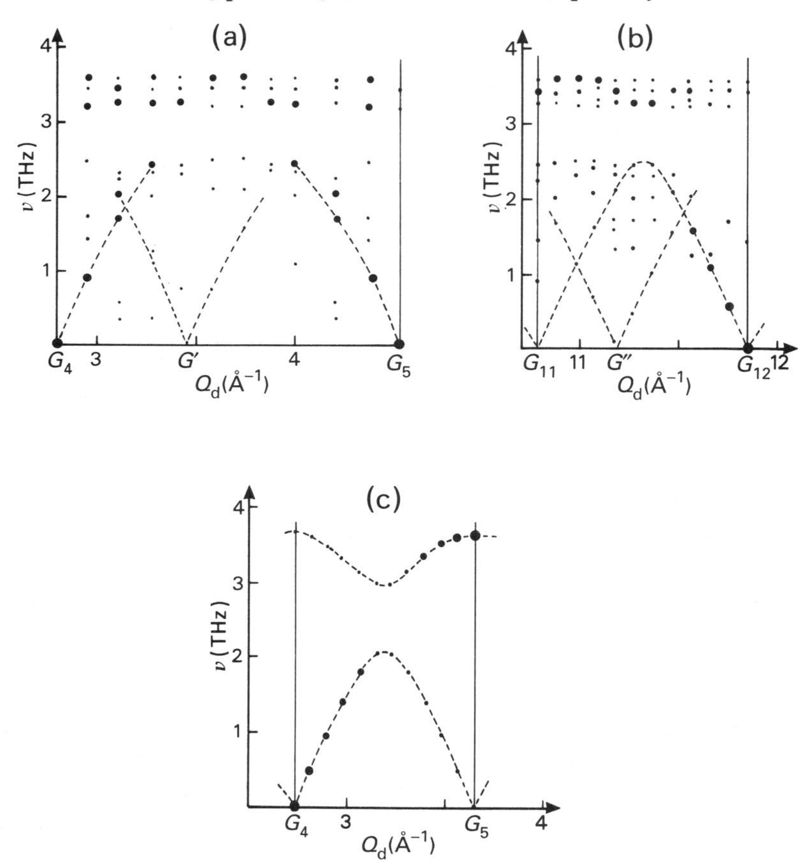

FIG. 5.27 Pseudo-dispersion curves as calculated for a Fibonacci chain [4] when scanning the reciprocal space between the strong Bragg peaks (a) G_4 and G_5 or (b) G_{11} and G_{12} in the notation of Table 5.1. G' and G'' indicate the positions of rather weak Bragg peaks (I less than 0.05 I_0). Intensity of the phonons is suggested by the dot sizes. The dispersion curves of the periodic diatomic chain, calculated within the same hypothesis, are shown for comparison (c).

can follow almost continuous branches, but with additional contributions coming from the weaker Bragg peaks that are densely distributed in between. Thus pseudo-dispersion curves, at least in the acoustic region, can be defined, as can pseudo-Brillouin zones: they are experimentally accessible. The diagrams are of course complicated (Fig. 5.27) by the superposition of many pseudo-dispersion curves attached to each Bragg peak. For a given

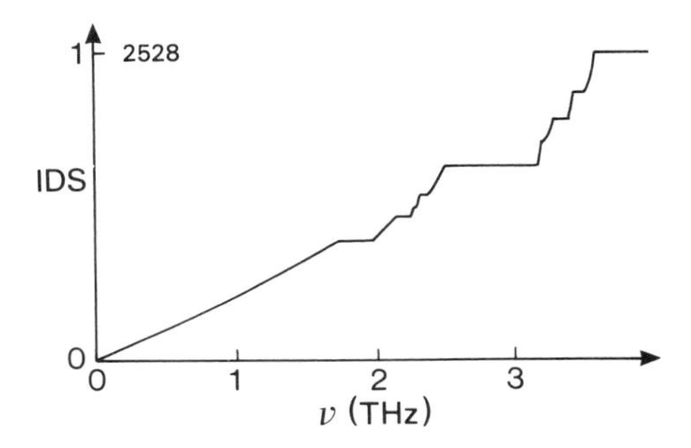

FIG. 5.28 Integrated density of state calculated for the Fibonacci chain.[4]

momentum transfer $\boldsymbol{k}$, there is in principle an infinity of phonon branches, but many of them have poor intensities. In other words, only a few of the Brillouin zones projected in the physical space from the high-dim reciprocal space are significantly 'activated'.

3. The 'optical' region is rather messy and forms a sort of flat continuous narrow band of frequencies.

4. The gap structure is more visible on the curve of Fig. 5.28, showing the calculated integrated density of states (IDS). There are gaps at all frequencies. Experimentally, as in the calculation reported in reference [4], the smaller gaps, at low frequencies, are smoothed by phonon lifetimes and/or experimental resolution effects. This generates quasicontinuous behaviour for the low energy part of the acoustic-like dispersion curve. Far from strong Bragg peaks, dispersion curves cannot be defined.

5. Displacement–displacement correlation calculations show that the atomic motions are always partially localized and the localization increases with the frequency. This is at variance with the behaviour of the perfect periodic chain which exhibits only fully propagating modes with no trace of localized motion.

Localization may be related to the presence of a very large number of gaps (more edge effects). Thus, the 'local' eigen mode of a given atom in the chain, namely $\omega_L \cong (2\alpha/m)^{\frac{1}{2}}$ as introduced at the end of Section 5.2.2, is never far from one of these gaps. The localization may, however, somewhat relax by some sort of **tunnelling** mechanism. Indeed, the Fibonacci chain is perfectly ordered and the number of neighbourhoods is restricted to only four different situations, i.e.

$$S \quad \text{between} \quad L \quad L$$

$$L \quad \text{between} \quad S \quad S$$

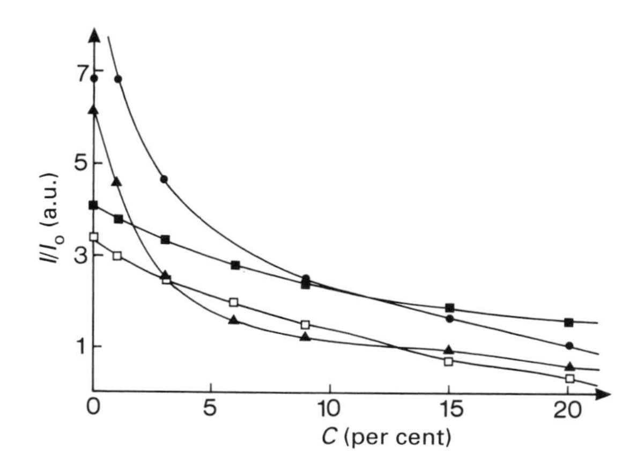

FIG. 5.29 Phonon intensity with respect to the corresponding Bragg peak, as a function of disorder (C defined in text).[4] •: 4.371 Å^{-1}, 0.94 THz; ▲: 4.210 Å, 1.70 THz; □: 4.058 Å, 2.44 THz; ■: 3.584 Å^{-1}, 3.54 THz.

$$L \quad \text{between} \quad L \quad S$$

$$L \quad \text{between} \quad S \quad L.$$

These situations may correspond to four different 'local' eigen modes. Some of these situations are distributed very regularly, within rather short distances of each other. For instance, *S* tiles are separated either by a single *L* tile or by two *L* tiles in sequence. Thus, the '*S* local mode' is expected to establish itself with actual phase relation from *S* site to *S* site, tunnelling through the occasional 'foreign' *LL* bonds. This may be a good illustration of what a so-called 'critical' mode might be.

6. The most interesting aspect is certainly that, in the Fibonacci chain, the slightest damping yields very strong localization of the atomic motion, even at low frequencies. Transport properties, as discussed in Chapter 6, are certainly affected by these localization effects.

7. Bond disorder due to (*LS*) bond inversion, slightly shifts the Bragg peak positions and drastically decreases their intensity (static Debye–Waller effect). As expected, weak peaks (large Q_{perp}) are more affected than strong peaks (small Q_{perp}). The consequences on the system dynamics are also quite drastic: phonon intensities are strongly reduced, the frequencies of a given mode are spread out, new modes are generated with energy in the gaps and also above the maximum frequency of the perfect Fibonacci chain, and the acoustic modes experience violent smearing effects. This suggests that quasicrystals have to be fairly perfect if one wishes to

investigate their dynamical properties. The curves of Fig. 5.29 illustrate these conclusions.

8. Phason-like defects act more or less as 'reciprocal space cleaners' in a sense that they somewhat erase the small intensity (large Q_{perp}) Fourier components. This results, in washing out some of the secondary pseudo-Brillouin zones, and the situation becomes closer to that of a regular crystal. In the Fibonacci scheme it can be viewed as longer strings of identical tiles (*LLL* . . . or *SS* . . .) generated when phason defects are produced. Again, this reduces the distance between the initial quasiperiodic structure and a true periodic crystal. Accordingly, and conversely to what is observed in crystals, some defects may improve the propagating aspects of phenomena in quasicrystals. We shall see in Chapter 6 that a perfect quasicrystal has a larger electrical resistivity than a defective one. Similarly, diffraction peaks with large Q_{perp} are washed out by increasing the temperature which should also generate more propagating modes.

More quantitative technical analyses of the problem of phonon localization in 1-dim or 2-dim quasiperiodic structure can be found in several publications.[5,6]

5.4.6 Lattice dynamics of three-dimensional quasicrystals: calculations and experiments

The problem we are faced with when dealing with the lattice dynamics of quasicrystals is that we have an infinite number of coupled equations of motion. In a normal crystal, the analysis of the motion in terms of Bloch waves reduces the number of these equations to three times the number of atoms in the unit cell. In principle, as a quasiperiodic structure is a 3-dim cut of a high-dim periodic structure, it is possible to make a Bloch-like analysis by using the superspace translational symmetry.[7] But this, unfortunately, does not reduce the number of coupled equations to a finite number, because in superspace the atomic objects contain an infinite number of points instead of point atoms.

Still, the archetypal Penrose tiling, even if it is not the best way to describe real quasicrystal structures, should provide a useful framework for lattice dynamic calculation. They are very well ordered indeed and this should generate selection rules to simplify the problem of finding the lattice vibrations. However, at the time of writing no one has so far succeeded in this.

The normal modes of vibrations for a set of N point-like particles are given by the analysis of any possible motion in terms of independent harmonic oscillators of frequency ω. The scientific case is very simple. When the particles are at their equilibrium positions in the system, they are by definition submitted to forces that sum to zero. When they are slightly displaced from these equilibrium positions, the particles are driven back toward these positions and oscillatory motions take place. In the harmonic approximations, the corresponding driving forces are assumed to be proportional to atomic displacements. Thus the basis equations of motion must be written as

$$m_j \frac{\mathrm{d}^2 \boldsymbol{u}_{nj}}{\mathrm{d}t^2} = -\sum_{n'j'} \phi(\boldsymbol{n}j; \boldsymbol{n}'j')\boldsymbol{u}_{n'j'}$$

where $\boldsymbol{u}_{nj}$ is the displacement vector of an atom of type j (mass equal to m_j) in a space position indexed by $\boldsymbol{n}$ (which represents the atomic coordinates) and $\phi(\boldsymbol{n}j;\, \boldsymbol{n}'j')$ is a force constant which operates between atoms $\boldsymbol{n}j$ and $\boldsymbol{n}'j'$.

Assuming that harmonic vibrations are good solutions, the equations of motion simplify

$$\omega^2 m_j \boldsymbol{u}_{nj} = -\sum_{n'j'} \phi(\boldsymbol{n}j; \boldsymbol{n}'j')\boldsymbol{u}_{n'j'}.$$

There are as many equations as atoms in the system under consideration. Formally, ω^2 and $\boldsymbol{u}_{nj}$ appear as the eigenvalues and the eigenvectors respectively of the dynamic matrix

$$D = [\phi(\boldsymbol{n}j; \boldsymbol{n}'j')].$$

The problem is not tractable for a macroscopic solid because of the large size of the D matrix. However, this simplifies for a regular crystal because of the lattice periodicity. Thus the Bloch theorem applies and the $\boldsymbol{u}_{nj}$ vectors propagate as plane wave, i.e.

$$\boldsymbol{u}_{nj} = m_j^{-\frac{1}{2}}\, U_j \exp(\mathrm{i}\,\boldsymbol{k}\cdot\boldsymbol{n}),$$

and the size of the dynamic matrix reduces to three times the number of atoms in the unit cell, i.e.

$$\omega^2 U_j = \sum_{j'} (m_j m_{j'})^{-\frac{1}{2}} \phi(jj')U_{j'}(\boldsymbol{k}).$$

The eigenvalues $\omega(\boldsymbol{k})$ are then distributed into branches according to the dispersion law. The wavevectors $\boldsymbol{k}$ are defined into the first Brillouin zone of the crystal and the corresponding eigenmodes are by definition extended.

In principle, there is no simple trick to reduce the number of equations of motion for non-crystalline matter. Moreover, the absence of periodicity, and hence of Brillouin zones, prevents the clear classification of the eigenmodes in terms of plane waves and vectors from the reciprocal space. However, this can be attempted for quasicrystals from both the theoretical (numerical calculations) and experimental points of view.

The formal basis for such an extension is contained in the hyperspace description of a quasiperiodic structure. All the derivations that are contained in the above analysis for crystals still hold for quasicrystals but they must be addressed in a very special way. The periodicity is now in the high-dimensional image of the real structure, say the 6-dim periodic image in the case of an icosahedral quasicrystal. This means that the Bloch theorem applies in the image space. In other words, the eigenmodes can still be expressed in terms of plane waves but with wavevectors in the first Brillouin zone of the 6-dim reciprocal lattice. The eigenvalues (frequencies) are also periodic in 6-dim reciprocal space. However, the energy components are in 3-dim physical space. For instance, atom displacements are vectors parallel to the physical space, but they are functions of the perpendicular coordinate attached to the atomic site of interest. Thus we must write $\boldsymbol{u}_{nj}(\boldsymbol{r}_{nj \cdot \mathrm{perp}})$, $U_j(\boldsymbol{r}_{j \cdot \mathrm{perp}})$ and $\boldsymbol{k}_{\mathrm{par}} \cdot \boldsymbol{n} + \boldsymbol{k}_{\mathrm{perp}} \cdot \boldsymbol{r}_{\mathrm{n} \cdot \mathrm{perp}}$ (instead of simply $\boldsymbol{k} \cdot \boldsymbol{n}$ in the phase factor). The only physical ingredients under control are the frequency ω and the propagation vector $\boldsymbol{k}_{\mathrm{par}}$. This means that in the high-dimensional modification of the dynamical matrix, a given eigenmode $(\omega, \boldsymbol{k}_{\mathrm{par}})$ is coupled with all $\boldsymbol{k}_{\mathrm{perp}}$ vectors of the 6-dim first Brillouin zone which give $\boldsymbol{k} = \boldsymbol{k}_{\mathrm{par}} + \boldsymbol{k}_{\mathrm{perp}}$ projecting on the same $\boldsymbol{k}_{\mathrm{par}}$ in physical reciprocal space. Conversely, a given $\boldsymbol{k}_{\mathrm{perp}}$ couples with a number of different $(\omega, \boldsymbol{k}_{\mathrm{par}})$ modes. If one tries to perform an inelastic neutron scattering experiment by looking for the ω values associated with a given $\boldsymbol{k}_{\mathrm{par}}$ (dispersion relation), the forced $\boldsymbol{k}_{\mathrm{par}}$ coupled to modes of various frequencies is associated with $\boldsymbol{k}_{\mathrm{par}} + \boldsymbol{k}_{\mathrm{perp}} = \boldsymbol{k}$ in which $\boldsymbol{k}_{\mathrm{perp}}$ scans the cross-section of the asymmetric unit of the first Brillouin zone at a distance $|\boldsymbol{k}_{\mathrm{par}}|$ from the origin. Thus the energy broadening of the 'phonon' at $\boldsymbol{k}_{\mathrm{par}}$ is expected to obey a power law $|\boldsymbol{k}_{\mathrm{par}}|^{\alpha}$. In particular, real extended phonons may appear in the acoustic regime where both $\boldsymbol{k}_{\mathrm{par}}$ and $\boldsymbol{k}_{\mathrm{perp}}$ remain small so that the energy broadening should vanish. If constant energy scans are performed (modes with the same ω but possibly different wavevectors), the 6-dim periodicity of ω must be accounted for. This means that the same ω appears at vectors $\boldsymbol{k}_{\mathrm{par}} + \boldsymbol{G}_{\mathrm{par}}$, where $\boldsymbol{G}_{\mathrm{par}}$ is the parallel component of any vector of the 6-dim reciprocal space. As it is well known that $\boldsymbol{G}_{\mathrm{par}}$ fills its space densely, this will obviously also happen for $\boldsymbol{k}_{\mathrm{par}} + \boldsymbol{G}_{\mathrm{par}}$, and the vibrational mode of frequency ω will appear somewhat extended in 3-dim reciprocal space. Consequently, localization of some sort must be observed in real space. Each contribution of $\boldsymbol{k}_{\mathrm{par}} + \boldsymbol{G}_{\mathrm{par}}$ to the ω response is related to the corresponding Fourier coefficient of the structure and thus is a decaying function of the $\boldsymbol{G}_{\mathrm{perp}}$ component associated with $\boldsymbol{G}_{\mathrm{par}}$. Hence only the small $\boldsymbol{G}_{\mathrm{perp}}$ components really contribute to ω broadening, if they are not too far away out of resolution in $\boldsymbol{k}_{\mathrm{par}}$ space. In particular, the ω broadening is expected to be completely invisible in the small $\boldsymbol{k}_{\mathrm{par}}$ acoustic regime.

Conclusively, well-defined phonons comparable to those observed in current crystals should show up in quasicrystals in the very-long-wavelength limit, particularly at $\boldsymbol{k}_{par}$ vectors very close to large $\boldsymbol{G}_{par}$ with small associated $\boldsymbol{G}_{perp}$ (large structure factor). At higher values of ω and/or $\boldsymbol{k}_{par}$, the (ω, $\boldsymbol{k}_{par}$) modes are no longer represented by a point in the dispersion law diagram but exhibit increasing ω and $\boldsymbol{k}_{par}$ broadening with spiky features. In addition, the high symmetry of the icosahedral point group reduces the dispersion law to one longitudinal and two degenerate transverse modes.

Some of the numerical calculations performed for low dimensional quasicrystals [4, 8] based on Fibonacci sequences can be qualitatively extended to three dimermsions. As explained in Chapter 1, a 2-dim Penrose lattice is a dual of a five grid pattern, each grid being a set of parallel lines arranged according to a Fibonacci sequence. A 3-dim Penrose lattice is also related to five sets of intersecting families of parallel planes, each family also arranged according to a Fibonacci sequence. Thus the conclusions as reported in the previous section may still qualitatively apply, to some extent, to the case of 3-dim Penrose tiling. The multigap structure of the dispersion curves may be even more chaotic, but measurements of acoustic phonons and of density of states may be possible; they have indeed been made for a few quasicrystals.

A numerical investigation of vibrational excitations in quasicrystals, as for crystals, requires that the structure is (i) sufficiently specified and (ii) is stable under realistic interatomic forces. If these conditions are satisfied, a calculation of the dispersion law density of states and eigenvectors can be attempted.

As reported earlier in this book, structures of existing quasicrystals are not specified at the detailed level that would lead to a complete set of atomic coordinates. Even so, reasonable 3-dim atomic models have been fitted to diffraction data and subsequently used for dynamic calculation [9–11] in the cases of AlMnSi and AlLiCu quasicrystals. For alloys of simple and noble metals such as AlLiCu, interatomic forces can be obtained via pseudopotential perturbation theory [10, 11] The normal modes of vibration of a solid are formally given in the harmonic approximation in terms of eigenvalues and eigenvectors of the real space dynamical matrix. This is just a 3-dim generalization of the secular equation, as introduced earlier in this chapter for 1-dim crystals. Then the straightforward approach is a direct diagonalization of this dynamic matrix. It is clear that this operation can only be achieved with a finite system. In crystals the periodicity of the structure reduces the matrix size to three times the number of atoms in the unit cell. This is not possible for quasicrystals and approximations must be made. The two types of approximation that have been made so far are the commensurate approximation and the cluster approximation. The cluster approximation is a rather 'brute force' approach which treats the system as a huge molecule with an increasing number of atoms, in the hope that some asymptotic behaviour will appear. In the commensurate approximation the quasiperiodic lattice is replaced by successive periodic approximants, with more and more atoms in the unit cells, and a possible evolution of the corresponding eigenmodes and density of states towards limis that would apply to the true quasicrystal is also sought. The method is certainly very useful for obtaining clues to the mystery of quasicrystal dynamics, and in any case is the only practical approach that has been proposed so far. However, it must be realized that such an approach contains several drawbacks, of various levels of importance, and they should be taken into consideration for correct appreciation of the results.

The first point to be considered is the basic aspect of looking for eigenmodes in the form of plane-like or Bloch-like waves. This kind of solution is fully adapted to periodic structures as they obey translational invariance. However, it is more doubtful for quasicrystals which have a self-similar structure. Therefore eigenmodes that respect some 'expansion invariance' (scaling law) are likely to be more appropriate.

The second comment that can be made is more marginal and concerns the density of states. Calculation of the density of states does not require full diagonalization of the dynamic matrix and hence is easier than the determination of the dispersion law and eigenvectors. It is also the quantity which is most easily accessible experimentally. Unfortunately, the density of states is mainly governed by short-range-order effects and short-range order does not vary significantly in a series of periodic approximants of a given quasicrystal. A possible evolution might show up as the appearance of more Van Hove singularities when the unit cell is expanded. However, in the case of structures that have more than 100 atoms in the unit cell this is beyond the resolution of both calculations and experiments.

The third difficulty comes from the practical limit on the acceptable size of the dynamic matrix. Taking the AlLiCu system as an example, the lowest-order periodic approximants of the quasicrystal,[10, 11] i.e. 1/1, 2/1, 3/2, and 5/3, have 160, 680, 2880, and 12 244 atoms respectively in their unit cells. This corresponds to dynamic matrices with dimensions of 480×480, 2040×2040, 8640×8640 and $36\,732 \times 36\,732$ respectively, and of course the dispersion laws will exhibit as many branches as lines or columns in the matrices! It is clear that, even with powerful modern computers, direct diagonalization and calculation of all eigenvectors is possible for, at most, the 2/1 approximant which is still far removed from a true quasicrystal. Density of states and dispersion law calculations are accessible up to the 5/3 approximants if recursion methods are used.

Typical results are presented showing the density of states for several approximants (Fig. 5.30), the specific density of states for the 5/3 periodic approximant of the AlLiCu quasicrystal compared with experimental results (Fig. 5.31), some dispersion curves for the same system (Fig. 5.32), and the participation ratio for a 2/1 periodic approximant (Fig. 5.33).

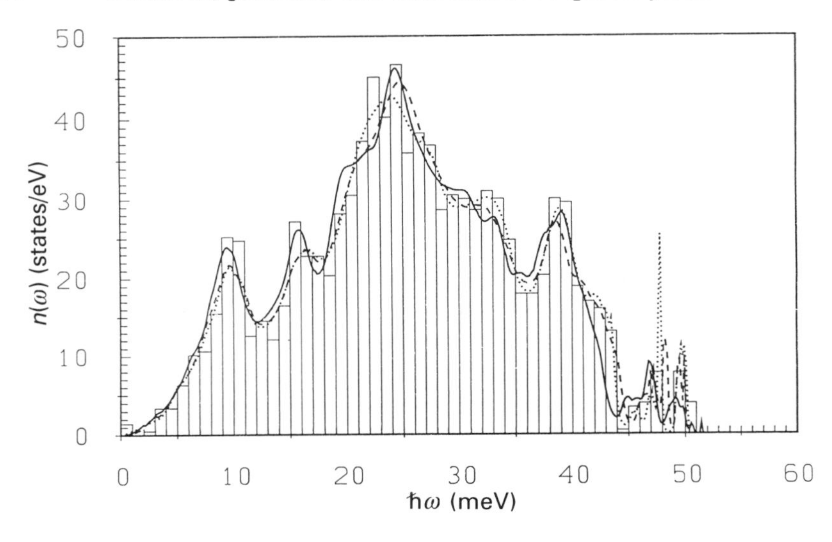

FIG. 5.30 Vibrational density of states for AlZnMg periodic approximants of order 2/1 (dotted line), 3/2 (dashed line), and 5/3 (full line)[10, 11] (courtesy of J. Hafner).

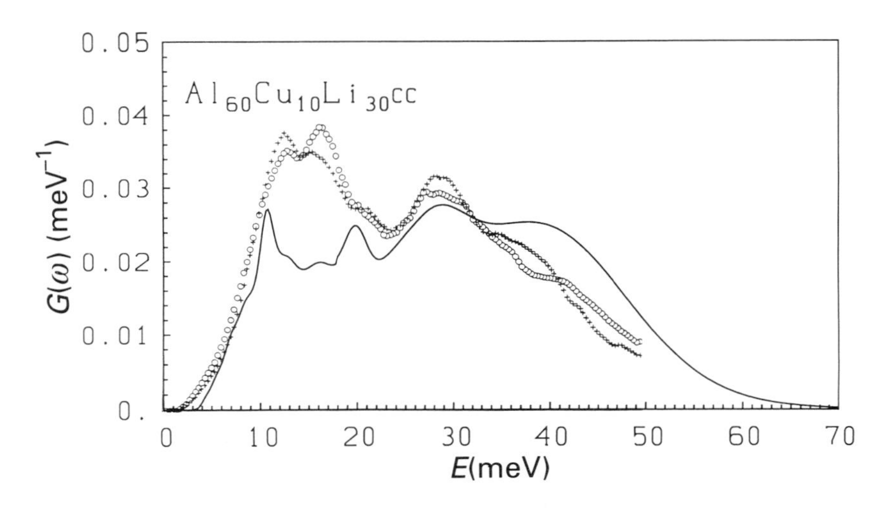

FIG. 5.31 Generalized density of states for the AlLiCu system: theoretical result, folded with a Gaussian resolution function which increases with energy (full line); experimental results for the R-phase 1/1 approximant (+) and the icosahedral quasicrystal (0).[10, 11, 13]

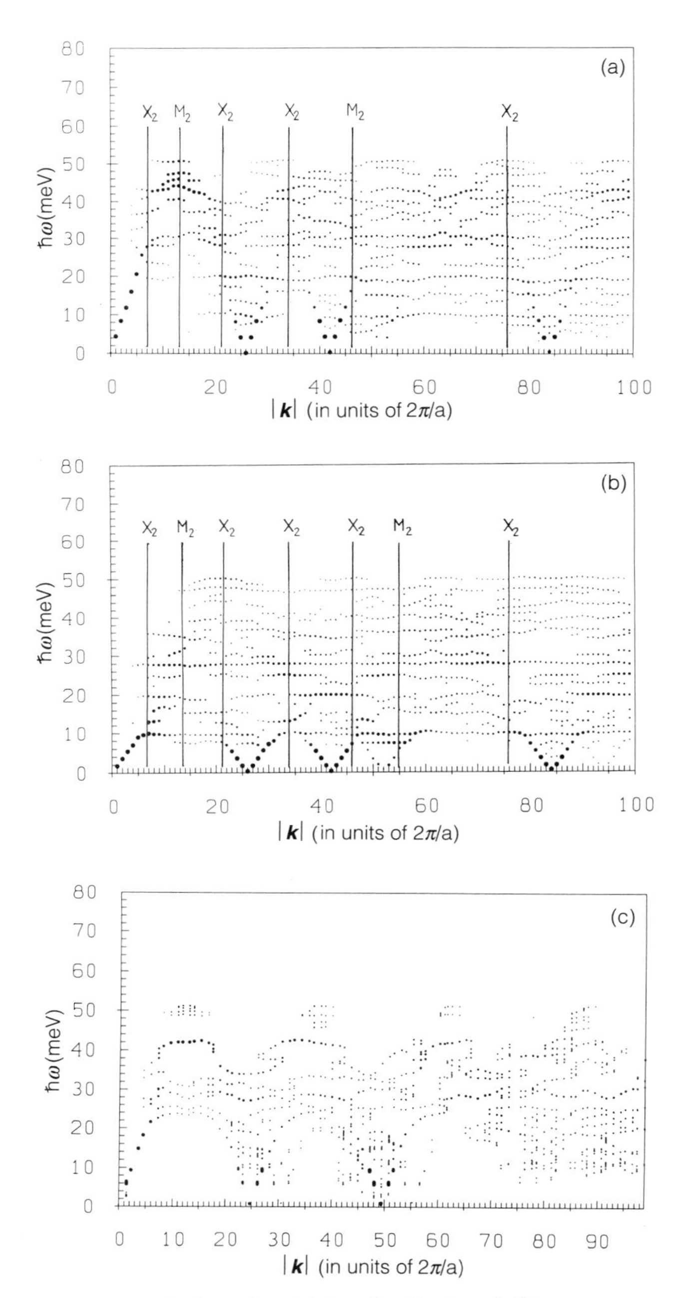

FIG. 5.32 Disperson relations for (a) longitudinal and (b) transverse excitations along a 2-fold axis, and (c) longitudinal excitations along a 5-fold axis in a 5/3 periodic approximant. Each dot corresponds to a peak in the spectrum; the dot size scales with the peak intensity.[10, 11] Isotropy is observed on (a) and (c) (courtesy of J. Hafner).

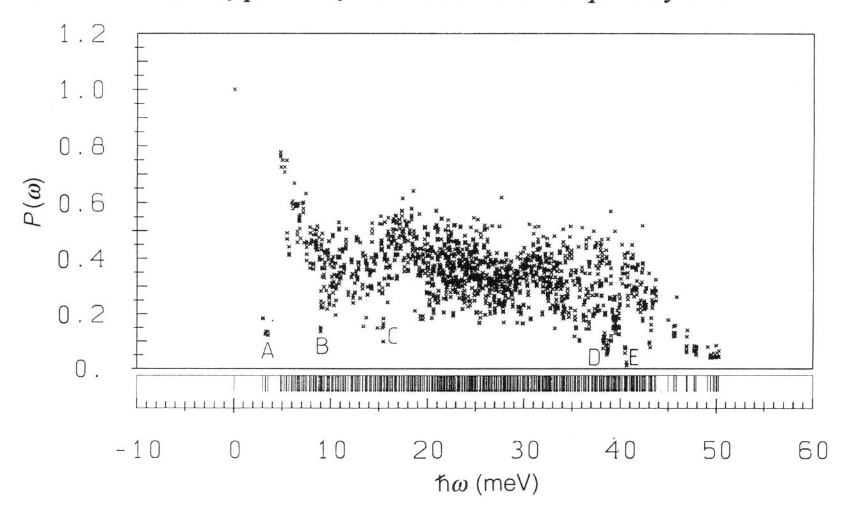

FIG. 5.33 Participation ratio $P(\omega)$ of the eigenmodes for a 2/1 periodic approximant of an Al–Zn–Mg alloy.[10, 11] (courtesy of J. Hafner).

These calculations[9–11] show the existence of well-defined propagating longitudinal and transverse collective excitations around the intense Bragg peaks, in the low-energy range (typically below about 5 meV), in agreement with sound velocity data. Calculations also confirm the elastic isotropy of icosahedral quasicrystals in this acoustic regime.

At higher energy, the vibrational spectrum is dominated by nearly dispersionless modes. These modes are particularly strong at points in reciprocal space corresponding to projections of Brillouin zone boundaries of the 6-dim hypercubic reciprocal lattice. The participation ratio (Fig. 5.33) of these 'high-energy' modes (above about 5 meV or 1.25 THz) oscillates around 0.3 for the 2/1 periodic approximant. In view of the analysis made in the 1-dim cases, we can anticipate that this participation ratio would be smaller for higher-order approximants, perhaps leading to strictly localized eigenmodes. The point seems to have been confirmed recently.[12] Dispersion curves, as shown in Fig. 5.32, are a good illustration of the complexity of the problem even when restricted to periodic approximants.

The calculated density of states (Figs. 5.30 and 5.31) are not particularly spiky, but show some structures with the appearance of pseudo-gaps and reinforcement of the density in the low-energy range (compared with a Debye ω^2-like behaviour). The corresponding spectrum possesses scaling properties and is singular continuous. As expected, periodic approximants of different order have rather similar calculated densities of states.

Both the density of states and the dispersion law have been measured experimentally using inelastic neutron scattering experiments.[13-16] As already summarized earlier, phonons or vibrational excitations can be investigated using inelastic neutron scattering measurements. The frequency and wavevector of the vibrations are deduced via conservation laws from neutron energies and momentum values measured before and after interacting with the sample. Experimentally, the sample is set on the spectrometer at a vector $\boldsymbol{q}$ from a Bragg position and an energy scan is performed to determine the corresponding frequency (maximum in the scattering profile). Typical results are shown in Fig. 5.34.[16] They correspond to different $\boldsymbol{q}$ vectors of the form $(0,q,0)$ originating at the Bragg reflection 4/6 0/0 0/0 (in the notation explained in Chapters 2 and 3). The particular dispersion curve deduced from these data is shown in Fig. 5.35 along with the width of the scattering profiles. Dispersion curves measured for $\boldsymbol{q}$ vectors ranging from the 4/6 0/0 0/0 to the 3/6 0/12 0/0 Bragg peaks, starting from one or the other, are also presented in these figures. All these data were obtained with an icosahedral AlPdMn quasicrystal which can be grown as a single grain of sufficient size for the purposes of inelastic neutron scattering measurements. Globally, the full set of observations can be summarized as follows:

1. The dispersion curves exhibit two pseudo-branches with transverse and longitudinal character respectively. This is consistent with the icosahedral symmetry.

2. In the long-wavelength limit phonons are well defined by both unbroadened energy and wavevector profiles, within experimental resolution. The isotropy of these ***acoustic modes*** is verified and the sound velocities deduced (slope of the $\omega(q)$ law) are in good agreement with ultrasonic measurements. This ***linear acoustic range*** extends up to about 0.3 Å^{-1} (wavelength) and 1.4 THz or 6 meV (energy). Beyond this range a clear relation between the scattering response function and a 'phonon branch' cannot be established unambiguously.

3. Between 1.4 and 2.8 THz (6–12 meV), corresponding to wavevectors ranging from about 0.3 to 0.6 Å^{-1} for the maximum response, the measured excitation progressively broadens. For a given energy in this range, the excitation finally extends over a q range of 0.40 Å^{-1} (full width at half maximum) and the energy width for a given wavevector reaches 1 THz ($\sim$4 meV). The broadening is isotropic. Simultaneously, the dispersion relation departs from linear behaviour.

4. Above about 3 THz (~ 12 meV) vibration modes become dispersionless and appear as flat branches.

The structure of the AlPdMn quasicrystal, as briefly discussed in Chapter 4, consists of a self-similar skeleton which results from inflation growth. The elementary cluster of atoms (pseudo-Mackay icosahedron (PMI)) has a diameter of 10 Å and the distances between PMI centres are about 20 Å. The PMI units seem to be very 'rigid', while interaction between them would be comparatively weak. In such structures, rigid unit soft modes are expected to propagate and to generate acoustic dispersion branches over wavevector values up to about 0.3 $Å^{-1}$ (π divided by the PMI diameter). At higher energies, the vibration should localized in the PMI as some sort of 'cavity modes'. In fact, these 'localised' modes are repeated in all PMI but with weak phase relations and hence are not strictly localized; they can be called 'critical' or 'confined' modes[17]. These results probably provide the basis for a derivation of the eigenmodes of a quasicrystal using scaling laws which are typical of self-similar structures. The idea would be to assume that eigenmodes are 'steady states' or 'standing modes' within the structural units generated at each inflation step. Thus wavelength corresponding to the eigenmodes would scale as the geometry, i.e.

$$\lambda\left(\frac{V}{b}\right) = \frac{1}{b^n}\lambda(V)$$

in which V is some size parameter (e.g. volume) of the macroscopic grain and b is the scale factor (inflation ratio). Each mode would be weighted by the frequency of the corresponding (inflated) unit in the piece of structure, which must also scale as b, i.e.

$$g[\lambda(V/b)] = bg[\lambda(V)]$$

or

$$g(\lambda/b^n) = bg(\lambda).$$

Derivation along these lines should lead to scaling laws for dispersion relations and density of states. However, it has not been attempted so far.

FIG. 5.34 Examples of inelastic neutron scattering response functions measured at vectors $\boldsymbol{q} = (0, q, 0)$ from the Bragg reflection 4/6 0/0 0/0. Dots show the data; the solid curves are fits with damped harmonic oscillators. The geometry of the measurements corresponds to transverse modes. The results are for AlPdMn quasicrystals.[16]

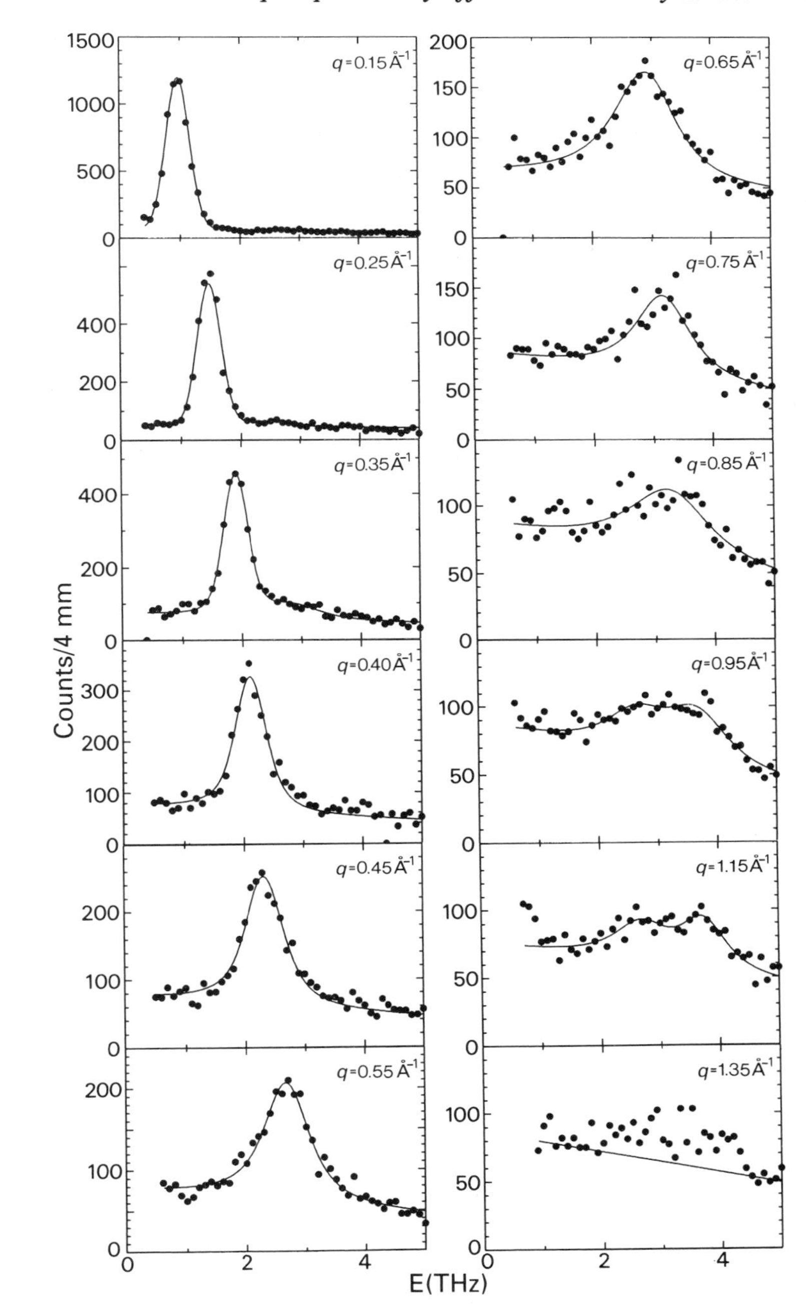
Counts/4 mm
E(THz)
q=0.15Å⁻¹
q=0.25Å⁻¹
q=0.35Å⁻¹
q=0.40Å⁻¹
q=0.45Å⁻¹
q=0.55Å⁻¹
q=0.65Å⁻¹
q=0.75Å⁻¹
q=0.85Å⁻¹
q=0.95Å⁻¹
q=1.15Å⁻¹
q=1.35Å⁻¹

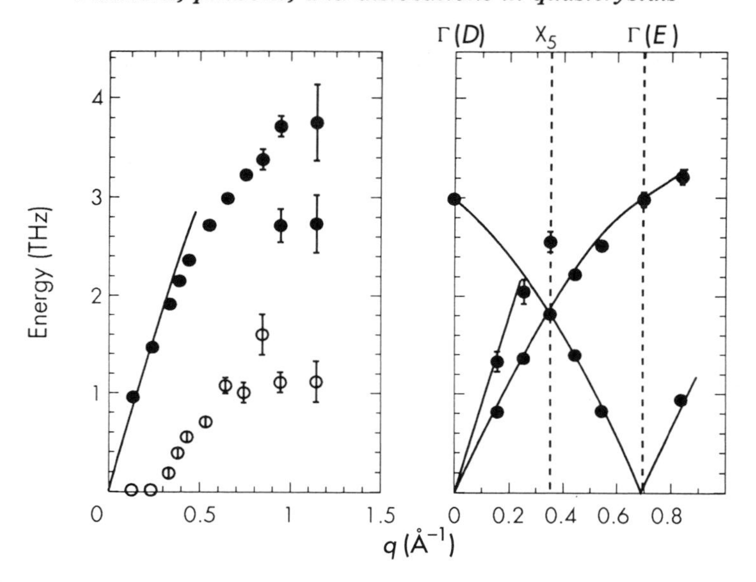

FIG. 5.35 (a) Dispersion curve (solid circles) and width (open circles) as deduced from the scattering response functions shown in Fig. 5.34; (b) dispersion curves measured between the intense Bragg peaks 4/6 0/0 0/0 and 3//6 0/1 0/0. There is no visible gap opening where the two acoustic pseudo-branches cross (within the 0.3 THz resolution of the instrument).[16]

5.5 Concepts of elasticity and defects in quasiperiodic structures

5.5.1 The density wave picture and the high-dim representation of quasiperiodic structures

As introduced in Section 5.2.6, the density wave picture of an ordered solid is expressed by a Fourier series expansion of the density, i.e.

$$\rho(\boldsymbol{r}) = \sum_{\boldsymbol{G}} \rho_G \exp(\mathrm{i}\boldsymbol{G}\cdot\boldsymbol{r}), \tag{5.41}$$

where $\boldsymbol{G}$ represents vectors of the reciprocal space. Each ρ_G is a complex number with an amplitude $|\rho_G|$ and a phase ϕ_G. Since $\rho(\boldsymbol{r})$ is a real quantity $|\rho_G| = |\rho_{-G}|$ and $\phi_G = -\phi_{-G}$. These ρ_G, with their amplitudes and phases, are nothing other than structure factors, or Fourier components of the ordered structure, for each vector $\boldsymbol{G}$ of the reciprocal lattice. All these remarks and formulae fully apply to the case of structures which are periodic in the 3-dim space, i.e. regular crystals.

The phases ϕ_G have values Φ_G^0 when the atoms are at their equilibrium positions (atomic sites), which are completely determined for a given vector

G and a given symmetry. The Φ_G^0 are the famous necessary quantities which are not directly obtained from diffraction data and must be reconstructed when structures have to be specified (see Chapter 4).

If the atoms are displaced from their equilibrium positions, say by a vector $\boldsymbol{u}$, the phases ϕ_G are modified by the quantities $\boldsymbol{G}\cdot\boldsymbol{u}$ and we write

$$\phi_G = \Phi_G^0 + \boldsymbol{G}\cdot\boldsymbol{u}. \tag{5.42}$$

A uniform shift in $\boldsymbol{u}$ corresponds to translation of the system. In crystals, spatial variation in $\boldsymbol{u}$ gives rise to propagating phonons, as explained in previous sections.

Now the question is: how should we use or modify such a description in the case of quasicrystals? Formally, everything still holds true if we consider the periodic n-dimensional description of the quasicrystal. Equations (5.41) and (5.42) can simply be rewritten to make explicit the n-dim character of the structure, i.e.

$$\begin{aligned} \rho(\boldsymbol{r}_n) &= \sum_{G_n} \exp(\mathrm{i}\boldsymbol{G}_n\cdot\boldsymbol{r}_n) \\ \Phi_{G_n} &= \Phi_{G_n}^0 + \boldsymbol{G}_n\cdot\boldsymbol{u}_n. \end{aligned} \tag{5.43}$$

But the physics of quasicrystals, as for any solid, is in 3-dim space and what is of interest is the density wave picture in three dimensions.

From the previous chapter, we know that the 3-dim quasiperiodic physical structure is obtained by a convenient cut of the n-dim periodic image. For the case of interest, this means that the $\rho(\boldsymbol{r}_3)$ density is precisely the cut of the $\rho(\boldsymbol{r}_n)$ density. In reciprocal spaces, we know also that the correspondence law is not a cut but a projection, i.e. the Fourier component $\rho_{G_{\mathrm{par}}}$ of the density in the 3-dim (or parallel) physical space, for a vector $\boldsymbol{G}_{\mathrm{par}}$ of the physical reciprocal space, is equal to ρ_{G_n}, with $\boldsymbol{G}_{\mathrm{par}}$ the component of $\boldsymbol{G}_n$ in the 3-dim physical reciprocal space. Amplitudes and phases are preserved in the projection procedure, so that eqn (5.43) can be easily altered to express the 3-dim physical density, i.e.

$$\rho(\boldsymbol{r}_{\mathrm{par}}) = \sum_{G_{\mathrm{par}}} \rho_{G_{\mathrm{par}}} \exp(\mathrm{i}\boldsymbol{G}_{\mathrm{par}}\cdot\boldsymbol{r}_{\mathrm{par}}),$$

with

$$|\rho_{G_{\mathrm{par}}}| = |\rho_{G_n}| \tag{5.44}$$

and

$$\Phi_{G_{\mathrm{par}}} = \Phi_{G_n} = \Phi_{G_n}^0 + \boldsymbol{G}_{\mathrm{par}}\cdot\boldsymbol{u}_{\mathrm{par}} + \boldsymbol{G}_{\mathrm{perp}}\cdot\boldsymbol{u}_{\mathrm{perp}}.$$

The displacement vector $\boldsymbol{u}_n$, which describes the shift of the atomic objects in the n-dim space, with respect to their equilibrium positions, has in general two components $\boldsymbol{u}_{\text{par}}$ and $\boldsymbol{u}_{\text{perp}}$, in the physical and complementary spaces respectively. The $\boldsymbol{u}_{\text{par}}$ vector field corresponds to real atomic displacements in the physical space and is the exact replica of the *phonon-like* strain field as described for a regular crystal. The $\boldsymbol{u}_{\text{perp}}$ vector field operates on the atomic objects by moving them in the complementary space, with the possible effect of exchanging those which are actually cut by the physical space. The point is illustrated in Fig. 5.36, again with the simple example of a Fibonacci chain related to a 2-dim square lattice structure. A uniform $\boldsymbol{u}_{\text{perp}}$ field is simulated by moving the R_{par} space. One can see that the effect of this simple $\boldsymbol{u}_{\text{perp}}$ field is to modify atomic positions without changing the unit cells; in the example of the figures some (L, S) or (S, L) bound pairs are reversed. We have demonstrated in Section 5.4.4 that in the picture of a Fibonacci chain as a regular periodic lattice plus incommensurate modulation of the atomic bonds with a periodicity τ, phason shifts, i.e. displacements of the origin of the atomic chain with respect to that of the modulation function, generate such inversions of (L, S) bound pairs. The $\boldsymbol{u}_{\text{perp}}$ vector fields are then *phason-like* modes. Any kind of strain fields that operates in the perpendicular complementary space exclusively are called phasons, for short, even when they do not correspond to dynamical processes and are simply related to static distortion fields. This is not fully appropriate, and has generated some confusion, but currency has conferred definitive authority on the terminology. For short also, phonon-like strain fields are usually written $\boldsymbol{u}$ for similarity with the crystal case, while phason-like fields are denoted $\boldsymbol{w}$.

The 1-dim example of Fig. 5.36 also shows clearly the main difference between $\boldsymbol{u}$ and $\boldsymbol{w}$ from the energy point of view. Spatially uniform shifts $\boldsymbol{u}$ correspond to a translation of the quasilattice as a single block; atomic pair distances remain unchanged. Thus the total energy of the system is kept constant, not only when initial and final states are compared, but also at any time of the translation shift. This justifies the label of phonon-like strain field, since non-uniform spatially variable shifts $\boldsymbol{u}$ will still take place within the same energetic scheme and give rise to propagating mode. Gradients in $\boldsymbol{u}$ produce distortions in the unit cell shapes without rearrangements, and the distortions can propagate when initiated at one point of the quasilattice.

On the other hand, as a rearrangement of unit cells, phason variation leads necessarily to diffusive movements of atoms, though maybe over rather short distances. This always requires that atoms cross over local energy barriers. Initial and final states have the same energy, but dissipating processes take place during the unit cell rearrangements.

These conclusions are generic and can be drawn from a mathematical analysis of the density wave expression as given by eqn (5.44). Introducing the notation $\boldsymbol{u}$, $\boldsymbol{w}$ and the explicit form for $\phi_{G_{\text{par}}}$ gives

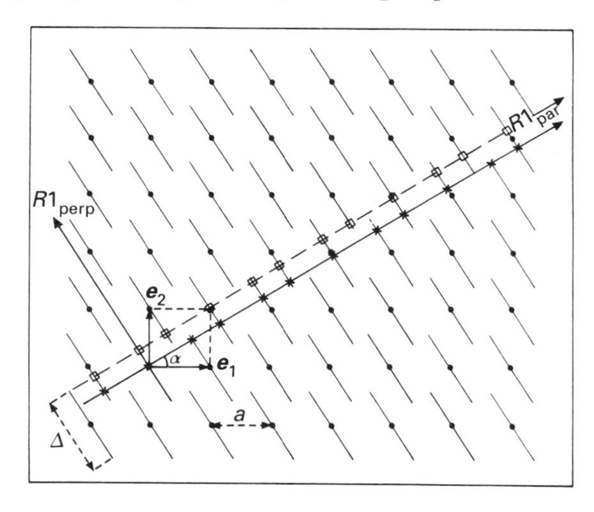

FIG. 5.36 A uniform phason strain of the 2-dim square lattice has been simulated by a shift of the 1-dim physical space R_{par}. The × and □ distributions of atoms differ locally by some large–short or short–large bond sequences being reversed. The two 1-dim structures are globally indistinguishable (same energy, same Fourier components).

$$\rho(\boldsymbol{r}_{\text{par}}) = \sum_{G_{\text{par}}} |\rho_{G_{\text{par}}}| \exp\{\mathrm{i}G_{\text{par}}(\boldsymbol{r}_{\text{par}} + \boldsymbol{u}) + \mathrm{i}G_{\text{perp}} \cdot \boldsymbol{w} + \mathrm{i}\Phi^0_{G_{\text{par}}}\}. \quad (5.45)$$

$\boldsymbol{G}_{\text{perp}}$ is of course defined in a one-to-one correspondence scheme with $\boldsymbol{G}_{\text{par}}$ but these two vectors are *independent* components of a n-dim vector $\boldsymbol{G}_n$. As such they induce different phase shifts effects in the density wave summations. Briefly, the phonon-like strain field $\boldsymbol{u}$ shifts the density waves relative to one another by an extra phase $\boldsymbol{G}_{\text{par}} \cdot \boldsymbol{u}$ which nicely combines with the position phase $\boldsymbol{G}_{\text{par}} \cdot \boldsymbol{r}$: the same Fourier component corresponding to the same space 'frequency' $\boldsymbol{G}_{\text{par}}$ still applies but is expressed at position vector $\boldsymbol{r} + \boldsymbol{u}$ instead of $\boldsymbol{r}$. A uniform (non-$\boldsymbol{r}$ dependent) field $\boldsymbol{u}$ keeps $\rho(\boldsymbol{r})$ unchanged; a gradient introduces simple distortions. The phason term $\boldsymbol{G}_{\text{perp}} \cdot \boldsymbol{w}$, even with uniform $\boldsymbol{w}$ fields, shifts the density waves relative to one another in a rather subtle way, since $\boldsymbol{w}$ applies to 'space frequencies' $\boldsymbol{G}_{\text{perp}}$ which do not follow the sequence of $\boldsymbol{G}_{\text{par}}$ vectors, although they do correspond to them (small $\boldsymbol{G}_{\text{perp}}$ is associated with small and large $\boldsymbol{G}_{\text{par}}$, and vice versa). Let us take again the 1-dim example of a Fibonacci chain. The $\boldsymbol{G}_{\text{par}}$ and $\boldsymbol{G}_{\text{perp}}$ vectors are given by eqns (1.33) and (1.34), that is:

$$\boldsymbol{G}_{\mathrm{par}} = \frac{2\pi}{a} \frac{1}{(2+\tau)^{\frac{1}{2}}} (h + \tau h')$$
$$\boldsymbol{G}_{\mathrm{perp}} = \frac{2\pi}{a} \frac{1}{(2+\tau)^{\frac{1}{2}}} (h - \tau h'), \tag{5.46a}$$

h and h' being any integer. The amplitudes of the density waves are also given in Chapter 1 (eqns 1.33):

$$\rho_{G_{\mathrm{par}}} = \Delta \left(\sin \frac{\boldsymbol{G}_{\mathrm{perp}} \cdot \Delta}{2} \right) \left(\frac{\boldsymbol{G}_{\mathrm{perp}} \cdot \Delta}{2} \right)^{-1}, \tag{5.46b}$$

with $\Delta = a(\tau + 1)$, where a is the lattice constant of the 2-dim square lattice. The expressions for $\boldsymbol{G}_{\mathrm{par}}$, $\boldsymbol{G}_{\mathrm{perp}}$, and $\rho_{G_{\mathrm{par}}}$ can be substituted into the density wave expansion (eqn 5.45). For the sake of simplifying the calculation, we restrict the series to two waves only, corresponding to ($h = 1$, $h' = 0$) and ($h = 0$, $h' = 1$). This keeps the translational symmetry features of the Fibonacci chain, since the two density waves that are combined have their periods in the irrational ratio τ. This is a description of the Fibonacci structure by its Landau model.

Going through the arithmetic results in a density expression given by

$$\rho(r) \simeq \frac{\sin A}{A} \exp\{\mathrm{i}\alpha(r + u) + aw\} + \frac{\sin \tau A}{\tau A} \exp\{\mathrm{i}\alpha\tau(r + u) - \mathrm{i}\tau\alpha w\}, \tag{5.47}$$

with

$$A = 2\pi(1+\tau)/(2+\tau)^{\frac{1}{2}},$$
$$\alpha = 2\pi/a(2+\tau)^{\frac{1}{2}}.$$

If $w = 0$ and $u \neq 0$, the two waves combine in a way which is exactly that corresponding to the no-strain field situation. The natural phase shift between the two waves, at equilibrium, comes from the factor τ in one of the wave phases. This factor τ applies in the same way for equilibrium position r and phonon-like phase shift u. Thus the density $\rho(r)$ at the same r position in space has indeed been modified, but in the simple way that now we find at r the density that was at $r - u$ in the absence of the strain field. The structure has kept the same topology but for weak distortions.

If now $u = 0$ and $w \neq 0$ the situation is quite different. Consider for instance, as an illustrative example, the position r which at equilibrium ($u = w = 0$) corresponds to two waves being perfectly in phase, i.e.

$$\alpha\tau r = \alpha r + 2\pi \ \text{or} \ \alpha r = 2\pi\tau.$$

The related density value is then

$$\rho(2\pi\tau/\alpha) \simeq \left(\frac{\sin A}{A} + \frac{\sin \tau A}{\tau A}\right) \cos 2\pi\tau$$

in the absence of phason strain, and

$$\rho(2\pi\tau/\alpha) \simeq \frac{\sin A}{A} \cos(2\pi\tau + \alpha w) + \frac{\sin \tau A}{\tau A} \cos(2\pi\tau - \alpha w)$$

with a phason strain. In a naïve numerical case with $A = \pi/2$ and $\alpha w = \pi/2$, the net density at the particular space point is proportional to 0.75 (1 being the expected maximum from the model) without phason strain and 0.05 with the phason strain. Thus the density has almost vanished out of this point, probably to reappear elsewhere, i.e. where the two waves were out of phase by π in the absence of a phason.

All these results can be generalized using a model based on the Landau theory approach and group theoretical properties.[18–20] The general idea is to express the Landau free energy F as a power series in the mass density. Truncations of the mass density wave terms are required to make the minimization of F a tractable problem. The truncated expression must however keep the basic symmetry ingredients of the system. Different methods have been proposed for this.[12] Basically, it has generally been assumed that the hydrodynamic degrees of freedom in the phases of the mass density waves can be parametrized by two vectors, $\boldsymbol{u}$ and $\boldsymbol{w}$, which are related to two different representations of the rotational symmetry group.[19, 20] The expression for the elastic energy as a function of gradients in $\boldsymbol{u}$ and $\boldsymbol{w}$ has been derived to quadratic order, again using group theory arguments. The basic results, whose demonstration we consider outside the scope of this introductory book, are as follows for icosahedral quasicrystals:

1. The elastic energy requires five independent elastic constants. Two are coefficients of terms quadratic in gradients of $\boldsymbol{u}$; two others are for terms quadratic in the gradient of w, and one is the coefficient for a cross-term containing both gradients of $\boldsymbol{u}$ and w.

2. The $\boldsymbol{u}$ mode (phonon) is propagating and the $\boldsymbol{w}$ mode (phason) is diffusive,[22] or in other words, the phonon mode couples to the momentum and the phason mode does not.

3. The terms that involve $\boldsymbol{u}$ are isotropic. Thus, in the absence of a phason gradient the phonon response to an external stress is isotropic. As the response of a uniform phason field to an external stress is diffusive, the response of the phonon field will be isotropic for quite a long time. Consequently, for all practical purposes, icosahedral quasicrystal may behave like *isotropic solids*,[22] except if there is a $\boldsymbol{w}$ gradient quenched during

solidification. This may be of interest to distinguish icosahedral quasicrystals from their periodic approximants.

4. Topological defects corresponding to dislocation exist in quasicrystals.[19] The analogue of the Burgers vectors for icosahedral quasicrystals is found to be 6-dim vectors containing both $\boldsymbol{u}$ and $\boldsymbol{w}$ contributions. There is no Burgers vector that would correspond to a $\boldsymbol{u}$ contribution alone, because translational periodicities are absent in the physical space: cutting a convenient piece of the quasicrystal, removing it and gluing the lips as for a crystal is still possible, but the result is not a linear defect but a stacking fault. A consequence of Burgers vectors having both $\boldsymbol{u}$ and $\boldsymbol{w}$ contributions is that dislocations cannot slip easily in icosahedral quasicrystals as they do in crystals. To move the phason field associated with a dislocation requires diffusive time-scales, equivalent to the climb mechanism in crystals. As crystals are believed to deform by pair production and rapid motion of dislocations, icosahedral quasicrystals should be brittle and difficult to deform plastically[23] except at high temperature.[24] This is indeed consistent with experimental observations.

The rest of this chapter will be now devoted to some more illustrations of the above statements.

5.5.2 Examples of phonon-like strain fields

The simple 1-dim example of the Fibonacci chain does not allow a proper illustration of the strain field effects on a quasiperiodic structure because of the usual 1-dim singularity problems. Though far from describing appropriately the structure of real quasicrystal (see Chapter 4), Penrose tiling schemes, especially with their 3-dim modifications, have more potential to give proper geometrical descriptions associated with phonons, phasons, and dislocations in these structures.

As we know from Chapter 1, a Penrose tiling in 2-dim can be obtained from a pentagrid (or Ammann quasilattice) either by decoration or by a generalized dual transformation. Similarly, a 3-dim Penrose tiling using the two canonical rhombohedra can be obtained from the quasilattice made of six sets of parallel planes: the distances between planes of a given set are in a Fibonacci sequence and the six directions are defined by the six-vector star of a regular icosahedra. Such a description of the quasiperiodic structure in terms of intersecting quasiperiodic families of planes has advantages:

(1) according to the experimental results presented in Chapter 4, this corresponds to a crude analysis of the structure of real quasicrystals in terms of rough atomic planes,

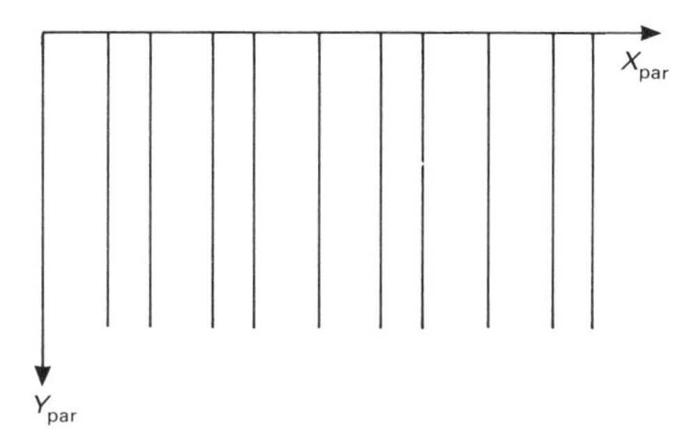

FIG. 5.37 Traces in a (X_{par}, Y_{par}) plane of a set of parallel planes whose separating distances are in Fibonacci sequence.

(2) the effects of strain fields on a quasiperiodic structure in three dimensions can be qualitatively approached as a combination of the effects on the individual families of planes.

An individual family of planes, or Fibonacci grid, can be easily generated by translating a Fibonacci chain in a plane. The result, as shown in Fig. 5.37, is a series of traces made by the Fibonacci plane family and is a pseudo-three-dimensional quasilattice. The n-dim (here 3-dim) periodic structure related to this quasilattice is also easily derived from that producing a Fibonacci chain. An X_{perp}-axis has to be added to the (X_{par}, Y_{par}) framework of Fig. 5.37. A square lattice, conveniently tilted and decorated with segment A_{perp}, is then drawn in $(X_{par}X_{perp})$ and continuously shifted parallel to Y_{par} (Fig. 5.38). The final n-dim periodic structure is thus a lattice of square cross-section channels decorated with parallel planes whose cut by the (X_{par}, Y_{par}) plane generates the expected Fibonacci grid. A section (X_{par}, X_{perp}) of the whole system is shown in Fig. 5.39, with arrows representing a uniform shifts $\boldsymbol{u}$ of the atomic object A_{perp} for a given value of the coordinate Y_{par}. Consider now that the shifts $\boldsymbol{u}$ are Y_{par}-dependent in the simplest way, namely $\boldsymbol{u}$ is parallel to X_{par} with a modulus proportional to the Y_{par} coordinate. This means that the 3-dim periodic structure is uniformly sheared, the plane atomic objects A_{perp} being now at an angle with respect to (X_{perp}, Y_{par}). Their traces in (X_{par}, Y_{par}) have of course the same behaviour, as represented in Fig. 5.39. This is indeed analogous to the elastic distortion of atomic planes in a regular crystal, and the associated dynamical modes for recovering the initial unstrained structure are obviously phonon-like.

If the considered strain field $\boldsymbol{u}$ is less simple, the net result is anyway a distortion field of the grid which remains confined in the (X_{par}, Y_{par}) plane, as illustrated in Fig. 5.40 for the case of a somewhat wavy phonon

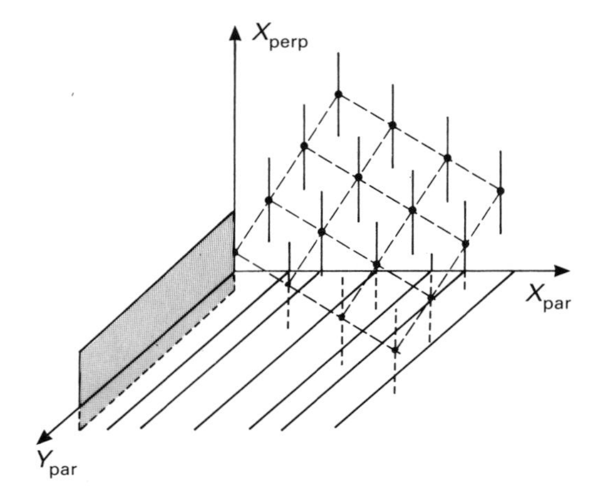

FIG. 5.38 A square lattice in the (X_{par}, X_{perp}) plane is 'decorated' by plane segments perpendicular to (X_{par}, X_{perp}) and extending along Y_{par} (the shaded area shows one of these plane segments). A cut of these pseudo-three-dim structures by (X_{par}, Y_{par}) generates an Ammann (or Fibonacci) grid.

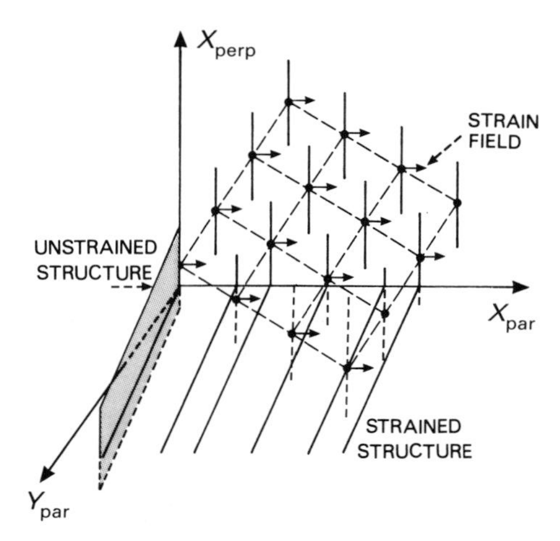

FIG. 5.39 A strain field parallel to X_{par} and increasing linearly along Y_{par} deviates the plane atomic objects of Fig. 5.38 and the resulting grid is now at angle with respect to Y_{par} ('phonon-like' strain).

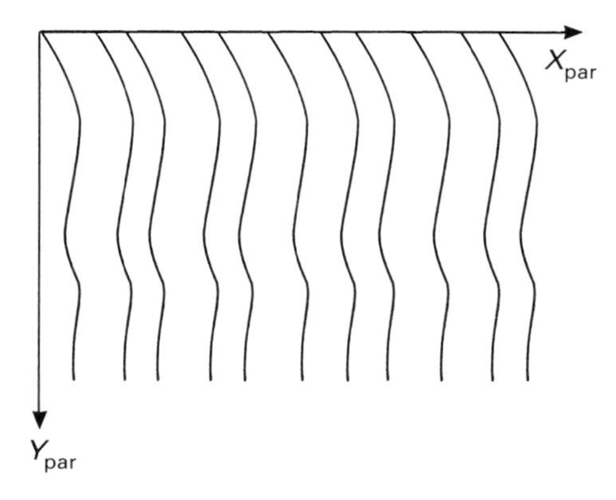

FIG. 5.40 Traces of the grid shown in Fig. 5.3 when submitted to a non-uniform phonon strain field. In this example, the strain field varies only along Y_{par}. Then, any section parallel to X_{par} reproduces a Fibonacci sequence, translated from the unstrained configuration.

strain field. Extension of the analysis to 2-dim and 3-dim Penrose tiling is qualitatively straightforward. Consider, for instance, the pentagrid equivalent to a 2-dim Penrose tiling (see Chapter 1). Each grid may be distorted by phonon strains. This is illustrated in Fig. 5.41. The topological distribution in space has not been altered by the strain field but the various 'tiles', or cells, of the quasilattice are distorted. Thus variation in $\boldsymbol{u}$, the variable associated with broken translational symmetry, leads to distortion of the unit cells (but without rearrangements) in a quasicrystal, as they do in a periodic crystal. The effect can also be visualized on the Penrose tiling

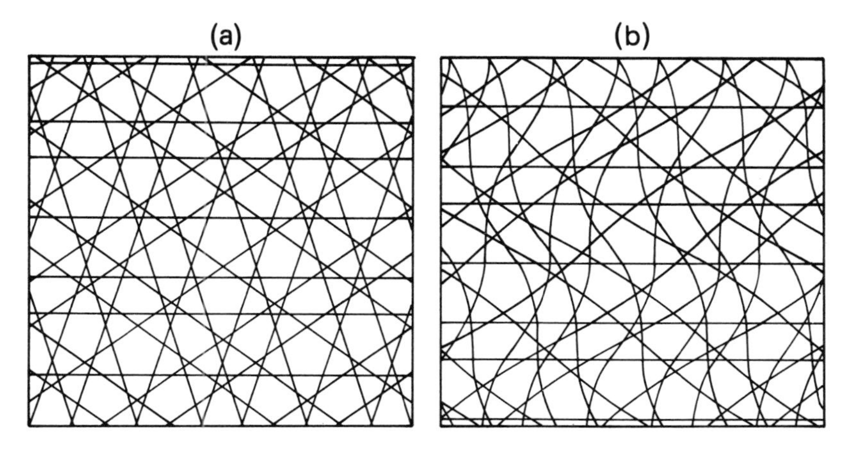

FIG. 5.41 Pentagrid either perfect (a) or submitted to a phonon strain field (b).[25]

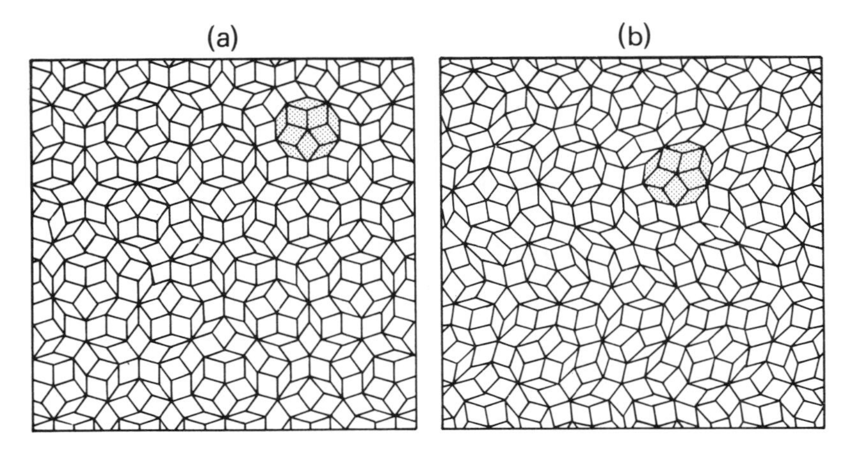

FIG. 5.42 Penrose tilings corresponding to the unstrained (a) and strained (b) pentagrids shown in Fig. 5.41. Shaded areas exemplify the distortion field, which, however, leaves the topology unaltered. The reader is encouraged to reproduce Fig. 5.41(b) and to superimpose it on to Fig. 5.42(b) in order to see the corresponding strain effects.[25]

dual of the pentagrid (see Chapter 1 for the correspondence scheme between a Penrose tiling and a pentagrid). The two tilings shown in Fig. 5.42 are the perfect Penrose tiling and the strained Penrose tiling equivalent to the two grids of Fig. 5.41. The phonon strain distortion of the tiles is clearly visible, especially around the decagonal clusters of tiles.

The n-dim periodic representation of the grid also provides an easy way to analyse the effects of phonon-like strain fields on the diffraction pattern. The strained atomic object A_{perp} now being curved, it has acquired a net cross-section (say a 'thickness') in the physical (parallel) space. Its Fourier transform in the n-dim reciprocal space is then a decaying function of $Q_{par} = Q_{exp}$ instead of being flat as before. Thus, again the variable associated with broken translational symmetry leads to a decrease of the Bragg peak intensities in a quasicrystal, especially at large Q values, exactly as it does in a periodic crystal. This is the so-called parallel Debye–Waller effect, already mentioned in previous chapters. Peak shifts and peak broadening increasing with Q_{par} are also observed.

5.5.3 Examples of phason-like strain fields

As already mentioned, uniform shifts and spatial variations in the $\mathbf{w}$ variable affect both the density wave images and the tilings in a rather more subtle manner.

The same simplification as that used in Section 5.5.2 will serve again for

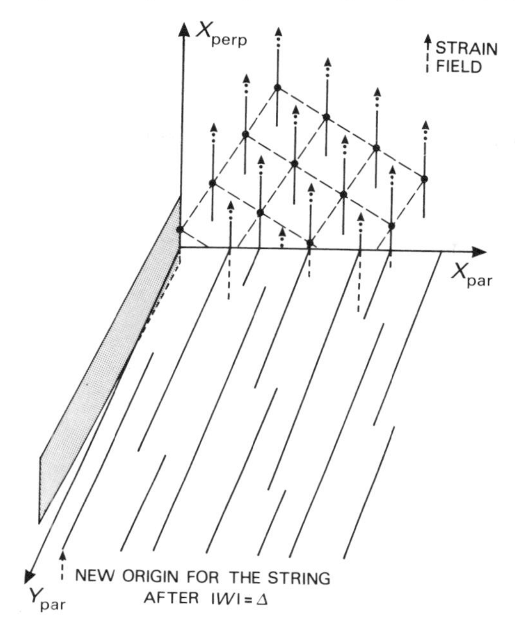

FIG. 5.43 A strain field parallel to X_{perp} and increasing linearly along Y_{par} produces a shear deformation of the plane atomic objects of Fig. 5.38. Then the subset of these plane which are cut by (X_{par}, Y_{par}) changes when going along Y_{par}, generating 'jags' and rearranging the lines in the grid.

the sake of illustration. Thus Figs. 5.37 and 5.38 still apply to this section, but the arrows of Fig. 5.39 must now be parallel to X_{perp}, with their modulus $\boldsymbol{w}$ still proportional to the Y_{par} coordinate (Fig. 5.43). Thus the atomic objects in the 3-dimensional periodic space are still portions of planes which remain parallel to the (X_{perp}, Y_{par}) plane, but their lower and upper boundaries are curved. In the (X_{par}, X_{perp}) section shown in Fig. 5.43, the uniform $\boldsymbol{w}$ displacements of the A_{perp} objects mean that some of them no longer intercept X_{par}, while new interceptions are induced. This corresponds to the already mentioned inversion (L, S) into (S, L) segments of a Fibonacci chain (Fig. 5.36). For the grid, the lines remain straight, but at each inversion one line fades out and a new one emerges. The point at which a line shifts is called a **jag**.

In the example, as $|\boldsymbol{w}|$ reaches a value equal to the length of the A_{perp} objects along X_{perp} all the lines of the grid have shifted by the same amount, equal to the difference $(L - S)$. The system has relaxed all its phason defects and is identical to its initial state, but for an overall translation. This of course does not happen for more complex phason strain fields.

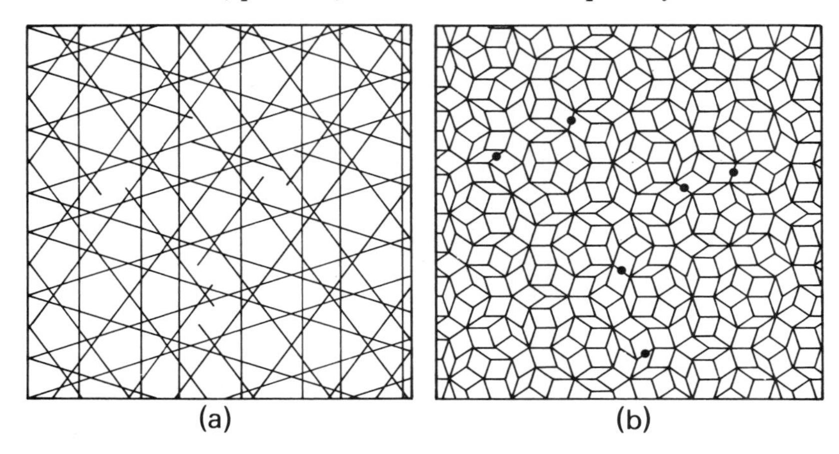

FIG. 5.44 Pentagrid (a) and corresponding Penrose tiling (b) as deformed by a rather complex phason strain field. The 'jags' regions in the grid correspond to isolated violation of the matching rules in the related Penrose tiling (black dots).[25]

The approach may be qualitatively generalized along the lines described in Section 5.5.2. This is illustrated in Fig. 5.44 showing the effects of variations in the variable $\mathbf{w}$ on a 'Fibonacci' pentagrid and the corresponding 2-dim Penrose tiling. The spatial variations in $\mathbf{w}$ produce jags in the grids which result in isolated violations of the matching rules in the tiling. Away from the 'phason defect', the matching rules are maintained, as illustrated in Fig. 5.45. Each defect corresponds to a simple flip of one 'atom' from its actual position into a 'virtual vacancy', without disrupting any other of the Penrose matching rules.

Such a phason-like disorder generates entropy in the system. This is the main ingredient of the so-called random tiling models, in which it is assumed that quasicrystals can only be in a high temperature equilibrium phase, the ground state being a periodic crystal.

Phason relaxation, or annihilation, can occur through a series of local rearrangements of unit cells. From Fig. 5.45 it appears that the location of the mismatch can be moved by flipping of either of the pseudo-hexagons that border on it. The jag is thus moved by one 'step'. The new configuration has the same energy as the old one, but one 'atom' had to cross an energy barrier in the flipping process. Then the relaxation is controlled by a diffusion rate. However, this relaxation is accomplished through short range motion of atoms: no mass must be transported over distances large compared to the unit cell's size. Each mismatch will execute a random walk along the relevant line of the related grid (no changes in the jump directions) until it walk off the sample or annihilates with another mismatch of the same family (a line

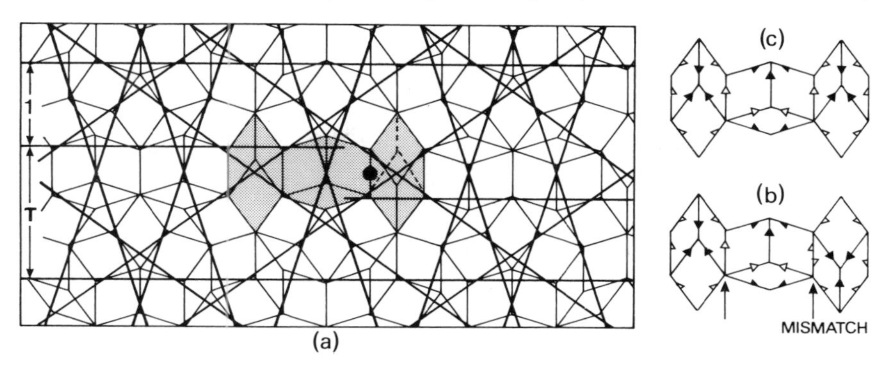

FIG. 5.45 Detail of the relation between a jag and a mismatch (black dot in part (a)). In the shaded area of (a) around the jag, the mismatch comes from the right-hand hexagonal set of three tiles: its configuration is as shown in (b) while it should be as in (c). Note that outside this hexagon, the matching rules are perfectly respected. We have here an example of an isolated phason defect.[25]

segment between two jags). A detailed picture may be more complicated, owing to the crossings of other jag paths. Globally, the mismatches may propagate at rates comparable to that of a vacancy in a regular periodic crystal; that is several orders of magnitude faster than expected for atomic diffusion in a solid. It seems that this has indeed been observed in a recent quasi-elastic neutron scattering experiment.[26]

Phason strains affect the perfect structure in the n-dim representation by changing and/or distributing the distances between the atomic objects and by increasing their effective cross-section in the perpendicular space. The latter enhances the Q_{perp} decay of the Fourier components. This is a perpendicular Debye–Waller effect. The former induces shifts of the Bragg reflections, in proportion to their Q_{perp} parameter. Mixed up shifts may simply result in peak broadening, also increasing with the Q_{perp} values. This has been indeed observed, in certain experimental conditions, with quasi-crystals of the AlFeCu system[27, 28] (see Fig. 5.46).

For a further and more pictorial illustration of effects of phason-like defects on the diffraction pattern, Fig. 5.47 shows a section of the diffraction pattern for the strained grid of Fig. 5.43 compared with that of the unstrained grid (Fig. 5.38). It can clearly be seen that phason defects destroy the spot alignment, with shifts along $Q_{Y_{\text{par}}}$ which are larger for weak reflections (large Q_{perp} component).

A density wave picture of a uniform phason strain is shown in Fig. 5.48. The figure was obtained by superimposing five cosine waves along the 5-axis 72° apart, and adding a linear phason strain. Looking at the glancing angle along the arrowed directions, it can be seen that the white dots are no

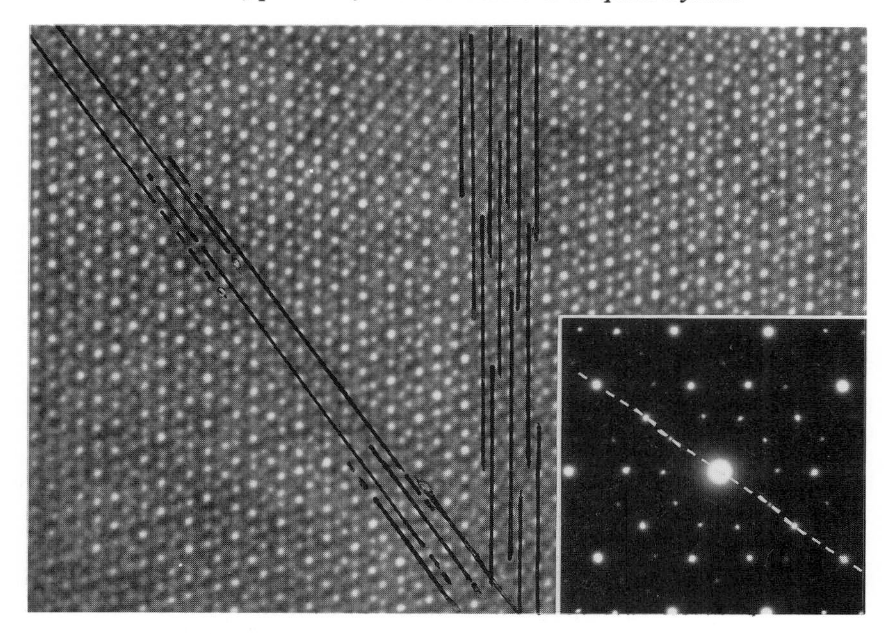

FIG. 5.46 High-resolution electron micrograph and the corresponding 5-fold zone axis diffraction pattern of an AlFeCu quasicrystal showing 'jags' in atomic rows (underlined) and shifted Bragg spots (courtesy of M. Audier).

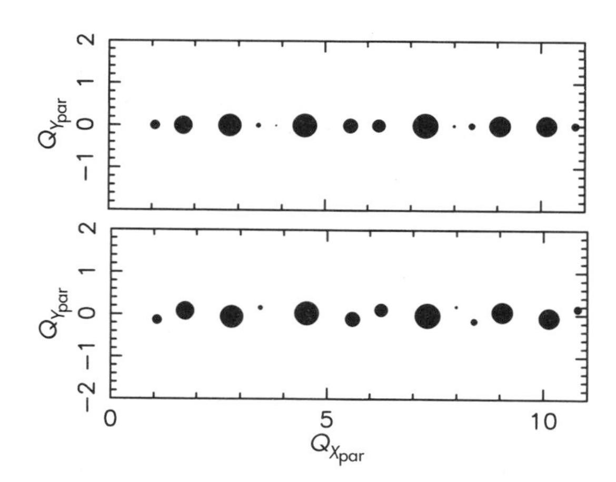

FIG. 5.47 Diffraction patterns of (a) a perfect Fibonacci grid and (b) a phason strained grid; spots are misaligned, particularly for weak reflections (large Q_{perp}). The sizes of the dots scale the diffracted intensity (courtesy of M. de Boissieu).

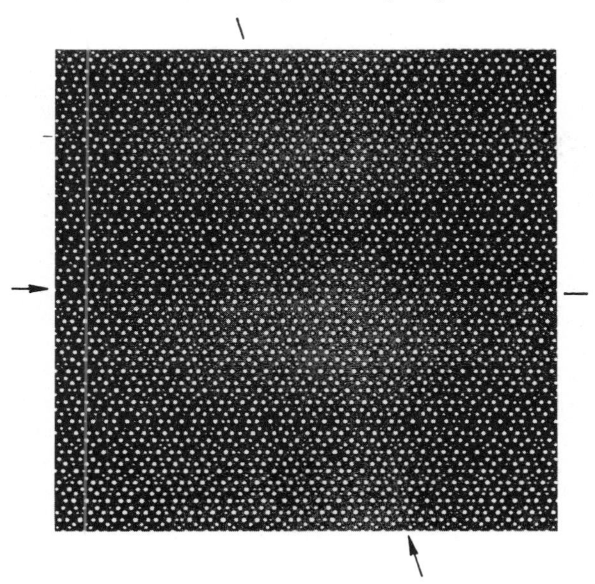

FIG. 5.48 Density wave pictrue of a linear phason strain. Five cosine waves oriented 72° apart are superimposed, plus a phase shift corresponding to a linear phason strain. Jags are visible at the glancing angle along the arrowed directions (courtesy of M. de Boissieu).

longer aligned. Instead, some 'jags' are visible as discontinuities in the lines of dots.

Phason defects are expected to play a major role in many problems such as phase transition, quasicrystal growth, or stability. One example is proposed in Fig. 5.49, with respect to phase transformation in the AlPdMn system. As reported in Chapter 2, the icosahedral quasicrystal in this system is stable for some composition only at high temperature. At lower temperatures, a decagonal phase precipitates in platelet-like morphology. The two phases have an orientation relationship with a 5-fold axis of the icosahedral phase parallel to the 10-fold axis of the decagonal phase. In fact, the decagonal phase forms almost via epitaxial growth on the icosahedral structure. This can occur at the expense of strain field effects which are relaxed via phason-like defects. Figure 5.49 shows an icosahedral phase domain which is squeezed between two slices of the decagonal phase. In the 5-fold epitaxy direction the structure is not visibly disturbed. However, phason jags clearly show up in inclined (arrowed) directions. This is also reflected in the corresponding diffraction patterns.

Other implications of phason defects in physical properties of quasicrystals will be discussed in the next chapter.

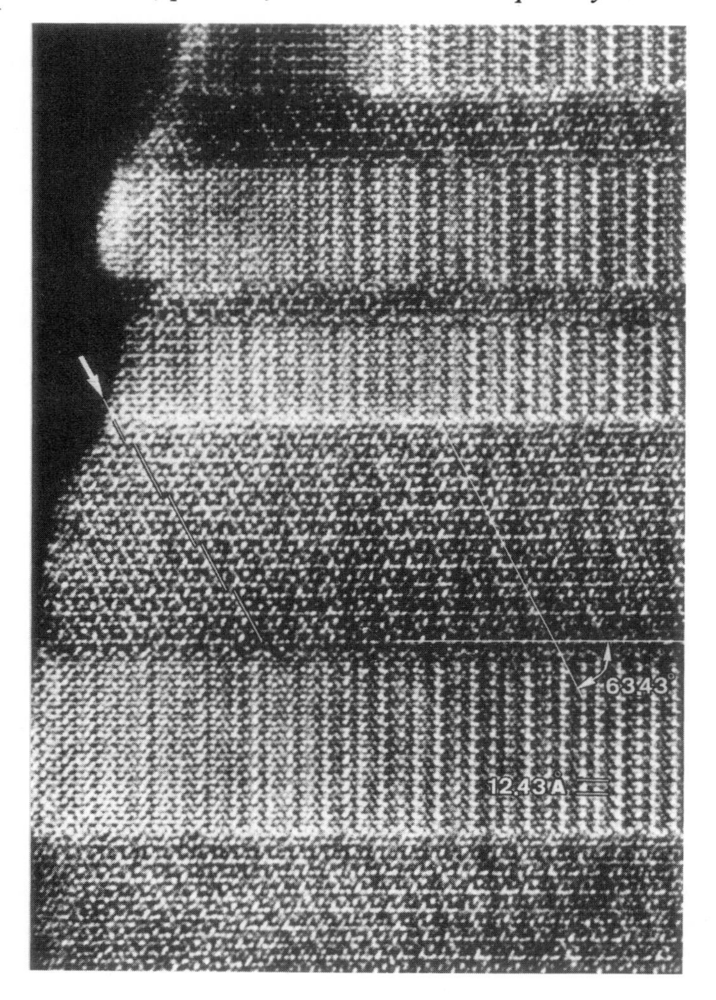

FIG. 5.49 High-resolution electron micrograph showing an AlPdMn icosahedral phase domain sandwiched between lamellar precipitates of decagonal phase. The 10-fold axis is vertical. Phason jags are visible (courtesy of M. Audier).

5.5.4 Examples of dislocation configurations in quasicrystals[25, 29, 30]

Dislocations are linear defects that for topological reasons cannot relax to the ground state. We have illustrated the point (Fig. 5.14) with the case of a 2-dim periodic crystal: annihilation of the edge dislocation results in a net shear of the crystal. Being a linear defect, the dislocation must leave the region far from its core without large distortions of the unit cells. In a periodic crystal, a dislocation is characterized by its Burgers vector $\boldsymbol{b}$ which is defined as the integral of the strain field $\boldsymbol{u}$ taken around a close loop containing the dislocation core (Fig. 5.13). This Burgers vector must be a real space lattice vector, geometrically illustrated in Fig. 5.13. In terms of the density wave picture, this is required for continuity of the phases around a closed loop containing the dislocation core. We remember that the phases of the density waves are expressed by eqn (5.42):

$$\phi_G = \phi_G^0 + \boldsymbol{G}\cdot\boldsymbol{u}.$$

For smoothness of these phases with respect to the polar angle about the dislocation core, one must fulfil

$$\phi_G(\theta = 2\pi) - \phi_G(\theta = 0) = 2\pi m$$

for some integers m, i.e.

$$\boldsymbol{G}\cdot\{\boldsymbol{u}(2\mathrm{n}) - \boldsymbol{u}(0)\} = 2\pi m$$

or

$$\boldsymbol{G}\cdot\boldsymbol{b} = 2\pi m. \tag{5.48}$$

This indeed imposes that $\boldsymbol{b}$ is a lattice vector.

For a quasicrystal, condition (5.31) cannot be satisfied with parameters taken only in the physical space. The phase expression for the density waves is given not by eqn (5.42), but rather by eqn (5.44), and then eqn (5.48) must be rewritten:

$$G_{\text{par}}\{\boldsymbol{u}(2\pi) - \boldsymbol{u}(0)\} + G_{\text{perp}}\{\boldsymbol{w}(2\pi) - \boldsymbol{w}(0)\} = 2\pi m \tag{5.49}$$

We conclude that the Burgers vector $\boldsymbol{b}$ of a quasicrystal must be n-dimensional (n = dimension of the related periodic image in the cut procedure), having both non-zero $\boldsymbol{u}$ and $\boldsymbol{w}$ components.

This statement can be readily illustrated, again using the example of a 2-dim Penrose tiling. By analogy with the crystalline case, one could remove a piece of one of the trails that can be seen in a Penrose tiling (Fig. 5.50). The lips of the cut can be glued without any problems. But all along the glued region, matching rules are violated, since this implies gluing edges which, beforehand, matched opposite edges of the removed lozenges. Thus, the result is not a point singularity (dislocation perpendicular to the plane of the figure), but a linear singularity (surface defect perpendicular to the plane of the figure) just equivalent to a stacking fault.

In the even more simplistic example of a Fibonacci chain, suppose that we remove an atom where an L and an S segment join together, along with the matter corresponding to $L/2$ and $S/2$ on one side and the other. After gluing the lips we are left with a strange unstable segment $(L + S)/2$.

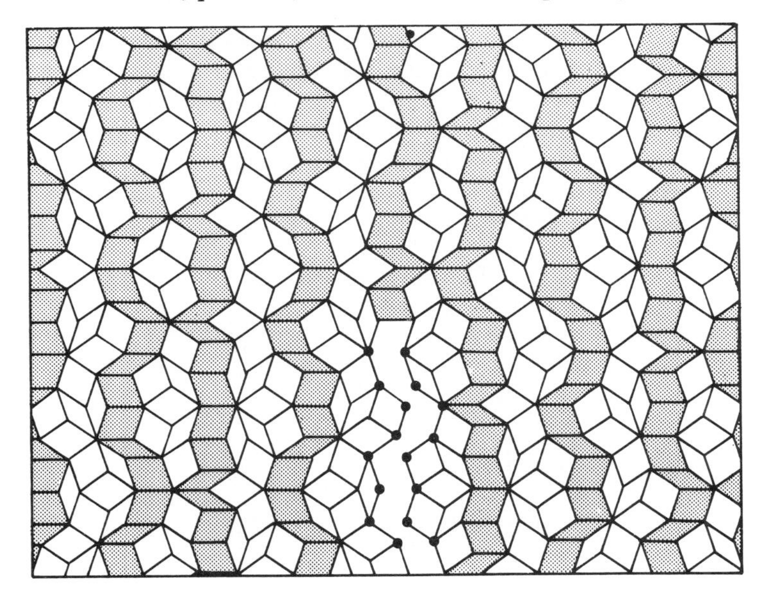

FIG. 5.50 If the shaded trails are considered as 'rough' atomic planes, removing a piece of a trail and gluing the lips would mimic the generation of an edge dislocation in a crystal. However, the resulting defect in the present case is not located at the termination point of the remaining part of the trail, but extends over all the lip region due to mismatches (parallel edges of the removed tiles do not 'match').[30]

This analysis transfers easily to the grid description as introduced in the previous sections.

One easy way to generate a strain field that satisfies eqn (5.49) is obviously to mimic the crystalline case, but where it can be applied, namely in the *n*-dim periodic representation. For instance, if we remove half a hyperplane in a decorated 6-dim cubic structure and glue the lips of the cut, we will create in the hyperspace the equivalent of a mere edge dislocation, with a 6-dim Burgers vector satisfying eqn (5.49). This would of course generate a defect in the related cut 3-dim quasilattice. Let us try to illustrate the point, again with the single grid scheme as already used for the pure phonon-like or pure phason-like strain fields. We still generate the grid by a plane cutting of 3-dim structure whose (X_{par}, X_{perp}) cross-section is reminiscent of the scheme used to generate a Fibonacci chain. But the figure is simply continuously copied along Y_{par} axis, perpendicular to the (X_{par}, X_{perp}) plane. Now, an 'edge dislocation' is introduced into this 3-dim periodic structure in the form of an additional 'half atomic plane', perpendicular to the cross-section (thus parallel to Y_{par}) and terminating at the core of the

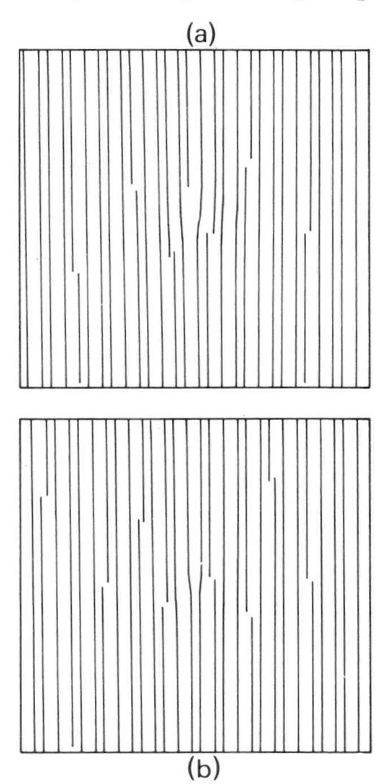

FIG. 5.51 Depending on the situation of the physical cut with respect to the core of the dislocation in the n-dim image, there may either be an extra half-line in the grid (a) terminating at the core of the 'physical' dislocation or no extra half-line (b). Both defects are stable.[24]

dislocation which is contined in the $(X_{\text{par}}, X_{\text{perp}})$ planes. The strain field produced vanishes in a region far from the dislocation core, is perpendicular to the added half-plane, and varies along Y_{par}. As both $\boldsymbol{u}$ and $\boldsymbol{w}$ (parallel and perpendicular) components participate in the total strain field, the single grid resulting from the plane cut of the 3-dim structure is expected to exhibit both curvature and jags. The additional half plane introduces additional A_{perp} planes, one of which may or may not generate a supplementary 'half line' in the grid, depending on the relative position with respect to the cut. Thus, two types of dislocation, with or without an extra line, must be considered in the single grid. In both types, a clear dislocation core is visible surrounded by a strong curvature gradient and numerous jags (Fig. 5.51). Far from the cores, the lines are straight and there is no jag.

In the pentagrid quasilattice scheme, some of the grids may have one

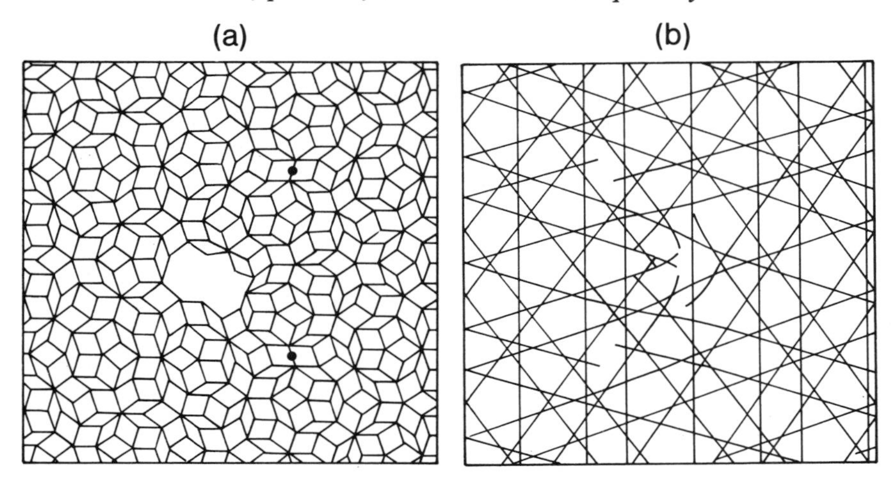

FIG. 5.52 (a) Example of a dislocation in a pentagrid and (b) the corresponding image in a Penrose tiling. Black dots are mismatches.[25]

type of (partial) dislocation while the rest has the other type. An example is shown in Fig. 5.52, which shows a dislocation in the pentagrid and its representation in the corresponding defective Penrose tiling, where a few mismatches are also visible. A more quantitative analysis reveals the following properties:

(1) the density of jags is inversely proportional to the distance from the dislocation core;

(2) the total number of jags is proportional to the length of the sample;

(3) the jags cannot be removed via phason relaxation because of criss-crossing of the jagged lines in a way that constrains the motion of the jag. Movements of the jags beyond a certain distance, linked to the Burgers vector, involve disruptions of the local structure. Quasicrystals can only be brittle in the absence of a grain boundary mechanism. Dislocations have indeed been observed in real quasicrystals.[31]

5.6 Conclusion

Quasicrystals, just as regular crystals, have basically $3N$ degrees of freedom that must be considered when elementary excitations and/or deformation modes are concerned. Contrary to what is sometimes wrongly stated, this number of degrees of freedom is not twice that of a crystal, simply because the quasicrystal structure is related to high-dimensional periodic image.

However, each degree of freedom has contributions from the physical and the complementary space. This yields different types of strain field corresponding either to topology conservation with cell distortion (phonon-like) or to subtle rearrangements of the unit cells (phason-like). Dislocations exist in quasicrystals. Their Burgers vectors are defined in the n-dim space and they induce strain fields with both phonon and phason-like contributions. They are expected to be hardly mobile.

Finally, the lattice dynamics of a quasicrystal is still a difficult subject. The problem cannot generally be tackled analytically, and numerical simulations, as well as experimental results, are still very scarce. However, a satisfactory qualitative analysis has been made, showing in particular localization effects. This is certainly an interesting aspect which deserves further investigation, as do other physical properties of the quasicrystals, as presented in Chapter 6.

5.7 Problems

5.1 Demonstrate that the allowed wave vectors for elastic waves in a monatomic chain of length L are of the form $k = n\, 2\pi/L$, n being any integer.

5.2 Calculate the density of states $g(\omega)$ in an isotropic solid within the continuous medium approximation. Deduce the maximum propagating frequency if the $3N$ degrees of freedom are restored.

5.3 Draw the first two Brillouin zones for a 2-dim square lattice.

5.4 Determine the dispersion curves of an atomic chain with three atoms of masses m_1, m_2, and m_3 in the unit cell and a single force constant α.

5.5 Use the Volterra process to generate all possible edge dislocations in hexagonal plane lattice.

5.6 Considering the Fibonacci chain as a regular periodic chain plus a modulation by a function taking values 0 and 1 with period τ, calculate the dispersion curves that would correspond to the approximants 1/1, 2/1, 3/2, 5/3, ... of τ.

5.7 Demonstrate that acoustic phonon intensities are proportional to intensity of the Bragg peak which is chosen for the measurement. Describe the shape of an acoustic phonon branch for a Fibonacci chain, when measured near a strong Bragg peak with a weak Bragg peak nearby (almost the same Q_{par} but small and large Q_{perp} respectively).

5.8 In a regular crystalline metal, atom diffusion occurs in particular via vacancy mechanisms. The activation energy (~2 eV) for diffusion contains a part corresponding to the formation of a vacancy (~1 eV) and a part corresponding to the energy barrier (~1 eV). Assuming that diffusion in a quasicrystal occurs via diffusive 'phason' defects (L, S inversion in a Fibonacci chain), what would be the ingredients to be included in the activation energy? Deduce an order of magnitude for the diffusion constant in quasicrystals, compared to that observed in crystals.

5.9 Write explicitly the density wave expression of the mass density in an icosahedral quasicrystal. Suggest truncated expressions to be used in a Landau model.

5.10 Consider a canonical 2-dim Penrose lattice as a proper irrational cut of a 4-dim primitive cubic lattice. Describe the possible edge dislocation by their Burgers vectors in 4-dim.

References

1. de Lange, C. and Janssen, T. (1981). Incommensurability and recursivity: lattice dynamics of modulated crystals. *J. Phys. C: Solid St. Phys.* **14**, 5269–92.
2. Currat, R. and Janssen, T. (1988). Excitations in incommensurate crystal phases. *Solid St. Phys.* **41**, 201–302.
3. Peyrard, M. and Aubry, S. (1983). Critical behaviour at the transition by breaking of analyticity in the discrete Frenkel–Kontorova model. *J. Phys. C: Solid St. Phys.* **16**, 1593–608.
4. Benoit, C., Poussigue, G., and Azougarh, A. (1990). Neutron scattering by phonons is quasi-crystals. *J. Phys.: Cond. Matter* **2**, 2519–36.
5. Steinhardt, P.J. and Ostlund, S. (1987). The physics of quasicrystals. pp. 513–652. World Scientific, Singapore.
6. Böttger, H. and Kasner, G. (1990). Electronic and vibrational properties of Fibonacci chains. In *Quasicrystals* (ed. M.V. Jaric and S. Lundqvist) pp. 459–67. World Scientific, Singapore.
7. Janssen, T. (1979). On the lattice dynamics of incommensurate crystal phases. *J. Phys. C: Solid St. Phys.* **12**, 5381–92.
8. Ashraff, J.A., Luck, J.M., and Stinchcombe, R.B. (1990). Dynamical properties of two-dimensional quasicrystals. *Phys. Rev.* **B41**, 4314–29.
9. Los, J. and Janssen, T. (1990). Lattice dynamics of three-dimensional quasicrystals. *J. Phys.: Cond. Matter* **2**, 9553–66.
10. Hafner, J. and Krajci, M. (1993) Propagating and confined vibrational excitations in quasicrystals. *J. Phys: Cond. Matter* **5**, 2489–510.
11. Windisch, M., Hafner, J., Krajci, M., and Mihalkovic, M. (1994). Structure and lattice dynamics of icosahedral AlCuLi, *Phys. Rev.* **B49**, 8701–17.
12. Poussigue, G. (1993). Private communication.
13. Suck, J.B., Janot, C., de Boissieu, M., and Dubost, B. (1990). In *Phonons*

(ed. S. Hunklinger, W. Ludwig, and G. Weis) pp. 576–8. World Scientific, Singapore.

14. Quilichini, M., Heger, G., Hennion, B., Lefebvre, S., and Quivy, A. (1990). Inelastic neutron scattering study of acoustic modes in a monodomain AlCuFe quasicrystal. *Le Journal de Physique* **51**, 1785–90.
15. Goldman, A. I., Stassis, C., Bellissent, R., Mouden, H., Pyka, N., and Gayle, F.W. (1990). Inelastic neutron scattering measurements of phonons in icosahedral AlLiCu. *Phys. Rev.* **B43**, 8763–6. (Short communication).
16. De Boissieu, M., Boudard, M., Bellissent, R., Quilichini, M., Hennion, B., Currat, R., *et al.* (1993). Dynamics of the AlPdMn icosahedral phase. *J. Phys.: Cond. Matter* **5**, 4945–66.
17. Janot, C., Magerl, A., Frick, B., and de Boissieu, M. (1993). Localized vibrations from clusters in quasicrystals. *Phys. Rev. Lett.* **71**, 871–4.
18. Steinhardt, P. J. and Ostlund, S. (1987). *The physics of quasicrystals*. pp. 137–74. World Scientific, Singapore.
19. Levine, D., Lubensky, T. C., Ostlund, S., Ramaswamy, S., Steinhart, P. J., and Toner, J. (1985). Elasticity and dislocations in pentagonal and icosahedral quasicrystals. *Phys. Rev. Lett.* **54**, 1520–3.
20. Bak, P. (1985). Symmetry, stability and elastic properties of icosahedral incommensurate crystals. *Phys. Rev.* **B32**, 5764–72.
21. Steinhardt, P. J. and Ostlund, S. (1987). *The physics of quasicrystals*. pp. 379–512. World Scientific, Singapore.
22. Lubensky, T. C., Ramaswamy, S., and Torner, J. (1985). Hydrodynamics of icosahedral quasicrystals. **32**, 7444–52.
23. Sainfort, P. and Dubost, B. (1988). Micro-mechanical properties of bulk crystalline and quasicrystalline AlLiCu compounds. In Quasicrystalline materials. (ed. C. Janot and J. M. Dubois) pp. 361–71. World Scientific, Singapore.
24. Bresson, L., Calvayrac, Y., and Gratias, D. (1991). Deformation plastique dans AlCuFe. In *Colloque quasicristaux* (ed. C. Berger and J. J. Préjean) p. 68.
25. Socolar, J. E. S. (1986). Phason strain in quasicrystals. *J. Phys. (Paris)* **47**, C3-217–28.
26. Coddens, G., Bellissent, R., Calvayrac, Y., and Ambroise, J. P. (1991). Observation of phason hopping in icosahedral AlFeCu quasicrystals. *Europhys. Lett.* **16**, 271–6.
27. Janot, C., Audier, M., de Boissieu, M., and Dubois J. M. (1991). AlCuFe quasicrystals: low-temperature unstability via a modulation mechanism. *Europhys. Lett.* **14**, 355–60.
28. Bessière, M., Quivy, A., Lefebvre, S., Devaud-Rzepski, J., and Calvayrac, Y. (1991). Thermal stability of AlCuFe icosahedral alloys. *J. Phys. (Paris)* **1**, 1823–36.
29. Socolar, J. E. S., Lubensky, T. C., and Steinhardt, P. J. (1986). Phonons, phasons and dislocations in quasicrystals. *Phys. Rev.* **B34**, 3345–60.
30. Bohsung, J. and Trebin, H. R. (1989). Defects in quasicrystals. In *Introduction to the mathematics of quasicrystals* (ed. M. V. Jaric) pp. 183–221. Academic, New York.
31. Zhang, Z. and Urban K. (1989). Transmission electron microscope observation of dislocations and stacking faults in a decagonal AlCuCo alloy. *Phil. Mag. Lett.* **60**, 39–102.

6

A little more about the physics of quasicrystals

Today, I feel no wish to demonstrate
that sanity is impossible
Aldous Huxley

6.1 Introduction

Quasicrystals pose a fundamental challenge to our understanding of the physical properties of condensed matter. As we have repeatedly stated, they lack the translational periodic order present in classical crystals. But they lack also the randomness present in truly disordered systems. They have, however, perfectly ordered structures that should generate proper selection rules for the formulation of their physical behaviour. Apparently, no one has so far succeeded in using this order, even within model descriptions such as the Penrose tiling scheme.

The easy option is to accept that quasicrystals have order 'intermediate' between periodic crystals and random materials, and then to apply methods that have previously been used in condensed matter. But this may be a blind and quite dangerous alley.

The degree of order in a material has indeed a profound effect on ideal properties. For instance, it is well known that there is only a single well-defined periodic structure which corresponds to a given unit cell with given rotational symmetries (point group). In a similar but random material short range order is mastered by chemical interaction and cluster packing obeys 'simple' sterical connectivity rules, whereas, for a fixed orientational symmetry (say, icosahedral point group), a fixed set of unit cells (say, two canonical rhombohedra) and a fixed quasiperiodicity (say, the one defined by Ammann grids) there are still an infinite number of non-equivalent quasilattices, related via 'phason induced' rearrangement of tiles. This obviously poses basic questions when addressing problems such as growing (perfect) quasicrystals and the stability of quasicrystals versus crystals in a given chemical system.

Another example of the influence of the degree of order on properties is the distribution of the eigenstates of the system. In one dimension, all eigenstates of a simple monatomic crystal are extended: they form a set of plane waves that fully propagate all along the chain. In a similar but random material, all states are localized. In higher (3)-dimensions, the eigenstates of the crystal remain extended, whereas both (weakly) localized and extended states may exist in a disordered system. A quasicrystal violates assumptions of both these systems and, hence, experience gained with both ordered and disordered systems must be carefully reconsidered before it is applied to quasicrystals.

Experimental evidence obtained from real quasicrystals about these problems must also be handle with great care, because of the actual difficulties in preparing and characterizing a perfect, single-phased sample of quasicrystal.

However, it is possible to get an interesting insight into the expected properties for quasicrystals by analyzing the physics of simple models and by qualitative derivations. These aspects will be briefly addressed in this final chapter.

6.2 Is perfect quasicrystal growth an acceptable physical concept?

6.2.1 The one-dimensional case as an illustration

Growing a piece of matter within certain rules for short- and long-range order is not necessarily an easy task. For regular periodic crystals, the sequence of atoms that exists in a seed cluster repeats exactly again and again, so it appears that the atom to be added must interact only with a small number of atoms at some place on the cluster surface. Moreover, there is a single ground state structure for a given space group which means that the structure is energetically stabilized and can be grown perfectly.

The growth problem of random material is not really relevant. Randomness in condensed matter is basically the result of non-ergodic transformations, such as rapid quenching from the melt or glass transitions, under the influence of short-range chemical interactions. They are metastable materials rather than a stable phase, with the exception of the equilibrium liquid state.

Conversely, it seemed apparent that the quasicrystal picture suffers some fatal flaws from these points of view and it is probably useful to explain, at least qualitatively that, contrary to common acceptance:

(1) long-range atomic interactions are not required to grow perfect quasicrystals;

(2) perfectly defined ground states can be identified and related to particular types of quasiperiodic order;

(3) rapid growth, comparable to that observed for periodic crystals, is not an idealistic view.

These points are not academic ones. Looking back at Chapter 1, the various mathematical procedures that have been described to generate quasiperiodic lattices are somewhat suggestive that growing a perfect quasiperiodic tiling would appear to be a daunting task. The sequence of atoms never exactly repeats, so that the atom added to the surface of a given cluster must interact with each atom in the seed cluster to ensure that it sticks at a site consistent with perfect quasiperiodic order. As the cluster grows, this requirement imposes arbitrary long-range interactions, which is physically implausible. As a consequence, perfect quasicrystals have been viewed by some people [1, 2, 3] as physically unrealizable and structural modelling for the icosahedral phases has sometimes turned towards random aggregation or glass models [1, 2] and/or entropically stabilized random tilings [3] (see Section 4.3.4). For several years, these theoretical arguments have resonated strongly with the observations of experimentalists, since a generic complaint was that all available icosahedral alloys exhibited (large) residual disorder as indicated by finite diffraction peak widths. Moreover, most of them did not seem to be stable phases with respect to their periodic crystal modifications. It is now accepted that this no longer holds true: an experimental breakthrough has taken place in which perfect icosahedral stable alloys have been discovered.[4–6] Even more recently,[7–11] it has been demonstrated that the common conclusion about long-range interactions being required to grow perfect quasicrystals had been misleadingly inspired by the treatment of the 1-dim case and restricted uses of the Penrose edge-matching rules.

Let us take again the Fibonacci 1-dim structure to illustrate the above statements. First, it is easy to show that a Fibonacci chain cannot be grown *ad infinitum* via a 'local interaction rule'. A 'local rule' means that in deciding whether to add L or S to the end of an existing finite string only 'atoms' of the end of the string should be known, as might be mimicked by some local atomic interactions. Unfortunately, the construction rules of the Fibonacci chain do not comply with this requirement. An example is the finite 'seed' chain:

$$LLSLSLLSLLSL; \quad \text{and then } SL \text{ or } LS?$$

'Local' features of the chain are that S is always between two Ls (Ss are never nearest neighbours) and/or that single L *and* pairs of Ls are sandwiched between two Ss. Accordingly, when a finite chain ends with S you can only add an L but when it ends, as in the example, with L, there are two apparent possibilities, S or L, or even SL or LS for the next atoms. The LS choice is actually illegal because it would generate a succession of (LLS) repeated three times, whereas ideal Fibonacci chains do not contain three repetitions of any sub-sequence in succession (see the inflation construction of the Fibonacci chain in Chapter 1). Obviously, to recognize the problem requires knowledge of the entire sequence, whatever its length. This implies that interactions extending over the length of the chain are required to ensure continued growth of the perfect Fibonacci chain. A seemingly reasonable conjecture would be that higher dimensional quasicrystals are, if anything, a worse problem. For example, the 2-dim Penrose tiling is equivalent to a five grid which forms *five* independent sets of parallel lines spaced according to a Fibonacci sequence of intervals (Chapter 1). Growing such a tiling from a seed cluster would require it to extend in each of the five directions simultaneously. The immediate feeling is that the 1-dim growth problem multiplies by five! Treated on the same footing, the 3-dim growing of quasicrystals would turn out to a dreadful nightmare! This is actually a little too simplistic an approach. Increasing dimensionality opens new ways to bypass otherwise blocked paths and generates criss-crossing effects that modify the degree of freedom, as will be explained in a forthcoming section.

Another drawback which can also be exemplified with the Fibonacci chain is the equal probability that different sequences may superimpose into the growing procedure. It other words, the infinite packing may not be uniquely specified given only a set of rules which fix the allowed unit cell clusters smaller than some bound. This is actually a corollary of the possibility that occasional short-range mistakes (e.g. *LS* instead of *SL* in the above illustration), even if not repeated very often, initiate new sequences which are energetically equivalent but geometrically distinguishable. This introduces the concept of 'local isomorphism' (*LI*) classes, which is not very useful for

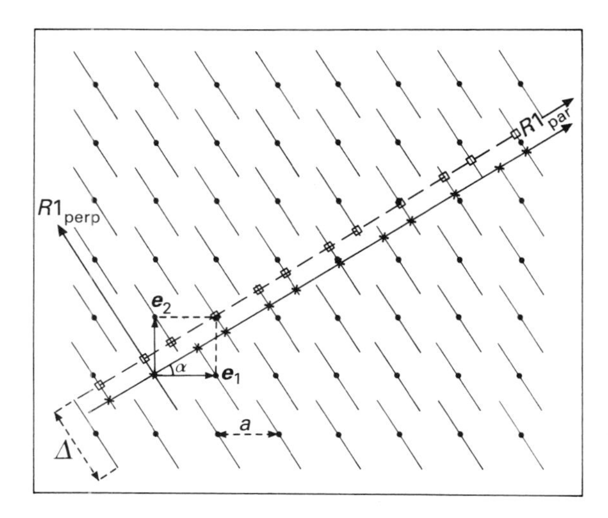

FIG. 6.1 The Fibonacci chain as generated via the cut procedure. Two chains are shown; they correspond to different 1-dim cuts within a phason translation. The strings, as visible in the figure, are different sequences of large–short segments. However, the two infinite chains are indistinguishable.

1-dim structures and will be illustrated further in a forthcoming section. For the moment, just consider again a Fibonacci chain as generated by a proper 1-dim cut of a 2-dim square lattice (see Sections 1.4.5 and 3.3.7 for properties used hereafter). The procedure is illustrated once more in Fig. 6.1. The sequence of (L, S) segments can be altered, with respect to the local features regarding the allowed nearest-neighbour positions, by reversing occasional *LS* pairs into *SL* or vice versa. As already mentioned, these point-like defects are improperly called 'phason-like' defects, because *LS* pair reversals are also produced by changing the offset between the physical space and the chosen origin of the 2-dim hyperspace. This is also illustrated in Fig. 6.1, where two *apparently* different Fibonacci strings are produced by selecting two different 2-dim sites as the origin of the 1-dim physical space. But the conclusion is not correct. The two *infinite* Fibonacci chains are actually identical (here the same LI class) up to a global translation of the figure. The demonstration is straightforward: the 2-dim square lattice has translation invariance properties that make all lattice sites perfectly equivalent and indistinguishable; any of these lattice sites can then be selected, at random, as the common origin of the 2-dim hyperspace and the related 1-dim physical and complementary spaces. All the corresponding generated Fibonacci chains will be indistinguishable and related by translations. Of course, finite chains corresponding to different origin choices are different when compared within the same portion of space (see Fig. 6.1).

In principle, non-equivalent distinguishable infinite Fibonacci chains would be generated if the offset between the square lattice and the origins of the associated spaces could be changed by a vector which is not a translation of the square lattice. That would define different LI classes. In order to quantify the idea, consider the Fibonacci chain as generated in Fig. 6.1 when the 1-dim physical space passes at a lattice site of the 2-dim square structure. With the notations already used in Chapters 1 and 2, we have:

$$
\begin{aligned}
L &= a\cos\alpha = \tau, \\
S &= a\sin\alpha = 1, \\
\Delta &= a(\cos\alpha + \sin\alpha) = \tau + 1.
\end{aligned}
$$

Sites of the square lattice are at vectors $\boldsymbol{r} = n_1\boldsymbol{e}_1 + n_2\boldsymbol{e}_2$ (n_i being integers). The components of $\boldsymbol{r}$ in physical and complementary space are r_{par} and r_{perp} respectively, with

$$
\begin{aligned}
r_{\text{par}} &= (n_1\cos\alpha + n_2\sin\alpha)a = n_1\tau + n_2, \\
r_{\text{perp}} &= (-n_1\sin\alpha + n_2\cos\alpha)a = -n_1 + n_2\tau.
\end{aligned}
$$

As analysed in Section 3.3.7, $|r_{\text{perp}}|$ must be smaller than $\Delta/2$ if a quasilattice site is to be generated from the $\boldsymbol{r}$ 2-dim site. It is easy to verify that for any n_1 integer, n_2 has then to be restricted to the values $\lfloor n_1/\tau \rfloor$ and $\lfloor n_1/\tau \rfloor + 1$ except if n_1 is a 'Fibonacci number' $(= 3, 5, 8, 13, \ldots)$ for which only $n_2 = \lfloor n_1/\tau \rfloor$ is permitted ($\lfloor x \rfloor$ stands for the largest integer contained in x) The latter case is of special interest, since the corresponding series of n_1/n_2 ratios contains all the successive rational approximants to the golden mean τ. The corresponding r_{perp} values are the successive smallest possible ones and define the 2-dim square lattice sites that will be crossed successively via a 'phason shift' γ of the physical space. We conclude that in the present example, all 1-dim quasiperiodic structures related by 'phason shift' γ of the form

$$\gamma = (-n_1 + n_2\tau),$$

with n_1/n_2 a rational approximant for τ, are equivalent and indistinguishable. The expression for γ can be simplified to

$$\gamma = \{n\tau\}, \tag{6.1}$$

with n a Fibonacci number (2, 3, 5, 8, 13, . . .) and $\{x\}$ being the fractional part of x. Actually, eqn (6.1) remains valid for any integer n. Phason translations not defined by eqn 6.1 would produce a different LI class of quasicrystals.

Now it is clear that the analysis of the 1-dim quasilattice case within the scope of physical growability has raised more questions than it has solved problems, and the overall situation can briefly be depicted as fairly hopeless. We are now going to inject, modestly, the dimensionality ingredient, via some features of the 2-dim Penrose tiling.

6.2.2 Criteria for the physical growability of a 2-dim Penrose tiling

It may be useful to repeat here that in order to reconstruct the atomic structure of a quasicrystal using a tiling description (2-dim or 3-dim), it is necessary to determine the symmetry, the atomic decoration of (two types at least of) unit cells, and the unit-cell packing scheme. Whereas only one perfect packing is possible for periodic crystals, there are, *a priori*, an uncountable infinite number of distinct, perfect quasicrystal packings for any given symmetry. They correspond to geometrically distinguishable arrangements of the same unit cells and are 'phason shift' related, as suggested by the 1-dim illustration. The point is also illustrated in Fig. 6.2 and 6.3. The former reproduces a result already presented in Section 5.5.3: phason translations produce alterations in some parts of the 5-grid associated with a 2-dim Penrose tiling; the dual procedure that attaches particular tiles to the 5-grid nodes is affected as a consequence; and the two tilings are Penrose-like, made with the same rhombus tiles arranged within pentagonal symmetries in both cases, but some local clusters do not show up in both structures.

This is exemplified further in Fig. 6.3, in which a simplified part of a 5-grid is represented. In the figure we observe a single line of a grid labelled 0 and

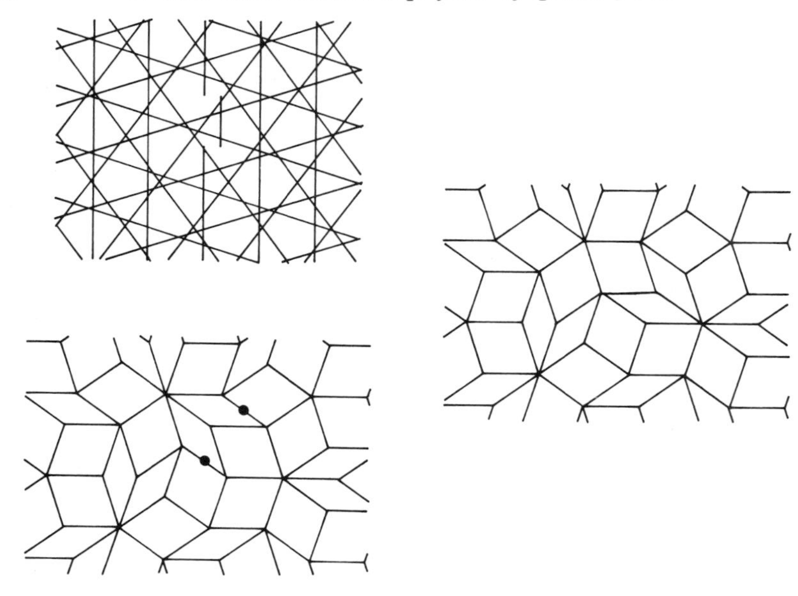

FIG. 6.2 Phason translations produce jags in the pentagrids and mismatch defects (dots) in the correlated Penrose tiling. The perfect Penrose tiling of the same part of the plane is shown on the right-hand side of the figure for comparison.

its intersections with the other four grids, labelled 1, 2, 3, and 4. Grids 1 and 4 intersect grid 0 at an angle $2\pi/5$, called type F intersections, whose dual modifications are fat rhombuses with two different orientations. Grids 2 and 3 intersect grid 0 at an angle $\pi/5$, called type S intersections whose dual modifications are skinny rhombuses. The result is a 'row' of rhombuses arranged in a given sequence according to the sequence of intersections along the associated line of grid 0. Assuming that now a proper 'phason translation' induces a parallel shift of this grid 0 line, the sequence of intersections F, S may be modified and, as a consequence, also the structure of the associated row (see Fig. 6.3(*b*).

Thus the first criterion to be satisfied for growability of a quasicrystal is that, among all the 'phason translated' structures, a finite set of the lowest energy clusters can be specified, that is that a ground state does exist. This criterion has been called 'restorability'.[11]

A 2-dim Penrose tiling can be generated via a planar cut of a 5-dim hypercubic lattice with rhombic icosahedral atomic volumes, as resulting from the cross-section of the hypercubic unit cell in the 3-dim complementary space.[12] The tiling is completely specified by the offset γ, defined as

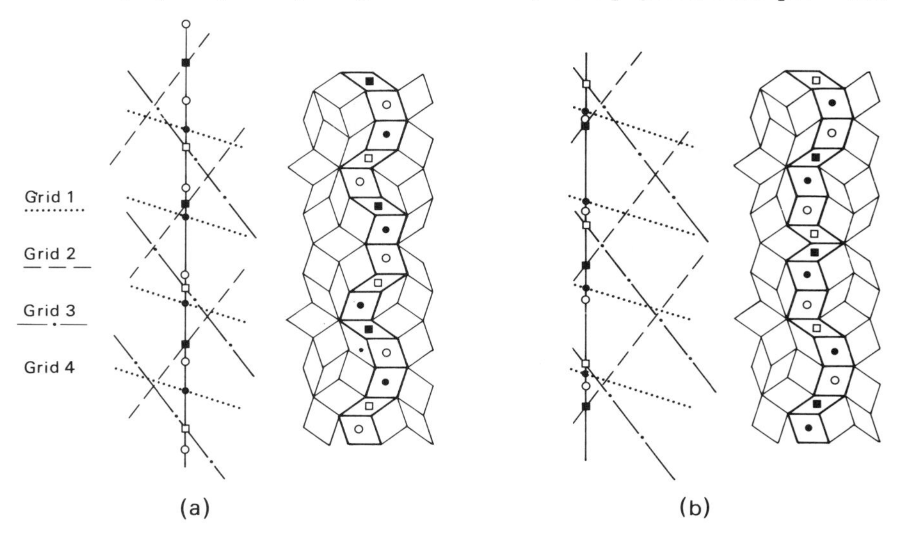

FIG. 6.3 The figure shows the position of a single line of the grid labelled ○, interesecting the four other grids of a pentagrid, (a) without and (b) with phason shift. The shift perturbes the sequence order of type F(circles) and type S(squares) intersections and produces different associated rows of dual tiles.

for the Fibonacci case in the preceding section. But contrary to the 1-dim quasicrystal, the offset $\gamma = \{n\tau\}$ (eqn 6.1) with different n integers generates a geometrically distinguishable arrangement of unit cells, and then different LI classes. This is due to the fact that 'phason translations' γ scan successively different possible origin sites, according to their r_{perp} distances. But these sites are distributed in different planar directions and two successive $\gamma = \{n\tau\}$ values do not necessarily affect the same grid of the associated 5-grid: $\gamma = 0$ is the classical Penrose LI class; shifts to γ between 0 and 1 produce a continuous spectrum of tilings belonging to other, non-Penrose, LI classes. A detailed analysis, beyond the scope of this book, would demonstrate that ranges of interactions with only modest values of n seem physically plausible.[13] Thus, determining the LI class for real quasicrystals may be reduced to a tractable problem. Yet it is certainly desirable that quasicrystal structures may be derived, e.g. from diffraction data, without *a priori* assumptions about their LI class. The point has been illustrated in Chapter 4, with the example of a structure forced to 3-dim Penrose tiling and the consequence of unphysical chemical decorations of the unit cells. (absence of Li vs. CuAl order in ico-AlLiCu).

Clearly, the conclusion is very positive, since a finite, even small, number of plausible LI classes means that structure and growth problems remain deterministic ones, within reasonably limited sets of free parameters.

The second criterion for growability is that the lowest energy clusters which specify a given LI class can be extended within an experimental time-scale via short-range atomic-like interactions. The question might appear to be not very relevant to Penrose tilings, since matching rules should offer a potential mitigating factor to growth problems. These matching rules, (more precisely, edge-matching rules) are typically indicated by placing different arrows on the edges of the tiles (Chapter 1) that constrain the way two tiles can match together edge-to-edge (Fig. 6.4). Penrose clearly showed that the only plane-filling tiling consistent with the matching rules is a perfect Penrose tiling. The statement only means that if you have obtained a perfect Penrose tiling, you can be sure that the edge matching rules are obeyed everywhere. Do these edge-matching rules also represent viable local rules for growing a tiling by aggregation, e.g. adding one tile at a time to a randomly chosen edge in accordance with the edge-matching rule? Unfortunately not.[14] Instead, following this procedure, mistakes are made every 10 tiles or so. The tricky aspect of those matching rules is that mistakes, when produced, are not revealed at once and you sometimes have to wait for many further building steps to appreciate the catastrophe. You have then to remove tiles and to make a different choice. The dismal failure of the edge-matching rules as a growth procedure (see an example in Fig. 6.5) is thus

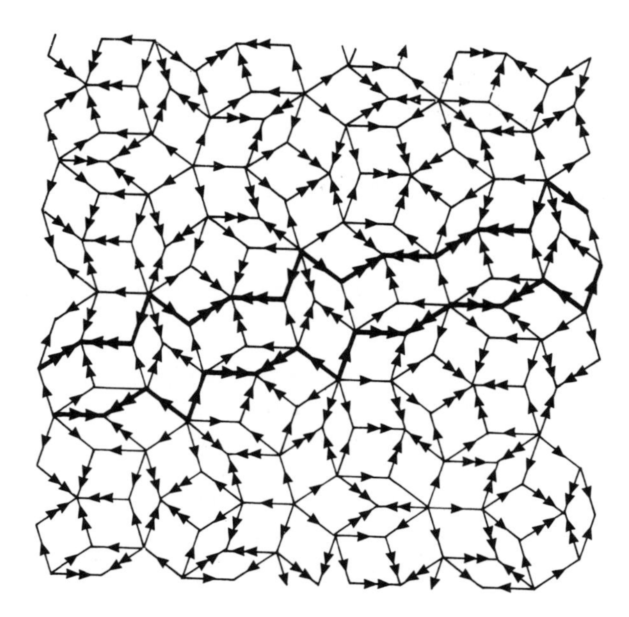

FIG. 6.4 The Penrose tiling showing the arrow edge-matching rules. A worm is highlighted.

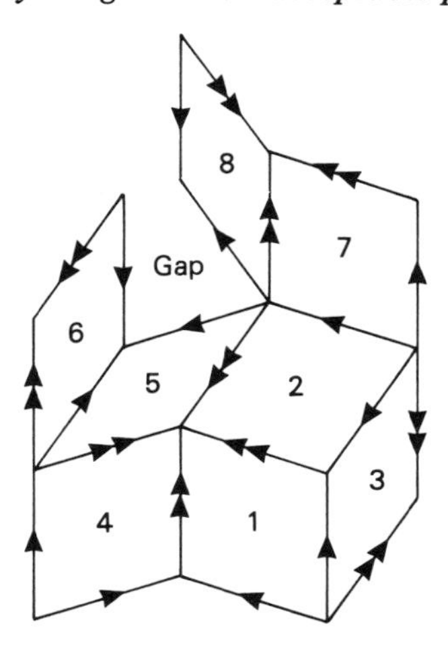

FIG. 6.5 Illustration of a typical failure of the edge matching rules as a growth procedure.[9]

obvious. The problem was indeed recognized in the original work of Penrose, as accurately reported by Gardner.[15] The important ingredient that was lacking there was the idea of 'forcing', implying that in many circumstances during growth there is no choice of whether to add a fat or a skinny tile, even though the arrow edge-rules always allow such a choice to be made.

The forcing rules had actually been only slighlty hidden. In fact, the required growth rules are remarkably short-range: edge matching rules must simply be replaced by **vertex matching** rules. This will be developed in the next section.

6.2.3 The vertex matching rules as local growth requirements for a quasicrystal

The basic principle in defining vertex matching rules is quite simple. They actually subsume the Penrose edge matching rules in the sense that any tiling obeying the vertex rules obeys the edge rules. Indeed, the vertex matching rules, demand that the vertex configurations be consistent with the types that appear in a perfect Penrose tiling.[7] A careful and simple scan of a perfect 2-dim Penrose tiling shows that there are only eight types of vertex configuration, as illustrated in Fig. 6.6. The growth procedure is now that, in deciding whether to add a given tile to a given vertex on the surface of a seed cluster, it is only necessary to determine the configuration of tiles already adjoined to that site. Such vertex matching rules are fairly short-range. By contrast, to grow a large unit cell periodic approximant with the same rhombus units requires local rules extending to the size of the unit cell. Thus, quasicrystal growth appears as a better alternative to the description of structures in terms of large unit cell crystals, simply because it requires shorter-range rules to grow.

Let us now describe how the corresponding growth algorithm may actually proceed.[7-13] Having the catalogue of the eight types of vertex con-

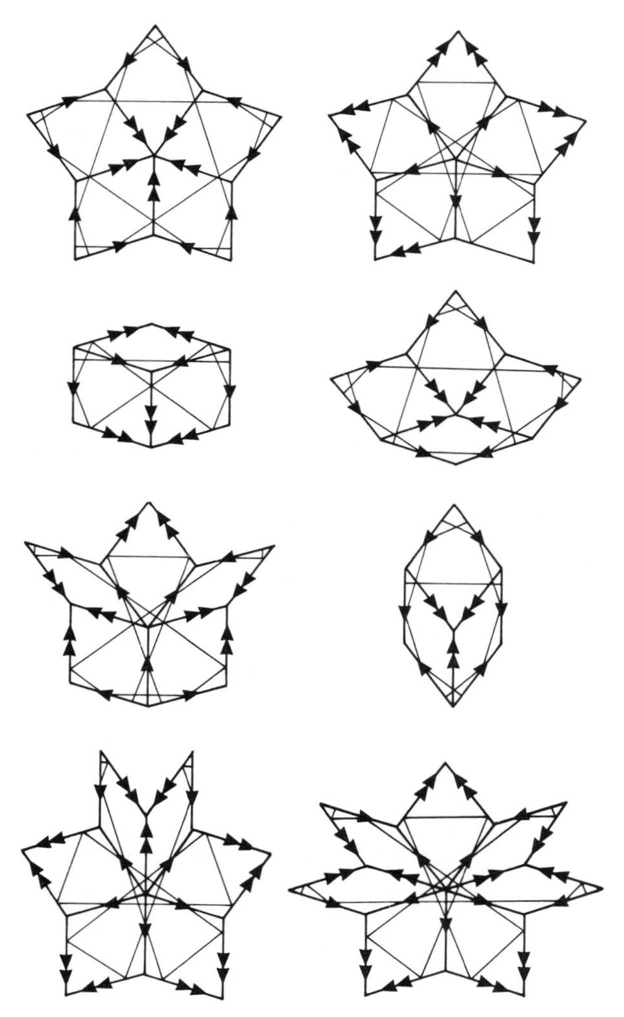

FIG. 6.6 The legal Penrose vertices, which can be used to define vertex matching rules and forced vertices. 'Decoration' of the tiles by the associated 5-grid is also shown.

figuration which show up in a 2-dim Penrose tiling (Fig. 6.6), a new tile can be added to an edge on the surface of a given cluster only if the configurations formed around each vertex of the new tile are consistent with at least one of these eight vertices. A vertex that is not yet completely surrounded is called a **forced vertex**[7] if, for at least one of the free edges sharing the vertex, there is only one way of adding the new tile, called also a **forced tile**. An example of a forced vertex is shown in Fig. 6.7(*a*) while the one in part (*b*) is not forced because the remaining gap can be filled either by an added fat rhombohedron or two skinny rhombohedra (see Fig. 6.6). If several vertices are forced at the surface of the seed cluster, the obvious growing rule

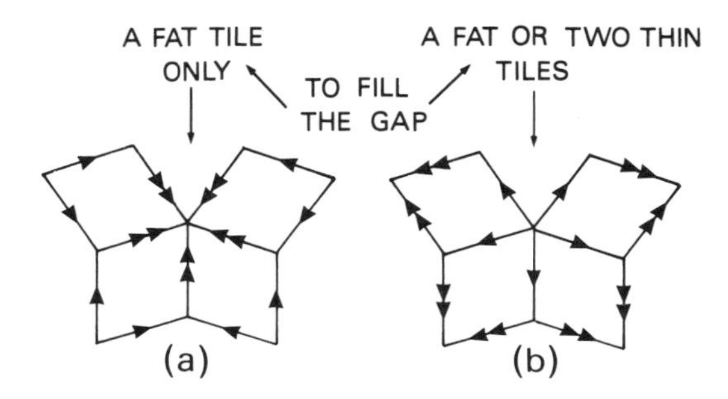

FIG. 6.7 Examples of a forced vertex (a) and a 'dead' vertex (b).

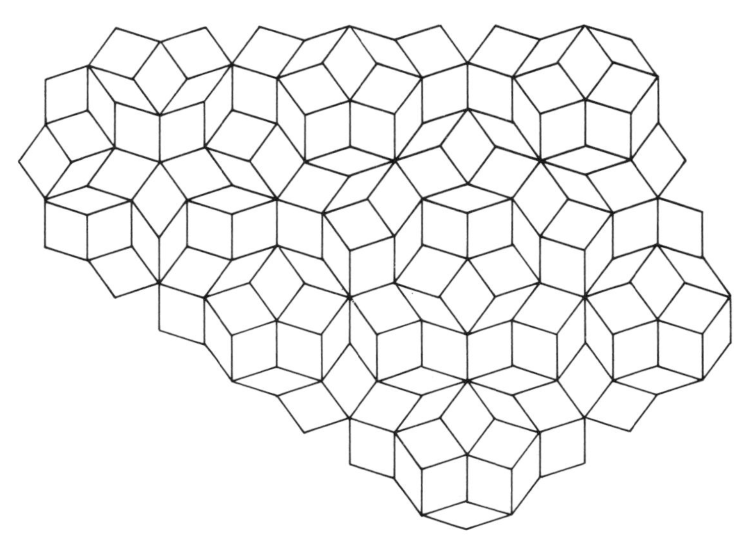

FIG. 6.8 Example of a dead surface: it is easy to verify that any free edge can be matched by either a fat tile or a skinny tile without violating the vertex matching rules.

is that a forced vertex is chosen at random to add a forced tile to it. Forced tiles being added in any order, it may happen that the entire surface of the seed cluster consists only of unforced vertices. The surface is then called a **dead surface**. An example of a dead surface is presented in Fig. 6.8. As explained further below, the dead surfaces, when looked at on a macroscopic scale, form a convex polygon whose edges joint at 108° or 72° corners and lie at the border of finite worm segments (see Fig. 6.9). Upon reaching a dead surface, the proposed growing rule is to choose a 108° corner at random and to add a fat rhombus to it, consistent with the vertex rules. A weaker form of this second rule could be to accept that either a fat or skinny tile is added to a randomly chosen vertex at the dead surface.

We shall now first justify some aspects of these growing rules, and second review their advantages and drawbacks. We know that a 2-dim Penrose tiling has an equivalent 5-grid description. The lines of the 5-grid 'decorate' the tiles, as represented once more in Figs. 6.6 and 6.9. The tiles that 'intersect' a given line form a **row**. The vertex rules obviously ensure the continuity of the lines in the tiles that are added around any vertex; the correct alignment of distant tiles is also ensured. When a forced tile is added to an edge, any lines in the seed cluster that pierce the edge are extended through the new tile.

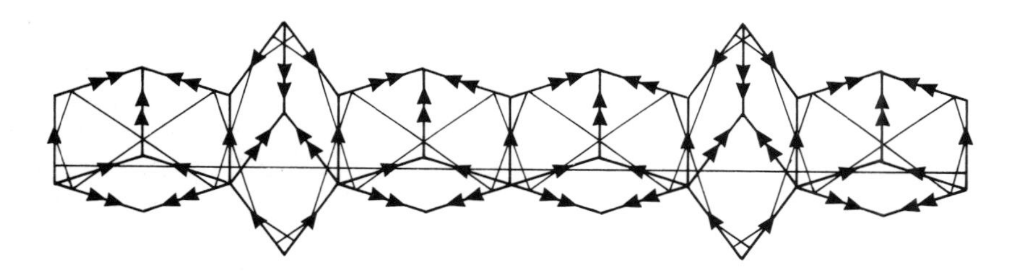

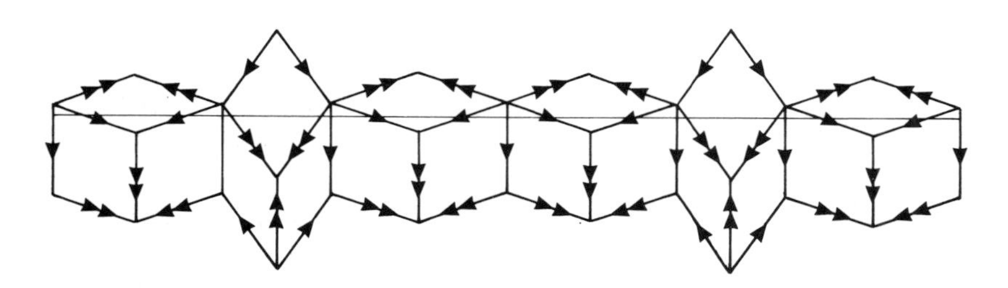

FIG. 6.9 A worm (top) and the flipped configuration (bottom). The lines and portions of lines of the associated 5-grids are also shown. The two worms match the upper part of the dead surface shown in Fig. 6.8.

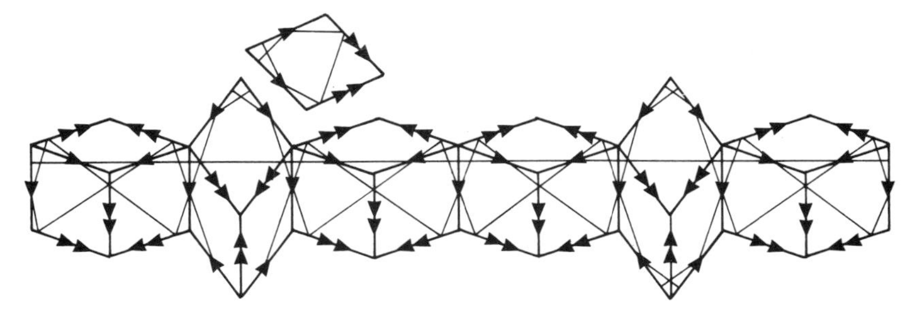

FIG. 6.10 Example of a forced tile as added properly to the edge of a row in order to initiate a new row.

The new tile, beyond extending existing lines in some directions, may initiate new lines in other directions. An example is shown in Fig. 6.10, where a new tile initiates a new line, and thus a new row, when added at a forced vertex of an existing line: row number 1 fixes four of the 5-grid lines which pierce the old *and* the new tiles, and the new tile fixes the fifth. The new tile produces further forced vertices where it adjoins row No. 1, etc., until a row No. 2 is grown by reaching the ends of row No. 1. The configuration of the tiles that are sited along a dead surface must be very special, in relation to the fact that at any edge of the dead surface, either a fat or a skinny rhombus can be added (within the vertex matching rules). Suppose that a skinny tile is added to a given edge; it initiates a new line of the 5-grid parallel to the dead surface and fixes the spacing (large or short) between the new line and the nearest parallel line in the cluster. This, in turn, forces a new row of tiles along the new line. Suppose instead that a fat rhombus is added to the same given edge; the other spacing is fixed and a different row of tiles is created along the dead surface. As the configuration of edges along the dead surface must match either of these two possible new rows, the only difference in the two lines that pass through the two alternative rows is the position of the new line. This imposes the rule that the two possible rows be related by a 'flip' reflection (Fig. 6.9) and that both of their sides have the same geometric shape and arrow directions. This unique property corresponds to a **worm segment**. In order to obtain a dead polygonal surface, dead surfaces that border the worm segments on the interior must meet to form **dead corners** and the only ways for worms to meet is for them to be associated with lines of the 5-grid that are tilted by 72° or 108°, as illustrated in Fig. 6.11. The only possible closed polygons with such corner angles have either two 72° and two 108° corners or five 108° corners. This also justifies that a 108° dead corner may be 'reactivated' by conveniently adding a fat rhombus whose angles are precisely 72° and 108°.

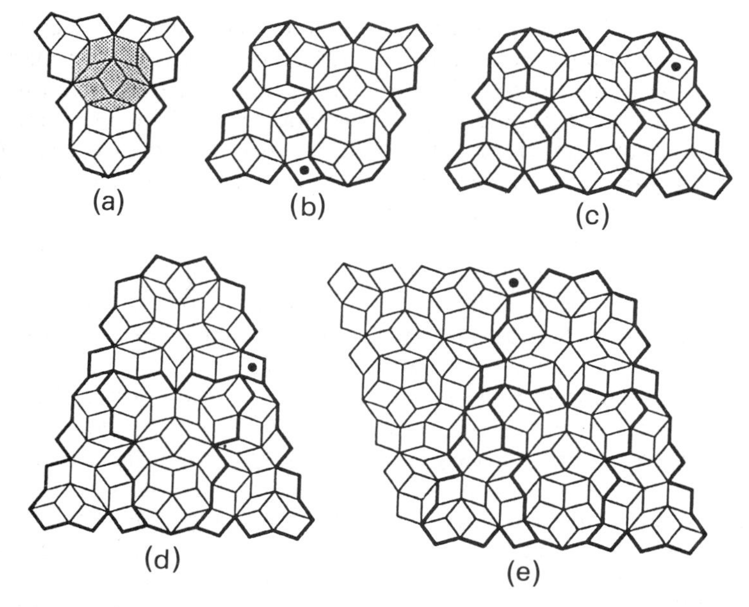

FIG. 6.11 Examples of polygonal dead surfaces with 108° and 72° corners, grown from a perfect seed cluster.[8]

We conclude that these growing rules, based on vertex matching rules, seem to be sound, at least from the purely geometrical point of view. Now, we must ascertain that they ensure rapid growability, within a physically reasonable time-scale. Obviously, the proposed growth rules impose at each step a search of the complete surface of the existing cluster to determine that there are or that there are not forced sites available, which somewhat hampers the true locality of the procedure. Actually, this is reminiscent of equivalent problems which are usually encountered in periodic crystal growth: the binding energy of an atom to a free site widely depends on the presences of ledges, kinks, screw dislocations, corners, and local atomic configuration. Thus, in a similar way, the rules governing binding energies to forced vertices, corners, and all other sites in quasicrystal growth can be mimicked by choosing 'sticking probabilities' that may minimize the effects of surface inspection. Let us call respectively P_F, P_C, and P_0 the sticking rates for forced, 108 ° corner, and generic 'dead' vertices. The condition that $P_F \gg P_C \gg P_0$ would cause growth to occur at forced sites first, as long as they exist at the cluster surface, and then at corner sites. Clearly, the sticking rate for a given site is determined only by the *local* arrangements of the atoms. In this way the algorithm becomes truly local, but unfortunately the ability to grow an infinite perfect Penrose tiling is lost. Let consider the case

$P_0 = 0$ and $P_C/P_F \neq 0$ but reasonably small. The prospect of growing a perfect Penrose tiling relies totally on tiles being added at 108° corners *only* when no forced vertices are left on the cluster surface, which requires that $P_C = 0$; but then the growth stops at the emergence of the very first dead surface. On average, with $P_C \neq 0$, it will be possible to grow perfect tilings only up to a size where the time needed to grow from a dead surface to the next becomes comparable with the inverse of the probability of adding tiles on a 108° corner. For any non-zero value of P_C, mistakes will appear, but the average size at which the first defect is formed increases as P_C decreases. On the other hand, P_C cannot be decreased too much if one wants to grow the tiling beyond the first dead surface. An example of the evolution of tiling size as time goes on is shown in Fig. 6.12.[16-18] The growth is characterized by brief intervals of time during which the growth proceeds rapidly on forced vertices from a dead surface to the next, where it stops for a long time, waiting for the unlikely event of sticking a tile on a 108° corner.

The case in which $P_C = 0$ and P_0 is very small is similar. The main difference is that there are many more generic vertices (dead vertices) on a cluster surface than 108° corners. Therefore, to get results similar to those generated by a given value of P_C one has to select a much smaller value of P_0. Also in this case, the value of P_0 fixes the average size of a perfect Penrose tiling which grows before defects are generated.

Besides understanding how large a perfect Penrose tiling can be grown, it is also of interest to know what happens when defects are generated. Basically, the emergence of a defect would correspond to the appearance of one or several vertices that cannot be completed according to the eight legal configurations (vertex mismatch). Such a defect, obviously, locally inhibits further growth. However, at a short distance from the defect, and all around

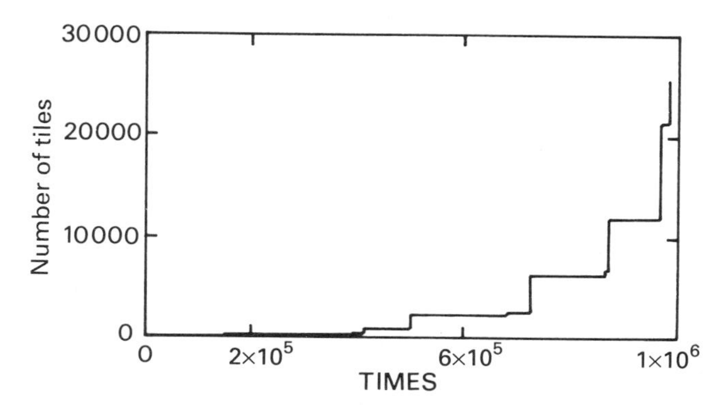

FIG. 6.12 Time evolution of tiling growth with the example of $P_c/P_F = 10^{-5}$.[18]

it, there may be possibilities for the sample to grow in the form of two bumps, along macroscopic directions defined by the defected worm:[14] two new pieces of perfect Penrose tiling have formed. Translational order is maintained within each single bump, but there is no guarantee that coherence is preserved between the two bumps. If it is lost, new defects are bound to be generated after a merging of the two regions. This generation will occur along their border and is responsible for a defect accumulation. Depending on the P_F/P_0 ratio, the grown structure is likely to range from random tiling to glassy arrangements. Accordingly, perfect or almost perfect quasicrystals should be, if anything, restricted to tiny particles.

At this stage it is worth pointing out that there is one ingredient in the growth process that we have not yet taken care of: the seed cluster used which has been arbitrarily chosen at the origin of the algorithm. In the next section we shall examine special defects which actually force the entire space around them: the decapods.[15]

6.2.4 Decapods acting as very efficient screw dislocations

In growing a perfect periodic crystal one leading parameter is the so-called supersaturation rate, which defines the concentration in the chemiostat surrounding the seed cluster with respect to what would be the equilibrium

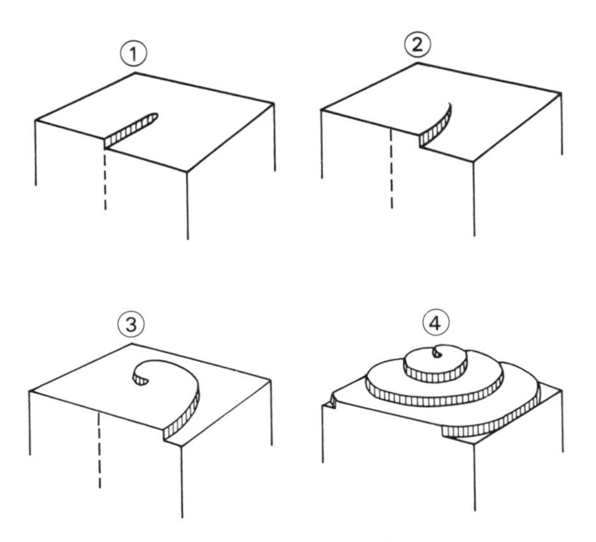

FIG. 6.13 New atoms attaching themselves to a nucleus already formed prefer to settle at a screw dislocation edge because there are more atoms there to attract them than there are on a flat surface. In the process, the screw dislocation fills up and spirals.

condition. Typically, seed crystals appear for supersaturation of about 10, liquid drops are formed out of vapour for values of 5 or so and values of 1.5 are required if one atomic place is to be added at the surface of a perfect crystal. However, in an historical example, iodine crystals were grown with a supersaturation ratio less than 0.01, corresponding in principle to a growth rate smaller than the smallest measurable rate by a factor $\exp(-3000)$! This enormous disagreement means that to start adding a new atomic plane on a perfect crystal is extremely difficult. If a screw dislocation (Fig. 6.13) exists, there is no need to 'start' a new plane: the 'edge' sites of the screw dislocation can usefully act as local seeds and the crystal grows in a spiral manner (Fig. 6.14). Crystals which have formed in low supersaturation conditions are expected to exhibit screw dislocation-like defects.

The mechanism is very simple: atoms are more easily added and fixed at ledge sites than at generic sites on a perfect plane; a screw dislocation offers such sites, and while they are successively used for spiral-like growing of the crystals, new such sites are automatically regenerated. The screw dislocation forces the entire space around it.

Now, the question is: is there any defect that can act like a screw dislocation in a quasicrystal? The answer is yes, but it cannot be a dislocation as has been defined in Chapter 5, since the phason-like component of a dislocation is not likely to help in growing perfect quasicrystals.

Looking back at the growing rules as they have been defined in Section 6.2.3, the main drawback seems to lie with the intermittent nature of the

FIG. 6.14 Spiral growth pattern in CdI_2 crystals.

growth, with rapid steps of accumulation of perfect vertices when forced vertices are available, separated by long standby periods ending with the use of an unforced vertex to go on further at the cost of generating defects and eventually imperfect tiling. To avoid such intermittent growth, and still achieve a perfect infinite tiling, the requisite in principle is to start with a seed cluster having only forced sites and keeping only forced sites upon growing. Is this possible? We are going to show that the answer is also yes.

Suppose that a complete plane tiling of properly matched Penrose tiles surrounds some empty region. What possible configuration of tiles can fill the empty space? If this empty space has no particular shape, or size, it can always be filled with properly matched Penrose tiles. An interesting particular situation is that of an empty space in the geometrical shape of a decagonal cluster occupying the space of 10 tiles (hence, decapod), five of each type. This situation is illustrated in Fig. 6.15. The problem was first introduced by Conway and presented by Gardner.[15] The empty decagon can be filled with a so-called 'cartwheel' decapod consistent with the Penrose matching rules (Fig. 6.16). But there are 61 other decapods that correspond to incorrect configurations and 60 of them can be used to force the tiling in the entire space: they are 'like an imperfection that solidifies a crystal'.[15] They even have better efficiency than screw dislocations in normal crystals, in that they allow the otherwise impossible growth of an *infinite perfect* Penrose tiling with *zero* defects and within *local* interaction rules.

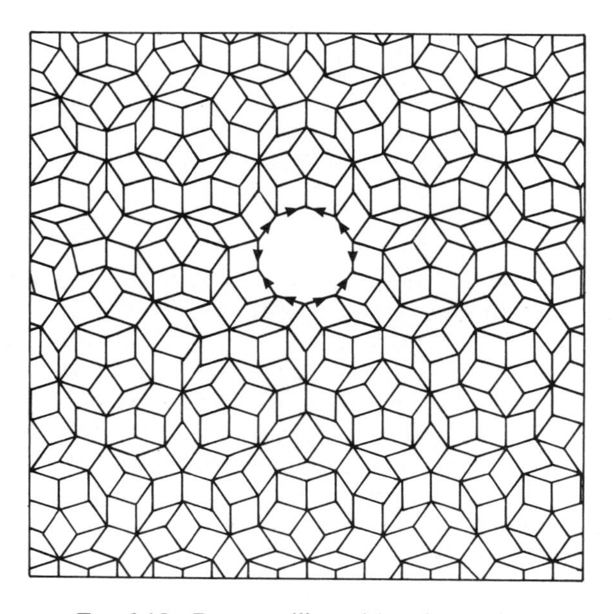

FIG. 6.15 Penrose tiling with a decapod hole.

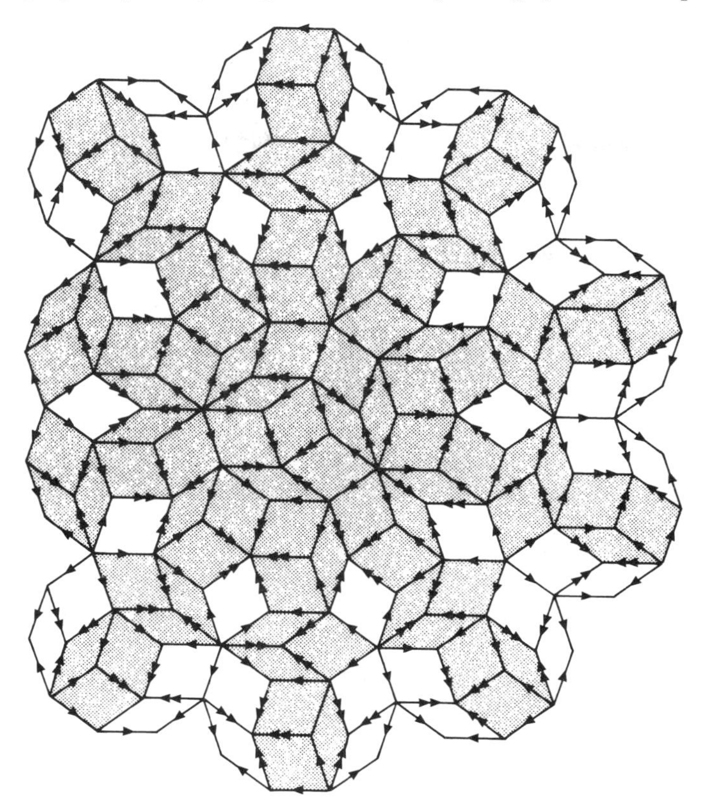

FIG. 6.16 The 'cartwheel' decapod consistent with the Penrose matching rules and the 10 semi-infinite worms that cross there (shaded).

As shown in Fig. 6.16, cartwheel decapods are associated with the crossing points of worms (two infinite plus six semi-infinite worms, or in other words 10 semi-infinite worms). An important property of a worm (see Fig. 6.9) is that it consists of a chain of hexagons made up of three tiles (two skinny plus one fat, or one skinny plus two fat tiles). Also, a worm can be 'flipped' without violating the matching rule configurations, except for the two edges at the end of the chain if a finite worm is concerned (Fig. 6.19). The decapod defects are formed by flipping one or more of the 10 semi-infinite worms which meet at a cartwheel decapod. The worms will still fit all surrounding tiles, but for a mismatch at the end of the semi-infinite worm inside the decagon (Fig. 6.17). If the decapod is left empty for convenience, it can be represented by a decagon with arrows properly oriented on its edges. Flipping

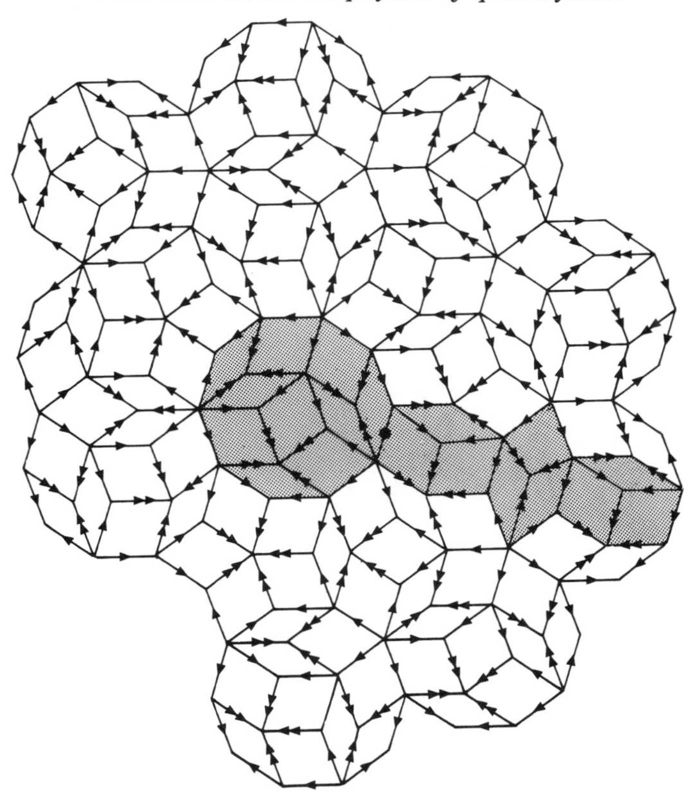

FIG. 6.17 Example of a decapod defect with one (shaded) flipped semi-infinite worm. The black dot indicates the resulting mismatch.

a semi-infinite worm then appears to reverse the corresponding arrow. Suppose that we run clockwise around the decapod along its edges, adding $+1$ if the arrow is clockwise oriented and -1 in the opposite case. The net total 'charge' is zero for cartwheel decapods and ranges from 2 to 10 (only even values) for defective decapods.

Now suppose that tiles are added to a given decapod consistent with the vertex rules in order to grow a perfect Penrose tiling around the decapod. The net 'charge', as defined above, along any closed path around the decapod at any distance of it somewhere in the Penrose tiling, is a constant of the tiling and equal to the net charge of the central decapod. It is indeed straightforward to understand that a Penrose tile (skinny or fat) brings a zero net charge when added consistently according to the matching rules (Fig. 6.18). Then any closed polygonal surfaces that bound a piece of

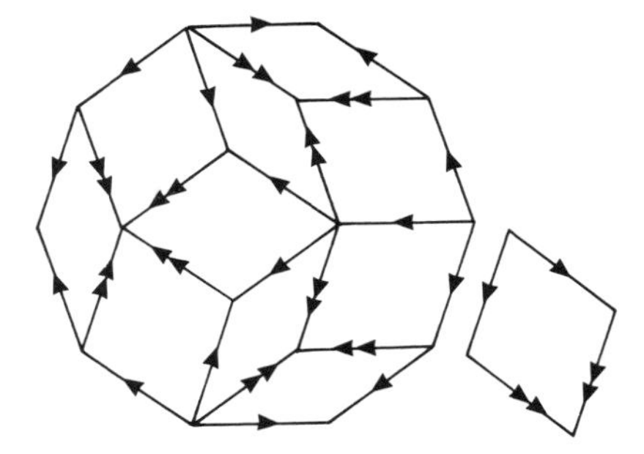

FIG. 6.18 When perfectly matched, an added tile leaves the 'arrow charge' of the Penrose cluster constant.

'perfect' Penrose tiling containing a central decapod (the only accepted defect), exhibit a net charge equal to that of the central decapod.

On the other hand, it is easy to verify that a closed polygonal dead surface can have a net charge of 0 or ± 2 only: 0 along their worm-like edge arrangement, 0 at $108°$ corners and ± 1 at $72°$ corners (with two of them at most). Thus, decapod defects with net charge greater than 2 can never be enclosed inside a dead surface; in other words, forced vertices keep on being available for ever if a Penrose tiling is grown from a decapod defect having a net charge equal to 4, 6, 8, or 10. This very important conclusion dispels definitively the myth that it is impossible to grow perfect quasicrystals when only short-range interactions are considered.

Qualitatively, the conclusion certainly holds true for the 3-dim icosahedral quasicrystals. The 3-dim rhombohedral tiles have many of the properties of the 2-dim rhombuses, including the generalization of line-grid to plane-grid. Quantitatively, the problem is complicated by the larger number of allowed distinct vertex types, which amount to 24 in perfect 3-dim Penrose tiling,[19, 20] with the occurrence of rings of up to five neighbours around a bond. The concepts of forced vertices and dead surfaces are still valid, and rapid growth can be expected around defected seed. Nevertheless, significant differences may arise in the analysis of defects, including seeds for perfect growth. For icosahedral structures, the analogues of Amman lines are Amman planes and a jag becomes a line of mismatches (a step in an Amman plane) rather than an isolated point mismatch. Thus the analogues of decapods consist of infinite lines of mismatches emanating from a central region which could be called an 'icosapod'. Defect lines can form closed loops or infinite curves, and it is not clear what sort of structures will typically be generated during growth. Although the use of local rules is obviously a resonable approach for growing perfect icosahedral quasicrystals without any relaxation processes whatsoever, it has not been firmly established so far and other mechanisms must be considered as possible alternatives.

The prerequisite for real perfect quasicrystals to be grown via a local rule mechanism is the existence of a finite number of local configuration in the structure. This forces constraints on the shape of the atomic surfaces in the high-dim periodic image and also on the composition/density parameters of the system. This is completely equivalent to the usual analysis which can be made for periodic crystals: local growth is mastered by atomic interactions (e.g. sp^3 bonds for carbon, germanium, or silicon) which forces the 'local symmetry', i.e. the point group (rotational invariance) which in turn forces the Bravais lattice and the unit cell (diamond cubic system); finally, the unit cell content defines the density and composition of the crystal. For icosahedral quasicrystals the local rules imply that the atomic surfaces are bounded by symmetry planes and that density/composition parameters are related to two components (e.g. tiles or chemical species) whose 'weight' ratio is a power of τ. As an illustration of this rule, it is interesting to consider compositions of some existing quasicrystalline systems. With a composition ratio of τ, we have the perfect stable $Al_{0.62}(CuFe)_{0.38}$ quasicrystal and the related system $Al_{0.64}(CuRu)_{0.36}$. The former is well characterized, but the composition of the latter has still to be established accurately. It should be noted that $\tau/(1 + \tau) = 0.618$ and $1/(1 + \tau) = 0.382$.

A second family may correspond to a composition ratio of $\tau^2(\tau^2/(1 + \tau^2) = 0.724$ and $1/(1 + \tau^2) = 0.276)$ with $Al_{0.721}(PdMn)_{0.279}$ and $Al_{0.71}(PdRe)_{0.29}$. Again, the latter is less well specified than the former (see Chapter 1).

A third family exhibits a composition ratio equal to $\tau^3(\tau^3/(1 + \tau^3) = 0.809$ and $1/(1 + \tau^3) = 0.191)$. This is rather well verified with $(AlSi)_{80}Mn_{20}$ which, apparently, is neither stable nor perfect.

Finally, the last of the achetypal quasicrystals, i.e. $Al_{0.57}Cu_{0.11}Li_{0.32}$ or $(AlCu)_{0.68}Li_{0.32}$, does not seem to obey any of τ^n composition laws and lies between the τ^1 and τ^2 families. However, while the quasicrystals quoted above are composed of atoms of almost equal sizes, the AlCuLi quasicrystal is such that the atomic volume Ω_{Li} of Li is larger than the average $(Al_{0.57}Cu_{0.11})$ atomic volume Ω_A. The product of Ω_A and the concentration of Al + Cu is practically τ times as large as the product of Ω_{Li} and the concentation of Li. This may be a loose way of obeying the τ^n law via the ratio of occupied volumes.

That a quasicrystal can truly form and be stable can also be demonstrated in a phenomenological description, a point that will be briefly addressed in the next section.

6.3 Can icosahedral order be grown out of disorder?

6.3.1 The Landau theory

When a liquid freezes to become a crystal there is ordering of the system. Suppose that there is a quantity ψ (the order parameter) which is zero in the disordered state and non-zero in the ordered state. The transition from one state to the other, and vice versa, can be the result of varying a suitable thermodynamic variable (e.g. temperature T or pressure P). In the Landau theory the transition is studied via the behaviour of the free energy F close to the transition temperature T_c.

The order parameter can be a complicated physical parameter. For the sake of simplicity, suppose for the moment that ψ is scalar (one component) and that F is not affected by the sign of ψ. A simple expression for F, in the vicinity of T_c, is

$$F = F_0 + \alpha\psi^2 + \beta\psi^4 + \ldots, \tag{6.2}$$

where F_0 is the value of F at T_c, β is a positive constant, and α is a smooth function of T of the form

$$\alpha = a(T - T_c) \qquad \text{with } a > 0. \tag{6.3}$$

Consequently, $\alpha > 0$ for $T > T_c$ (disordered state) and $\alpha < 0$ for $T < T_c$ (ordered state). General behaviours of F are shown in Fig. 6.19. The stable state ψ_0 is found by minimizing F with respect to ψ:

$$\left(\frac{\partial F}{\partial \psi}\right)_{\psi_0} = 0, \quad \left(\frac{\partial^2 F}{\partial^2 \psi}\right)_{\psi_0} > 0. \tag{6.4}$$

In the example, F is invariant under the transformation $\psi \to \psi$ and $\psi \to -\psi$,

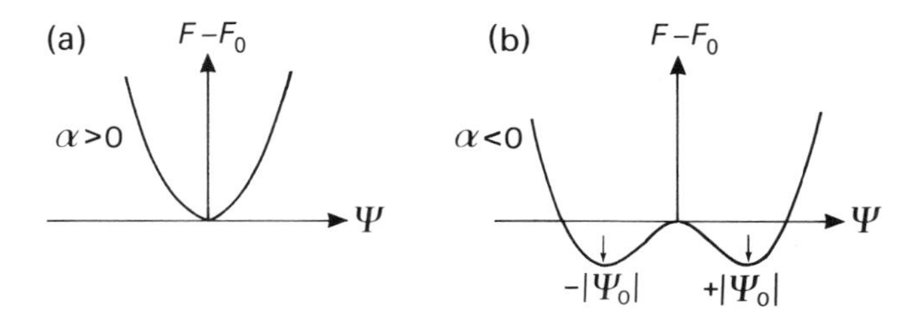

FIG. 6.19 Schematic plots of the free energy function (6.2) for (a) $T > T_c$ and (b) $T < T_c$.

i.e. under the discrete two-element group $\mathbb{Z}_2 = \{E, I\}$ (identity plus inversion). In the disordered state $(T > T_c)$, $\psi_0 = 0$ and the conditions

$$E\psi_0 = \psi_0 \text{ and } I\psi_0 = \psi_0$$

are trivially satisfied.

For the non-trivial states $\pm\psi_0$ below T_c:

$$E(+\psi_0) = (+\psi_0), \qquad E(-\psi_0) = (-\psi_0),$$
$$I(+\psi_0) = (-\psi_0), \qquad I(-\psi_0) = (+\psi_0).$$

They do not have the $\mathbb{Z}_2$ symmetry. Thus, below T_c the states lack the full symmetry of F: this is called a state of **broken symmetry**.

6.3.2 Liquid–solid transitions

The expression for F can be more complicated than the one shown in eqn 6.2. Third-order terms may be added to account for first-order (discontinuous) transitions, or even higher order terms when the transitions are not driven by symmetries.

If the transition implies atomic rearrangements, the density function $\rho(x, y, z)$ is a relevant order parameter. Let $\rho = \rho_0 + \delta\rho$ characterize the distribution in the ordered state and ρ_0 be the uniform density function in the liquid. If G and H are the symmetry groups of the liquid $(G[=E(3)])$ and in the broken symmetry state respectively, ρ_0 is invariant under the operation of G whereas ρ is invariant under H.

According to Landau:[21]

$$\delta\rho = \sum_{i=1}^{n} \eta_i^{\Gamma} \phi_i^{\Gamma}, \tag{6.5}$$

where ϕ_i^{Γ} are basis functions transforming to the n-dimensional irreducible representation of G, and η_i^{Γ} are the expansion coefficients. They depend on P, T, . . . and constitute order parameters. It is common to write

$$\eta_i = \eta\gamma_i \qquad \text{with } \sum_i \gamma_i^2 = 1.$$

Thus $\{\gamma_i\}$ describes the symmetry of the ordered state, whereas the scale factor η is a measure of the degree of order. We have seen that in practice $\delta\rho(\boldsymbol{r})$ is expressed in terms of a restricted set of density waves referred to as **a star:**

$$\delta\rho(\boldsymbol{r}) = \sum_i \rho(\boldsymbol{k}_i)\exp(\mathrm{i}\boldsymbol{k}_i \cdot \boldsymbol{r}) \tag{6.6}$$

Generally the $\rho(\boldsymbol{k}_i)$ are complex quantities and they must encode the orientational symmetry of the ordered state. The general expression of the free energy becomes

$$F = F_0 + \int A(\boldsymbol{k}', P, T)\rho(\boldsymbol{k}')\rho(-\boldsymbol{k}')\mathrm{d}\boldsymbol{k}'$$

$$+ B(|\boldsymbol{k}|, P, T)\sum_{ijl}\rho(\boldsymbol{k}_i)\rho(\boldsymbol{k}_j)\rho(\boldsymbol{k}_l)\delta(\boldsymbol{k}_i + \boldsymbol{k}_j + \boldsymbol{k}_l) \quad (6.7)$$

$$+ \sum_{ijlm} C(\bar{\boldsymbol{k}}_i\cdot\bar{\boldsymbol{k}}_j, \bar{\boldsymbol{k}}_l\cdot\bar{\boldsymbol{k}}_m, P, T)\rho(\boldsymbol{k}_i)\rho(\boldsymbol{k}_j)\rho(\boldsymbol{k}_l)\rho(\boldsymbol{k}_m)$$
$$\times\,\delta(\boldsymbol{k}_i + \boldsymbol{k}_j + \boldsymbol{k}_l + \boldsymbol{k}_m) + \ldots .$$

$\bar{\boldsymbol{k}}_i$ denotes a unit vector in the direction of $\boldsymbol{k}_i$ and $|\boldsymbol{k}| = |\boldsymbol{k}_i|$.

Using such an expression it is possible to understand the wide prevalence of bcc structures, especially at high temperature, among the elements of the periodic table.[22] Icosahedral structures can also be described by the free energy (6.7), under the condition that higher order terms are included.

6.3.3 *Liquid to icosahedral solid transitions*

Actually, there is no fundamental reason to limit the expansion in eqn (6.7) to the third-order term. Let us rewrite the expression for F in a shorter form

$$F = F_0 + \mu(\rho_i\rho_{-i})^2 + \ldots + v_3\rho_1\rho_2\rho_3\delta(\boldsymbol{k}_1 + \boldsymbol{k}_2 + \boldsymbol{k}_3) + \ldots$$
$$+ v_5\rho_1\rho_2\rho_3\rho_4\rho_5\delta(\boldsymbol{k}_1 + \boldsymbol{k}_2 + \boldsymbol{k}_3 + \boldsymbol{k}_4 + \boldsymbol{k}_5) + \ldots . \quad (6.8)$$

Basically, a liquid to solid transition can be viewed as the emergence of a 'multi-$\boldsymbol{k}$' structure out of the disordered isotropic liquid. In the expression for F, products of the density wave amplitudes are formed by combining those waves with wavevectors adding to zero. This suggests strongly that the n-order terms in F favour multi-$\boldsymbol{k}$ structures with wavevectors forming an n-star or, equivalently, an n-regular polygon. For instance, the third-order term favours multi-$\boldsymbol{k}$ structures with wavevectors forming regular triangles (Fig. 6.20). This may correspond to a 2-dim hexagonal phase or a 3-dim rod-like phase. Subsets of wavevectors arranged in regular triangles are also found among the eight vectors of an octahedron or the six vectors of a tetrahedron, representing the reciprocal lattices of bcc and fcc structures, respectively.

The fifth-order term in F is then expected to favour multi-$\boldsymbol{k}$ structures with a set, or subsets, of wavevectors forming a regular pentagon (Fig. 6.20)[23, 24] which satisfies the δ-function constraint in the free energy expansion. If the order parameter r_i is written $a\exp(\mathrm{i}\theta_i)$, the fifth-order term in eqn (6.8) becomes

$$F_5 = v_5 a^5 \cos(\theta_1 + \theta_2 + \theta_3 + \theta_4 + \theta_5). \quad (6.9)$$

The resulting structure,

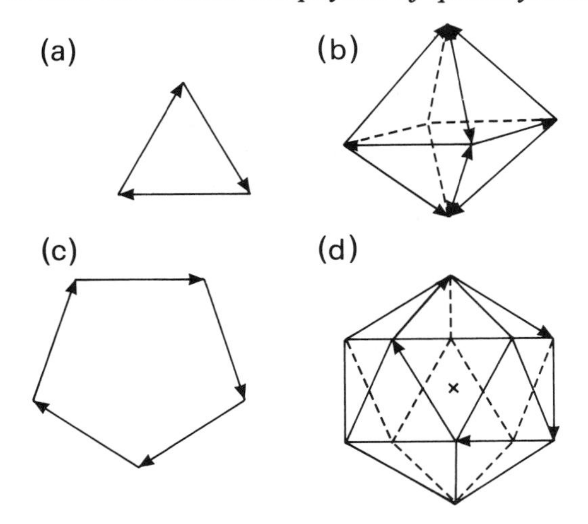

FIG. 6.20 Wavevector combinations representing (a) a rod-like structure, (b) a bcc structure, (c) a 2-dim Penrose structure, and (d) an icosahedral 3-dim structure.

$$\rho(\boldsymbol{r}) = \sum_{i=1}^{5} a\cos(\boldsymbol{k}_i\cdot\boldsymbol{r} + \theta_i) + \text{higher harmonics}, \tag{6.10}$$

obviously has the symmetry of a 2-dim Penrose lattice (equivalent to a pentagonal quasilattice).

Subsets of wavevectors arranged in regular pentagons are also present in a regular dodecahedron or a regular icosahedron (see Figs. 1.18 and 6.20). The latter case is particularly interesting, since one can take advantage of both the third-order term and the fifth-order term with, respectively, the 10 pairs of triangles and the six pairs of pentagons that can be formed with the vectors. Normalizing the density waves to $a = \rho/(15)^{\frac{1}{2}}$ *gives*:

$$\begin{aligned} F_3 + F_5 = &-\frac{\rho^3 v_3}{(15)^{\frac{3}{2}}} \sum_{10\ \text{triangles}} \cos(\theta_i + \theta_j + \theta_k) \\ &-\frac{\rho^5 v_5}{(15)^{\frac{5}{2}}} \sum_{6\ \text{pentagons}} \cos(\theta_i + \theta_j + \theta_k + \theta_l + \theta_m). \end{aligned} \tag{6.11}$$

Minimization of the free energy gives rise to nine constraints on the 15 phases, and the resulting energy is

$$(F_3 + F_5)_{\text{ico}} = -2\rho^3 v_3/3(15)^{\frac{1}{2}} - 2\rho^5 v_5/75(15)^{\frac{1}{2}}. \tag{6.12}$$

The calculation for the bcc structure with $a = \pi/(6)^{\frac{1}{2}}$ classically gives

$$(F_3)_{\text{bcc}} = -2\rho^3 v_3/3(6)^{\frac{1}{2}}. \tag{6.13}$$

The free energy of eqn (6.12) can become favourable compared with the $(F_3)_{\text{bcc}}$ free energy in system with small enough v_3/v_5 values, resulting in the formation of icosahedral quasicrystals out of the liquid phase. The transition is first-order because the order parameter and the Landau expansion are the same as for regular 3-dim solidifications, with third-order invariants.

Superficially, it may seem that the Landau theory can be generalized to produce crystalline structure with arbitrary rotational symmetry. This is not actually the case because there is only a finite number of regular polyhedra. For instance, there is no polyhedron that can be obtained by combining regular heptagons. Thus, quasicrystals with 7-fold axes should not exist.

6.4 The alternative ways for a quasicrystal to grow

6.4.1 Random accretion models or relaxation processes

In the perfect quasicrystal growth model described in Section 6.2, additional atoms or clusters of atoms are assumed to be attached to the growing piece of matter at exactly the right sites for the structure to assume perfectly. This means that the corresponding growth proceed without relaxation processes, i.e. according to rules which do not incorporate any mechanism for repositioning or removing an atom once it has been attached to the growing patch.

The alternative descriptions are either to accept random accretion rules, still without subsequent relaxation effects, or to assume a mechanism in which atoms are repositioned into a quasiperiodic lattice by relaxation effects from other structures which have grown more rapidly.

All these models are based on the idea that growth can be described by mechanisms governing the aggregation of geometric units that represent clusters of atoms. Details of the formation of the units are not included in the models and are implicitly assumed to be related to the precursor in the very early stage of solidification, if not in the liquid state.

The observation that some crystalline phases related to quasicrystals contain icosahedral units suggested, very early in the history of the subject, that simple rules for the accretion of icosahedra should be considered. In the simplest case, a structure is grown via the addition of icosahedra to the growing cluster according to the following four rules.

1. Orientations of the units are identical.
2. Linkages between nearest-neighbour units are identical (same length, same direction).
3. Units are not allowed to overlap.

4. New units are added randomly to the set of possible sites according to rules 1–3.

Although these rules produce longer translational correlation lengths than might have been expected (as measured by the widths of the simulated diffraction peaks), they do not generate enough correlations to account for experimental observations with the perfect quasicrystals currently produced.

Modifications of the random accretion model have been proposed in which new units are not added randomly.[25,26] Before a cluster sticks at the growth front, linkage constraints within a finite region of the growing piece of matter are considered. For instance, the sticking cluster may be forced to accept two linkages simultaneously, say along the 3-fold and 2-fold directions, with two different clusters at the front. Units which are already attached may also be removed. In a way this is reintroducing types of local rules which certainly produce structures which have a better relation with experimental observations. However, the growth mechanism may be at least as difficult as that described in Section 6.2. A systematic search for the appropriate site where a new unit can stick results in standby periods that preclude rapid growth. Branching defects cannot be avoided, with the related occurrence of diffraction peaks whose width scales with the Q_{perp} power law.

One family of the most sophisticated random accretion models is the **random tiling model** (RTM).[27] In subsequent sections the RTM is analysed with respect to growth mechanisms, after which some possible relaxation processes are presented.

6.4.2 *What about random tiling growth?*

It may be useful to clarify the description of a tiling. By definition a tiling is a filling of space using tiles taken from a finite set, *without overlaps and without gaps*. This is at variance with other random rigid networks, such as glass models, which can have arbitrarily large space gaps with no nodes. A tiling is defined by *packing rules* which specify linkages between tiles. These include the obvious constraints that faces adjoin, as well as symmetry invariances and possibly some other less trivial restrictions. Packing rules correspond to the strict matching rules in the special cases that force perfect quasiperiodicity (see previous chapters of this book). Therefore, packing rules in the RTM are types of relaxed matching rules in the sense that matching rules are fully obeyed for a single-space tiling configuration while RTM packing rules may allow a large number of tiling configurations. Thus RTM structures may be entropically favoured with respect to perfect quasiperiodicity, particularly at high temperature:[27] each additional tile is offered quite a large distribution of allowed sticking places with equal energy which obviously may increase the growth rate. In maximally random tiling,

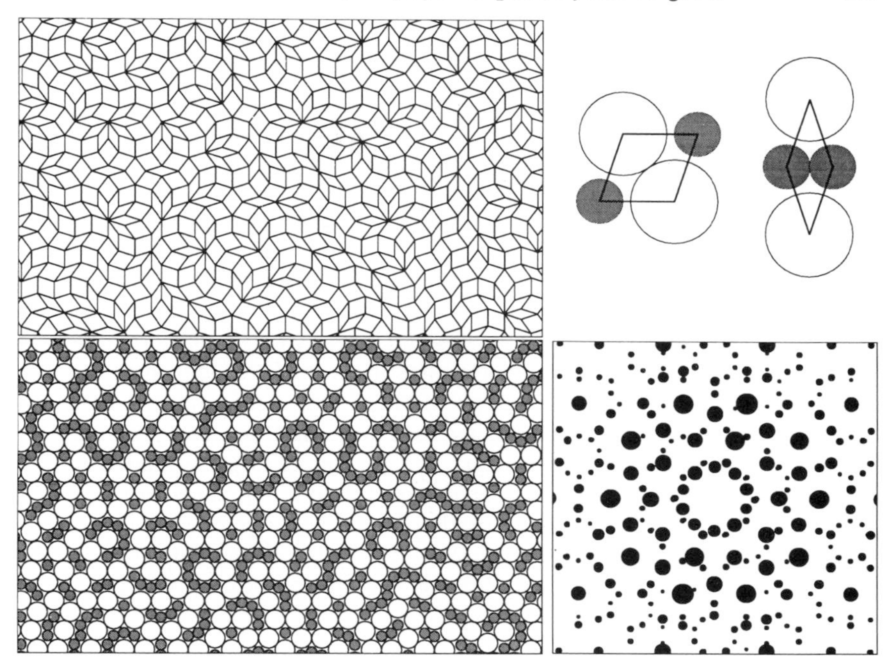

FIG. 6.21 Random binary tiling obtained via the Monte Carlo method (top left) and a possible atomic decoration (top right). The corresponding configuration of particles (bottom left) and its diffraction pattern (bottom right) are also shown. The dotted area is intensity related. Cut-off at 0.02 I_{max} (courtesy of F. Lançon and L. Billard).[28, 29]

all configurations are equally probable:[27] matching rule violation costs no energy and all the properties of the system are driven by its configurational entropy.

As an illustration, let us consider simulation studies for the stability of binary tiling.[28, 29] In the example shown in Figure 6.21, the binary tiling consists of the decoration of two Penrose rhombi by two types (small and large) of atoms. Molecular dynamic calculations, using simple Lennard–Jones interatomic potentials and Monte Carlo methods, result in a stable tiling with a simulated diffraction pattern which exhibits sharp peaks (Fig. 6.21). However, it is not perfect Penrose tiling: matching rules are violated on many occasions resulting in numerous 'phason flips' that reverse hexagon-like configurations as illustrated earlier in Fig. 5.45 (Chapter 5). Interestingly, the structure is stable at temperatures in a finite range below the melting point. We shall return to this point.

To understand the main properties of a RTM it is useful to consider its high-dim periodic image. For instance, if a 1-dim structure is made of two types of segments, i.e. L and S, the perfect quasicrystal (the Fibonacci chain) is obtained via a flat irrational slice through a 2-dim square lattice conveniently decorated with A_{perp} segments. Position fluctuations in the perpendicular space of the A_{perp} segments modify frequencies and/or sequences of the L and S segments in the structure. These position fluctuations of the A_{perp} segments can be roughly simulated by turning the flat slicing into 'undulating', faceted, tilted, etc. slicing (see Fig. 3.16 in Chapter 3). A general random tiling would correspond to slicing which randomly fluctuates into the square lattice image. Accordingly, the corresponding slicing can be described by two parameters related to the perpendicular coordinates, i.e. an average uniform phason strain E, which, in the simple 1-dim example, is the average slope of the slicing with respect to that of the perfect quasicrystal, and the average phason fluctuation or perpendicular 'cross-section' of the slicing. The former is directly related to the 'tile concentration' (relative proportion of L and S segments in the 1-dim case), and $E = 0$ for all structures that keep the same tile concentration as in the perfect quasiperiodic structure. The latter allows for distributions of matching-rule violations around the average phason strain. It can be defined as

$$(\Delta \boldsymbol{r}_{\text{perp}})^2 = \langle\, |\boldsymbol{r}_{\text{perp}}(\boldsymbol{r}_{\text{par}} = 0) - \boldsymbol{r}_{\text{perp}}(\boldsymbol{r}_{\text{par}})\,|^2 \rangle \tag{6.14}$$

and it compares the perpendicular components at any site of the structure with that of a point arbitrarily chosen as origin. For perfect quasiperiodic structures the $(\Delta \boldsymbol{r}_{\text{perp}})^2$ are bounded quantities and have a simple relation with the perpendicular cross-sections of the atomic surfaces A_{perp}. As examples, we have demonstrated in Chapter 3 (Section 3.3.7, eqns (3.28) and (3.29)) that for a Fibonacci chain the r_{par} and r_{perp} coordinates of the relevant square lattice sites are

$$\begin{aligned} r_{\text{par}} &= \frac{a}{(2+\tau)^{\frac{1}{2}}}\,(H\tau + H') \\ r_{\text{perp}} &= \frac{a}{(2+\tau)^{\frac{1}{2}}}\,(H'\tau - H) \end{aligned} \tag{6.15}$$

with the condition $|H'\tau - H| \leqslant \tau^2/2$, which means that any integer H can be associated with two and only two integers H' equal to $\lfloor H/\tau \rfloor$ and $\lfloor H/\tau \rfloor + 1$. If H is a 'Fibonacci number', i.e. 3, 5, 8, 13, . . ., H' reduces to a single value which is also a Fibonacci number 2, 3, 5, 8, . . . respectively. In any case, it appears clearly that $(\Delta \boldsymbol{r}_{\text{perp}})^2 \leqslant (\tau^4/4)[a^2/(2 + \tau)]$ remains bounded upon expansion of r_{par}. For the 6-dim cubic image of a perfect icosahedral quasicrystal, the derivation is similar (eqn (3.29) in Chapter 3)

$$|r_{par}|^2 = \frac{a^2}{2(2+\tau)}(S+\tau T)$$
$$|r_{par}|^2 = \frac{a^2}{2(2+\tau)}(S\tau - T) \tag{6.16}$$

with $|\Delta r_{perp}|^2 \leqslant a^2/2$ (in the simple case of a first-approximation spherical A_{perp}). $|\Delta r_{perp}|^2$ is also bounded in 3-dim RTM, but diverges like r_{par} and $\ln(r_{par})$ in 1-dim and 2-dim RTM respectively.

Considering further the 1-dim case for qualitative illustration, the integers H and H' in eqn (6.15) are just the number of long and short tiles (L and S) respectively in a Fibonacci strip of length r_{par}. Let us call n_1 and n_2 the 'concentration' of tiles L and S respectively. If these tiles are allowed to connect randomly, the number of possible configurations is simply derived from the binomial coefficient:

$$\Omega = \frac{(n_1 + n_2)!}{n_1!n_2!}.$$

Using the Stirling approximation for a very large number of tiles leads to the expression of the configurational entropy $\mathcal{S} = -k_B \ln \Omega$, i.e.

$$\mathcal{S}(n_1, n_2) = -k(n_1 \ln n_1 + n_2 \ln n_2).$$

This entropy is maximum for $n_1 = n_2$ which corresponds to $H'\tau = H$ and $r_{perp} = 0$, the precise concentration of a perfect unlimited Fibonacci chain. Changing the (H, H') concentration is equivalent to introducing a phason or perpendicular strain E (see again eqn (6.15)):

$$E_{\lim r_{par} \to 0} = \frac{\Delta r_{perp}}{\Delta r_{par}} \quad (|E| \leqslant 1)$$

with the related consequence that $\mathcal{S}(n_1, n_2)$ is reduced. In other words, eliminating (H, H') between $\mathcal{S}(n_1, n_2)$ and E gives the expression of the entropy density as a function of E for a maximally random tiling structure. Since this function is symmetric in (n_1, n_2), it must be quadratic in E and can be expanded in powers of E as follows:

$$\mathcal{S}(E) = \mathcal{S}(0) - \frac{1}{2}KE^2 + \frac{1}{12}K'E^4 - \ldots. \tag{6.17}$$

Exact demonstrations[30] and numerical simulations[31–33] suggest that the formulation of eqn (6.17) is generic and applies to any maximally random tiling:

$$\mathcal{S}(E) = \mathcal{S}(0) - \frac{1}{2}\sum K_{iljm}E_{il}E_{jm} + O(|E|^n). \tag{6.18}$$

This clearly means that the phason free energy $F = U - TS$ is a function of the gradient-squared form of the perpendicular coordinates. Formally, phasons must then obey an elastic theory and can be treated in the same way as phonon elasticity. For instance, there is a perpendicular Debye–Waller factor and the phason coordinates show a linear response.

This may be a surprising result at first sight since a phason move corresponds to a diffusive process and it is hard to imagine that such a process could easily be modelled by a harmonic potential. In fact, the perpendicular gradient-squared behaviour reflects that, in a RTM, the structure explores statistically a very large number of different configurations. One of the non-trivial problems is to identify a phason move which allows all possible configurations of the tiling to be explored. In a 1-dim random Fibonacci-like structure, this is obviously obtained via L–S flips. There is also the 'umbrella' reverse of a hexagonal-like set of tiles in a 2-dim random Penrose tiling (Fig. 5.45 in Chapter 5). It has been shown[33] that such possible phason moves also occur in 3-dim random Penrose tiling; they are located on 2 fold planes, inside dodecahedral clusters of fat and thin rhombi. This strongly suggests that a RTM should be addressed with respect to its dynamic aspects rather than as a static structure. This is partly the purpose of the next section.

6.4.3 Quasicrystal growth via 'relaxation processes'

As previously discussed, perfect quasiperiodic structures and random tiling models can be considered as two irreconcilable descriptions of quasicrystals. Indeed, while the former invokes energetic stabilization, the latter is based on entropy effects. In fact, as is often the case in both life and science, the truth may lie in between. Assuming that the ground state (0 K configuration) of a structure is a perfect quasiperiodic tiling means that this quasiperiodicity is enforced by the equivalent of some local matching rules. However, in any of the current cases of quasicrystals, 'phason jumps' or 'tile flipping', which may produce matching-rule violation, always correspond to atom displacements made to small distance shifts (L–S reverse, umbrella-like reverse of tiles, etc.) which are shorter than the atomic bonds in the structure. Thus the energy cost of a matching-rule violation is certainly much smaller than the formation energy of a vacancy in a periodic crystal. Treating the two defects on the same formal footing suggests that 'isolated phason defects' form in the perfect quasicrystal as soon as its temperature is no longer zero. Extending the analysis leads to a formulation of the concentration d_{pha} of 'isolated phason defects':

$$d_{\text{pha}} = \exp\left(\frac{S_{\text{f}}}{k_{\text{B}}}\right) \exp\left(-\frac{E_{\text{f}}}{k_{\text{B}} T}\right)$$

where S_f and E_f are the formation entropy and energy respectively. The entropy factor comes from the local changes produced by the defect in the vibration modes. This must be considered as negligible in the phason flip situation since quasicrystals are not supposed to exhibit so many propagating modes. Thus we are left with

$$d_{\text{pha}} \propto \exp\left(-\frac{E_f}{k_B T}\right) \tag{6.19}$$

and there is certainly a range of temperature where the quasicrystal can be considered as perfect quasiperiodic tiling with some isolated phason defects. Increasing the temperature increases d_{pha}. If E_f is sufficiently weak, the number of phason defects may become so large that they can no longer be isolated. They may even percolate into a continuous network above some temperature T_c. The site percolation threshold for the close-packed 3-dim structure is about 20 per cent, which would yield $E_f \approx 0.13$ eV for $T_c \approx 1000$ K if eqn (6.19) is applied even in the percolated state. Above T_c the quasicrystal would become a RTM structure, exploring all the possible configurations, and entropically stabilized. Several properties of the best quasicrystals, such as those of the AlFeCu or AlPdMn systems, support these estimates. For instance, they clearly show a fragile-to-ductile transition at about 1000 K, and this temperature is also that required to anneal frozen phason defects in rapidly quenched samples.

Thus a 'random tiling scenario' for quasicrystal growth may be sensible. It would start at high temperature with a true RTM structure as defined above, stabilized by entropy effects, and somewhat easier to achieve owing to the randomness ingredient. Then, reducing the temperature induces relaxation with phason strain annihilation and a transition at T_c into a state where the phason defects would no longer percolate and would become isolated.

An alternative growth model via a relaxation process is to approach the problem from the opposite direction. Among the many configurations that a RTM is supposed to explore are some which correspond to periodic strips of approximant structures. In a regular RTM all configurations are equally probable. However, according to current experience and crystal growth history this is very unlikely. It is reasonable to envisage that periodic configurations may be (slightly) more stable than the others, particularly those with short periods which would form more easily at high temperature. Thus the structure could be some sort of multilevel system with a quasiperiodic ground state and many excited states corresponding to the various periodic approximants whose free energy would be a decreasing function of the cell size. In such a mechanism, phason defects would permit exploration of the perfect quasiperiodic and periodic configurations only, again with some sort of percolation threshold above T_c when essentially the periodic

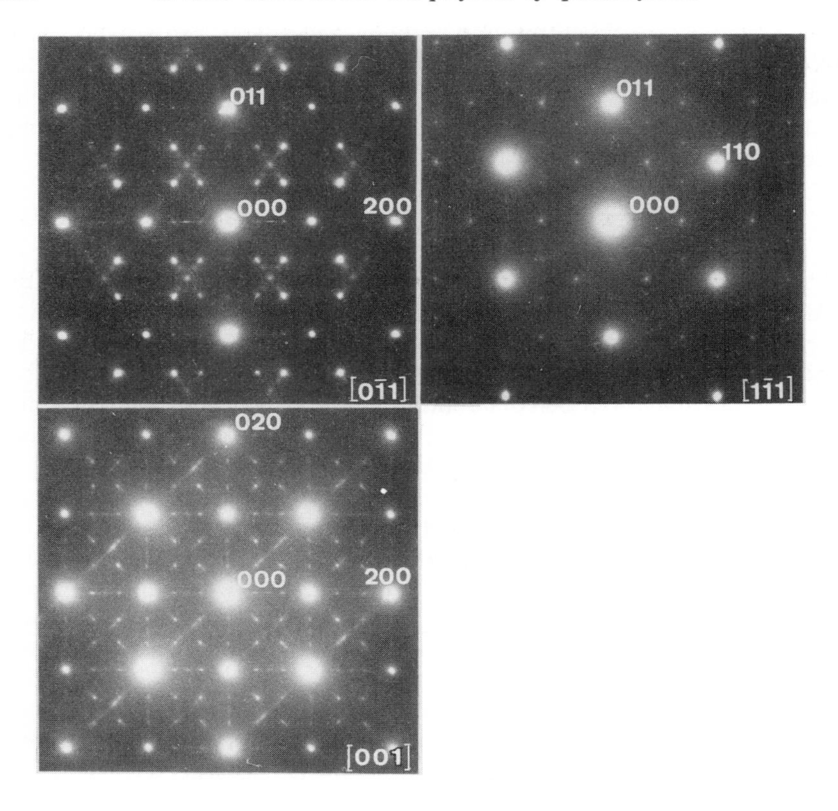

FIG. 6.22 A high temperature cubic phase (π phase) as observed in the AlFeCu system. Superstructure reflections are observed, indicating a transition via expansion of the cell size (courtesy of M. Audier).

configurations are explored. Continuous transition from the periodic structure to the quasicrystals along a decreasing temperature have actually been observed,[34–39] with the presence of small cell cubic phases at high temperatures (Fig. 6.22). Thus, initially grown periodic configurations, including small period structures, would relax to a perfect quasicrystal, again via progressive annihilation of phason defects and with a threshold transition at some temperature. However, this must be seen in its dynamic aspects, with all possible configurations being statistically explored but with 'frequencies' that depend on temperature. Moreover, the periodic approximant configurations must be considered as topological frameworks to rearrange fixed numbers of tiles, i.e. the composition is that which stabilizes the perfect quasiperiodic structures and then is different from those of equilibrium stable periodic approximants.[40] We shall return to this point later to explain how phason defects can participate in electrical conductivity.

6.5 Where are the electrons and how do they move?

6.5.1 The basic concepts as introduced for regular crystalline structures

Figure 6.23 summarizes the basic three situations that may encounter one electron. In a free isolated atom, the electrons move in a potential well and their energies belong to a series of discrete **levels** called 1s, 2s, 2p. . . . Now, consider the situation in which two atoms form a molecule. The electrons then move in a double well, with an energy spectrum made of discrete

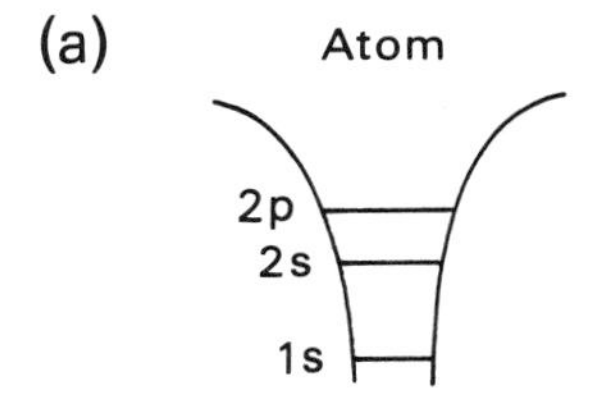

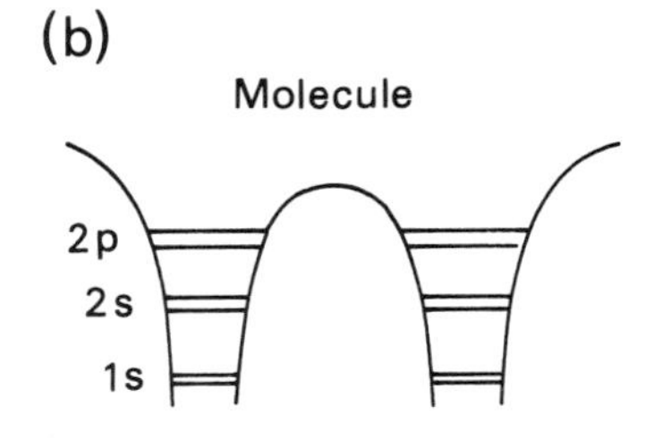

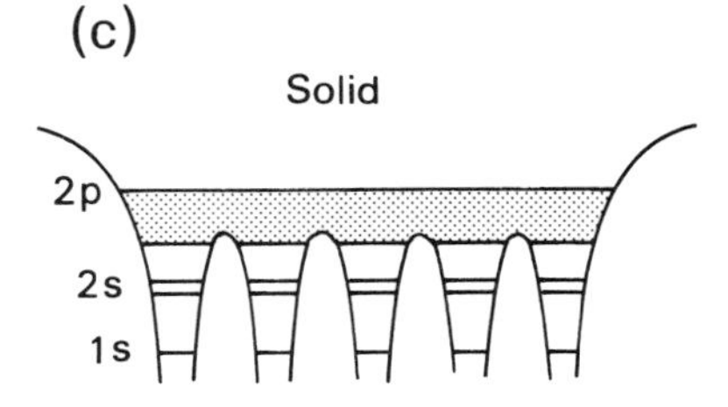

FIG. 6.23 The evolution of the energy spectrum for electrons from an atom (a) to a binary molecule (b) and a solid crystal (c).

doublets, due to double degeneracy lifting of the atomic levels by the atom interaction. A solid may then be viewed as the limit case of a very big molecule, with a large number N of atoms. Each of the atomic levels is now split into N closely spaced sublevels (about 10^{23}), so closely spaced that they coalesce and form an **energy band**. The width of the band is greater for the higher bands because low energy atomic states correspond to tightly bound orbitals which are affected only slightly by the presence (perturbation) of the other atoms of the structure.

From a more quantitative point of view, the behaviour of an electron in a crystalline solid is determined by studying the appropriate Schrödinger equation:

$$\left(-\frac{\hbar^2}{2m}\nabla^2 + V(\boldsymbol{r})\right)\psi(\boldsymbol{r}) = E\psi(\boldsymbol{r}), \tag{6.20}$$

where $V(\boldsymbol{r})$ is the crystal potential acting on the electron and $\psi(\boldsymbol{r})$ and E are the state function and energy of this electron, respectively. In a regular crystal $V(\boldsymbol{r})$ is periodic and has the same translational symmetry as the lattice, i.e.

$$V(\boldsymbol{r} + \boldsymbol{R}) = V(\boldsymbol{r}), \tag{6.21}$$

where $\boldsymbol{R}$ is any lattice vector.

According to the well-known **Bloch theorem**, the solution of eqn (6.14) within a (6.15) potential is of the form:

$$\psi_k(\boldsymbol{r}) = u_k(\boldsymbol{r})\exp(\mathrm{i}\,\boldsymbol{kr}), \tag{6.22}$$

where $u_k(\boldsymbol{r} + \boldsymbol{R}) = u_k(\boldsymbol{r})$.
This is due to the observable quantity $|\psi(\boldsymbol{r})|^2$ being subjected to the symmetries of the system, including the periodicity. Thus, $\psi_k(\boldsymbol{r})$ is a travelling plane wave with periodic modulation of its amplitude and a wavevector $\boldsymbol{k}$ such that the momentum of the electron is $\boldsymbol{p} = \hbar\boldsymbol{k}$. The Bloch function ψ_k describes a fully delocalized state of the electron. Substituting $\psi_k(\boldsymbol{r})$ into eqn (6.20) gives

$$\left(-\frac{\hbar^2}{2m}(\nabla + \mathrm{i}\boldsymbol{k})^2 + V(\boldsymbol{r})\right)u_k(\boldsymbol{r}) = E_k u_k(\boldsymbol{r}). \tag{6.23}$$

Since $u_k(\boldsymbol{r})$ and $V(\boldsymbol{r})$ are periodic functions with the period of the crystal, they can be expanded in a Fourier series:

$$V(\boldsymbol{r}) = \sum_G V_G \exp(\mathrm{i}\,\boldsymbol{G}\cdot\boldsymbol{r})$$
$$u_k(\boldsymbol{r}) = \sum_G U_G \exp(\mathrm{i}\,\boldsymbol{G}\cdot\boldsymbol{r}) \tag{6.24}$$

where the $\boldsymbol{G}$ vectors are those of the reciprocal lattice. Substituting eqn (6.24) into eqn (6.23) and assuming that all (V_G, U_G) Fourier coefficients are small except for U_0 (nearly free electron approximation) leads to

$$U_G \approx -\frac{2m}{\hbar^2}\frac{U_0 V_G}{2\boldsymbol{k}\cdot\boldsymbol{G} - |\boldsymbol{G}|^2}. \tag{6.25}$$

This defines each Fourier coefficient of the wavefunction with respect to the corresponding Fourier coefficient of the crystal potential. For the eigenstate $\boldsymbol{k}_{\mathrm{BZ}}$ and some $U_{G'}$ coefficients, such as $2\boldsymbol{k}_{\mathrm{BZ}}\cdot\boldsymbol{G}' - |\boldsymbol{G}'|^2 = 0$, the above derivation no longer applies because the expression diverges. This occurs for the eigenstate whose wavevector ends on the Brillouin zone attached to the $\boldsymbol{G}'$ site of the reciprocal lattice. Then U_0 and $U_{G'}$ are very large compared with the other Fourier coefficients and the wavefunction given by eqn (6.24) can be approximated by

$$\begin{aligned}\Psi_{k\mathrm{BZ}}(\boldsymbol{r}) &\approx U_0 \exp(\mathrm{i}\boldsymbol{k}_{\mathrm{BZ}}\cdot\boldsymbol{r}) + U_{G'}\exp[\mathrm{i}(\boldsymbol{k}_{\mathrm{ZB}} + \boldsymbol{G}')\cdot\boldsymbol{r}] \\ &\approx U_0[\exp(\mathrm{i}\boldsymbol{k}_{\mathrm{BZ}}\cdot\boldsymbol{r}) + \exp(-\boldsymbol{k}_{\mathrm{BZ}}\cdot\boldsymbol{r})].\end{aligned} \tag{6.26}$$

Substituting $\Psi_{k\mathrm{BZ}}(\boldsymbol{r})$ into he Schrödinger equation (6.20) demonstrates that non-trivial solutions for $U_{G'}$ and U_0 are obtained provided that the energy satisfies the condition

$$E = \frac{\hbar^2 k_{\mathrm{BZ}}^2}{2m} \pm |V_{G'}|.$$

Equation (6.23) has to be solved for each value of $\boldsymbol{k}$, which corresponds to several eigenstates with discrete energy E_{1k}, E_{2k}, ... Each level leads to an **energy band**, as shown in Fig. 6.24. We shall henceforth write the energy eigenvalues as $E_n(\boldsymbol{k})$ and refer to the subscript $_n$ as the **band index**. Energy bands are separated by **gaps**. The $E_n(\boldsymbol{k})$ bands have symmetry properties which can be described in the Brillouin zones of the reciprocal lattice:

(1) $E_n(\boldsymbol{k} + \boldsymbol{G}) = E_n(\boldsymbol{k})$,

(2) $E_n(-\boldsymbol{k}) = E_n(\boldsymbol{k})$,

(3) $E_n(\boldsymbol{k})$ has the rotational symmetry of the lattice.

It follows that we may confine our attention to the asymmetric unit of the first Brillouin zone.

Consider now the simple one-dimensional case in which the Bloch function has the form

$$\psi_k(x) = u_k(x)\exp(\mathrm{i}kx).$$

k extends from $-\pi/a$ to $+\pi/a$ in the first Brillouin zone if a is the period of the 1-dim system. If the potential $V(x) = 0$ (free electron or empty lattice model) the solution of the Schrödinger equation is straightforward and gives

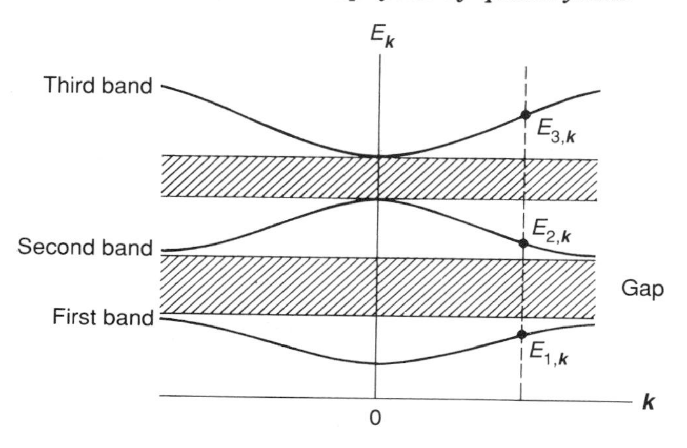

FIG. 6.24 Energy band structure. The cross-hatched areas indicate energy gaps.

$$\begin{aligned} \psi_k^{(0)}(x) &= \frac{1}{L^{\frac{1}{2}}} \exp(\mathrm{i}\, kx) \\ E_k^{(0)} &= \hbar^2 k^2/2m \end{aligned} \tag{6.27}$$

The energy bands are pieces of parabola when folded in the first Brillouin zone (reduced zone scheme) (Fig. 6.25). There is no energy gap. The effect of switching on the crystal potential is essentially to open gaps at the zone limits by smoothing over the sharp 'corners' present in the band structure of the empty lattice. If $V(x)$ remains relatively weak, this can be demonstrated within the frame of the perturbation theory. For instance, the perturbed energy $E_1(k)$ up to the second order of the potential is given by

$$E_1(k) = E_1^{(0)}(k) + \langle \psi_{1,k}^{(0)} | V | \psi_{1,k}^{(0)} \rangle + \sum_n \frac{|\langle n, k | V | 1, k \rangle|^2}{E_1^{(0)}(k) - E_n^{(0)}(k)} + \ldots \tag{6.28a}$$

The first-order correction is equal to the average value of the potential and is independent of k: we can ignore it. In the second-order correction, the denominator increases rapidly as the band index u rises: the major effect on band 1 arises from its coupling to band 2. We may therefore write

$$E_1(k) = E_1^{(0)}(k) + \frac{|V(-2\pi/a)|^2}{E_1^{(0)}(k) - E_2^{(0)}(k)}, \tag{6.28b}$$

where $V(-2\pi/a)$ is the Fourier component of the potential, $E_1^{(0)}(k) = \hbar^2 k^2/2m$, and $E_2^{(0)}(k) = \hbar(k - 2\pi/a)^2/2m$. For a weak enough potential, the second-order perturbation is negligible, i.e. $E_1(k) \simeq E_1^{(0)}(k)$,

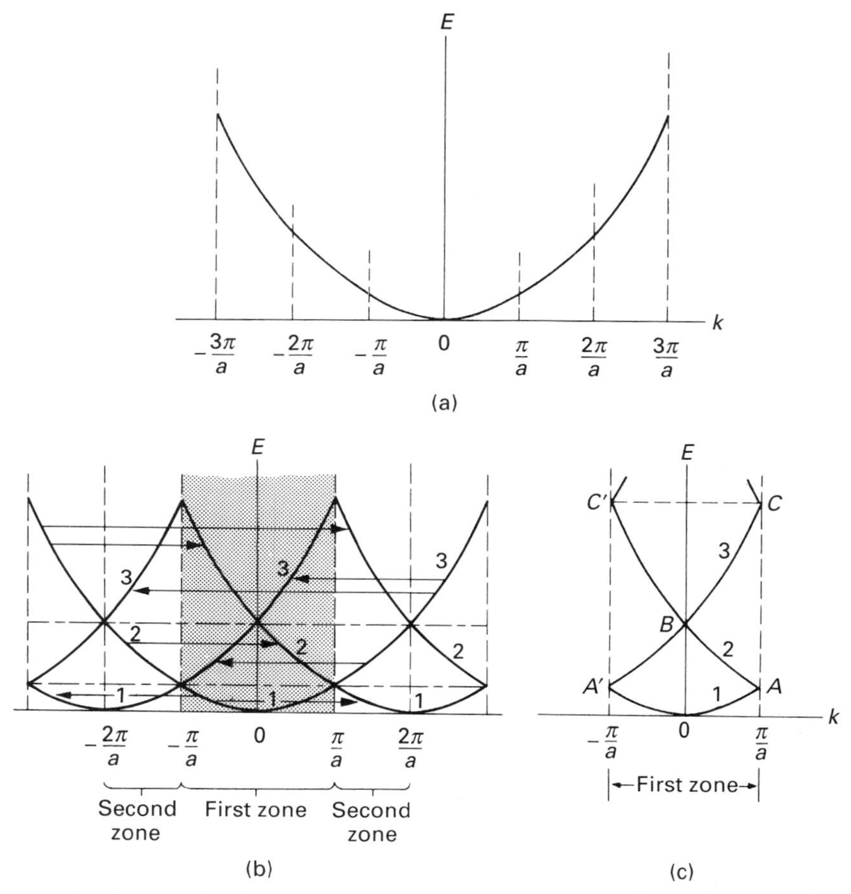

FIG. 6.25 (a) The familiar parabola representing the energy dispersion curve for a free particle. (b) The curves for the same particle in the empty lattice model, showing translational symmetry and the various bands. (c) Dispersion curves folded in the first Brillouin zone (reduced zone scheme).

except at the zone edge ($k = \pi/a$ and $E_1^{(0)}(k) = E_2^{(0)}(k)$) where the perturbation theory cannot hold true.

The degenerate perturbation theory, in which both bands 1 and 2 are treated simultaneously on an equal footing gives:

$$E_{\pm k} = \frac{1}{2}\,[E_1^{(0)} - E_2^{(0)} \pm \{ (E_2^{(0)} - E_1^{(0)})^2 + 4|V(-2\pi/a)|^2\}],$$

where the $\pm$ sign refers respectively to deformed bands 2 and 1. The band structure is shown in Fig. 6.26. The energy gap is equal to twice the Fourier component of the potential, i.e.

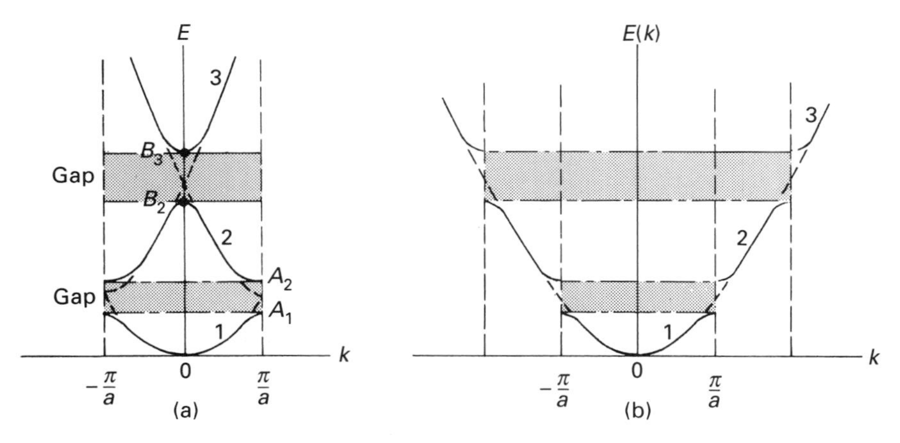

FIG. 6.26 Band structure in the nearly free electron model (a) in the reduced zone scheme and (b) in the extended zone scheme.

$$E_g = 2|V(-2\pi/a)|.$$

These gaps becomes $2|V(-4\pi/a)|$ between bands 2 and 3, $2|V(-6\pi/a)|$ between bands 3 and 4, etc., and thus decrease as we move up the energy scale, while the bands become wider. Explicit calculations of the energy suggest that electrons at the top and bottom of the bands behave formally as free electrons ($E \sim k^2$) but with an effective mass m^* that can be either much larger than the true mass m or even negative. This is consistent with the state functions being mixed states at the zone edge, in the form of standing waves (consistent with eqn (6.26)):

$$\psi_{\pm}(x) = \frac{1}{(2L)^{\frac{1}{2}}}\{\exp(i\pi x/a) \pm \exp(-i\pi x/a)\}.$$

Thus, at the zone edge the scattering is so strong that the reflected wave has the same amplitude as the incident wave and the electron has a zero velocity (Bragg-like situation).

If the crystal potential is not weak enough, the perturbation approach must be disregarded. Other methods, such as the tight binding model, the augmented plane wave technique, or pseudopotential calculations are then more relevant. But the above conclusions remain qualitatively valid, with band structures exhibiting many features.

In higher dimension lattices, the Bragg conditions is satisfied along all boundaries of the Brillouin zone and this results in the appearance of energy gaps along these boundaries, in agreement with the above conclusions.

The *density of states* $g(E)$ for electrons yields the number of states in a

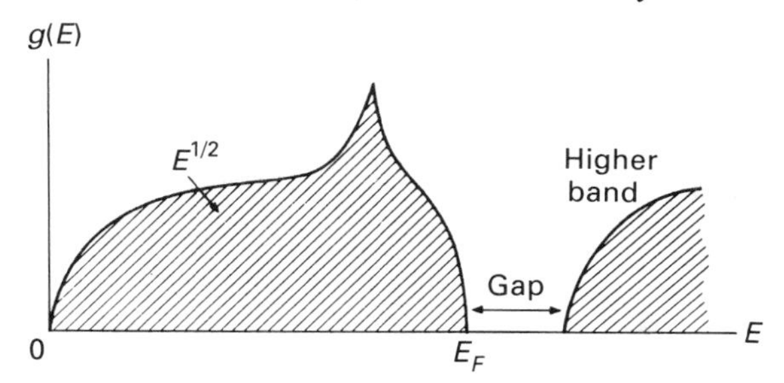

FIG. 6.27 Schematic representation of the density of states.

certain energy range. This function is important in transport phenomena and other properties; $g(E)$ is defined as $g(E)\mathrm{d}E$, being the number of electron state per unit volume in the energy range $[E, E + \mathrm{d}E]$. $g(E)$ is intimately related to the band structure. For a standard parabolic form, $g(E)$ is

$$g(E) = \frac{1}{2\pi^2}\left(\frac{2m^*}{\hbar^2}\right)^{\frac{3}{2}} E^{\frac{1}{2}}, \tag{6.29}$$

as represented in Fig. 6.27 for the low energy part. The $g(E)$ function deviates from the increasing parabolic shape as zone edge effects take over (Fig. 6.27) where the parabola is inverted.

At $T = 0\mathrm{K}$, the states are occupied up to the Fermi level E_F defined as:

$$\int_0^{E_\mathrm{F}} g(E)\,\mathrm{d}E = n, \tag{6.30}$$

n being the number of electrons per atom. Inserting eqn (6.29) into eqn (6.30) gives

$$E_\mathrm{F} = \frac{h^2}{2m^*}(3\pi^2 n)^{\frac{2}{3}}.$$

For higher-dimensional lattices the Fermi level defines a Fermi sphere if isotropy holds true, and a Fermi surface in the more general case.

In the definition of the density of states $g(E)$ it is assumed that surfaces of equal energy states can be identified inside the Brillouin zones. Far from zone boundaries the energy is fairly well described by the free-electron formula $E^\circ = \hbar^2k^2/2m$ and the corresponding surfaces of equal energy are almost spherical. On approaching a zone boundary, surfaces of equal energy are first deformed by protuberances directed towards the boundary. This increases the area of the surfaces of equal energy with respect to the free-electron sphere. On closer approach, these protuberances are intercepted by the boundary and the efficient area of the surface of equal energy shrinks

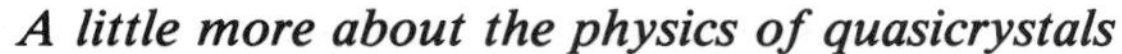

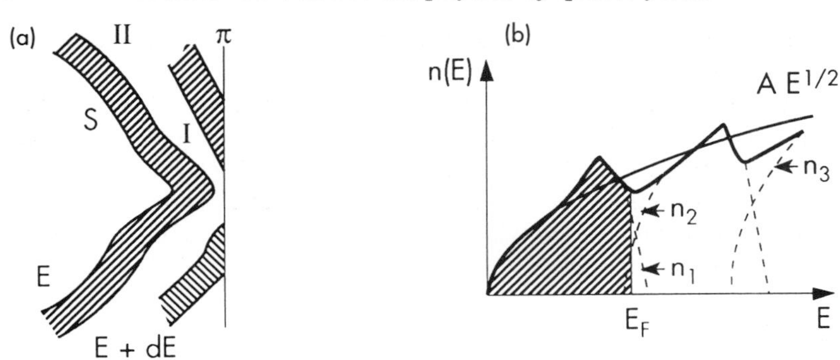

FIG. 6.28 (*a*) Volume between two surfaces of equal energy (E and E + dE) in the first Brillouin zone when approching the zone boundary π and (b) the corresponding density of state for three overlapping bands: broken lines show the partial densities of states for the individual bands; the solid line gives the total density of states. The Fermi level indicated corresponds to a divalent metal.

progressively. The consequence for the density of states is a departure from the $E^{1/2}$ law, first with a spiky increase followed by a rapid decay to zero. If several energy bands overlap due to anisotropy of the Brillouin zones combined with a relatively weak potential, many such singularities are observed, each corresponding to the cross-over of a zone boundary by the surfaces of equal energy. The complete mechanism is illustrated schematically in Fig. 6.28 and is obviously responsible for the existence of depressed energy ranges in the density of states.

These ranges of relatively smaller densities of states are called 'pseudo-gaps'. When the energy level falls into such a pseudo-gap, the corresponding structure is electronically stabilized with respect to the free-electron situation since the protuberance in $g(E)$ below E_F allows accommodation of more electrons with lower energies than is the case with the $E^{1/2}$ profile. Thus in metallic alloys with a given electron concentration (average number of valence electrons per atom) the structure will tend to build up in such a way that a Brillouin zone boundary is generated in the right place to induce a pseudo-gap at the Fermi level. This is called a Hume-Rothery alloy. The mechanism is particularly efficient if the pseudo-gap is very deep, which is the case for almost isotropic materials with the zone boundary situated at about the same distance from the origin in all directions in reciprocal space.

For a spherical Fermi surface, the electrical conductivity of the material is given by

$$\sigma = \frac{1}{3} e^2 v_F^2 \tau_F g(E_F), \tag{6.31}$$

where v_F and τ_F are the velocity and the collision time for an electron at the Fermi surface.

In periodic crystalline metals $g(E_F)$ is large and the Fermi surface is not at a zone edge. Then σ is large. In any insulator, the Fermi level lies within a gap and $\sigma = 0$. In a disordered material σ is small, even if $g(E_F)$ is large, because τ_F is small (very short mean free path for the electron). The mobility edge being zero, an electron at the zone boundary does not really participate in conduction.

6.5.2 Effect of disorder on electron behaviour

In the conventional nearly-free-electron theory of the band structure of crystalline metals, the potential is treated as a perturbation of the plane-wave functions and the energy of a Bloch state of wavevector $\boldsymbol{k}$ can be expanded as a series (see eqn (6.28)). In principle, a disordered system can formally be treated the same way. Modifications come from the fact that both the structure factor $S(\boldsymbol{q})$ and the Fourier coefficient $V(\boldsymbol{q})$ of the potential are not restricted to a reciprocal lattice but are continuous functions in momentum space. Thus the analogue of eqn (6.28) is

$$E(\boldsymbol{k}) \simeq E^{\circ}(\boldsymbol{k}) + \frac{\Omega}{8\pi^3}\int \frac{S(\boldsymbol{q})\,|V(\boldsymbol{q})|^2\,\mathrm{d}\boldsymbol{q}}{E^{\circ}(\boldsymbol{k}) - E^{\circ}(\boldsymbol{k}+\boldsymbol{q})}\,. \tag{6.32}$$

Since $S(\boldsymbol{q})$, $V(\boldsymbol{q})$ and $E^{\circ}(\boldsymbol{k})$ must all be spherically symmetrical functions and the integral converges, the above equation can be solved numerically at least. It is tempting to regard the result as a dispersion function for an electron propagating in the disordered metal. However, this is unsound in principle. Since there is no true translation symmetry, momentum is not a good quantum number and the Fourier transform of the eigenfunction cannot have a perfectly sharp spectrum in reciprocal space. The correct procedure is to calculate the spectrum of the system via Green function formalism, which is not developed in this book. Only the main results are summarized.

On the $(E, \boldsymbol{k})$ planes, the spectral density of states $\mathfrak{N}(E, \boldsymbol{k})$, of energy E and apparent momentum $\boldsymbol{k}$, looks like a ridge, as if a sharply defined dispersion function had been broadened in energy and/or momentum by incoherent scattering from the disordered assembly of atoms (Fig. 6.29). The physical notion of a dispersion function for the electrons in a disordered metal might be represented by the trajectory $E_L(\boldsymbol{k})$ of the apex of $\mathfrak{N}(E, \boldsymbol{k})$. However, this curve has no special significance and is not, for instance, the solution of the nearly-free-electron equation.

The density of states itself is obtained classically by integrating overall values of $\boldsymbol{k}$. An example of the calculated spectrum is shown in Fig. 6.30. At some energy, the density of states rises above the free-electron curve and then drops below it in a manner reminiscent of the density of states for

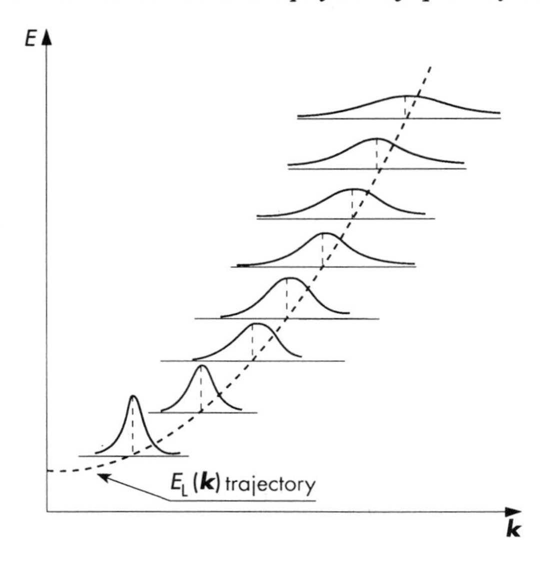

FIG. 6.29 Schematic representation of the energy–momentum distribution for an electron in a randomly disordered metal (redrawn from reference [41]).

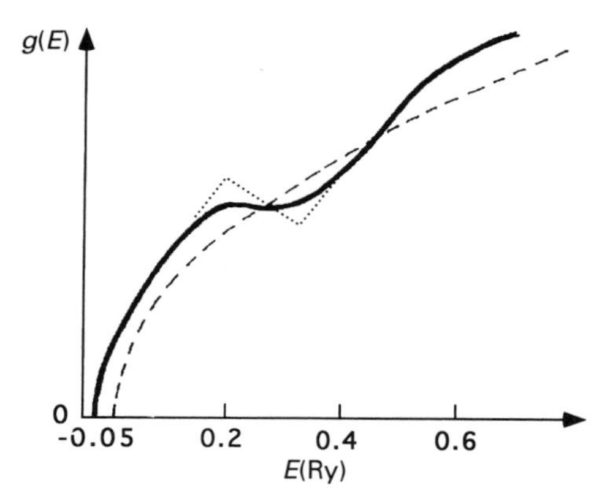

FIG. 6.30 Typical density of states for a randomly disordered metal (full line) compared with the free-electron parabola (broken line) and a crystalline solid with Van Hove singularities (dotted line).

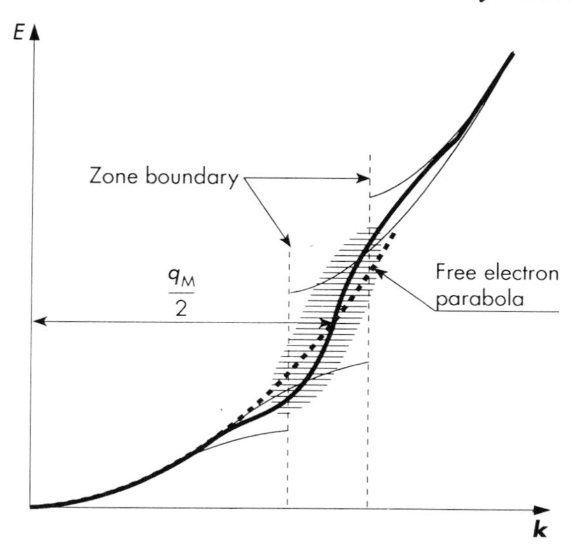

FIG. 6.31 The ridge top of the energy–momentum distribution as shown in Fig. 6.29 exhibits kink and broadening in the region where there would be jump discontinuities of $E(\boldsymbol{k})$ in a crystalline solid (redrawn from reference [41]).

overlapping bands in a regular periodic crystal (Fig. 6.28). Thus a pseudo-gap also appears, in this case at an energy corresponding to $|\boldsymbol{k}| \approx |\boldsymbol{q}_M|/2$ where $\boldsymbol{q}_M$ is the position for the maximum structure factor $S(\boldsymbol{q})$. This also gives rise to a kink and a broadening in the $E_L(\boldsymbol{k})$ curve (Figure 6.31) which can be interpreted as a smudged version of the discontinuity in $E(\boldsymbol{k})$ at the zone boundary in a crystalline solid. As an interesting consequence, Hume-Rothery stabilization can also be evoked in disordered alloys to justify the particular composition range in which they form. Finally, as readily observed in Fig. 6.29, it can be roughly stated that the eigenstates are rather extended in momentum space. This induces localization in physical space and will corrupt transport phenomena.

It is easy to prove that, in a one-dimensional system, all eigenstates are localized by disorder. However, this cannot be generalized to any two- or three-dimensional models. This point was addressed by Anderson some time ago (1958).[41] The essence of the Anderson localization model is that it is a tight-binding approximation calculation where the disorder variable is taken from a continuous probability distribution. This is equivalent to treating the problem of electrons in a randomly distributed potential. The main result is that the eigenfunctions all become localized if the 'strength' of the disorder exceeds some definite value, and below this value the spectrum can be partitioned by the so-called 'mobility edges' into ranges where either

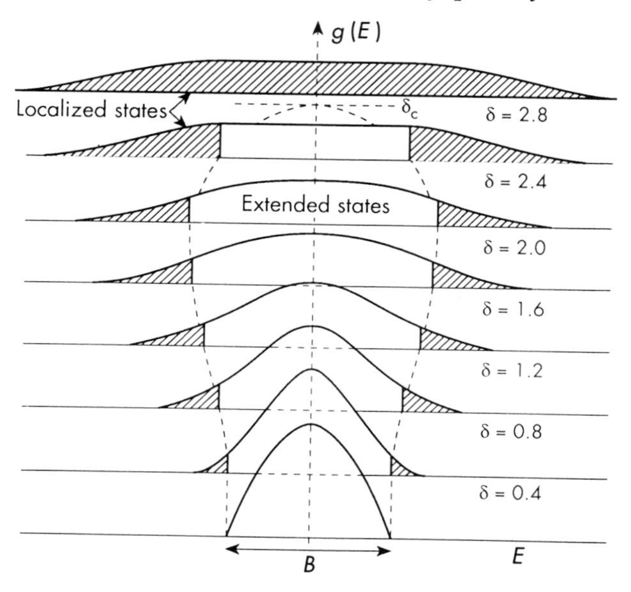

FIG. 6.32 As disorder increases ($\delta = W/B$ (see text)) more states are localized. For a certain critical disorder ($\delta \geqslant \delta_c$), the mobility edges (limit between localized and extended states) coalesce at the centre of the band. This is the Anderson transition (redrawn from reference [41]).

all the states are localized or all are extended. The 'strength' parameter of the disorder is usually labelled δ and is equal to W/B, where W is the width of the probability distribution function and B is the width of the prospective valence band for a perfect crystal of the same system. Extended and localized states with mobility edges are shown in Fig. 6.32.

Obviously, Anderson localization may be responsible for weak electrical conductivity at low temperature. However, hopping mechanisms via tunnelling processes can force the localized states to contribute to conductivity.

Many mechanisms have been proposed to describe hopping conduction via diffusive transport of electrons between localization sites. Generally speaking, an energy eigenfunction in a spherically symmetrical potential decays exponentially at large distances:

$$\Psi(r) \approx \exp(-r/a) \tag{6.33}$$

where a is a characteristic radius for the localized state in question.[41] States centred at $\boldsymbol{R}_i$ and $\boldsymbol{R}_j$ must interact through an overlap integral of the type

$$\int \psi^*(\boldsymbol{r} - \boldsymbol{R}_i) \cdot \psi(\boldsymbol{r} - \boldsymbol{R}_j)\, \mathrm{d}^3 \boldsymbol{r} \approx \exp\left(-\frac{|\boldsymbol{R}_j - \boldsymbol{R}_i|}{a}\right).$$

Thus small currents may be due to electron tunnelling from site to site through this overlap of wave functions. To a first approximation the hopping probability can be assumed to be proportional to the square of the overlap integral. However, the variation of energy from site to site inhibits the hopping transition. For the overlap between wavefunctions to have its effect, we must find a means of neutralizing the energy difference $\Delta E_{ij} = E_i - E_j$. In practice, this is taken care of by the thermal vibration and introduces a factor $\exp(-\Delta E_{ij}/k_B T)$ which reduces the transfer probability. Conclusively, the hopping probability P_h for electron-tunnelling transfer from a site i to a site j, at a distance $\boldsymbol{r}_{ij}$ from each other, corresponding to two bound-state energies E_i and E_j can be expressed as

$$P_h \propto \exp\left(-\frac{\Delta E_{ij}}{k_B T}\right) \exp\left(-\frac{|\boldsymbol{r}_{ij}|}{R}\right) \tag{6.34}$$

where R is a characteristic distance depending on the potential barrier to be overcome. At high temperatures, P_h and hence the hopping conductivity will vary as $\exp(-\Delta E_{ij}/k_B T)$. At low temperatures, the transfer may involve more distant localization sites if this reduces the energy barrier. Indeed, the number of Fermi states in the ΔE_{ij} range in a sphere of radius r_{ij} around site i is

$$n(r_{ij}, \Delta E_{ij}) = \tfrac{4}{3}\pi r_{ij}^3 g(E_F)\Delta E_{ij}$$

where $g(E)$ is the density of localized states. The term $n(r_{ij}, \Delta E_{ij})$ must be equal to unity if the proper localized state is to be found at site j. Thus

$$\Delta E_{ij} = [\tfrac{4}{3}\pi g(E_F)]^{-1}/r_{ij}^3 = \alpha/r_{ij}^3$$

and expression (6.28) becomes

$$P_h \propto \exp\left(-\frac{\alpha}{r_{ij}^3 k_B T}\right) \exp\left(-\frac{r_{ij}}{R}\right).$$

P_h has maximum value ($\mathrm{d}P_h/\mathrm{d}r_{ij} = 0$) for $r_{ij} = (3\alpha R/k_B T)^{\frac{1}{4}}$, which gives

$$P_h(\max) \propto \exp(-T_0/T)^{\frac{1}{4}}. \tag{6.35}$$

Such a $T^{-\frac{1}{4}}$ power law (the Mott law) is often observed in poorly organized matter. The last condition for hopping conduction to be achieved is dominated by the possibility of finding a continuous path of infinite length linking the localization sites.

In contrast with hopping mechanisms which tend to drive localized states back into conduction, quantum interference may modify the contribution of the extended states. Quantum interference has roughly two different effects which either operate via a modification of the electron diffusivity (***weak***

localization) or induce changes in the density of states at the Fermi level **(electron–electron interactions)**.

1. *Weak localization* In disordered structures, travelling electrons are randomly scattered via numerous elastic collisions. The electron transport is then a random diffusion process from collision site to collision site. The average travelling distance is $L = (Dt)^{\frac{1}{2}}$ after a travelling time t; D is the electron diffusivity. More generally, the probability of finding an electron at a distance r after a time t is a Gaussian function of r with half-width at half-maximum equal to $(Dt)^{\frac{1}{2}}$. Successive collisions may result in closed-loop trajectories with quantum interference between wavefunctions corresponding to the two 'traffic ways' for electrons along the loop. In the absence of inelastic collisions, these interferences produce additive probabilities. Electrons diffuse back to their original sites twice as often as expected from random walk calculations. This is equivalent to some (weak) localization and will reduce conductivity. The weak localization is efficient at low temperatures only. Thermal vibrations and any inelastic processes destroy interferences. The resulting $\delta\sigma_{\mathrm{WL}}(T)$ conductivity is then expected to be an increasing function of T roughly linear up to the Boltzmann conductivity.

2. *Electron–electron interactions* This effect is intended to take into account the Coulomb interactions between electrons. The Coulomb interactions between electrons in disordered metals are less efficiently screened by the positive ions than in crystals because of the random fluctuation density. The resulting influence on the density of state propagates up to the Fermi level. This produces an additional contribution to conductivity $\delta\sigma_{\mathrm{e\text{-}e}} = c\sqrt{T}$; c is almost negligible and negative for a crystalline metal but becomes significant and positive for disordered (amorphous) metals. Electron–electron interactions also become ineffective away from the very low temperature range.

6.5.3 Experimental electronic properties of quasicrystals

Transport properties and specific heat studies of icosahedral phases were carried out soon after the discovery of these materials.[42] The results were rather disappointing, in the sense that they seemed to indicate that quasicrystals were similar to their corresponding crystalline or amorphous counterparts.[43] However, it has since been realized that quasicrystals like those of the AlMn system are intrinsically disordered and may not be able to reveal novel properties. As already stated, the discovery of stable perfect quasicrystals,[44–47] notably those of the AlCuRu and AlFeCu systems, has dramatically changed the situation and has provided an impetus for exploring the intrinsic properties of quasiperiodic alloys. Indeed, studies of the new systems in the few past years have indicated the precursor to electron localization.[48–55] The main features can be summarized as follows:

1. Low-temperature resistivities (ρ at 4.2 K) that are two orders of magnitude larger than those seen in amorphous metals and more than four orders of magnitude larger than those observed in current crystalline metals have been measured. This is closed to the metal–insulator transition[55] ($\rho_M = 5000\ \mu\Omega$ cm).
2. A strong temperature dependence of the resistivity ρ, as seen in the resistivity ratio $\rho(4.2\text{ K})/\rho(290\text{ K}) \approx 10$ (this ratio is ~1.2 for amorphous metals and less than 0.1 for crystalline metals) is also measured.
3. Analysis based on Boltzmann transport, which takes into account the carrier density,[52, 53] shows that the electron mean free path is much shorter than the Fermi wavelength and is of the order of 20–30 Å.[50]
4. The optical conductivity is not Drude like.[56]
5. Symmetry breaking due to either defects or non-optimized compositions has dramatic effects on these features and operates in the 'wrong way' when compared with crystalline material. For example, both composition mismatches and structural defects due to poor annealing treatment reduce the resistivity drastically.

Some of these features are illustrated in Figs 6.33–6.36.

It is not completely clear whether the ultimate perfection and compositon have been reached so that we can take the measured parameters for granted. To date, with the resisitivity data measured for AlPdRe quasicrystals,[48] it seems reasonable to accept the idea that quasiperiodicity produces *insulators at 0 K*. Indeed (see Fig. 6.37), in this experimental record,[48] the low-temperature resistivity culminates above 1 Ω cm and the temperature dependence is that of an insulator. The electron density at the Fermi level is that of a *semimetal* (~ 10^{20}cm^{-3}). Clearly, the large resistivity observed, particularly at low temperatures, indicates that even a very small inclusion (less than 1 per cent by volume) of a comparatively better conductor (10 to 100 $\mu\Omega$ cm) can potentially 'short circuit' the intrinsic electron transport. This makes both sample preparation and resistivity measurements very difficult and very critical. It appears that, upon varying the composition of i-AlPdRe, the conductivity at 0.5 K is as low as 0.03 Ω^{-1} cm^{-} and the temperature dependence seems to be in the variable-range hopping regime.

Some other properties have also been measured and deserve discussion. Specific heat data confirm to some extent the low density of states at the Fermi level, but are not always completely reliable because of strong corruption effects, particularly at low temperatures, owing to impurities, defects, magnetic inclusion, etc. The Hall effect coefficient R_H may also show a very typical behaviour for perfect quasicrystals which seem to be sensitive to the

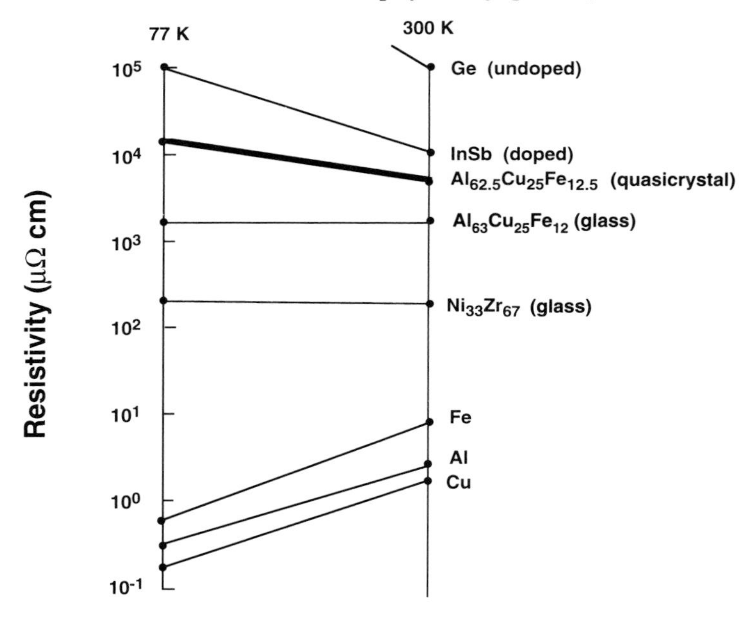

FIG. 6.33 An overview of the resistivity scale showing typical temperature behaviour and orders of magnitude for crystalline metals, amorphous alloys, quasicrystals, and semiconductors (courtesy of C. Berger).

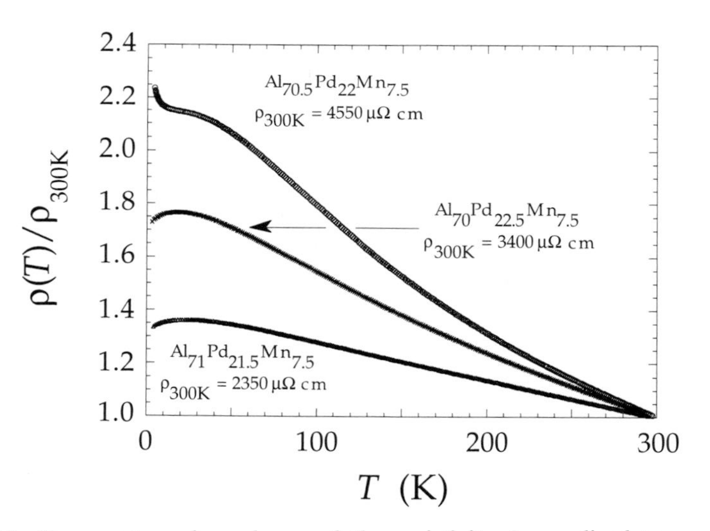

FIG. 6.34 Temperature dependence of the resistivity (normalized to $\rho_{300\,\mathrm{K}}$) for quasicrystals of the AlPdMn system. Criticality of the composition is clearly observed (courtesy of C. Berger).

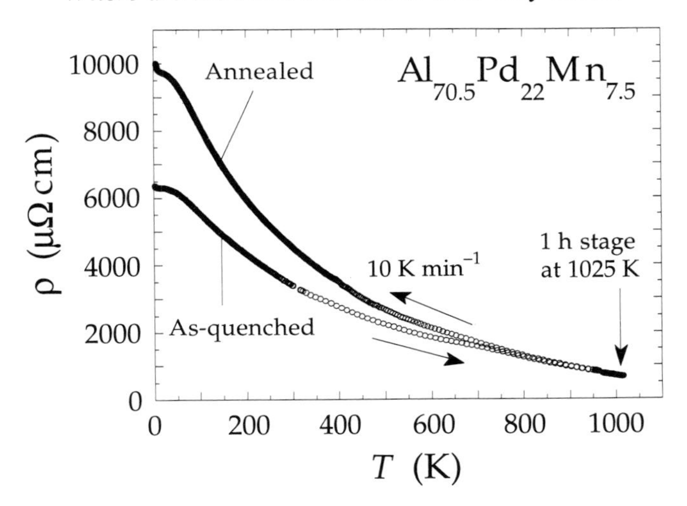

FIG. 6.35 Temperature dependence of the electrical resistivity, in a large temperature range, for the perfect quasicrystal $Al_{70.5}Pd_{22}Mn_{7.5}$ (larger resistivity) compared with that of the same sample before annealing (smaller resistivity). Structural defects reduce the resistivity (courtesy of C. Berger).

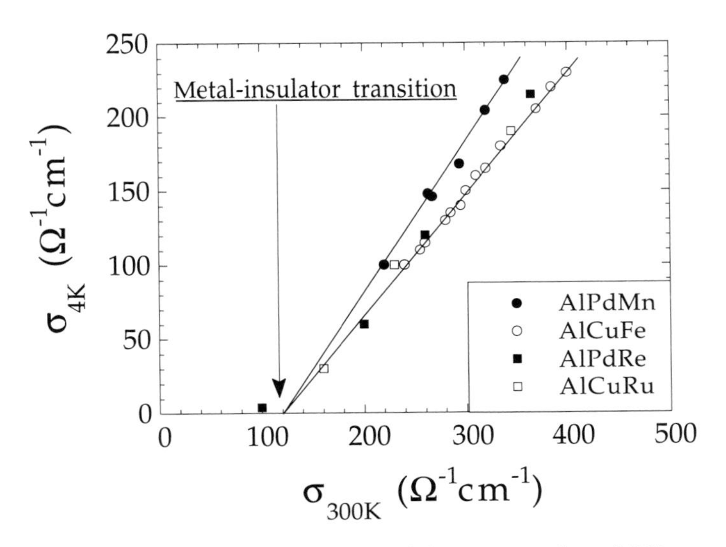

FIG. 6.36 Correlation plot for the conductivity measured at 4.2 K versus that measured at 300 K for several icosahedral quasicrystals, suggesting the proximity of a metal–insulator transition (courtesy of C. Berger).

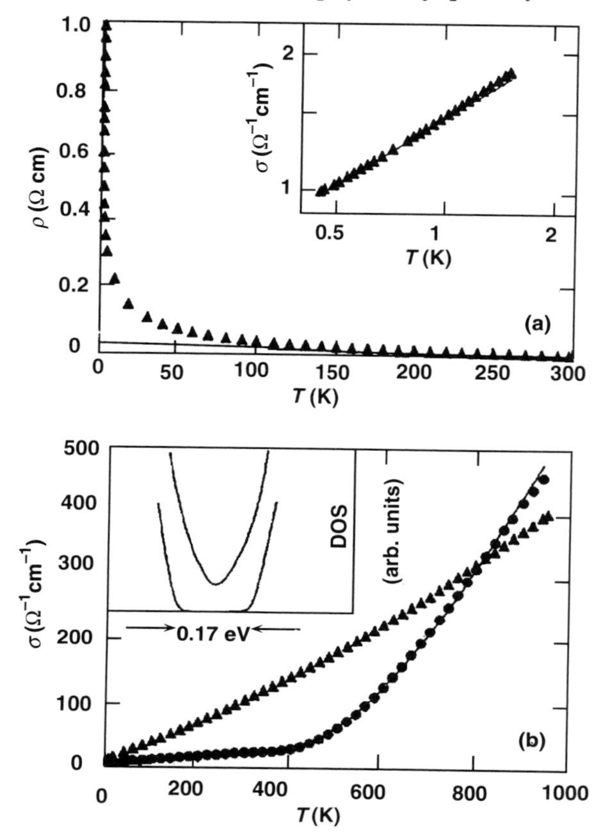

FIG. 6.37 (a) resistivity and (b) conductivity of the $Al_{70}Pd_{20}Re_{10}$ quasicrystal. The conductivity of the semiconductor Al_2Ru is shown for comparison (•) The pseudo-gap of the quasicrystal is also compared with the true semiconductor gap (redrawn with permission from reference [48]).

various causes of imperfections. This is illustrated in Fig. 6.38. Alloys which are considered as the more perfect have R_H values essentially weakly negative, with a change of sign at some temperature.

Thermoelectric power data,[53] as shown in Fig. 6.39, are also unusual. For metallic glasses or free-electron-like metals, the thermopower $S(T)$ is dominated by the electron diffusion contribution, which is proportional to the temperature. It is clear from Fig. 6.39 that the $S(T)$ behaviour in the icosahedral phases is fundamentally different, with strongly non-linear temperature dependence and a change of sign at some temperature.

Photoemission and photoabsorption spectra of the AlCuFe quasicrystal

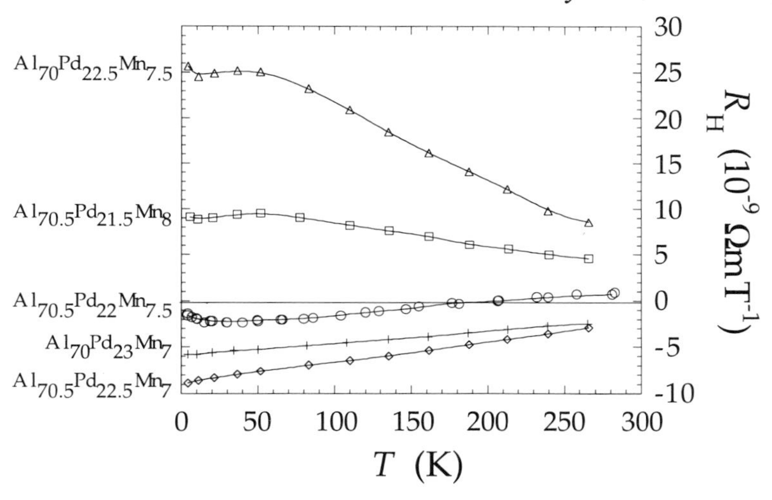

FIG. 6.38 Temperature dependence of the Hall coefficient for AlPdMn quasicrystals of various compositions. The perfect icosahedral alloys are those with weakly negative R_H at low temperatures. (courtesy of C. Berger).

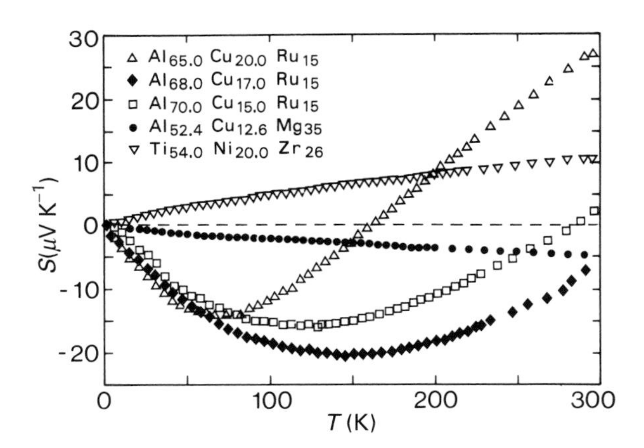

FIG. 6.39 Thermoelectric power data. The glass-like material (● and ∇) shows the linear temperature behaviour expected from a weak localization effect. The quasicrystals (△, □, ♦) instead exhibit an unusual behaviour, with a pronounced minimum in the $S(T)$ curves.[53]

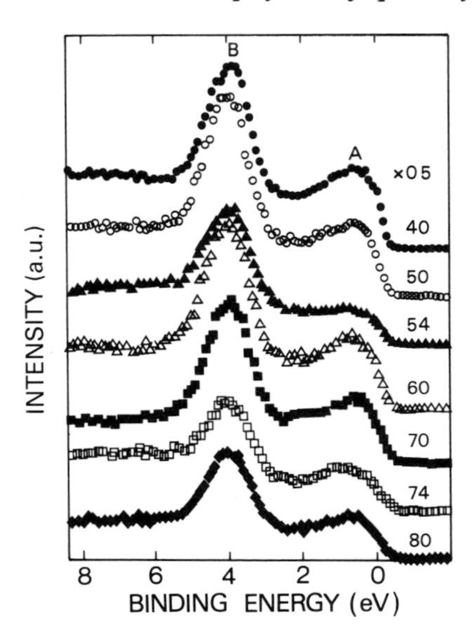

FIG. 6.40 Photoemission spectrum of the AlFeCu ico-phase alloy measured at roon temperature with various values of the photon energy as indicated on the right-hand side of the curves.[58]

have confirmed the low density of state at the Fermi level.[58, 59] The shape of the observed spectrum can roughly be described as a broad peak culminating on a plateau with a cut-off at E_F (Fig. 6.40). The peak at 4 eV comes mainly from the 3d states of the copper atoms. The 3d states of Fe atoms show up less markedly as a bump at about 1 eV below E_F. A high-resolution analysis of the region close to the Fermi edge demonstrates that this Fermi edge is not well defined in quasicrystals as compared with normal metal (Fig. 6.41) and that there is a dip valley in the electronic density of states at E_F, with half-width 0.3–0.4 eV and a depth 70 per cent of the regular value. This is perfectly consistent with the density of states as deduced from specific heat data.[52]

As already mentioned, intrinsic electronic properties of quasicrystals are less easily observed in other systems, due to the lack of perfection in the quasiperiodic order formed. However, the AlLiCu system, as imperfect as it may be, is especially interesting for the double reason that it has been the subject of detailed structural analysis (see Chatper 4) and that both quasi-crystalline icosahedral and crystalline bcc R-phases have the same compositions and density. Thus, we may be sure that differences between them, if any, can be attributed to modifications from periodic to quasiperiodic order, within imperfections for each of them. The point has indeed been fully

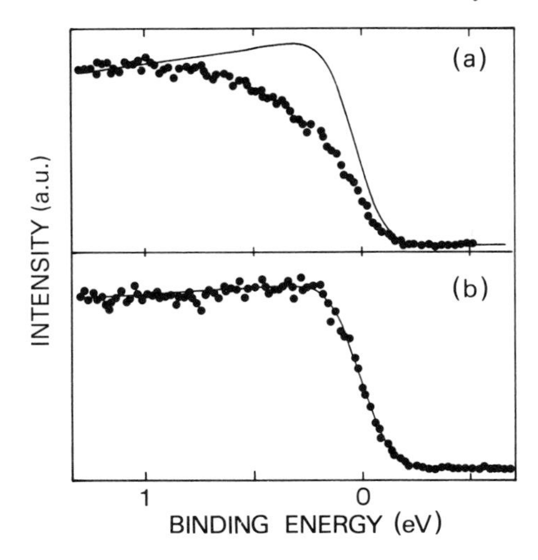

FIG. 6.41 High resolution (≅0.14 eV) photoemission spectra near the Fermi edges for the AlFeCu ico-phase (a) and pure aluminium (b). The 'dots' are experimental data and the solid curves are fits assuming a regular density of state for (b) and using data corresponding to energies smaller than 0.75 eV for (*a*).[58]

investigated,[60] and the results are summarized in Fig. 6.42. Basically the large resistivity of the quasicrystalline phase with respect to its crystal counterpart is well confirmed (Fig. 6.42(a)). Around room temperature the resistivity ratio amounts to about 4 and increases to abut 6 at low temperature. These values are far below those observed with the AlFeCu and AlRuCu systems, even if there is less significance for them, in as much as comparison with crystals has to be done with different compositions. The results are consistent with the idea that the AlLiCu system does not produce perfect quasicrystals, but rather some sort of random tiling structure, stabilized via an entropy mechanism with phason-like defects. The figure also shows that the quasicrystal resistivity has a decaying behaviour with increasing temperature, while the R-phase exhibits the classical temperature dependence of a crystalline solid. The low temperature regime is $(T)^{\frac{1}{2}}$-like, while it turns to T behaviour above 20 K or so.

Another very elegant demonstration of the differences in transport properties between periodic and quasiperiodic structures has been obtained with $Al_{65}Cu_{15}Co_{20}$ and $Al_{20}Ni_{15}Co_{15}$ decagonal quasicrystals.[61] As already mentioned, decagonal quasicrystals can be viewed as two-dimensional quasiperiodic atomic arrangements which are then periodically stacked in the third dimension. Thus, decagonal quasicrystals have the privilege of offering simultaneously periodic and quasiperiodic ingredients. Accordingly, the temperature-dependent resistivity of the decagonal phases has

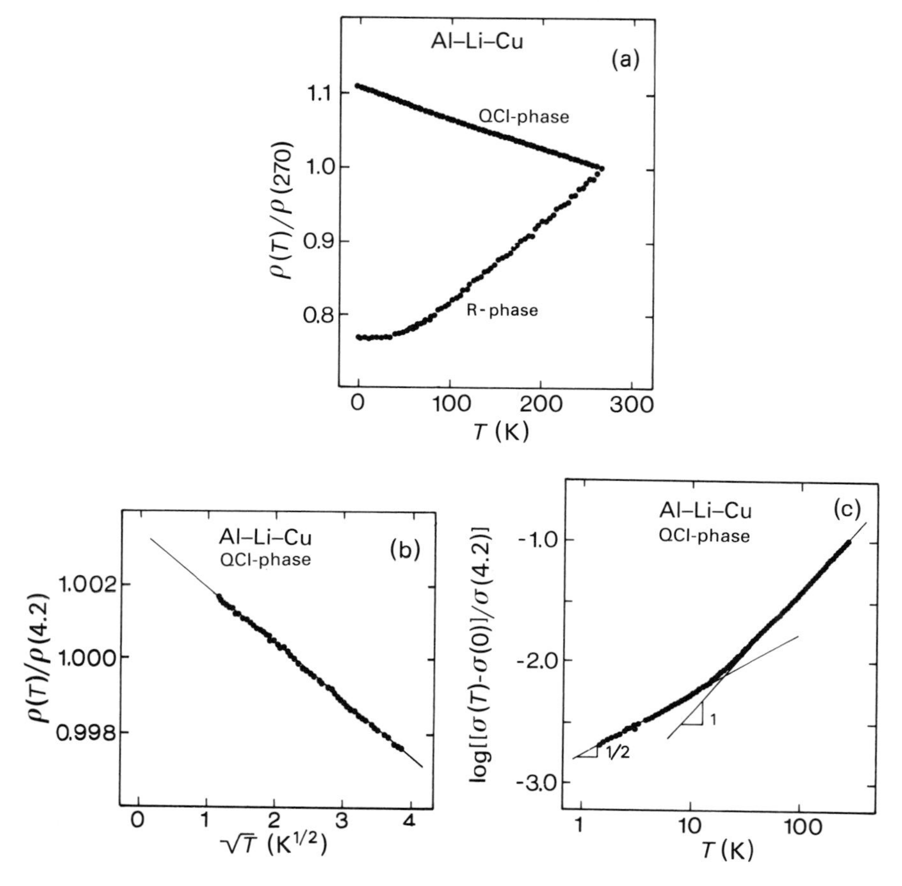

FIG. 6.42 (a) Resistivity data for the icosahedral phase and the bcc R-phase of the AlLiCu system, normalized to the values measured at 270 K (800 and 200 $\mu\Omega$ cm, respectively). (b) $\sqrt{T}$ behaviour of the resistivity at low temperature of the quasicrystal. (c) Observation of two conductivity regimes for the icosahedral AlLiCu phase.[60]

been studied along the periodic direction (ρ_p) and in the quasiperiodic plane (ρ_q) from 4.2 to 600 K.[61] In both materials ρ_p is metallic, can be described by a semi-classical electron–phonon scattering model, and increases roughly linearly with temperature from about 30 to 60 $\mu\Omega$ cm. For ρ_q, a non-metallic behaviour has been found; ρ_q is about 10 times as large as ρ_p and shows a plateau at 340 $\mu\Omega$ cm for $T \simeq 650$ K. Such a transport anisotropy is certainly the best evidence of the expected peculiar behaviour of quasiperiodic structures compared with regular periodic arrangements.

Magnetism in quasicrystals has mainly been measured with systems

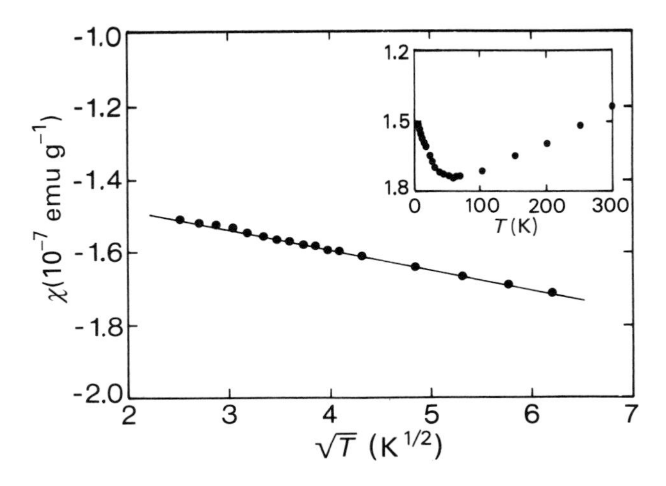

FIG. 6.43 $\sqrt{T}$ behaviour of the low-temperature magnetic susceptibility for AlFeCu icosahedral alloys. The inset shows the overall temperature dependence of χ up to 300 K.[52]

which are now accepted as very imperfect quasicrystals, such as, for instance, those related to the AlMn family. Most of these materials exhibit classical paramagnetism and, for some of them, a spin-glass-like behaviour at low temperature.[62, 63] The latter is obviously a very exciting phenomenon, but does not seem to be due to intrinsic aspects of quasiperiodicity. This has been revealed by crossed studies of the magnetic susceptibility in quasicrystals of the AlFeCu system. When measured in an $Al_{63.5}Cu_{24.5}Fe_{12}$ alloy[64] the magnetic susceptibility appears as basically paramagnetic, though with a strong departure from Curie behaviour at low temperatures and a spin-glass-like peak near 1.6 K. The composition of the alloy measured in this study is now known not to be optimized and gives rise to imperfect quasicrystals containing defects and a parasitic phase. In another more systematic study,[55] the same paramagnetic behaviour has been also observed in 'imperfect' samples, while the true quasicrystal $Al_{62.5}Cu_{25}Fe_{12.5}$ has shown a *diamagnetic* behaviour from 4 to 300 K, with $\chi(4\text{K}) = -4 \times 10^{-7}$ emu g^{-1} (to be compared to the -2.3×10^{-7} ion core contribution); χ has a $(T)^{\frac{1}{2}}$ dependence up to 40 K or so (Fig. 6.43), with a slope about one order of magnitude larger than the free electron estimate; above ~50 K, $|\chi|$ increases and is roughly proportional to T. Diamagnetism has also been recently observed in AlPdMn quasicrystals when their composition is optimized, while paramagnetism appears in imperfect samples.[65]

6.5.4 The theoretical aspects of electrons in quasicrystals: the critical states

In the absence of specific models to give a proper description of electron behaviour in quasicrystals, it has been tempting to use the available theories, i.e. those which apply to either periodic crystals or amorphous alloys. Indeed, some of the experimental observations can be partially explained by conventional effects.

For instance, many low-temperature conductivity data[50, 55, 65] have been reasonably interpreted within the scheme of quantum interference effects, such as weak localization and/or electron–electron interactions as commonly expected for amorphous solids. A combined $\sqrt{T}$ and T dependence of the conductivity $\sigma(T)$ can actually be fitted to data for some alloys in the 0–100 K range. However, this approach does not apply universally and a large number of drawbacks, including the following, make it questionable.

- It does not explain why the 0 K conductivity can be so low (see the AlPdRe alloys[48]).
- It is unexpected in alloys with such a low density of states at the Fermi level and a very short mean free path.
- It does not provide any clues to understanding why $\sigma(T)$ should improve under disorder effects (structural disorder due to bad annealing treatment and/or chemical disorder due to composition mismatches).
- It does not account for the continuous increase of $\sigma(T)$ up to temperatures (1200 K!) well above the limit which is accessible to quantum interferences.

To continue with the possible description of electrons in quasicrystals in terms of amorphous-like systems, Anderson localization must also be ruled out. Again, this would not explain the almost zero conductivity at 0 K of AlPdRe, the strong $\sigma(T)$ increase above room temperature, or the structure–composition effects. Moreover, exponentially localized states, as expected from usual Anderson localization, normally contribute to the specific heat but not to the electrical conductivity. Experimental data[48, 66] show that the density of states at the Fermi level for quasicrystals is equally low whether deduced from specific heat (Fig. 6.44) or electrical conductivity. This demonstrates that the density of localized states at E_F is very low, if not zero, in contradiction with Anderson localization. The associated variable-range hopping mechanism is not observed either, since $\sigma(T)$ is not simply thermally activated at high temperature and does not follow the $T^{-\frac{1}{4}}$ Mott law at low temperature.

In we now turn to descriptions of quasicrystal electrons in terms of band structures, using the models currently applied to periodic alloys, the conclu-

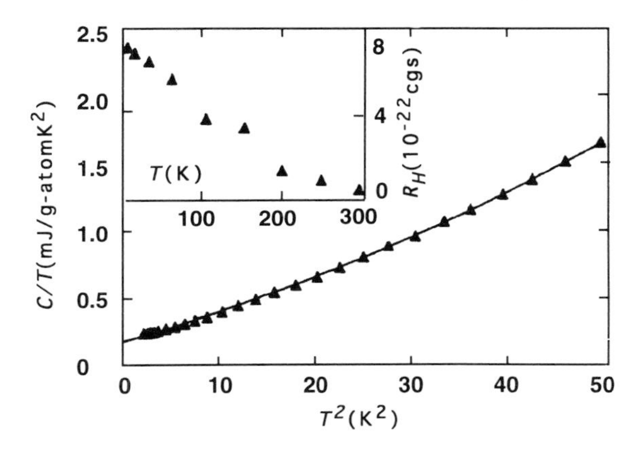

FIG. 6.44 Specific heat data in a C/T vs. T^2 plot for an AlPdRe quasicrystals. The inset shows Hall coefficient vs. temperature (redrawn with permission from reference [48]).

sions are rather similar. The almost zero conductivity at 0 K and the increasing $\sigma(T)$ associated with very low density of states at the Fermi level are features reminiscent of semiconductor behaviour. However, the occurrence of a strict gap is excluded in quasicrystals as again there is no sign of any thermally activated contribution to $\sigma(T)$ or to the dynamic conductivity $\sigma(\omega)$. Moreover, spectroscopy data[58, 59] confirm the presence of a mere pseudogap with partial overlapping between filled and empty states, even if the valley appears to be rather deep (Fig. 6.45). Thus interband transitions of this type are certainly not the key to the quasicrystal mystery. This is particularly clear in Fig. 6.45 where the density of conduction states appears to be dramatically low in the quasicrystal compared with Al metal.

However, the low density of states at the Fermi level (about 20 per cent that of A1) can be understood in terms of a Hume-Rothery stabilization mechanism,[67, 68] which is just the equivalent for higher dimensions of the Peierls instability. It has been explained in previous sections devoted to a summary of electrons in crystals that the periodicity of the potential is responsible for gap opening in the dispersion law and in the density of states at the Brillouin zone borders. This is due to the 'total reflection' (diffraction) of any plane wave which tries to propagate with the corresponding momentum (wavevector). The natural anisotropy of the crystals (and hence of their Brillouin zones) associated with finite values of the potential Fourier coefficient generally results in partial overlap of the successive bands, and the gaps transform into pseudo-gaps with non-zero densities of states in their valleys. Higher symmetry of the point group, i.e. less anisotropy of the Brillouin zones, reduces the band overlap and produces deeper pseudo-gaps. When the energy level falls into such pseudo-gaps, the kinetic energy of the

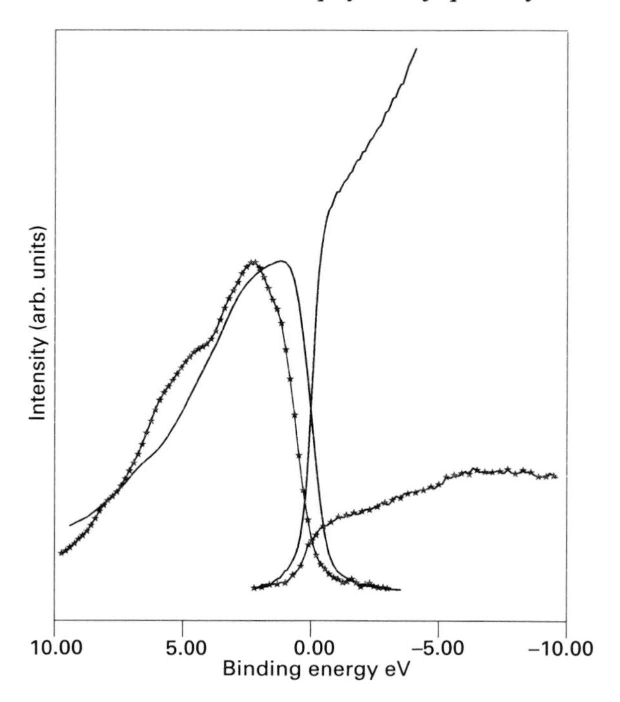

FIG. 6.45 Electronic density of states Al 3p and Al p in AlPdMn quasicrystals (starred line) measured using X-ray spectroscopy compared with that of Al metal (solid line) (courtesy of E. Belin who kindly provided it prior to publication).

conduction electrons is reduced with respect to that of the free electron gap.

Icosahedral quasicrystals seem to be good candidates for the Hume-Rothery stabilization mechanism. Indeed, spectroscopy and conductivity data demonstrate that the Fermi level lies at the bottom of a deep valley in the density of states. This deep valley can be explained in a way similar to the introduction of gap opening and Brillouin zone effects in crystals. Strong Fourier components of the structure which produces strong Bragg peaks in the X-ray or neutron diffraction pattern are also responsible for total backscattering of any other type of plane wave. This suggests that pseudo-Brillouin zones can be defined in quasicrystals as the volume of the reciprocal space enclosed between mediator planes of the links from the origin to a set of sites in reciprocal space. As the icosahedral point group is the most isotropic that can be envisaged, these pseudo-Brillouin zones are expected to be almost spherical. This is illustrated in Fig. 6.46 which shows the pseudo-Brillouin zone whose size matches, at best, the Fermi sphere in AlFeCu or AlPdMn quasicrystals. Indeed, it looks very 'spherical' and possesses all the necessary ingredients to produce a deep pseudo-gap in the density of states. Conclusively, treating quasicrystals as Hume-Rothery alloys provides

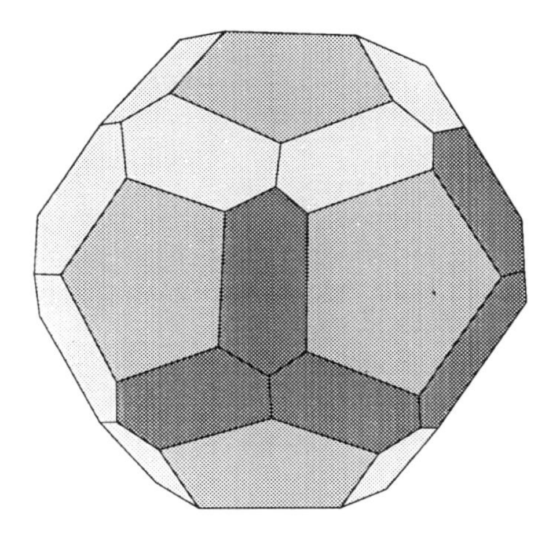

FIG. 6.46 Pseudo-Brillouin zone for icosahedral symmetry constructed for the $M/N = 18/29$ and $M/N = 20/32$ sites in reciprocal space (30 + 12 multiplicities).

obviously acceptable explanations for the stability of quasicrystals, for the existence of a pseudo-gap at E_F, and, to some extent, for its consequences (sign reversal for Hall coefficient and thermoelectric power for instance, as well as low electron mobility). However, main difficulties are far from being resolved through this approach. The very low conductivity at 0 K cannot be quantitatively inferred from the Hume-Rothery mechanism. Moreover, σ should decrease with increasing temperature and when removing structural and/or chemical disorder. This is just the opposite of what is observed. The electron/atom dependence of the structures and properties, the optical behaviour, and the diamagnetic behaviour, among others, are parameters which cannot be explained by a Hume-Rothery mechanism. Finally, it should be remembered that features such as pseudo-gaps in the density of states are not directly related to long-range order of the structure and are actually due to local order effects. Thus the Hume-Rothery rules are certainly not the best route towards the understanding of quasiperiodicity and self-similarity in atomic structures.

So far, *ab initio* calculations of the electronic structures of quasicrystals have been just extensions of methods currently applied to periodic structures.[69–76] Basically, quasicrystals are replaced by rational approximants of the highest possible order compatible with computer limitations. The rational approximant approach offers the double advantage of restricting the system to a finite size and restoring periodicity. The electronic structure can then be treated classically. Bloch waves are constructed using the tight-binding linear muffin-tin orbital method.[48, 70, 71] Typical results of such calculations are shown in Fig. 6.47, 6.48, and 6.49 for the Bloch spectral function $f(\boldsymbol{k}, E)$, the pseudo-dispersion relation $E(\boldsymbol{k})$, and the density of

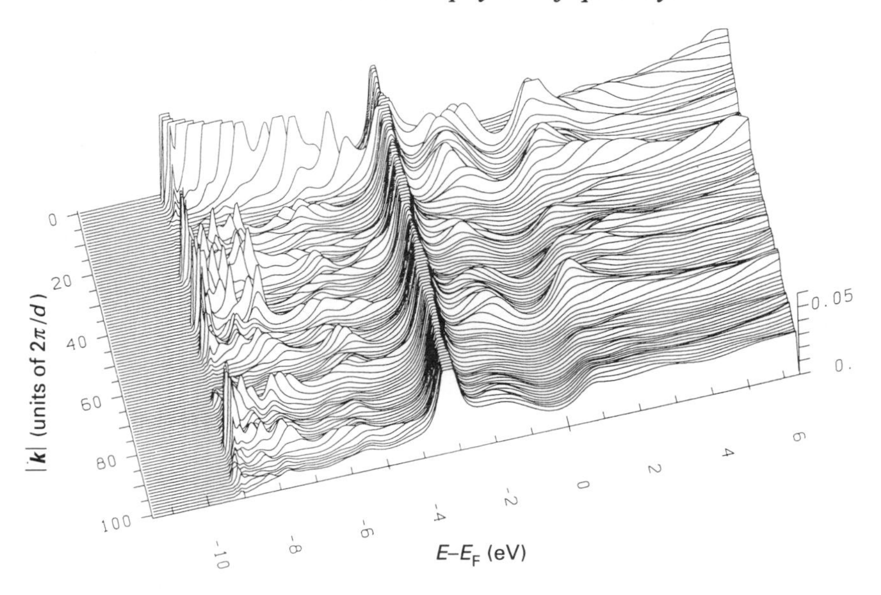

FIG. 6.47 Bloch spectral function $f(\boldsymbol{k}, E)$ for wavevectors along a 2-fold axis ($\boldsymbol{k}$ is in units of $2\pi/d$ where d is the approximant period) in a 5/3 approximant of the AlLiCu quasicrystal[70, 71] (courtesy of J. Hafner).

states. Close to sites of the reciprocal space (Γ points), the electronic states are extended. But everywhere else the energy bands are almost dispersionless and the Fermi level falls into a well defined pseudo-gap. Thus, both Fermi velocity and density carriers are very low. The conclusions are very similar to the deductions made from the phenomenological Hume-Rothery rules. Again here, the results are mainly mastered by short and medium range effects. The quasiperiodicity ingredient is completely rubbed out by the periodic approximation. Similar calculations have been performed to evaluate phason strain effects on conductivity.[77] Depending on the Fermi energy, i.e. on the average density of conduction electron, random phason strain destroy or improve the system conductance. The latter occurs when the energy level of the perfect structure falls in a narrow energy gap which is somewhat smeared out upon phason strain effects.

The electronic structure of quasicrystals can also be tentatively analysed within the formalism of nearly-free-electron systems. As described in Section 6.5.1, the energy dispersion law $E(\boldsymbol{k})$ is obtained using second-order perturbation of the free-electron states by the crystal potential. Divergence problems occur for eigenstates whose wavevector ends on Brillouin zone borders, i.e. satisfies the diffraction conditions for total backscattering effects. Then the currently extended Bloch waves couple to degenerate into standing waves. As Fourier components of a quasiperiodic structure densely

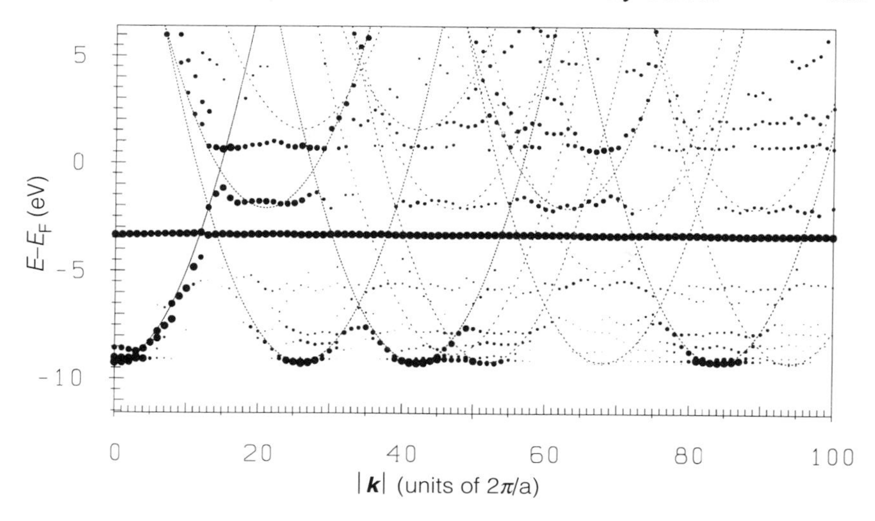

FIG. 6.48 Pseudo-dispersion relation derived from the maxima of the Bloch spectral function shown in Fig. 6.47[70, 71] (courtesy of J. Hafner).

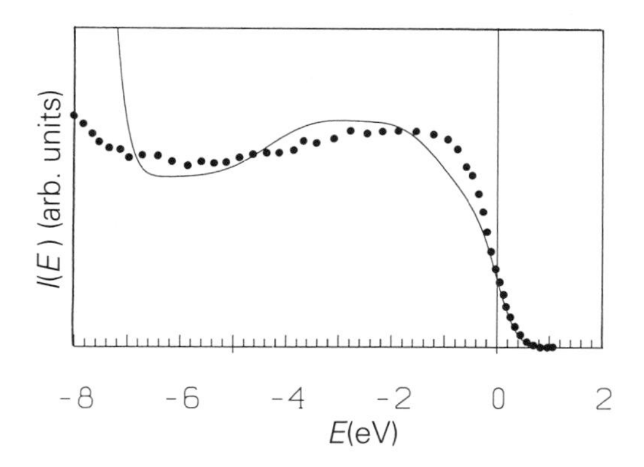

FIG. 6.49 Density of states of the 3/2 approximant (solid line) compared with the densities of states of the liquid (dotted line) in the AlZnMg alloy[70, 71] (courtesy of J. Hafner).

fill the reciprocal space, any prospective extended Bloch wave may be confronted with the above divergence situation. Assuming that the difficulty can also be overcome by mixing extended states into almost 'standing' (critical?) eigenwaves is the foundation stone of this shivering free-electron building! The use of eqn (6.28) to calculate energy bands in a periodic crystal requires a knowledge of the Fourier components of the potential, or any form of pseudo-potential. It is not actually necessary to know the pseudo-potential components everywhere, but only at a few reciprocal lattice vectors, since the low-energy part of the band structure (E_1, E_2) is the only one to contribute to the useful density of states up to the Fermi level. If we use the same procedure for quasiperiodic structures, the reciprocal lattice vectors must be replaced by 'quasireciprocal lattice vectors'. In practice, these are defined as those vectors of the reciprocal space which show scattering in diffraction experiments, particularly those which display 'strong' scattering. The procedure has been implemented for a hypothetical Al quasicrystal[78] using diffraction data from AlMn-like quasicrystals as input. The resulting density of states, shown in Fig. 6.50, exhibits very pronounced deviations from the free-electron parabola and the singularities are much stronger than those found in the fcc Al calculation. For valences at which the Fermi level is close to a minimum in the density of states, as observed in Fig. 6.50, the quasicrystal should be the most stable phase, its resistivity should be large, and its composition very strictly defined. Although the results seems to be promising, the method is actually very difficult to implement from a fundamental point of view. There is no Hamiltonian that we can solve to provide a basis for perturbation theory and the apparent success of the above derivation is due to a forced artificial injection of extended states into the analysis; the quasicrystal has been treated as if it were a crystal but with more Fourier components, which are unusually distributed.

At this point, it is clear that the basis question about electronic states in quasicrystals is whether they are extended as in periodic crystals or localized as in amorphous materials. Looking back to the beginning of Section 6.5.1, the expression (6.20) for the Schrödinger equation is still valid for a quasicrystal. However, now the 'quasicrystal potential' $V(\boldsymbol{r})$ acting on the electrons of the system is not a space periodic function. Instead, we must

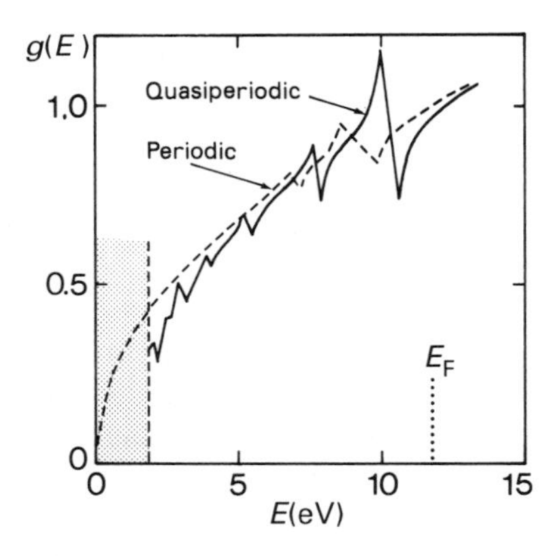

FIG. 6.50 Density of states calculated for a nearly-free-electron model of Al atoms on a 3-dim Penrose lattice (solid line) compared with the result for a periodic fcc Al lattice (broken line). [78]

consider $V(\mathbf{r})$ as having the same symmetry as the quasicrystal, i.e. in particular changing with $\mathbf{r}$ as a quasiperiodic function. Obviously, the Bloch theorem as expressed by eqn (6.22) is no longer true. However, the statement that the observable $|\psi(\mathbf{r})|^2$ must have the symmetry of the system suggests that a 'modified' revisited Bloch theorem may be considered. Suppose that the plane wave $\psi_k(\mathbf{r})$ solves the eigenvalue problem of eqn (6.20) for a quasicrystal; then $\psi_k(\mathbf{r})$ satisfies the relation

$$\psi_k(\mathbf{r}) = f_k(\mathbf{r}) \exp(\mathrm{i}\mathbf{k}\cdot\mathbf{r}). \tag{6.36}$$

In the regular Bloch theorem (eqn 6.22), the modulation function $u_k(\mathbf{r})$ was periodic and hence had Fourier coefficients on a set of discrete points sited at the vectors of the pertinent reciprocal lattice. Now, $f_k(\mathbf{r})$ is quasiperiodic and has the form

$$f_k(\mathbf{r}) = \sum_{\mathbf{G}} f_k(\mathbf{G}) \exp(\mathrm{i}\mathbf{G}\cdot\mathbf{r}) \tag{6.37}$$

where $\mathbf{G}$ scans all the Fourier components of the quasicrystal potential. Hence $f_k(\mathbf{G})$ is defined on a countable dense set of points. In most cases this actually has very little value: the set of $f_k(\mathbf{G})$ functions is as dense as the set of $\psi_k(\mathbf{r})$ functions and we have simply reordered the quantities to be determined by changing the base vectors on which the wavefunctions are expressed. We conclude that eqn (6.36) has almost no useful predictive value, in contrast with its counterpart in periodic crystals. The procedure may help in some very specific cases when the potential remains very weak *and* has significant Fourier components only for small vector $\mathbf{G}$($V_\mathbf{G}$ decreases rather rapidly as $|\mathbf{G}|$ increases).

Globally, one can say that the electronic states of a regular periodic perfect crystal are extended in real space because they are strongly localized in reciprocal space (periodicity of $u_k(\mathbf{r})$). Conversely, the plane wave states in quasicrystals cannot extend very far in real space because they densely occupy the reciprocal space ($f_k(\mathbf{r})$ is quasiperiodic). As a special case, a weak $V(\mathbf{r})$ may aditionally have sufficiently few Fourier components to mimic a localization in reciprocal space and allow extended plane waves.[79]

In order to push the analysis a little further, we may remember that gap opening for periodic crystal is a mere consequence of diffraction-like effects due to a full sampling of one of the structure Fourier components by the prospective propagating wave. This happens very occasionally since the Bragg condition $2\mathbf{k}\cdot\mathbf{G} \pm |\mathbf{G}|^2 = 0$ can be satisfied only for very special wave vectors $\mathbf{k}$ (those pointing along a Brillouin zone limit). More specifically, the Bragg condition expresses a strict selection of a pair ($\mathbf{k}$, $\mathbf{G}$). This means that a given incoming plane wave, with a given wavevector $\mathbf{k}$, either propagates almost freely if $\mathbf{k}$ does not fulfil the condition or forms a standing wave (eqn (6.26)) by coupling with the single $\mathbf{G}$ Fourier component of the potential. In the former case we obviously have an extended state and in the latter the state is also extended in real space since it is localized at a single $\boldsymbol{G}$ site in reciprocal space.

Of course, this is a little too simplistic. Owing to the strong electron–lattice interaction, dynamic aspects such as multiple scattering are more efficient than in X-ray or neutron diffraction and are responsible for perturbations of the plane-wave regime at any $\boldsymbol{k}$ vectors. However, the main conclusion is rather robust.

If we try to apply a similar analysis to a completely random structure (gas-like solid), we must consider a flat continuous Fourier spectrum where $\boldsymbol{G}$ has any value in an isotropic distribution and all Fourier coefficients of the potential are the same. Now, any incoming plane wave of wavevector $\boldsymbol{k}$ is going to be scattered by a subset of the structure Fourier components. Indeed, a Bragg-like relation is satisfied for a given $\boldsymbol{k}$ and any $\boldsymbol{G}$ vector of modulus ranging from 0 to $|2\boldsymbol{k}|$ in a continuous distribution. This can easily be deduced from an application of the Ewald construction (Fig. 1.5). All the corresponding scattered wavelets have the same amplitude and combine inside the defined flat window function of the reciprocal space. Thus the resulting state is extended flatly in the reciprocal space over a $\boldsymbol{G}$ vector range of full width $|2\boldsymbol{k}|$ and then will localize in real space over distances roughly equal to $\pi/|\boldsymbol{k}|$. Small-$\boldsymbol{k}$ incoming waves are expected to be almost extended while a large $|\boldsymbol{k}|$ value will force localization.

For quasiperiodic structures, the situation lies between those of crystals and random disorder. The Fourier spectrum is called singularly continuous, which means that it exhibits a dense set of sharp features at $\boldsymbol{G}$ vectors defined by more independent integers than the dimensions of the system (see Chapters 1 and 2). As in random disorder structure, any incoming plane wave with wavevector $\boldsymbol{k}$ couples with a subset of the structure Fourier components, in agreement with the Bragg-like relation, with $0 \leqslant |G \leqslant |2\boldsymbol{k}|$. Here, multiple scattering is certainly the rule. Beyond the fact that the $\boldsymbol{G}$ ensemble does not form a continuous distribution, the Fourier spectrum is not uniform and the Fourier coefficients of the potential, like the structure factors, are a decaying slightly oscillatory function of the $\boldsymbol{G}_{\perp}$ vectors associated with $\boldsymbol{G}$ in the periodic image of the structure. This has the consequence of 'smoothing away' the localization effects. For instance, taking the case of a very large $\boldsymbol{k}$ such that the situation is approximated well by assuming full coupling of the incoming wave with all Fourier components of the potential would result in strict localization for random disorder structure but in localization domains of the A_{perp} size for a quasicrystal. Indeed, the Fourier coefficients of a quasicrystal structure are just the Fourier transform of the atomic surfaces in the high-dimensional periodic image. This means that the length scale for localization is about 10 Å for the current perfect icosahedral quasicrystals. But this is not the end of it. One of the typical properties of quasicrystals, in addition to their quasiperiodicity, is the existence of a uniform (self-similar locally isomorphic) distribution of local atomic arrangements. For instance, if one takes a region of diameter d in the quasicrystal, Conway's theorem states that there exists an identical piece of matter within a distance of less than $2d$. For example, when considering the Fibonacci sequence $L\ L\ S\ L\ S\ L\ L\ S\ L\ L\ S\ L\ S\ L\ L\ S\ L\ S\ L\ L\ S\ldots$ it is straightforward to check that the LSL sequences have their middle S segments separated by distances equal to $2S + 4L$, $LSLSL$ strips are $4S + 6L$ apart, $LSLSLLS$ are $6S + 8L$ apart, etc. We have also reported (Chapter 4, Section 4.2.4) that the experimental structure of AlPdMn quasicrystals is based on a self-similar skeleton of 10 Å atomic clusters separated by distances of 20 Å. As the coupling of a plane wave with Fourier components of the structure operates like a window function, the resulting localization domain in real physical space will convolute the structure, i.e. it will repeat itself at each equivalent site according to the above self-similarity. In other words, a state with pseudo-wavevector $\boldsymbol{k}$ will appear in atomic domains of size $d \approx \pi/|\boldsymbol{k}|$ separated by distances $2d \approx 2\pi/|\boldsymbol{k}|$. (Such a convolution effect is not efficient in random disorder structures because identical clusters of size d are on average separated by distances exponentially large as a function of d.) Thus, in quasicrystals, the 'localization domains' are sufficiently close to each other with respect to their size and it is not too difficult to tunnel from one cluster to its copy, to the price of a finite damping factor. This is exactly the description of the so-called ***critical states***.[72, 80–82] An equivalent description of these critical states of wavefunction ψ_i is contained in the question: on how many sites $N(\varepsilon)$ do we have $|\psi_i|^2$ larger than any small value ε? For a localized state $N(\varepsilon)$ is bounded and does not increase as the system size is expanded. For an extended state, $N(\varepsilon)$ is proportional to the system size d^3 since all sites are identical. For a critical state, $N(\varepsilon)$ scales like $d^{3\beta}$ with $0<\beta<1$.

As the electron transport operates via a tunnelling mechanism, the wavefunctions must essentially be identical on each cluster within a finite damping factor, i.e.

$$\psi_{2d}(x) \sim z(x)\psi_d(x)$$

whose solution is of the form

$$\psi_d(x) \sim 1/d^{\alpha(x)} \tag{6.38}$$

with $\alpha(x) = -\ln|z|/\ln 2$. (This derivation is also valid for a crystal in which $\psi(x)$ transfers without damping, i.e. $z(x) = 1$ and $\alpha = 0$ as it should be for an extended state of same order of magnitude everywhere).

Although this analysis is qualitative, it contains the necessary ingredients for more exact calculations of the electron eigenmodes in a quasiperiodic structure without the unjustified prerequisite that the wavefunctions should the Bloch-like or plane-wave-like.[83] The tunnelling conductivity can also be estimated by the overlap between two similar states ψ_i and ψ_j (eqn (6.38)) on two self-similar neighbouring sites, which scale as $1/d^{2\alpha}$ (with $3/8<\alpha<5/8$),[83] for a tunnelling distance which scales as d. Thus we end with a conductivity which vanishes slowly according to a $d^{1-2\alpha}$ power law, and the quasiperiodic structures are then expected to behave as *insulators at 0 K*.

The conclusion obviously implies that the structure is rigorously self-similar and extends quasiperiodically *ad infinitum* without any disruption of the long-range order.

If the alloy contains some impurities due to improperly matched composition, or static structural defects due to bad processing, the quasiperiodic structure appears as rather periodically coarse-grained by the average distance d_i between defects and/or impurities. Thus the conductivity is limited only by this coarse-graining size and will increase as a power law of the defect concentration in principle up to the current metal conduction. As reported in Section 6.4, rising temperature also produces fluctuations in a quasiperiodic structure in the form of isolated phason defects in the low and medium temperature range which degenerate into a random-tiling-like structure of some sort when phason defects interact and percolate. It is easy to see that this thermal phason strain generates coarse-graining of the structure via fluctuations of wavelength λ_{phason} which decreases with increasing temperature. This can be modelled in the high-dimensional periodic image by replacing the flat slicing by undulating cuts. The result is again that the conductivity will have a positive slope versus T, up to the metal conduction limit, with a power-law dependence.

Another way of reaching the same conclusion is to remember (see Section 4.24) that the phason fluctuation acts as a Debye–Waller like factor but is coupled with the perpendicular component of the scattering vector and is more than 100 times as large as the regular (parallel) Debye–Waller factor. Thus static and/or thermal disruptions of the quasiperiodic long-range order erase the Fourier (large $Q_\perp$) component to some extent. Then incoming plane waves sample a restricted portion of the reciprocal space and the corresponding eigenmode becomes more extended in real space.

Conclusively, any defect (static and dynamic) in a quasiperiodic structure acts as a periodic disruption of the long-range order and reinjects some extended modes in the otherwise critical eigenstates of the perfect structure. We shall end the chapter—and the book—with a particular approach to the quasicrystal properties via a tentative description in terms of hierarchical aggregates.

6.5.5 Quasicrystals as a hierarchy of clusters[84]

It has been reported in Chapter 4, Section 4.2.4, that the diffraction data strongly suggest a description of some real icosahedral quasicrystal in terms of a skeleton made of τ^3-inflated clusters (pseudo-Mackay icosahedra (PMI) or triacontahedra of different sizes).

The occurrence of icosahedral short-range order (ISRO) in condensed matter is not surprising. Computer simulation on simple Lennard-Jones potentials confirmed that ISRO should be energetically favoured. Further evidence comes from the identification of special numbers in the mass spectra of *free*-atomic-cluster beams corresponding to icosahedrally ordered atomic aggregates. In these free-atomic-cluster beams the total numbers of 'free' electrons must also be certain 'magic numbers'. A semiquantitative interpretation of these electron magic numbers has been deduced from a very simple model. If the aggregates are stable, this means that electrons are trapped in a deep potential well induced by the positive ions of the cluster. At the extreme simplification such an aggregate can be viewed as a deep spherical square well whose Schrödinger equation reduces to a Bessel equation. The energy eigenvalues are then given by

$$E_{n,l} = \frac{\hbar^2}{2\mu a^2} \chi^2_{n,l} \tag{6.39}$$

in which n, l are the usual quantum numbers, μ is the electron mass, a is the well radius, and $\chi_{n,l}$ are the zeros of the Bessel functions $J_{l+1/2}(kr)$. The occupation numbers define the so-called electron 'magic numbers' M which have indeed been observed in free atomic aggregates (Fig. 6.51). The cluster geometry induces a roughly Lorentzian broadening of the level, leading to densities of state which are in good agreement with the results of tight-binding calculations[85] (see the first large feature of $n(E)$ in Fig. 6.52). If magic numbers of electrons give rise to stable rare-gas-like free aggregates, we can assume that slightly different numbers of electrons per aggregate will produce 'connecting clusters' for the same reasons that non-rare-gas atoms do. Thus we just consider that atom aggregates behave like giant atoms.

A special situation may arise if the number of electrons per cluster is equal to a magic number plus the exact average number n of electrons per atom. The inflated cluster (a cluster of N clusters, where N is the number of atoms in the elementary cluster) will then require the same magic number of electrons to stabilize itself. In order to go beyond this, we must dispose of the same number of excess electrons after each inflation step. Thus we have enough electrons to stabilize the currently inflated cluster *and* initiate the forthcoming step of inflation *ad infinitum*. The condition can be expressed as

$$n = \frac{M}{N}\left(1 + \frac{1}{N} + \frac{1}{N^2} + \ldots\right)$$

or (6.40)

$$n = \frac{M}{(N-1)}$$

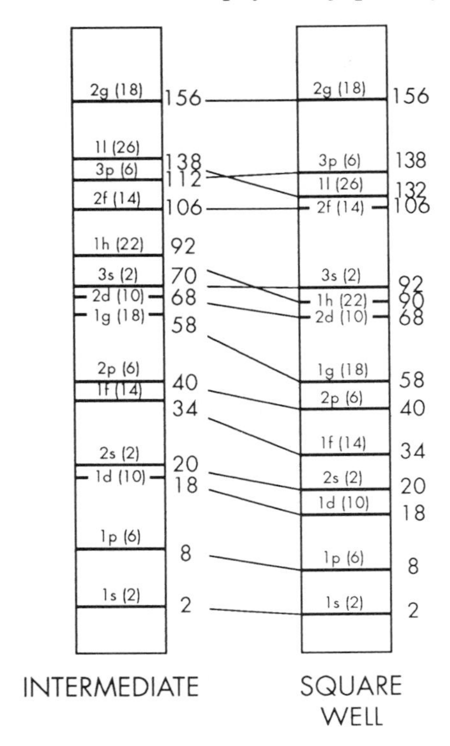

FIG. 6.51 Energy level occupations for three-dimensional spherical wells. Geometrical effects due to the icosahedral symmetry regroup some levels together and the global 'magic numbers' M become 2, 8, 20, 34, 58, 92, 138, . . .; $M = 92$ is a particularly stable aggregate.

Thus the composition for the structure to grow is very critical. Actually the model must be considered in its real dynamic image. It is assumed that the elementary PMI clusters confine electrons like spherical square wells surrounded by potential barriers. Electrons tunnel through these barriers. This can be approximated as hopping within the cluster of the next stage of inflation which in turn confines electrons in an inflated spherical square well, and so on. At each inflation stage electrons are confined by a barrier across which tunnelling is harder and harder. In other words, the effective mass increases at each stage of the inflation hierarchy. The model proposes to account for such a 'hierarchical tunnelling' by stating that a hierarchically decaying fraction of electrons per atom only is allowed to leave freely the well states of a given stage of inflation to fill those of the next stage. At any stage of inflation, each cluster constantly keeps a time-averaged number of electrons equal to M to stabilize itself via intercluster exchanges.

The electron density of states $n(E)$ of the built hierarchical structure can

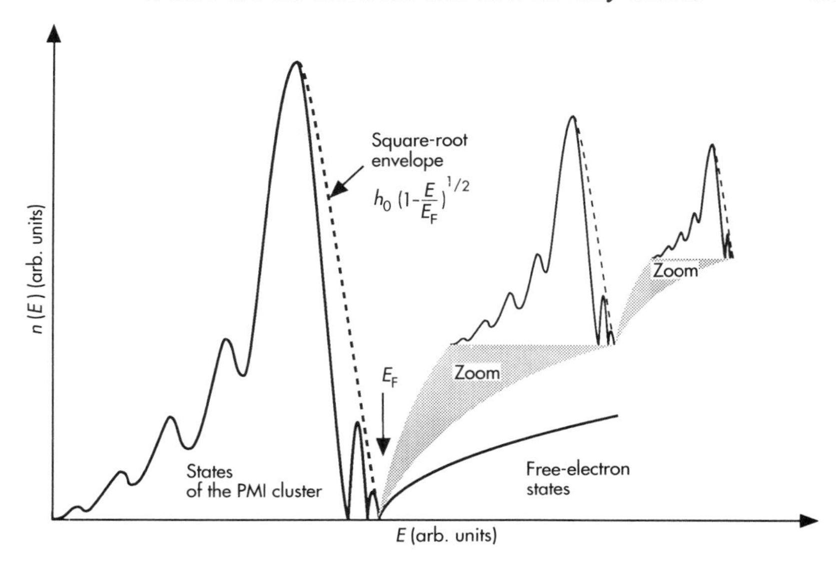

FIG. 6.52 Schematized density of electronic states. The large feature on the left-hand side is the contribution of the elementary PMI. Features due to the successive inflated PMI are simplified for the drawing. Zooming at E_F gives a flavour of the self-similar decay of the density of states. The broken line is a square-root envelope of the main maxima in the successive cluster feature of $n(E)$.[84]

be roughly inferred by considering the succession of inflated spherical square wells attached to successive generations of aggregates. The resulting atomic potential can then be viewed as a deep well whose bottom is a three-dimensional surface with hierarchical roughness; the narrowest and deepest wells are gathered into groups of N on the floor of wells which are τ^6 wider and τ^3 shallower which, in turn, are similarly distributed into even wider and shallower wells, and so on. The τ^6 and τ^3 scale factors come from the inflation growth. The cluster, and then the attached spherical square well, has a radius which inflates with the scale factor τ^3. This is turn forces the number of electronic states per atom specifically attributed to a cluster to decrease with the scale factor τ^9 and the corresponding energy 'sub-band' to extend over a width decaying with the scale factor τ^6 (see eqn (6.39)). Then the average number of electronic states per atom and per energy unit decreases with the scale factor τ^3. The total density of states is qualitatively obtained by adding the 'bonded' states of each well generation, with the unbounded states of one generation falling into the bounded states of the next generation. Thus this total density of states will show the density of states of the basic clusters on the low energy side, completed by successive similar features with surfaces reduced by a cumulative scale factor τ^9 (τ^6 in width and τ^3 in height) with deep pseudo-gaps between them. The density of states roughly converges as τ^{-3} up to a Fermi-surface-like cut-off given by the geometric sum

$$E_{\mathrm{F}} = \Delta E_0 \sum_{n=0}^{\infty} \frac{1}{\tau^{6n}} = \Delta E_0 \frac{\tau^6}{\tau^6 - 1} = 1.0590\,\Delta E_0 \tag{6.41}$$

in which ΔE_0 is the full extension of the density of states for the basic PMI. It is easy to see that only the very first features have significant surfaces, as illustrated in Fig. 6.52. Beyond that, the density of states degenerates into a dense piling up of very narrow features, which is reasonably represented by a short tail decaying toward E_{F} following the equation

$$n(E) = n_0/\tau^{3p}$$

$$E = \Delta E_0 \sum_{n=0}^{p} \frac{1}{\tau^{6n}} \tag{6.42}$$

i.e.

$$n(E) = n_0 \left[1 - \frac{E}{E_{\mathrm{F}}}\right]^{\frac{1}{2}}.$$

Empty states at 0 K are distributed into a free-electron band ($n(E) \approx (E - E_{\mathrm{F}})^{\frac{1}{2}}$) (Fig. 6.52). On both sides of the Fermi level $n(E)$ increases as a square-root function of energy changes with respect to E_{F}. However, the $n(E_{\mathrm{F}})$ valley is asymmetric as it separates free-electron states on one side (extended over a large energy range with a rather flat density) from semibonded states with a sharply increasing density on the other side. The width of the pseudo-gaps is zero because the inflation mechanism continuously reduces the well deepness and then progressively degenerates the energy level into free states.

In summary, the global electron density of states exhibits a self-similar geometry and the Fermi surface has a fractal character. If the lower energy pseudo-gap corresponding to the top of the first sub-band (electron states for elementary clusters) is typically situated at 10 eV, the Fermi level is only some 0.5901 eV beyond. The full width of the successive sub-bands rapidly falls from 10 eV to 0.5573, 0.0310, 0.0017, 0.0001, . . . eV. Close to the Fermi level there is an infinity of decaying pseudo-discrete levels within an energy range smaller than 5×10^{-5} eV. Thus, at rather low energy excitation, say below about 0.05 eV, one should observe some sort of semimetal-like behaviour with square-root distribution of electronic states in both filled valence and empty conduction bands. For a slightly different composition that would allow cluster states to be filled up to a certain stage of inflation only but not *ad infinitum*, the hierarchy would be interrupted at some energy and the Fermi level would fall somewhere within the valence band which becomes conductive. This generates an approximant periodic structure with a density of states at E_{F} scaling as $1/\tau^{3n}$.

The predictions of the models seems to be reasonably consistent with experimental observation. At 0 K the present model is a perfect insulator and exhibits diamagnetic behaviour since all the electrons are in saturated states of the well hierarchy. As the temperature increases electrons are transferred from the bonding states to the conduction band. For $kT \leqslant 0.05$ eV, i.e. $T \leqslant 600$ K, the number of free carriers (electrons in the conduction band and holes in the filled well band) can easily be calculated using the square-root shape of the density of states. Then the number of electrons in the conduction band increases as $T^{\frac{3}{2}}$. In this temperature range only the conduction band is conducting (electron conduction) since holes are created in flat subbands with quite poor mobilities. At higher temperatures, electrons from the states of the smaller clusters are also excited, and both valence and conduction bands are conducting. Thus $\sigma(T)$ should increase roughly as $T^{\frac{3}{2}}$ up to $T \approx 600$ K and more rapidly above this temperature.

The analysis can also explain the sign changes in the Hall constant R_H and thermal power S. At low temperatures, free electron carriers (in small numbers) give weakly negative R_H and S. As T increases the small flat subbands between the Fermi level and the upper part of the larger features of the density of state are progressively stripped of their electrons, generating an effective pseudo-gap of increasing width. Thus the behaviour becomes that of a poor semiconductor with increasing energy gap $E_g(T)$ and positive increasing $R_H(T)$ and $S(T)$ values.

Using the same scheme, one can calculate the temperature dependence of the total electron energy and the electron contribution to heat capacity. This gives $\Delta E_{el} \propto T^{\frac{5}{2}}$ and consequently $C_{el} \propto T^{\frac{3}{2}}$ (instead of T^2 and T respectively as obtained for a metal) in the low-temperature range ($T \leqslant 600$ K). If this is the case, the usual analysis of the heat capacity data in terms of a C/T vs. T^2 curve would be irrelevant.

Finally the experimental behaviour of the optical conductivity is also explained rather well by the present model. At 0 K, the filled envelope of subbands just below the empty conduction band can produce optical absorption via interband transition only. The optical conductivity σ_{op} can be expressed as

$$\sigma_{op} = \frac{c}{\omega^2} M^2 \text{ (JDS)}$$

where c is a constant, M is a matrix element for the electronic transition, and the joint density of states JDS can be approximated by the product of the number of states found in each of the two bands within an energy range equal to the photon energy $\hbar\omega$ on each side of E_F. A straightforward calculation, still with square-root envelopes, results in $\sigma_{op} \propto \hbar\omega$. Beyond a certain energy threshold, the two bands are expected to have enough empty and filled states in appropriate proportions to permit intraband transition and the recovery of a Drude-like behaviour.

Now, it may be interesting to test eqn (6.40) with respect to the composition of real quasicrystals. We have mentioned in Section 6.2.4 that compositions of quasicrystals should obey a τ^n law which, when applied to real systems, results in chemical formulae such as $Al_{0.68}(CuM)_{0.382}$ where M is Fe or Ru (τ law) or $Al_{0.724}(PdM)_{0.276}$ where M is Mn or Re (τ^2 law). These icosahedral quasicrystals have a structure based on averaged PMI clusters, each containing 50.5 atoms. The valences of Al and Cu can confidently be taken as equal to +3 and +1 respectively. Transition metals in alloys have negative valences (they behave as acceptors for completing their d states). Accounting for geometrical mixing of states (Lorentzian broadening), the successive magic numbers M (Fig. 6.51) are 2, 8, 34, 58, 92, 138, etc. The only realistic value for the above PMI is obviously $M = 92$, which gives an average of 1.858 electrons per atom. This forces the final compositions of the above alloys to be $Al_{0.618}Cu_{0.256}Fe_{0.126}$ with $Z_{Fe} = -2$, $Al_{0.618}Cu_{0.256}Ru_{0.126}$ with $Z_{Ru} = -2$, and $Al_{0.724}Pd_{0.200}Mn_{0.076}$ and $Al_{0.724}Pd_{0.200}Re_{0.076}$ with $Z_{Pd} = -0.5$ and $Z_{Mn} = Z_{Re} = -2.8$, which are in good agreement with measurements.

The predictions of the hierarchical cluster model also apply to the $Al_{0.570}Cu_{0.108}Li_{0.322}$ quasicrystals which contain 2.14 electrons per atom, i.e. they also contain 92 electrons in a triacontahedral cluster of the structure. Owing to the very strong criticality, it may also be conjectured that exact chemical composition and local homogeneity cannot strictly be satisfied in real alloys, which may result in 'experimental quasicrystals' actually being mixtures of approximants rather than pure perfect quasicrystals and explain the non-typical behaviour of the properties, particularly electrical resistivity, at low temperatures where the pure quasicrystal state can easily be corrupted by minute amounts of metallic phases.

6.6 Conclusion

This chapter may leave the reader somewhat unsatisfied, as also is the author. Most of the derivations that have been reported here are, indeed, limited to rather poor qualitative approaches. This actually reflects the state of the art, even if quite a large number of studies dealing with numerical simulations have been neglected. However, basic questions have been addressed which are essential if quasicrystals are to be considered as a special state of matter in its own right. The structure specification exercises which have been developed, under their various aspects in Chapters 1, 2, and 4, would have been of very little interest without the assurance that such structures could actually be grown and that they had specific properties. The points have now been made clear: (i) quasicrystal growth within physical time-scales is a realistic, well-described prospect; (ii) the electronic band structure of quasicrystals, as mysterious as it still is, cannot be less than very peculiar, and the related observed properties have already anticipated this point.

Many other properties of quasicrystals have not been reported in this chapter, despite their prevalence in today's interests. Some of them are even on the verge of making obsolete the popular inquiry: 'What are quasicrystals useful for?'. As a good example, let us note here the recent suggestion, already tested, that coatings containing a significant amount of quasi-crystalline phases may offer an interesting alternative to the surface reinforcement of soft metallic materials.[86] They are hard ($Hv > 500$), but become ductile at high temperature. Their friction coefficient is less than half that of normal aluminium-based alloys. They are poorly adhesive to organic material under certain conditions. They are bad heat conductors. They are also thermally stable and they can be chemically adjusted to meet corrosion problems. Accordingly, they are good candidates for mild wear applications, ranging from non-stick frying pans to a self-lubricated combustion chamber in engines.[87–90]

For quasicrystals, the age of effectiveness has already started to overlap the dreaming period.

6.7 Problems

6.1 Calculate the probability that a Fibonacci string of n segments is grown without mistakes with respect to the perfect sequence, if the following two rules are applied:

(1) After S only L can follow;

(2) After L, either S or L can follow, but after two Ls only S can follow.

Take any n value you wish.

6.2 Using the cut procedure to generate the Fibonacci chain, demonstrate eqn (6.1) using the formulae given in Section 6.2.1

6.3 Explain how 'flipping' a worm in a 2-dim Penrose tiling is equivalent to reverse L, S sequences in the Fibonacci chain.

What would happen in the structure of a perfect Penrose tiling if, in a region of the associated 5-grids where the five lines almost cross at the same point (star-like crossing), two lines are permuted in each grid so that large–short or short–large spacings are reversed?

6.4 Starting from a single fat rhombus, grow a 2-dim Penrose tiling cluster up to the first dead surface, using the vertex matching rules.

Do the same when starting with a perfect 'cartwheel' decapod as a seed cluster.

Do the same when starting with the defect decapod having 'arrow' charge equal to two.

Compare the sizes of the three grown clusters.

6.5 Using any dead corner you like, as taken from the results of problem 6.4, explain why a 108°-corner is very convenient for 'reactivating' a dead surface.

6.6 Decapods with defects are generated by flipping some of the 10 semi-infinite worms of the Penrose tiling that cross inside the original 'cartwheel' perfect decapods. Any combination of flipped worms can be arbitrarily chosen. Some of the generated defect decapods are, however, identical, due to the symmetry of the tiling. Demonstrate that there are, in total, 62 distinct combinations and classify them by the rate of possible occurrence within total randomness.

6.7 Taking any example you like, convince yourself that any defect 'mismatch' in a 2-dim Penrose tiling can be attributed a 'defect charge' or 'defect arrow charge' which may be reminiscent of Burgers vectors for dislocations in crystals.

6.8 Minimizing the free energy given by eqn (6.2), demonstrate that the general form of F is described by Fig. 6.19.

6.9 Demonstrate that the free energies at equilibrium for icosahedral and bcc structures are given by eqn 6.12 and 6.13. Express the condition for a prevalence of the ico-phase.

6.10 Copper has a mass density $\rho_m = 8.95\ \mathrm{g\,cm^3}$ and an electrical resistivity $\rho = 1.55 \times \mu\Omega\,\mathrm{cm}$ at room temperature. Assuming that the effective mass $m^* = m_0$ calculate:

(a) the concentration of conduction electrons
(b) the mean free time for collision
(c) the Fermi energy E_F
(d) the Fermi velocity
(e) the mean free path at the Fermi level
(f) the density of state at the Fermi level.

Calculate the same quantities for a quasicrystal, where the mass density is $3.80\ \mathrm{g\,cm^{-3}}$ and electrical resistivity $\rho = 6500\ \mu\Omega$ at room temperature.

6.11 Given a Bloch function at point $\boldsymbol{k}$ in the reciprocal lattice, such as

$$\psi_k(\boldsymbol{r}) = u_k(\boldsymbol{r}) \exp(\mathrm{i}\boldsymbol{k}\cdot\boldsymbol{r}),$$

express the Bloch function at point $\boldsymbol{k} + \boldsymbol{G}$ with $\boldsymbol{G}$ a vector of the reciprocal lattice.

Deduce that $E_n(\boldsymbol{k} + \boldsymbol{G}) = E_n(\boldsymbol{k})$.

6.12 Demonstrate eqn (6.23) for the state function and energy values of a free electron one-dimensional model.

Calculate the gap openings in the nearly free electron model.

Do the same for a Fibonacci chain assuming that the models are valid and choosing an arbitrarily restricted set of vectors in the reciprocal space, corresponding to the (strongest) Fourier components of the structure. How is the qualitative aspect of the density of states modified when more and more (medium and weak) Fourier components are considered within the same $\boldsymbol{k}$ range?

References

1. Stevens, P.W. and Goldman, A.I. (1985). Sharp diffraction maxima from an icosahedral glass. *Phys. Rev. Lett.* **56**, 1168–71.
2. Stevens, P.W. and Goldman, A.I. (1986). Errata. *Phys. Rev. Lett.* **57**, 2331.
3. Henley, C.L. (1988). Random tiling with quasicrystal order: transfer matrix approach *J. Phys.* **A21**, 1649–77.
4. Tsai, A.P., Inoue, A., and Masumoto, T. (1987). A stable quasicrystal in Al-Cu-Fe system. *Japan. J. Appl. Phys.* **26**, L1505–7.
5. Calvayrac, Y., Quivy, A., Bessière, M., Lefebvre, S., Cornier-Quinquandon, M., and Gratias, D. (1991). Icosahedral AlCuFe alloys: towards ideal quasicrystals. *J. Phys. (Paris)* **51**, 417–31.
6. Tasi, A.P. Inoue, A., Yokoyama, Y., and Masumoto, T. (1990). Stable icosahedral AlPdMn and AlPdRe alloys. *Mat. Trans. JIM* **31-2**, 98–103.
7. Onoda, G.Y., Steinhardt, P.J., DiVicenzo, D.P., and Socolar, J.E.S. (1988). Growing perfect quasicrystals. *Phys. Rev. Lett.* **60**, 2653–6.
8. Steinhardt, P.J. (1990). Growing perfect quasicrystals. In *Quasicrystals* (ed. T. Fujiwara and T. Ogawa) pp. 91–9. Springer, Berlin.
9. DiVincenzo, D.P. (1990). Physical models of perfect quasicrystal growth. In *Geometry and thermodynamics on common problems of quasicrystals, liquid crystals and incommensurate systems* (ed. J.C. Toledano) pp. 133–9. NATO-ASI Series, Series B: Physics, NATO.
10. Ingersent, K. and Steinhardt, P.J. (1990). Matching rules and growth rules for pentagonal quasicrystal tilings. *Phys. Rev. Lett.* **64**, 2034–7.
11. Dotera, T., Jeony, H.C., and Steinhardt, P.J. (1990). Properties of decapod defects. In *Methods of structural analysis of modulated structures and quasicrystals* (eds J.M. Perez-Mato, F.J. Zuniga, G. Madariaga and A. Lopez-Echarri), pp. 660–6. World Scientific, Singapore.
12. Katz, A. and Duneau, M. (1986). Quasiperiodic patterns and icosahedral symmetry. *J. Phys. (Paris)* **47**, 191–6.
13. Ingersent, K. (1991). Matching rules for quasicrystals. In *Quasicrystals: the state of the art* (ed. P.J. Steinhardt and D.P. DiVincenzo) World Scientific, Singapore.

14. Penrose, R. (1989). Tiling and quasicrystals: a non-local growth problem? In *Introduction to the mathematics of quasicrystals* (ed. M. Jaric) pp. 53–80. Academic, New York.
15. Gardner, M. (1977). Mathematical games *Sci. Am.* **236**, 110–21.
16. Ronchetti, M., Bertagnolli M., and Jaric, M. V. (1990). Generation and dynamics of defects in 2-dim quasicrystals. In *Geometry and thermodynamics on common problems of quasi-crystals, liquid crystals and incommensurate systems* (ed. J. C. Toledano) pp. 141–57. NATO ASI Series, Series B: Physics, NATO.
17. Jaric, M. V. and Ronchetti, M. (1989). Local growth of quasicrystals. *Phys. Rev. Lett.* **62**, 1209–12.
18. Ronchetti, M. and Jaric, M. V. (1990). Defect generation during quasicrystal growth. In *Quasicrystals* (ed. M. V. Jaric and S. Lundqvist) pp. 227–41. World Scientific, Singapore.
19. Henley, C. L. (1986). Sphere packings and local environments in Penrose tilings. *Phys. Rev.* **B34**, 797–816.
20. Duneau, M. (1985). Quasiperiodic patterns. *Phys. Rev. Lett.* **54**, 2688–91.
21. Lifshitz, E.M. and Pitaevskii, L.P. (1980). In *Landau-Lifshitz course of theoretical physics: statistical physics*, Part I. Pergamon, Oxford.
22. Jaric, M. V. (1985). Long-range icosahedral orientational order and quasicrystals. *Phys. Rev. Lett.* **55**, 607–10.
23. Bak, P. (1985) Phenomenoligical theory of icosahedral incommensurate (quasiperiodic) order. *Phys. Rev. Lett.* **54**, 1517–19.
24. Bak, P. (1985). Symmetry, stability and elastic properties of icosahedral incommensurate crystals. *Phys. Rev.* **B52**, 5764–72.
25. Robertson, J.L. and Moss, S.C.Z. (1991). Random cluster models for icosahedral phase alloys, *Z. Phys. B.* **83**, 391–405.
26. Elser, V. (1989). The growth of icosahedral phase. In *Aperiodicity and order*, Vol. 3, *Extended icosahedral structures* (ed M.V. Jaric and D. Gratias), pp. 105–34. Academic Press, New York.
27. Henley, C. L. (1991). Random tiling models. In *Quasicrystals: the state of the art* (ed D.P. Di Vincenzo and P.J. Steinhardt), pp. 429–518. World Scientific, Singapore.
28. Lançon, F., Billard, L., and Chandari, P. (1986). Thermodynamical properties of two-dimensional quasicrystal from molecular dynamics calculation. *Europhys. Lett.* **2**, 625–9.
29. Lançon, F. and Billard, L. (1988). Two-dimensional system with a quasicrystalline ground state. *J. Phys. Fr.* **49**, 249–56.
30. Widom, M. (1993). Bethe *Ansatz* of the square-triangle random tiling model. *Phys. Rev. Lett.* **70**, 2094–7.
31. Widom, M., Deng, D.P., and Henley, C.L. (1989). Transfer matrix analysis of a two-dimensional quasicrystal. *Phys. Rev. Lett.* **63**, 310–13.
32. Strandburg, K.J., Tang, L.H., and Jaric, M.V. (1989). Phason elasticity in entropic quasicrystals. *Phys. Rev. Lett.* **63**, 314–17.
33. Tang, L.H. (1990). Random-tiling quasicrystals in three dimensions. *Phys. Rev. Lett.* **64**, 2390–3.
34. Yacaman, M.J. and Torres, M. (ed) (1993). *Crystal–quasicrystal transitions.* North-Holland, Amsterdam.
35. Donnadieu, P. (1992). Defects in the approximant crystal of the quasicrystal

Al_6Li_3Cu: from π-boundaries to a near triacontahedron polyhedron. *Phil. Mag. B* **65**, 15–28.
36. Donnadieu, P. and Degand, C. (1993). Evidence of intermediate states between approximant crystal and quasicrystal. *Phil. Mag. B* **68**, 317–28.
37. Donnadieu, P. (1994). The deviation of the Al_6Li_3Cu quasicrystal from icosahedral symmetry: a reminiscence of the cubic phase. *J. Phys. Fr.* in press.
38. Kang, S. S. and Dubois, J. M. (1992). Lattice transformations in the approximant and decagonal phase. *J. Phys.: Cond. Matter* **4**, 7025–40.
39. Menguy, N. (1993). Thesis, Institut National Polythechnique de Grenoble.
40. Janot, C. (1994). *Int. J. Mod. Phys. B.*, in press.
41. Ziman, J.M. (1979). *Models of disorder: the theoretical physics of homogeneously disordered systems* Cambridge University Press.
42. Berger, C., Gozlan, A., Lasjaunias, J.C., Fourcaudot, G., and Cyrot-Lackmann, F. (1991). Electronic properties of quasicrystals. *Physica Scripta* **T35**, 90–4.
43. Mizutani, U., Kamiya, A., Matsuda, T., Kishi, K., and Takeuchi, S. (1991). Electronic specific heat measurements for quasicrystals and Frank–Kasper crystals. *J. Phys.: Cond. Matter* **3**, 3711–18.
44. Tsai, A.P., Inoue, A. and Masumoto, T. (1988). New stable icosahedral Al–Cu–Ru and Al–Cu–Os alloys. *Japan. J. Appl. Phys.* **27**, L1587–90.
45. Tsai, A.P., Inoue, A., and Masumoto, T. (1990). Chemical order in an AlPdMn icosahedral quasicrystal. *Phil. Mag. Lett.* **62(2)**, 95–100.
46. Tsai, A.P., Inoue, A., Yokoyama, Y., and Masumoto, T. (1990). New icosahedral alloys with superlattice order in the AlPdMn system prepared by rapid solidification. *Phil. Mag. Lett.* **61(1)**, 9–14.
47. Faudot, F., Quivy, A., Calvayrac, Y., Gratias, D., and Harmelin, M. (1991). About the Al-Cu-Fe icosahedral phase formation. *Mat. Sci. Eng.* **A133**, 383.
48. Pierce, F.S., Poon, S.J., and Guo, Q. (1993). Electron localization in metallic quasicrystals. *Science* **261**, 737–9.
49. Poon, S.J. (1992). Electronic properties of quasicrystals. An experimental review. *Adv. Phys.* **41**, 303–63.
50. Mayou, D., Berger, C., Cyrot-Lackmann, F., Klein, T., and Lanco, P. (1993). Evidence for unconventional electronic transport in quasicrystals. *Phys. Rev. Lett* **70**, 3915–18.
51. Pierce, F.S., Poon, S.J., and Biggs, B.D. (1993). Band-structure gap and electron transport in metallic quasicrystals and crystals. *Phys. Rev. Lett.* **70**, 3919–22.
52. Klein, T., Gozlan, A., Berger, C., Cyrot-Lackmann, F., Calvayrac, Y., and Quivy, A. (1990). Anomalous transport properties in pure AlFeCu icosahedral phases. *Europhys. Lett.* **13**, 129–34.
53. Biggs, B.D., Poon, S.J., and Munirathman, N.R. (1990). Stable AlCuRu icosahedral crystals: a new class of electronic alloys. *Phys. Rev. Lett.* **65**, 2700–3.
54. Biggs, B.D., Li, Y., and Poon, S.J. (1991). Electronic properties of icosahedral approximant and amorphous phases of the AlFeCu alloy. *Phys. Rev.* **B43**, 8747–51.
55. Klein, T., Berger, C., Mayou, D., and Cyrot-Lackmann, F. (1991). Proximity of a metal–insulator transition in icosahedral phases of high structural quality. *Phys. Rev. Lett.* **66**, 2907–10.
56. Homes, C.C., Timusk, T., Wu, X., Altounian, Z., Sahnoune, A., and

Ström-Olsen, J.O. (1991). Optical conductivity of the stable icosahedral quasicrystal AlCuFe. *Phys. Rev. Lett.* **67**, 2694–6.

57. Poon, S.J. (1994). Private communication.
58. Mori, M., Matsuo, S., Ishimasa, T., Matsuura, T., Kamiya, K., Inokuchi H., and Matsukawa, T. (1991). Photoemission study of an AlCuFe icosahedral phase. *J. Phys.: Cond. Matter* **3**, 767–71.
59. Belin, E., Dankhazit, Z., Sadoc, A., Calvayrac, A., Klein, T., and Dubois, J.M. (1992). Electronic distribution of states in crystalline and quasicrystalline Al–Cu–Fe and Al–Cu–Fe–Cr alloys. *J. Phys.: Cond. Matter* **4**, 4459–72.
60. Kimura, K., Iwahashi, H., Hashimoto, T., and Takeuchi, S. (1990). Electrical properties of Al-based high-quality quasicrystals. In *Quasicrystals and incommensurate structures in condensed matter* (ed. M.J. Jacaman, D. Romeu, V. Castaño and A. Gomez) pp. 532–57. World Scientific, Singapore.
61. Martin, S., Hebard, A.F., Kortan, A.R., and Theil, F.A. (1991). Transport properties of $Al_{65}Cu_{15}Co_{20}$ and $Al_{70}Ni_{15}Co_{15}$ decagonal quasicrystals. Phys. Rev. Lett. **67**, 71922.
62. Berger, C. and Préjean, J.J. (1990). Evidence of a spin-glass transition in the quasicrystal $Al_{73}Mn_{21}Si_6$. *Phys. Rev. Lett.* **64**, 1769–72.
63. O'Handley, R.C., Dunlap, R.A., and McHenry, M.E. (199). Magnetism and quasicrystals. In *Ferromagnetism materials* (ed. K.H.J. Buschow) North-Holland, Amsterdam (in press).
64. Wagner, J.L., Wong, K.M., and Poon, S.J. (1989). Electronic properties of stable icosahedral alloys. *Phys. Rev.* **B39** 8091–5.
65. Lanco, P. (1993). Thesis, Université Joseph Fourier de Grenoble.
66. Klein, T. (1992). Thesis, Université Joseph Fourier de Grenoble.
67. Friedel, J. and Denoyer, F. (1987). Les quasicristaux de AlLiCu comme alliages de Hume–Rothery. *C.R. Acad. Sci. Paris* **305**, 171–4.
68. Vaks, V.G., Kamyshendo, V.V., and Smolyuk, G.D (1988). Possible effects of bandstructure in properties of quasicrystals. *Phys. Lett. A* **132**, 131–6.
69. Fujiwara, T. and Tsunetsugu, H. (1991). Electronic structure and transport in quasicrystals. In *Quasicrystals: the state of the art* (ed D.P. DiVincenzo and P.J. Steinhardt), pp. 343–60. World Scientific, Singapore.
70. Hafner, J. and Krajci, M. (1993). Electronic structure of quasicrystalline Al–Zn–Mg alloy and related crystalline amorphous and liquid phases. *Phys. Rev. B* **47**, 11795–11809.
71. Windisch, M., Krajci, M., and Hafner, J. (1994). Electronic structure of AlLiCu quasicrystal. Submitted to *J. Phys.: Cond. Matter*.
72. Kim, Y.J., Lee, M.H., and Choi, M.Y. (1990). Novel transition between critical and localized states in one-dimensional quasiperiodic system. In *Quasicrystals* (ed. M.V. Jaric and S. Lunqvist) pp. 442–58. World Scientific, Singapore.
73. Odagaki, T. and Nguyen, D. (1986). Electronic and vibrational spectra of two-dimensional quasicrystals. *Phys. Rev.* **B33**, 2184–90.
74. Choy, T.C (1985). Density of states for a two-dimensional Penrose lattice: evidence of a strong Van Hove singularity. *Phys. Rev. Lett.* **B55**, 2915–18.
75. Fujiwara, T. and Yokokawa, T. (1991). Universal pseudo-gap at the Fermi energy in quasicrystals. *Phys. Rev. Lett.* **66**, 333–6.
76. Kitaev, A.Y. (1990). On the electronic properties of a three-dimensional quasicrystal with weak potential. In *Quasicrystals* (ed. M.V. Jaric and S. Lunqvist) pp. 409–24. World Scientific, Singapore.

77. Fujiwara, T., Yamamoto, S., and Trambly de Laissardière, G. (1993). Band structure effects of transport properties in icosahedral quasicrystals. *Phys. Rev. Lett.* **71**, 4166–9.
78. Smith, A.P., and Ashcroft, N.W. (1987). Pseudopotentials and quasicrystals. *Phys. Rev. Lett.* **59**, 1365–8.
79. Dinaburg, E.J. and Sinai, Y.G. (1976). *Funct. Anal. Appl.* **9**, 279.
80. Kollar, J. and Süto, A. (1986). The Kronig–Penney model of a Fibonacci Lattice. *Phys. Lett.* **A117**, 203–9.
81. Ostlund, S., Pandit, R., Rand, D., Schellenhuber, H.J., and Siggia, E.D. (1983). One dimensional Schrödinger equation with an almost periodic potential. *Phys. Rev. Lett.* **50**, 1873–6.
82. Kohmoto, M. and Sutherland, B. (1986). Electronic states on a Penrose lattice. *Phys. Rev. Lett.* **56**, 2740–3.
83. Sire, C. (1994). Properties of quasiperiodic Hamiltonian spectra, wavefunctions and transport. In *Ecole d'hiver sur les quasicristaux* (ed D., Gratias and F. Hippert). Editions de Physique, Paris, in press.
84. Janot, C. and de Boissieu, M. (1994). Quasicrystals as a hierarchy of clusters. *Phys. Rev. Lett.* **72**, 1674–7.
85. Carlsson, A.E. and Phillips, R. (1991). Electronic structure and total-energy calculations for quasicrystals and related crystals. In *Quasicrystals: the state of the art* (ed D.P. DiVincenzo and P.J. Steinhardt), pp. 361–401, World Scientific, Singapore.
86. Dubois, J.M., Kang, S.S., and von Stebut, J. (1991). Quasicrystalline low-friction coatings. *J. Mat. Sci. Lett.* **10**, 537–41.
87. Kang, S.S. and Dubois, J.M. (1992). Pressure-induced phase transitions in quasicrystals and related compounds. *Europhys. Lett.* **18**, 45–51.
88. Kang, S.S. and Dubois, J.M. (1992). Compression testing of quasicrystalline materials. *Phil. Mag. A* **66**, 151–63.
89. Dubois, J.M., Kang, S.S., Archambault, P., and Colleret, B. (1993). Thermal diffusivity of quasicrystalline and related crystalline alloys. *J. Mater. Res.* **8**, 38–43.
90. Massiani, Y., Ait Yaazza, S., Crouzier, J.P., and Dubois, J.M. (1993). Electrochemical behaviour of quasicrystalline alloys in corrosive solutions. *J. Non-Cryst. Solids*, **159**, 92–100.

Index